Agricultural Medicine

Rural Occupational and
Environmental Health
for the Health Professions

Agricultural Medicine
Rural Occupational and Environmental Health for the Health Professions

Kelley J. Donham, M.S., D.V.M., D.A.C.V.P.M.
Anders Thelin, M.D., Ph.D.

Blackwell Publishing

© 2006 Blackwell Publishing
All rights reserved

Blackwell Publishing Professional
2121 State Avenue, Ames, Iowa 50014, USA

Orders: 1-800-862-6657
Office: 1-515-292-0140
Fax: 1-515-292-3348
Web site: www.blackwellprofessional.com

Blackwell Publishing Ltd
9600 Garsington Road, Oxford OX4 2DQ, UK
Tel.: +44 (0)1865 776868

Blackwell Publishing Asia
550 Swanston Street, Carlton, Victoria 3053, Australia
Tel.: +61 (0)3 8359 1011

First edition, 2006

Library of Congress Cataloging-in-Publication Data

Donham, Kelley J.
    Agricultural medicine : occupational and environ-
mental health for the health professions / Kelley J.
Donham, Anders Thelin.— 1st ed.
       p. cm.
     ISBN 978-0-8138-1803-0 (alk. paper)
     1. Agricultural laborers—Health and hygiene.
2. Agricultural laborers—Diseases. 3. Agriculture—
Health aspects. 4. Medicine, Rural. 5. Rural health.
I. Thelin, Anders. II. Title.
    RC965.A5D66 2006
    362.196′98—dc22

                                          2005019313

Printed and bound in Malaysia by Vivar Printing Sdn Bhd
The last digit is the print number: 9 8 7 6 5 4

Contents

Dedication

We dedicate this book to the millions of farmers, ranchers and farm workers around the world who have died, or suffered disabling injuries or illness over the years in their work of producing food and fiber for all the people on this Earth. Furthermore, we dedicate this book to the many committed rural health care professionals, veterinarians, agricultural health and safety professionals, relevant non-governmental organizations, and the many volunteers around the world who care for and work to prevent injuries and illnesses among the farming communities.

We are thankful to the Scientific Working Group of the International Commission of Occupational Health who charged us to write this text. We thank our many colleagues who have reviewed and provided advice on the text. For financial and human resources support, we thank Iowa's Center for Agricultural Safety and Health, the U.S. National Institute for Occupational Safety and Health, who provided funding to the project through two University of Iowa centers; the Great Plains Center for Agricultural Health, and the Heartland Center For Occupational Health and Safety.

Finally, we thank our families for their support and understanding through this effort, which like farming is to so many, and to us—a labor of love.

Kelley J. Donham
Anders Thelin

About the Authors

Kelley J. Donham, M.S., D.V.M., D.A.C.V.P.M.
Professor and Associate Head for Agricultural
  Medicine, Department of Occupational and
  Environmental Health,
College of Public Health, The University of Iowa

Kelley was born and raised on a swine and beef cow farm in Johnson County, Iowa, where he was actively involved in the farm operation for many years. He still is actively involved in agriculture and owns and manages a farm in Mahaska County, Iowa. Kelley obtained a B.S in Premedical Sciences, and an M.S. in Preventive Medicine and Environmental Health from the University of Iowa, and a Doctorate of Veterinary Medicine degree from Iowa State University. He was in a rural veterinary practice before returning to the University of Iowa in 1973, achieving the rank of full Professor in 1984.

Dr. Donham has developed the first—and one of the few—didactic teaching programs today in

Kelley J. Donham

agricultural medicine, which provides specialty training for health care professionals, occupational health professionals, and veterinarians in occupational and environmental health for agricultural communities. Dr. Donham is also the founder of the AgriSafe Network, a group of specialty clinics who attend to the occupational and environmental health issues of farm families and workers in their communities. Dr. Donham directs Iowa's Center for Agricultural Safety and Health (I-CASH) and the Agricultural Health and Safety Training Core of the Heartland Education and Research Center, and is Deputy Director of The Great Plains Center for Agricultural Health (GPCAH).

Dr. Donham's research has focused on diseases of agricultural workers, particularly respiratory diseases, zoonotic infectious diseases, and intervention methods of prevention. He conducted the original studies in regard to air quality and respiratory illnesses in workers and swine in intensive housing. He has published over 100 articles, books, and chapters in the areas mentioned above.

Dr. Donham is a Diplomate and past president of the American College of Veterinary Preventive Medicine, and in 2002 received the Helwig-Jennings Award for sustained and lasting contributions to the field of veterinary preventive medicine. In 2003, he received the outstanding faculty award for service from the University of Iowa College of Public Health. Also in 2003, he received the Stange outstanding alumni award from the Iowa State University College of Veterinary Medicine.

Anders Thelin, M.D., Ph.D.
Former Head of Research and Development,
The Swedish Farmers' Safety and Health
    Association (Lantbrukshalsan), Stockholm,
    Sweden

After medical studies at the University of Lund, Sweden, Dr. Thelin obtained his M.D. in 1970. He was active at the hospital of Wexiö for some years and as a G.P. in a rural area, followed by some years as an occupational health service physician. A rising interest in farmers' lives and working conditions brought him back to the University of Lund for further studies, resulting in a Ph.D. with a thesis focusing on the panorama of farmers' occupational health and risks.

In the 1970s, Dr. Thelin was a principal person in establishing the first occupational health service in the world specifically for farmers (the Swedish Farmers Health and Safety Association, or Lantbrukshalsan). Later he was head of research and development of Lantbrukshalsan and was active in designing innovative service programs to promote health and safety in dangerous working conditions.

Dr. Thelin's research has focused on diseases of agricultural workers, especially rheumatic disorders and injuries of the musculoskeletal system. He conducted several original studies and is currently responsible for a large prospective research program established in 1990 involving more than 1000 farmers in an ongoing study. Dr. Thelin noticed early on the significant risk of osteoarthritis among farmers and has written a number of publications over the years focusing on hip and knee joint osteoarthritis.

Dr. Thelin lives in the countryside in Southern Sweden and, together with his family, operates a farm. The Thelin family is active in horse breeding, sheep production, and development of a novel fodder for horses (a special silage stored in a plastic wrap). His special knowledge and connection to the farm community provide him with a unique background for writing about farmer's health.

Anders Thelin

Murray Madsen, B.S.Ag.E., M.B.A.
Adjunct Assistant Professor,
Department of Occupational and Environmental Health
College of Public Health, The University of Iowa

Mr. Madsen was born and raised on a farm in southwest Minnesota. He obtained his Bachelors degree in Agricultural Engineering from the University of Minnesota. He has served as product safety engineer for the farm machinery division of John Deere Inc. for nearly 30 years and has been on the staff of the University of Iowa since 2002. He also serves as Coordinator of the Great Plains Center for Agricultural Health.

LaMar Grafft, M.S.
Associate Director,
Agricultural Health Training Program
Iowa Center for Agricultural Safety and Health

Mr. Grafft was born and raised on a farm in northeast Iowa. He was the principal operator of that farm for a number of years. He received his B.S. and M.S. degrees from Iowa State University in Animal Science, his EMT-Paramedic certification in 1985, his Farm Medic Certification in 1989, and his certificate in Agricultural Safety and Health from the University of Iowa in 2002. Mr. Grafft has been a farm extension specialist and a part-time paramedic-EMT since 1985. Since 1993, Mr. Grafft has served as an agricultural health and safety educator and Associate Director for the training program of Iowa's Center for Agricultural Safety and Health.

Danelle Bickett-Weddle, D.V.M., M.P.H.
Associate Director,
Center for Public Health and Food Security,
Iowa State University

Dr. Bickett-Weddle was born and raised on a farm in South Dakota. She received her Doctor of Veterinary Medicine degree from Iowa State University and served in the practice of dairy veterinary medicine for several years. She received her Masters of Public Health and Certificate in Agricultural Safety and Health from the University of Iowa in 2004.

Greg Gray, M.D., M.P.H.
Professor of Epidemiology and Director,
Center for Emerging Infectious Diseases,
College of Public Health, The University of Iowa

Dr. Gray received his B.S. degree from the U.S. Naval Academy, his M.D. degree from the University of Alabama, and his M.P.H. from Johns Hopkins University. He also has completed a residency in Preventive Medicine. He served in various scientific and administrative positions in infectious diseases epidemiology for over 10 years in the U.S. Navy. Since 2001, he has been professor of epidemiology and Director for the Center for Emerging Infectious Diseases, the University of Iowa College of Public Health.

Foreword

This long-awaited volume by Kelley J. Donham and Anders Thelin will be an essential primer and reference for the thousands of existing and future scientists and health professionals who make the emerging science of health and safety issues in agriculture the focus of their study and professional development. With each having served a professional lifetime in the science and practice of agricultural medicine, the authors bring to life a cohesive statement of the spectrum of the issues involved leading to the practical application of measures to mitigate occupational and environmental health and safety risks for farmers, family members, workers, and the public at large. The student will find this volume an easy read that is logical in sequence and well organized in substance.

With a focus on agricultural production in industrialized countries, the work draws considerably on the research, practice, and experience of the authors in the North American and European context, but the message is equally applicable to current industrialized practices in general. An important element of the work is the attention that is given to placing agricultural medicine within a human and economic framework that is so essential to understanding not only the risks but also the opportunities and impediments to managing such risks according to the kind of modern principles of occupational and environmental health that are now fully established in modern mature industries where such issues as uncontrolled exposures, or use of children in the workplace would not be acceptable by either local custom or regulatory oversight.

Commencing in chapter 1 with an introduction and overview, the authors outline the special populations at risk in addition to the farm owner-operator, these being children, women, the elderly, migrant and seasonal workers, and various religious groups. What follows is a succinct treatment of the principal risks, including respiratory and skin diseases, cancer, the health effects of agricultural pesticides, and issues related to clean air and water with an emphasis on confined animal feeding operations. Chapters 8, 9, and 10 explore the important areas of musculoskeletal problems, physical factors such as heat, cold, vibration and noise, and the important and heretofore largely untreated area of psychosocial problems in farmers, including depression, stress, and suicide. Chapter 11 deals with the still unchecked and shocking epidemic of accidental death of and serious injury to agricultural workers. The section on the use of pharmaceuticals and the implications of their use in animal rearing in North America provides an overview of an area that is just emerging as a field of science and action, and the issue of transmission of infectious diseases from animals to humans provides a very current backdrop to the work.

The fact that both authors have been pioneers in prevention and knowledge translation in this area is evident in the final chapter, which deals with the public health tools at our disposal to meet the challenges that they have described, these being regulations, engineering, education, personal protection, and ergonomics. These principles and the descriptions that form the balance of this text make this book a welcome and necessary resource for all of us.

James A Dosman, SOM. MD, FRCPC
Director, Canadian Centre for Health
and Safety in Agriculture
University of Saskatchewan, Saskatoon, Canada

Preface

Kelley J. Donham and Anders Thelin

Our major objective for this book is to provide in-depth information on occupational and environmental illnesses, injuries, and prevention associated with production agriculture to a broad range of health and safety professionals. Furthermore we aim to motivate the reader to apply this information to serve the health and safety needs of the agricultural population.

The industry of agriculture provides essential needs for the world's population. Essential needs of a society were put into perspective in 1943 by the famous sociologist Abraham Maslow in his well-accepted theory of hierarchy of needs entitled "A Theory of Human Motivation" (Psychological Review 50:370–396.) . He used a pyramid metaphor to describe how basic human needs must first be met in order to advance to a higher order of human function. At the base of his pyramid (Figure below, adapted from Maslow) are physiological needs, such as food, clothing and shelter; these are products of the commodities of the industry of agriculture. The secure and productive agriculture of industrialized countries has allowed these societies to develop and progress in many areas of human endeavor. On the other hand, developing countries must first achieve a productive, secure, and sustainable agriculture in order to facilitate their development.

Agriculture is an essential industry too often taken for granted in industrialized countries, especially when it comes to its human resources. The agricultural work force must be protected to assure a sustainable domestic and international food source. The theme for this book is ". . . help-

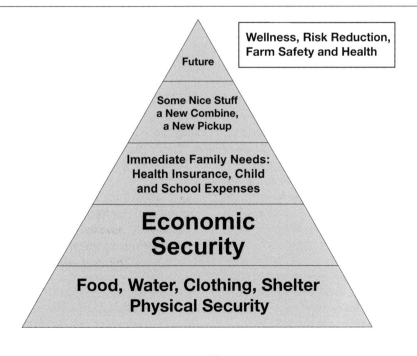

ing keep the rural community alive and well in agriculture." This book is modeled after a course (Rural Health and Agricultural Medicine) that was started by one of us (KJD) in 1974 at the University of Iowa. It was the first such course that could be documented in agricultural medicine. This book is intended to be used as a textbook for courses in this field at other institutions. This book is also intended for use as a reference book for rural health care professionals and health and safety practitioners of all backgrounds.

The geographic focus of this book is those countries defined in 2005 by the United Nations as more developed (or industrialized) countries: the United States of America (U.S.), Canada, countries of the European Union Community (EU), countries in line to become EU members, Australia, New Zealand, and Japan (Absolute Astronomy Encyclopedia, http://www.absoluteastronomy.com/encyclopedia/d/de/developed_country.). The reason for the geographic boundaries of this book is to maintain a focus on a very large amount of diverse information. The industrialized countries have similar agricultural industries, and thus similar occupational and health safety concerns. Agriculture in developing countries is very different from that of industrialized countries. General environmental health and economic issues in developing countries, such as sanitation, water quality, infectious diseases, and nutritional problems, outweigh their occupational health problems. Even though a much higher portion of the population in developing countries work in agriculture compared to industrialized countries, related health problems are so different that it is difficult to deal with them to any degree in this same text.

Although there may be a greater amount of information in this book that pertains directly to North America, we have made attempts to cover information that pertains to agriculture in industrialized countries generally. One of us (AT) is from Sweden, and one of us is from North America. We are co-authors on all chapters for the purpose of assuring that we have input and perspective from the two continents. Furthermore, we have chosen multiple reviewers that are international experts in their fields. The reviewers and their backgrounds are listed elsewhere in this text.

Definitions or terms used for people in industrialized countries working in agriculture have statistical as well as cultural significance. The agricultural work force is not all farm workers. The majority of the work force in agriculture consists of members of family farming operations. These are operations where all members of the family may participate in the work or management of the operation. One of the parents is usually the principal operator or owner-operator, although spouses are often co-principal operators. They are management and labor combined. The terms used for the latter in this book are farm owner-operator, farmer, rancher, or producer. These terms can generally be used interchangeably; however, rancher refers more commonly to a cattle or sheep producer who raises his or her animals on pasture or open range. The term farm worker is reserved for those wage earners who are employed by the owner-operator, farmer, rancher, or producer. (More details of these definitions are seen in Chapter 1, "Introduction and Overview," under the heading "Breakdown of the Production Agricultural Work Force and Types of Farms.")

AUDIENCE FOR THIS BOOK

The primary audience for this book includes health science students with an interest in rural and agricultural health and safety. Specific health science students targeted include physicians, nurses, physician assistants, nurse practitioners, veterinarians, emergency medical technicians, and paramedics, among others. Students in industrial hygiene, safety, and other occupational health fields are also targets of this book. This book is meant to be a reference text for practicing health care professionals in all the fields mentioned. There is an emphasis on a medical approach to the terminology, but the detail and terminology are modified so as not to exclude comprehension by non-health care professionals.

The principal authors of this book are both from an agricultural background as well as a health profession background. The authors are still involved in agriculture today. Furthermore, both authors have been professionally involved in agricultural medicine for more than 30 years. Our aim is to bring our practical experience to the pages of this textbook, so that the exposures, risks, and preventive measures are based on practical and realistic scenarios and based on firsthand involvement with agricultural production and its people, rather than just a reporting of the published literature.

ACKNOWLEDGMENTS

The initial impetus for this book came from the Agricultural Health Scientific Working Committee of the International Commission of Occupational Health. The members of this committee represent occupational health professionals from a broad array of backgrounds and countries.

This book is intended to cover issues in the rural and agricultural environments of North America, countries of the European Union, Australia, and New Zealand. To achieve international coverage, we have two reviewers for most chapters: one from North America and one from a country of the European Union. These reviewers were selected for their international recognition and experience in rural and agricultural health and safety. These highly qualified reviewers were charged with addressing the topic matter for broad geographic coverage of industrialized western countries, as well as addressing factual accuracy. Following is a list of the chapters of the book and the respective reviewers and their titles and locations.

Chapter 1: Introduction and Overview
- Jacek Dutkiewicz, Ph.D.
  Professor and Head
  Department of Occupational Biohazards
  Institute of Agricultural Medicine
  Lublin, Poland

Chapter 2: Special Risk Populations in Agricultural Communities
- Ken Culp, R.N., Ph.D.
  Associate Professor and Director
  Occupational Health Nursing
  College of Nursing
  University of Iowa
  Iowa City, Iowa USA

- Barbara Lee, R.N., Ph.D.
  Director
  Farm Medicine Center
  Marshfield Medical Foundation
  Marshfield, Wisconsin USA

- Kirsti Taattola, Agricultural Engineer
  Director

National Centre for Agricultural Health
Kuopio, Finland

Chapter 3: Agricultural Respiratory Diseases
- Susanna Von Essen, M.D., M.P.H.
  Professor
  Pulmonary Medicine
  University of Nebraska Medical Center
  Omaha, Nebraska USA

Chapter 4: Agricultural Skin Diseases
- Christopher Arpey, M.D.
  Associate Professor
  Department of Dermatology
  College of Medicine
  University of Iowa
  Iowa City, Iowa USA

- Paivikki Susitaival, M.D., Ph.D.
  Department of Dermatology
  North Karelia Central Hospital
  Joensuu, Finland

Chapter 5: Cancer in Agricultural Populations
- Charles Lynch, M.D., M.S., Ph.D.
  Professor
  Department of Epidemiology
  College of Public Health
  University of Iowa
  Iowa City, Iowa USA

Chapter 6: Health Effects of Agricultural Pesticides
- Laurence Fuortes, M.D., M.S.
  Professor
  Department of Occupational and
    Environmental Health
  College of Public Health
  University of Iowa
  Iowa City, Iowa USA

Chapter 7: General Environmental Hazards in Agriculture
- William Field, Ph.D.
  Associate Professor
  Department of Occupational and
    Environmental Health
  College of Public Health
  University of Iowa
  Iowa City, Iowa USA

- Ragnar Rylander, M.D., Ph.D.
  Professor Emeritus
  Department of Environmental Medicine
  University of Gothenburg
  Gothenburg, Sweden

Chapter 8: Musculoskeletal Diseases in
Agriculture
- Steve Kirkhorn, M.D., M.P.H.
  Medical Director
  National Farm Medicine Center
  Marshfield Medical Foundation
  Marshfield, Wisconsin USA

- Dan Anton, P.T., Ph.D.
  Assistant Professor
  Department of Occupational and
      Environmental Health
  University of Iowa
  Iowa City, Iowa USA

- Eva Vingård, M.D., Ph.D.
  Professor
  Department of Medical Sciences,
      Occupational and Environmental
      Medicine
  Academic Hospital, Uppsala, Sweden

Chapter 9: Physical Factors Affecting Health in
Agriculture
- Gösta Gemne, M.D., Ph.D.
  Professor
  Bygdøy Allé
  Oslo, Norway

Chapter 10: Psychosocial Conditions in
Agriculture
- Lennart Levi, M.D., Ph.D.
  Professor Emeritus
  National Institute for Psychosocial Medicine
  Stockholm, Sweden

Chapter 11: Acute Agricultural Injuries
- Risto Rautiainen, Ph.D.
  Assistant Professor (Clinical)
  Department of Occupational and
      Environmental Health
  College of Public Health
  University of Iowa
  Iowa City, Iowa USA

Chapter 12: Veterinary Biological and
Therapeutic Occupational Hazards
- Danelle Bickett-Weddle, D.V.M., M.P.H.
  Associate Director
  Center for Public Health and Food Security
  College of Veterinary Medicine
  Iowa State University
  Ames, Iowa USA

Chapter 13: Agricultural Zoonoses
- Danelle Bickett-Weddle, DVM MPH
  Associate Director Center for Public
      Health and Food Security
  College of Veterinary Medicine
  Iowa State University
  Ames, Iowa USA

- Gregory Gray, MD MPH
  Professor
  Department of Epidemiology
  College of Public Health
  University of Iowa
  Iowa City, Iowa USA

Chapter 14: Prevention of Illness and Injury in
Agriculture
- Risto Rautiainen, Ph.D.
  Assistant Professor (Clinical)
  Department of Occupational and
      Environmental Health
  College of Public Health
  University of Iowa
  Iowa City, Iowa USA

- Mark Purschwitz, Ph.D.
  Research Engineer
  National Farm Medicine Center
  Marshfield Medical Foundation
  Marshfield, Wisconsin USA

Agricultural Medicine
Rural Occupational and Environmental Health
for the Health Professions

1
Introduction and Overview

Kelley J. Donham and Anders Thelin

INTRODUCTION

Several terms have been used to describe fields of endeavor aimed at the health of our rural and agricultural communities. These terms are often used interchangeably although they have distinct histories and meanings. Figure 1.1 illustrates the various terms used, and their relationships. The two primary terms used to describe health-related activities in rural areas are **rural health** and **agricultural health and safety. Rural health** is defined by the National Rural Health Association as **a field of endeavor aimed at the development and support of health care services (providers and facilities) that are accessible and appropriate for all rural residents** (NRHA 2005). The field of rural health does not focus on any particular diseases, occupation, ethnic group, or prevention. Rural health focuses on provision of services (health care personnel and facilities) aimed to take care of the usual episodic illnesses of rural residents (Mutel and Donham 1983). **Agricultural health and safety** on the other hand is a broad term that is used to describe a **field of practice and associated endeavors aimed at reducing occupational injuries and illnesses in agricultural populations**. Underneath this latter umbrella term, there are several interrelated fields (**agricultural medicine, agromedicine, and agricultural safety**). Although each term is associated with activities aiming to decrease injuries and illness in agricultural populations, each term has a slightly different history, concept, focus, professional makeup, and culture.

Within the broader term of agricultural safety and health, the term agricultural medicine has been used for at least 50 years, describing a specialty discipline of the broader field of occupational medicine and occupational health. Table 1.1 summarizes the key dates and events in the development of agricultural medicine and agromedicine. Bernardo Ramazzini, an Italian physician of the early 1700s, has been generally recognized as the father of occupational medicine. His book (*Diseases of Workers,* translated from Latin) describes in detail many occupation-related diseases—many of which we still recognize today—that he observed in his farm patients (Ramazzini 1713). The history of occupational medicine and occupational health generally (and agricultural medicine specifically) can be traced to his writings.

In more modern times, a physician named Toshikazu Wakatsuki in Japan developed a strong outreach program to his farming patients following World War II. Wakatsuki began his tenure at Saku Central Hospital in the Nagano Prefecture of central Japan in 1945. He spent his professional lifetime transforming the care of the rural farming community from what may have been considered benign neglect to a world model outreach and prevention program. He established the Japanese Rural Medicine Association, and was one of the principal founders of the International Association of Agricultural Medicine and Rural Health. His humble, dedicated, humanitarian approach to his mission earned him the Ramon Magsaysay Award (the Asian version of the Nobel Peace Prize) in 1976 (Wakatsuki 2003). In Europe, the Institute for Rural Occupational Health was initiated at Lublin, Poland, in 1951, the first

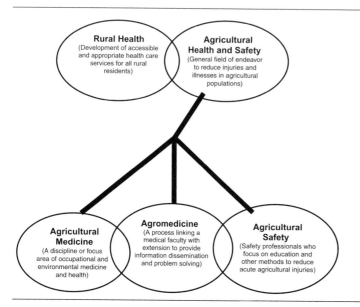

Figure 1.1. Terminology/fields of endeavor addressing the health of rural residents, owners/operators, and workers in production agriculture.

Table 1.1. History of Agricultural Medicine

- 1713—Bernardo Ramazzini published his book *Diseases of Workers.*
- 1945—Toshikazu Wakatsuki established an outreach medical and prevention program for the farming community at Saku Central Hospital, Japan.
- 1951—Institute of Agricultural Medicine, Lublin, Poland, established.
- 1955—Institute of Agricultural Medicine, University of Iowa, established.
- 1961—Founding of the International Association of Agricultural Medicine and Rural Health in Tours, France.
- 1965—Founding of the *Journal of the International Association of Agricultural Medicine and Rural Health.*
- 1973—Institute of Rural Environmental Health (Occupational Health and Safety Section), Colorado State University, established.
- 1973—The term *Agromedicine* first used by John Davies (James 1994).
- 1974—Beginning of the Agricultural Medicine Training program at the University of Iowa.
- 1976—Peer-reviewed article published in *Minnesota Medicine* outlining the didactic areas of Agricultural Medicine (Rasmussen and Cole 1976).

- 1979—Article published in the *Journal of the Royal Society of Medicine,* which includes the first definition of Agricultural Medicine (Elliott 1979).
- 1982—Article published in the *Journal of Family Practice,* which includes a more detailed definition of Agricultural Medicine and differentiates it from the field of Rural Health (Donham and Mutel 1982).
- 1984—Establishment of the first Agromedicine Program as a consortium of the Department of Family Medicine at the Medical University of South Carolina and Clemson University.
- 1986—Institute for Agricultural Medicine and Rural Environmental Health, University of Saskatchewan, Canada, established.
- 1988—Founding of the North American Agromedicine consortium.
- 1988—National Farm Medicine Center, Marshfield, Wisconsin, established.
- 1994—Founding of the *Journal of Agromedicine.*
- 1994—Founding of the *Annals of Agricultural and Environmental Medicine.*
- 1995—Founding of the *Journal of Agricultural Safety and Health.*

apparent research institute to focus on the occupational health of farmers. The Institute at Lublin had a name change to the Institute of Agricultural Medicine in 1984. This Institute houses a multidisciplinary team of some 150 scientists studying occupational health and care of Poland's rural and farming community (WSI IM 2004). The Institute at Lublin founded a new journal in 1994 titled the Annals of Agricultural and Environmental Medicine, which publishes peer-reviewed scientific articles on a wide variety of occupational and environmental health problems among agricultural workers (AAEM 2005).

The first use of the term **agricultural medicine** can be traced to 1955 with the founding of the Institute of Agricultural Medicine (IAM) in the U.S. The Institute of Agricultural Medicine was founded at Iowa in 1955 within the College of Medicine. This Institute was organized with a multidisciplined faculty that included a physician, an industrial hygienist, a veterinarian, a microbiologist, an anthropologist, an agricultural engineer, and a toxicologist (Berry 1965; Knapp 1965). Existing primarily as a research institute early on, a training program in agricultural medicine and rural health was initiated in 1974 for health care professional students and graduate students in occupational health specialties. This institute exists today as the Institute for Rural and Environmental Health within the College of Public Health at the University of Iowa. The research, training, and outreach programs have a principal focus on occupational health problems of the agricultural community. The term and concept of agricultural medicine became more sanctioned with the founding of the International Association of Agricultural Medicine and Rural Health in 1961 at Tours, France (Elliott 1979; IAAMR 2005). This organization is a multidisciplined group of medical professionals, health care practitioners, and agriculturalists, aimed at identifying and controlling health and environmental problems in rural and agricultural communities. This organization has regional meetings and a major congress every 3 years. The organization also has connections with the World Health Organization (WHO), International Labor Organization (ILO), and other international health organizations.

Other national/international organizations that have developed along the agricultural medicine model include the Agricultural Health Scientific Committee of the International Commission of Occupational Health and the French *Agricole Mutualite*.

A variant on the term **agricultural medicine** was used with the founding of the National Farm Medicine Center in 1981 at Marshfield, Wisconsin. This medical research and outreach group to the farm community was developed within a private multispecialty physician group, but it has evolved into a multidisciplined group, focusing on occupational illnesses in the farming community (Marshfield Clinic Research Foundation 2005). The University of Saskatchewan developed the Center for Agricultural Medicine in the early 1990s, and since 2001 has been known as the Institute for Agricultural, Rural, and Environmental Health (IAREH 2005).

Franklin Top (1962) set out the didactic basis of agricultural medicine, which included the importance of understanding the processes and work environment of agriculture, acute injuries, sanitation, allergies, farm chemicals, zoonoses, and social and mental health. Rasmussen and Cole (1976) expanded on Top's comments and further established the didactic content of agricultural medicine. Berry (1965; Berry 1971) established the first published articles regarding the peculiarities of agricultural employment relative to other industrial employment. He also commented on the occupational health of the farm community, and suggested a research agenda for agricultural medicine (Berry 1965). Elliott (1979) published the first attempt at a definition of agricultural medicine, which used a variation of the definition of industrial hygiene. One of us (KJD) expanded on Elliott's definition, to include clinical medicine and public health: ***Agricultural medicine is ". . . the anticipation, recognition, diagnosis, treatment, prevention, and community health aspects of health problems peculiar to agricultural populations"*** (Donham and Mutel 1982). Agricultural medicine is a discipline, a specialty area of occupational and environmental medicine and public health. It is multidisciplinary in its approach and involves professionals from all the clinical and basic health sciences and veterinary medicine. Agricultural medicine has a research base and a core of didactic information. This didactic core of information serves as the basis for training programs for health or safety professionals who work in the area. There is an international professional organization and two journals that have agricultural medicine in their title (the *Journal of the International Association of Agricultural Medicine and Rural Health,* and the *Annals of Agricultural and Environmental Medicine*).

Another variant to the term agricultural medicine, **agromedicine,** was first used in 1976 and defined in more detail in 1978 (Davies and others 1978). Dr. John Davies expressed his concern over the public and agricultural producers' fears regarding the health and environmental effects of pesticides. He felt there was a need for the medical and agricultural health communities to work more closely together on this issue, and called for an "agromedicine approach." Davies descried that pesticides were clearly a boon for agricultural production, but the occupational, environmental, and public's concern created problems for their use and expansion with concerns of fear rather than science driving regulatory efforts. Dr. Stanley Schumann at the medical school of South Carolina expanded on the agromedicine concept. He observed that the Cooperative Extension Service of the Land Grant College had agricultural specialists in every county of the state to disseminate information from the research campuses to the farmers in the countryside. Extension agents are located in rural areas to assist with problem solving that might arise regarding production issues. Dr. Schumann thought that medical and health information and problem solving could and should be disseminated in a similar manner, but in full collaboration and in context with agricultural production information (James 1994). The S. Carolina Agromedicine Program was initiated in 1984 as a joint collaboration between the Division of Family Medicine at the Medical University of South Carolina and the Agricultural Extension Service of Clemson University (MUSC 2005). The program grew, gaining first regional then national interest, and in 1988, the National Agromedicine Consortium (NAC) was established. This organization holds annual professional conferences, and publishes a scientific, peer-reviewed journal (the *Journal of Agromedicine*) (NAC 2005).

The differences between the history and concepts of agricultural medicine and agromedicine in their beginnings were clear. However, their basic goal of improved health of the agricultural community was common. The practice of both

areas has moved toward one another, and in general the similarities today are greater than the differences. Both are fields of agricultural health and safety, with the objective of controlling or preventing agricultural occupational and environmental illnesses and injuries. They are both multidisciplinary in their approach. Both promote the importance and understanding of agriculture and the culture of its people. The basic difference is that agricultural medicine is a health/medical **discipline** (a subspecialty of occupational and environmental health), and agromedicine is a **process** linking the medical faculty (usually Family Medicine) with the agricultural college (Extension).

Agricultural safety has had a long history starting in North America in the 1940s. The first individuals involved were a combination of extension agents, insurance loss control personnel, and Farm Bureau representatives among others. Professionals who became involved in this effort brought an orientation from the developing field of industrial safety. The methods have focused on promoting awareness of safety problems resulting in acute injuries in the farm community. The National Institute for Farm Safety is the primary professional organization for this group of professionals (NIFS 2005). The principal scientific journal of this group is the *Journal of Agricultural Safety and Health*.

Although the three terms described above have different histories, professional makeup and culture, they are tied together by the common goal of reducing illnesses and injuries among agricultural populations. Furthermore, the related professional journals and societies provide a common ground for these groups. For example, professionals working in any of the three areas of agricultural health and safety publish in all of the agricultural health and safety journals mentioned above. Furthermore, the North American Agromedicine Consortium is also the North American Congress affiliate of the International Association of Agricultural Medicine and Rural Health. In the U.S., The National Institute of Occupational Safety and Health (NIOSH) funds an agricultural health initiative that supports research and outreach activities that pull together professionals aligned with all three sectors of Agricultural Safety and Health. Often professionals in the fields of occupational safety and health attend professional meetings held by other groups. For example, in 2004 and 2005 combined national meetings were cohosted by the National Institute of Farm Safety, the National Institute of Occupational Safety and Health Agricultural Health Centers Group, and the Agromedicine Consortium.

WHAT IS AN AGRICULTURAL HEALTH AND SAFETY PROFESSIONAL?

There are many individuals who can make a difference in the health and safety of our agricultural communities, including health care practitioners who may serve the agricultural population as only one part of their patient or client responsibilities. There are also professionals who deal with agricultural health and safety on a full-time basis, who have basic core knowledge and competencies in agricultural health and safety. Figure 1.2 describes the roles and background education needed to fill these different niches.

The Primary Care Physician, Nurse, Allied Health Professionals, and Veterinary Practitioners

Health care professionals and veterinarians in rural communities have an excellent opportunity to address the issues of agricultural health and safety with their patients and/or clients. They have frequent contact with farmers and their families. They are in a position of respect and credibility within their community. The agricultural health and safety training program at the University of Iowa (2005) has had the following health professionals as students who are contributing to the health and safety of their respective communities: 1) primary care physicians, 2) nurses, 3) nurse practitioners, 4) physician assistants, 5) veterinarians, 6) respiratory therapists, 7) emergency medicine technicians and paramedics, 8) occupational therapists, 9) physical therapists, and 10) public health practitioners. Although these professionals may not consider themselves agricultural health and safety professionals, if equipped with some specific agricultural medicine training, they can play an important role in addressing the agricultural health and safety issues in their communities. The authors' goal for this text is to arm health professionals with knowledge that can facilitate their practice of fundamental agricultural medicine, including anticipation, recognition, evaluation or diagnosis, treatment, and prevention of occupational and environmental illnesses in our agricultural communities.

Mutel and Donham (1983) have proposed an expanded role for the rural practitioner that includes clinical work, community service, health education, and research. Issues of agricultural health and safety can fit into each of these areas. The physician in the community often serves as the health leader in the community, and therefore his or her actions may have a profound effect on their community's health activities. Nurses have found and developed important roles in agricultural health (Fleming 2004; Gay and others 1990; Lundvall and Olson 2001; Walsh 2000). Veterinarians have an excellent opportunity to play a role in the health and safety of their farm clients. They frequently visit their clients' farms, they know their clients' exposures, and they have the medical background to make a connection to those exposures and possible health risks. Furthermore, veterinarians are one of the most trusted professionals for the farm population in regard to health and medical issues (Thu and others 1990).

The Agricultural Health and Safety Specialist

There is a cadre of professionals in North America, the EU, Australasia, and South East Asia that devotes the majority of their professional time to activities dealing with the health and safety of agricultural populations. They are associated with extension services, university or hospital research programs, or occupational health programs for the farming community. These programs provide the research basis and training for health professionals to practice in their field. They provide outreach and service to the agricultural community. There are many nongovernmental agricultural health and safety organizations that provide outreach to the agricultural communities and advocacy for their cause. These people and organizations are the core of the professional field of agricultural safety and health. They make up the professional organizations, write and publish the scientific manuscripts in the field, and advocate public policy to address farmers' health issues. Chapter 14 lists the organizations in which these agricultural health and safety specialists serve.

TRAINING OF AN AGRICULTURAL HEALTH PRACTITIONER AND AGRICULTURE HEALTH AND SAFETY SPECIALIST

The Health Professional Practitioner

There are very few formal programs in the world that train physicians or other health professionals or veterinary medicine practitioners regarding agricultural illnesses and injuries. Most health professionals who end up practicing in rural areas have to learn about these issues through experience or through informal continuing education. There are exceptions. The University of Alabama provides an agromedicine training experience by linking medical students with extension agents to visit farms, review the hazards, and discuss issues directly with farmers and their families (Wheat and others 2003). The University of Iowa provides case-based learning experiences on agricultural injuries and illness, in addition to an elective 3–credit-hour course in agricultural medicine for medical students and physician assistant students (University of Iowa 2005).

There are several graduate programs and continuing education programs that provide training for nurses (Jones and others 2004; University of Minnesota 2005; University of Iowa 2005). The University of Iowa provides a certificate program in agricultural occupational health for health care practitioners.

Lantbrukshalsan in Norway is developing an online interactive educational program in agricultural health and safety. The first level is for farm-

ers, the second level is for farm schools and localized advisers, and the third level is for researchers and teachers at universities.

The Agriculture Health and Safety Specialist

There are few formal training programs for agricultural health and safety specialists. Some organizations have workshops; some universities have a course or two in agricultural health and safety, plus research experience that can enhance the training in this area. Only a few universities have a formalized didactic training specialty in agricultural medicine or agricultural safety and health that leads to a degree or certificate. Table 1.2 provides information about several training programs in agricultural health and safety.

Individuals interested in formalized agricultural health and safety training may come from several different backgrounds, including any of the health sciences, occupational medicine and health, veterinary medicine, public health, agricultural engineering, or agriculture. The key is to have the combination of an agricultural training—or at least a background in agriculture—along with a related health, engineering, or educational training. Also essential to the training of an agricultural health and safety specialist is core knowledge in occupational exposures, resulting illnesses and injuries, and prevention. Figure 1.2 provides a diagram of this author's (KJD) concept of a recommended background and training to make up an agricultural health and safety specialist.

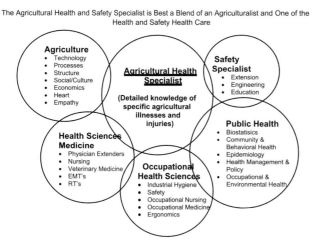

Figure 1.2. The components of the field of agricultural medicine (agricultural health and safety and the specialist).

Table 1.2. Training Programs in Agricultural Health and Safety for Health Professionals

	Audience	Agricultural Health Program Name or Courses Available	Comments	Contact Information/URL
UNIVERSITIES—CANADA				
University of Saskatchewan	Graduate students	Public Health and the Agricultural Rural Ecosystem (PHARE)		**Institute of Agricultural Rural & Environmental Health** I.ARE.H. P.O. Box 120, RUH 103 Hospital Dr. Saskatoon, SK, S7N 0W8
UNIVERSITIES—U.S.				
University of Iowa	Graduate students and Health Sciences students	Agricultural Health and Safety Training program	M.S., Ph.D. with Agricultural health focus. Also Certificate course in Agricultural Safety and Health	**Kelley Donham, DVM or LaMar Grafft, M.S.** Institute for Rural Environmental Health University of Iowa Iowa City, Iowa 52242 Kelley-Donham@UIOWA.edu
University of Kentucky	Undergraduate students with Biological and Agricultural Engineering, Technology or related field **Course BAE 432** Undergraduate Industrial Engineering students **Course IE 544**	**No recognized Agricultural Health and Safety degree or program.** **Course BAE 432:** Agricultural and Environmental Safety and Health 3 credits **Course IE 544:** Occupational Biomechanics is a related, but not exclusively Agricultural course.	Undergraduate Minor in Agricultural and Environmental Technology available for students interested in safety and the application of engineering technology analysis in agricultural and environmental systems.	**Larry Piercy, M.S.** lpiercy@bae.uky.edu **Julia Storm, M.S.** N.C. State University Department of Environmental and Molecular Toxicology Box 7633 Raleigh, NC 27695-7633 919-515-7961 julia_storm@ncsu.edu
University of Eastern North Carolina				
Pennsylvania State University	Undergraduate students in Agricultural and Industrial Health and Safety	**No recognized Agricultural Health and Safety degree program.**	Offers a BA level Occupational Safety and Health Degree program.	**Dennis Murphy, Ph.D.** 221 Agricultural Engineering Building

(continued)

Table 1.2. Training Programs in Agricultural Health and Safety for Health Professionals (*continued*)

	Audience	Agricultural Health Program Name or Courses Available	Comments	Contact Information/URL
	Course ASM 326 Senior/Beginning graduate **Course ASM 426**	**Course ASM 326:** Hazard Identification and Control in Production Agriculture and Related Business **Course ASM 426:** Management of Safety and Health Issues in Production Agriculture and Related Business	Most students who take these two courses are seniors.	University Park, PA 814-865-7157 djm13@psu.edu
Ohio State University	Primarily Agricultural students (ASM majors) or Construction students (CSM majors); other students include ASM minor, Landscape Construction, Agricultural Business, Animal Science, Agricultural Education, Horticulture, Turfgrass Management, Agronomy	**No recognized Agricultural Health and Safety degree program.** **Course ASM 600:** Agricultural Health and Safety 3 credits **Course CSM 600:** Construction Health and Safety 3 credits	Mostly undergraduate students. Courses count as graduate credit.	**Margaret Owens, Ph.D.** 204 Agricultural Engineering 590 Woody Hayes Dr Columbus, OH 43210 614-292-1731 owens.162@osu.edu
Purdue University				**William Field, Ph.D.** Department of Agricultural & Biological Engineering 225 South University Street Office: ABE 218 West Lafayette, IN 47907-2093 765-494-1162 field@purdue.edu
University of Illinois	All programs award both undergraduate and graduate credit.	**Minor in Agricultural Safety and Health.** Minor at college level for approval.	Primary goal is to provide students with background in Agricultural safety and health for use	**Robert (Chip) Petrea, Ph.D.** Agricultural Engineering Department University of Illinois

Table 1.2. Training Programs in Agricultural Health and Safety for Health Professionals (*continued*)

	Audience	Agricultural Health Program Name or Courses Available	Comments	Contact Information/URL
	Course TSM 422: Agricultural Health and Illness Prevention 3 credits **Course TSM 423:** Agricultural Safety and Injury Prevention 3 credits **Course TSM 425:** Agricultural Safety and Health Interventions 3 credits Agricultural Safety and Health Traineeships 3–4 credits	Will be offered as Minor within the College of Agriculture, Consumer and Environmental Sciences, UIUC	directly in graduate studies (e.g., Industrial Hygiene) or indirect use within Secondary Agricultural Education, Cooperative Extension, Public Health, etc.	1304 W. Pennsylvania Urbana, IL 61801 217-333-5035 rcp@sugar.age.uiuc.edu
University of Minnesota	Course offered to Juniors, Senior Graduate students No degree available here **Agricultural Safety and Health** 3 credit hours (offered only web-based)	**No recognized Agricultural Health and Safety degree.**		
Utah State University				**William Popendorf, Ph.D.** popendorf@biology.usu.edu
NON-UNIVERSITY ORGANIZATIONS **National Institute for Farm Safety**	Extension Safety Specialists. Insurance risk managers. Farm Bureau Insurance risk managers. Machinery manufacturer product safety specialists. Academic health and safety specialists.	An annual meeting in June provides continuing education sessions. Occasionally a day-long preconference agricultural health and safety workshop is provided.	NIFS is a professional organization for agricultural health and safety specialists. One of the organization's objectives is to educate the membership. However, there is no formalized academic training or certification program.	Contact Chair of the NIFS Professional Improvement Committee, which can be found at http://www.ag.ohio-state.edu/%7Eagsafety/NIFS/officers.htm#Standing.
Norwegian Lantbrukshalsan				**Anne Heiberg** anne.marie.heiberg@lhms.no

DEMOGRAPHICS OF THE AGRICULTURAL WORK FORCE

Worldwide, agriculture is the largest employer. Over 40% of the world's 3 billion workers (including self-employed owner-operators and hired farm workers) are employed in agriculture (ILO 2004). Relative to developing countries, where as high as 70% of the total population is engaged in agriculture, a much smaller percentage of the total population (around 3%) are engaged in agriculture as owner-operators. (However, it should be noted that even in industrialized countries, the agricultural sector is a very significant portion of the total work force.) Compared to over 190 million farms in China with a median acreage of under 2 hectares (Food and Agricultural Organization of the United Nations Economic and Social Department The Statistics Division 1997), the EU has about 7 million farms (European Commission 2004). Mexico has about 4 million farms (Mexico Agricultural Census 1991), the U.S. has about 2 million farms (Bureau US 2005), Canada has about 246,000 farms (Canada 2001), and New Zealand has about 70,000 farms (Food and Agricultural Organization of the United Nations Economic and Social Department The Statistics Division 2002).

It is important to understand the evolving structure of agricultural enterprises and the terminology used to understand the demographics of the agricultural work force. The agricultural work force includes principal operators (also called *owner-operators*), nonwage family members, and wage-earning employees or farm workers (indigenous and foreign-born nationals). Additionally, large and corporate-style farms employ farm managers. The vast majority of the agricultural work force across industrialized countries is involved in family-style operations. These operations include a principal operator, who is also the owner-operator, and nonwage earning family members. Large family farms may have hired either full-time workers or seasonal labor.

The principal features of these family operations are that residence, ownership, management, and labor are all unified in the family operations. Global economic forces have created economic stress on the traditional family farm operations, enhancing their decline in numbers that began in the 1940s. On the other hand, there is a growing component of large and corporate-style operations that include large family corporations or nonfamily corporations. These types of operations separate labor and management, and often the farm and residence are separated. Some operations are connected to large multinational vertically integrated food conglomerates. The corporate-style farms employ the majority of hired farm workers. These workers may come from the local area or foreign countries, and they travel for employment purposes (migrant seasonal workers).

There is another growing style of agricultural enterprise called *niche farming*, which is a new variant of the traditional family farm operation. These operations produce and market products (e.g., organic foods, exotic food crops, and livestock) that have small, often localized, markets not met by the traditional family or corporate family operations (US Department of Labor 2002–2003). Figure 1.3 illustrates the breakdown of these types of operations for the U.S. This pattern is similar in all the developed countries discussed in this book. However, the large corporate-style agriculture has developed more rapidly in the U.S., compared to other industrialized countries. Although these large farms make up less than 5% of the total farms, they contribute about 50% of the total U.S. commodity production (US Department of Labor 2002–2003).

BREAKDOWN OF THE PRODUCTION AGRICULTURAL WORK FORCE AND TYPES OF FARMS

Family Farms

Family farming historically developed in the Western industrialized countries as Northern Europe evolved from a feudal system to democratic nation states in the late 1700s and early 1800s. Opportunities for land ownership, farming opportunities, and religious freedom in the 1800s drew many Northern Europeans to Australia, New Zealand, and North America. They developed family farming in those countries with the encouragement of the colonial and early indigenous governments, as a method to "settle the land." Millions of emigrants came to these countries and found the "New World dream." During the same time periods, greater democratization, reformed land tenure, and socialization of Europe allowed family farming to develop in those coun-

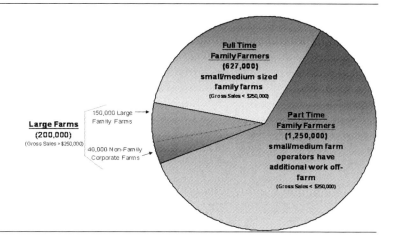

Figure 1.3. Demographics of farm types (U.S.). (This figure also is in the color section.)

tries. History, political and social change, and perhaps genetics, created the dominant culture of the agricultural work force called *family farming*. Family farms are the dominant type of farm still today in industrialized countries, making up more than 95% of the total farm operations in the developed countries. The trend in industrialized regions has been a decrease of about 10% of the farms over the previous decade and a corresponding decrease in the number of farm residents and relative increase in farm size.

There is an important culturally relevant point as to how one should refer to a member of a family farm or ranch unit. Farm family members consider themselves to be owner-operators, managers, and self employed (not farm workers). A farm worker, to them, is someone who is an employee, a person with less socioeconomic status. Family farm members are proud and may feel being referred to as a farm worker as an indication of disrespect, or naiveté (although they would not likely make mention of that because they are also humble). Acceptable terms for adult farm family members would include *farmer, producer* (with a prefix of specific commodity such as pork, wheat, etc.), *rancher, grower, owner-operator,* or *principal operator*. As gender equality has evolved in the past two decades, women also want to be recognized in a similar fashion. "Farm wife" may not always be an acceptable term, as women also are principal operators or joint principal operators. Children under teenage years may be referred to as "farm children," but adolescents who are actively involved in the oper-

ation would be proud to be referred to as a farmer, rancher, etc., just as an adult would.

Principal Operator

Approximately 80% of the principal operators of family farms in developed countries are male. In most operations, they put in the majority of hours of work on the farm. They typically are older than the general mean of the work force in their country. For example, the average age (in 1997) of the male principal operators in the U.S. was 54 years, and the trend is an increasing age (Bureau US 2005). This compares to a mean of about 45 years in the general work force (US Department of Labor 2002–2003). This same trend of aging in the family farm work force is seen in other industrialized countries, including the European Union, Canada, and Australia (Canada 2001; Economics 2005; European Commission 2004). A total of 28% of the farmers in the EU are over 65 years and only 9% are under 35 (European Commission 2004).

The vast majority (over 95%) of the U.S. family farm work force is Caucasian, primarily of Northern European descent. In the U.S., about 1.5% of the principal farm operators is of Hispanic origin, 1% is black, and about an additional 1% includes Native American Indians and Asian or Pacific Islanders. Over 70% of the principal operators live on their farm (Bureau US 2005; USDA Agricultural Statistics 2004).

Family Members

Women on farms have long been a vital component in the family farming enterprise. Their role varies from part-time assistance with bookkeeping, to full partner in the operation, to principal operator. Women principal operators have been increasing in both North America and Europe. Although only 5% of principal farm operators in Canada are women, 30% are jointly operated by a man and a woman (Canada 2001). Nearly 25% of farms in the U.S. are operated by women. In Finland, a high percentage of women consider themselves to be full-time farmers. In Poland, over 60% of the farms are operated by women (European Commission 2004) as they tend to stay on the farm if their husband dies or takes a job off the farm.

A high percentage of both men (30–50%) and women (45–60%) on family farms in North America have additional employment off the farm (US Department of Labor 2002–2003). This has been an increasing trend over the past three decades in all industrialized countries, as profit margins have decreased, making it a necessity to have additional income for the family. In the U.S., taking an off-farm job is also motivated by the possibility of obtaining health insurance from the employer, as insurance may otherwise cost a family up to $8,000 per year. This additional employment has increased the total work load and stress on modern family farm operations, and it increases risk for adverse mental and physical health outcomes, as is discussed in Chapters 10 and 11 of this book (Spengler and others 2004).

Children typically begin more independent work on farms at about the age of 10 years. Boys are typically more involved in the farm work than girls, and their farm injury rates bear that out; they are about twice as high as for farm girls (Frank and others 2004). In Sweden girls 14–16 have the same high injury rate as boys, but it is related to contacts with horses.

Farm Workers

Farm workers are those who are hired and receive wages for farm work. Of the estimated 1.3 billion persons employed in agriculture worldwide, nearly half are farm workers; 38% are migrant or seasonal farm workers, and 50% of these workers are women (ILO 2004). Farm workers make up an important part of the agricultural labor force in all industrialized countries. Farm workers may come from the local area (indigenous), or they may come from distant and often foreign countries (migrant and seasonal farm workers).

INDIGENOUS FARM WORKERS

Indigenous farm workers may have the same culture and socioeconomic status of the owner operator. They in fact may be farm youth who work seasonally or part time on another person's farm. This group of workers may be exposed to the same hazards as other workers but do not have some of the same inherent socioeconomic risks as do the nonindigenous workers. Generally speaking in the U.S., indigenous farm workers are similar in number to foreign-born migrant or seasonal workers.

MIGRANT AND SEASONAL FARM WORKERS

The developed countries depend highly on foreign-born workers to conduct farm work. The general term used for these workers is *migrant and seasonal workers,* as they tend to move from place to place depending on seasonal work, returning to their home place in the "off season." The EU employs 4.5 million migrant and seasonal workers; about 500,000 of these are from outside the EU. U.S. farm operators hire (at some time during a year) about 2.5 million foreign-born migrant and seasonal workers (the highest among industrialized countries) (ILO 2004). In North America, they make up nearly a quarter of the agricultural work force. Many of these workers have found more permanent employment in larger industrial-style farms and have or are in the process of "settling out" in the community. For Europe and Australasia, the percentage of migrant and seasonal workers of the agricultural work force is somewhat lower than in the U.S. (ILO 2004). The vast majority of these workers are foreign born. In North America, they are largely Hispanics from Mexico. However, Central and South America contribute workers as well as Bosnia, Asia, Africa, and the Caribbean Islands. The US Department of Labor Statistics (2002–2003) indicates that only 30% of U.S. farmers hire one or more employees, and just over 8% of the farms hire more than 10 employees. The latter figure is significant in the U.S. because federal worker protection laws can be enforced only on those farms with more than 10 employees. Australian orchardists often depend on student labor from the EU or North America, who work part time to help pay for their vacation travel expenses.

Corporate-Style Farms

Previously mentioned, one of the megatrends in industrialized countries' agriculture has been an increase in size of operations, along with a loss of family farms and an increase of corporate-style farms. The reasons for this trend is discussed in more detail later. However, it has developed farms that are not only large, but take on the general structure and work organization of an ordinary private industry or factory with emphasis on high productivity based on routine and tightly managed work processes. Hired labor, not family labor, is essential. Labor and management are separated. Residence and the farm business are usually separated. The operations may be involved to some extent in vertical integration with large regional or international corporations. They may be large family corporations, or large industrial corporate operations. These farms are often less diversified than family farms. They are sometimes referred to as "factory farms" by those who prefer traditional family farms. There may be stockholders involved in these corporations. They may be regional or international companies that are involved in large vertically integrated food systems. They may rely on funds from stockholders or venture capitalists to expand their operations.

Family Corporations

Family corporations are usually enterprises that have grown from family farms over the years to involve multiple generations and extended family members. Their management scheme is little different from private corporate farms.

OTHER OCCUPATIONS EXPOSED TO THE AGRICULTURAL ENVIRONMENT

Although this book is primarily about the occupational health of those who work producing food and fiber, there are many other workers that have similar exposures to those of farmers and ranchers (US Census Bureau 2001). These include many workers who provide services, sales, and processing of agricultural commodities. Many of these workers have similar exposures as farmers and ranchers because they may work in the same environment, or they may be exposed to the same hazardous agents. These occupations and potential exposures are included in, but are not limited to those listed in Table 1.3.

GENERAL HEALTH STATUS OF THE AGRICULTURAL POPULATION

Table 1.4 summarizes numerous studies from several different countries, suggesting that the rural and agricultural populations generally have better health status compared to urban populations (based on the major causes of death and morbidity) (Blondell 1988; Doll 1991). Furthermore, it is clear that the agricultural population has better general health when compared to the overall population and the general rural population, and when compared to other occupational groups (Stiernström and others 1998). The health advantage of being a member of the farm population is a result of lower cancer rates and cardiovascular disease rates, including ischemic heart disease and stroke (Burmeister and Morgan 1982; Pomrehn 1982). Details of cancer risks are seen in Chapter 5. In Sweden Thelin (Thelin 1991) has found lower morbidity in the farm population due to lower rates of cancer, alcohol-related diseases, psychological conditions, cardiovascular conditions, and urinary conditions. The "farmer health effect" for these health benefits ranges from 20–50% lower relative risk, depending on the study and the country.

The reasons for these health benefits are thought to be lifestyle factors, including 1) decreased smoking, 2) less alcohol consumption, 3) more exercise, and 4) healthier diet (Stiernström and others 1998). It is clear that farmers smoke significantly less (approximately 50% less) compared to the general population (Thelin and Höglund 1994). They also appear to consume less alcohol (Thelin and Höglund 1994). Increased exercise might also be a benefit for farmers, primarily from their work, as they expend as much as 30% more calories per day than the general population (Stiernström and others 1998). However, their leisure time is less likely to include vigorous exercise (Stiernström and others 1998). Farmers' body mass index is similar to the general rural population (Thelin and Höglund 1994). Although there are no large-scale definitive dietary studies relative to health outcomes in the farming population, there is slight evidence diet may be a health benefit. One study in Italy suggested that farmers who primarily eat food grown on their

Table 1.3. Agricultural Support and Service Occupations That May Have Similar Occupational Exposures to Farmers and Ranchers

Agricultural–Exposed Occupations	Potential Exposures
Veterinarians • Veterinary assistants	• Animal-related injuries • Rural roadway crashes • Organic dusts (e.g., livestock confinement buildings) • Zoonotic infections • Antibiotics and resistant organisms • Veterinary biologicals and therapeutics • Insecticides used on animals • Excessive noise inside livestock buildings (mainly swine)
Livestock and poultry production/ animal health technicians (employees of large integrated livestock production companies who service the companies' animals and contract growers)	• Animal-related injuries • Rural roadway crashes • Organic dusts (e.g., livestock confinement buildings) • Zoonotic infections • Antibiotics and resistant organisms • Veterinary biological and therapeutics • Insecticides used on animals • Excessive noise inside livestock buildings (mainly swine)
Livestock auction sale employees	• Animal-related injuries • Organic dusts (e.g., inside livestock sales buildings) • Antibiotic-resistant organisms • Zoonotic infections • Excessive noise inside livestock sales buildings (mainly swine)
Meat and poultry processing plant workers • Those handling live animals • Those killing and processing animals	• Animal-related injuries • Organic dusts • Antibiotic-resistant organisms • Zoonotic infections • Excessive noise (mainly swine)
Livestock transporters	• Animal-related injuries • Organic dusts • Antibiotic-resistant organisms • Zoonotic infections • Excessive noise (mainly swine)
Crop production service workers • Pesticide formulators, mixer/ loaders, applicators • Crop scouts	• Pesticides • Fertilizers • Stinging/biting arthropods
Grain elevators/animal feed mixing, loading, and delivery	• Rural roadway crashes • Organic dusts (grain dust) • Antibiotics and other growth-promoting feed additives

own farms have significantly reduced risk for cancer (Blondell 1988).

OCCUPATIONAL HEALTH STATUS OF THE AGRICULTURAL WORK FORCE

Although the general health status of the agricultural work force appears to be better than compar-

ison populations (Table 1.4), their occupational health appears to be one of the worst among all occupations (McCrane 1999; Reed 2004; Reed and Wachs 2004). Since the writings of Olaus Magnus of Sweden in the 1500s and Ramazinni (1713) on occupational diseases, there have been numerous reports, review articles, books, and

Table 1.4. Overall Health Status (Morbidity and Mortality) of the Farm Population Relative to Comparison Populations

Comparison Population	Location	Findings	Reference
Rural compared to urban populations	New York, U.S. Kentucky, U.S. U.S. overall	Lower overall mortality Lower cancer mortality	Stark and others 1987 Blondell 1988; Doll 1991
	Poland	Lower cardiac mortality and risk factors	Rywik and Kupsc 1985
	Costa Rica U.S.		Campos and others 1992; Zhai and McGarvey 1992
Agriculture compared to the general population	Finland Scandinavia	Lower mortality (9%) Lower mortality (10%)	Notkola and others 1987 The Nordic Statistical Secretariat 1988
	New York, U.S.	Lower mortality (40%)	Stark and others 1987
Agriculture compared to the rural population	New York, U.S. Sweden Iowa, U.S.	Lower overall mortality Lower morbidity Lower mortality overall and cardiovascular (20%)	Stark and others 1987 Stiernström and others 1998 Burmeister and Morgan 1982; Pomrehn 1982
Agriculture compared to other occupations	Italy	Lower mortality: cardio-vascular (50%), cancer (28%), overall (46%)	Figa-Talamanca 1993
	Sweden Sweden	Lower mortality Lower morbidity (15%–70%)	Vagero and Norell 1989 Thelin 1991

book chapters documenting the low occupational health status of the agricultural work force (Frank and others 2004; Mutel and Donham 1983; Rasmussen and Cole 1976). Agriculture in every industrialized country is one of the most hazardous occupations, based on occupational fatality rates, nonfatal occupational injury rates, and occupational illness rates.

OCCUPATIONAL INJURY AND ILLNESS STATISTICS

Agriculture in the U.S. appears to be relatively more hazardous compared to other industrialized countries. The fatality rate for agricultural workers in the U.S. is 22.5/100,000 (Frank and others 2004; Rautiainen and Reynolds 2002). This compares to 3.8/100,000 for all occupations (making the U.S. rate over five times higher). Fatality rates for youth under 20 years of age are about 8/100,000 (over twice the fatality rate for all occupations). There is large variation of reporting regarding rates of nonfatal injuries. For the U.S., rates vary from about 0.5/100 to 17/100 (McCurdy and Carroll 2000). Studies that had an

active injury surveillance process recorded 42/100 (Rautiainen and others 2004). Most studies fall somewhere in the middle of this range, around 10/100. In other words, one out of 10 farmers suffers a disabling injury every year.

Reported fatality rates among farmers in Canada are 11.6/100,000 (Pickett and others 1999) and 6.5/100,000 in Finland (Notkola and others 1987; Rautiainen and Reynolds 2002). Injury rates in Canadian farmers are reported at 390/100,000, compared to 320/100,000 for all occupations (Canada 2001). Data from Finnish insurance indicates that occupational illnesses in agriculture occur at a rate of 640/100,000. Respiratory illnesses make up nearly 40% of these conditions, followed by skin (21%) and joint illnesses (31%).

Occupational disease rates are much more difficult to quantify, because such illness in self-employed people would rarely be identified as occupation related (there is no mechanism to create a need to report). Therefore objective data is based only on employed workers. Other data is based on self-reported surveys, which have inherent sensitivity and specificity problems. Given these caveats, work-related illness reported in the U S Bureau of Labor Statistics (2002–2003) indicates an injury rate of 309/100,000. The top three conditions causing these illnesses included skin conditions (56%), cumulative trauma (14%), and respiratory diseases (13%). There are many more occupation-related illness and injury conditions among the agricultural work force. These categories of specific conditions are included in this text as fundamental core information in agricultural medicine.

Not only is there a high cost in human resources from agricultural injuries and illness, but there is also an enormous economic consequence. Leigh (2001) found that each agricultural fatality creates an expense for the family on the average of $20,700 direct and $422,069 indirect costs. When multiplied by all the fatalities that occur annually in the U.S., this adds up to over $37 million. Adding the estimated costs for disabling injuries brings the annual total to over $3 billion.

SPECIFIC OCCUPATIONAL HEALTH AND SAFETY RISKS AND CONDITIONS

The specific conditions included here were chosen based on reported research data. Furthermore, they are included based on the combined nearly 70 years of experience of the authors having dealt directly with agricultural populations on three different continents regarding the clinical and preventive aspects of their occupational health exposures. The following paragraphs provide a brief overview of each of the conditions and injuries and the referent chapter where they are discussed in depth. References for this overview are found in the following reviews (Donham and Mutel 1982; Frank and others 2004; Horvath and Donham 1987; Popendorf and Donham 1999).

Special Risk Populations in Agriculture (Chapter 2)

Children, women, the elderly, migrant and seasonal farm workers, and Anabaptist religious groups are populations at increased risk for illness or injury from exposure to the farm environment. Children are at risk because they live, play, and may work in a very hazardous environment (Gwyther and Jenkins 1998). Children may not have the developmental capacity to play or work safely in this environment without close informed

parental supervision. Pregnant women exposed to certain zoonotic infectious agents or veterinary pharmaceuticals (oxytocin or prostaglandins), carbon monoxide, and certain insecticides or herbicides may risk damage to an unborn fetus. Elderly farmers are at increased risk because their physical and mental status might not be sufficient to keep them safe in a hazardous working environment. Furthermore, they often use the older equipment they are used to (which is often less safe than newer equipment), and may have comorbidities such as poor eyesight, hearing loss, arthritis, and diabetes that may increase their risk for injury. Migrant and seasonal workers are at risk because they are generally of a poorer economic and social circumstance (one of the most common links to general health status is socioeconomic status), have language barriers, lack education, and have physically hard jobs in hot environments. Anabaptist groups are at increased risk because the equipment they use is often old or homemade and without usual safety features. Furthermore, their beliefs include minimal use of mainstream health care, which decreases their probability of having up-to-date immunizations, prenatal care, and early diagnosis of chronic disease.

Agricultural Respiratory Conditions (Chapter 3)

Based on research and surveillance data, the combined high rate and severity of these conditions, and the authors' experience, we think that respiratory illnesses are the most important occupational illness of agricultural workers. Available data reveal that 10–30% of agricultural workers experience one or more occupational respiratory conditions (Gomez and others 2004). The most frequent cause of respiratory illness is organic (agricultural) dust from livestock production and handling grain or hay. There is a syndrome of respiratory conditions caused by agricultural dust that includes bronchitis, asthma-like conditions, irritation of the mucosa of the upper airways and eyes (mucous membrane irritation or MMI), and organic dust toxic syndrome. The former two conditions are usually chronic, and the latter is an acute influenza-like condition lasting one to several days.

Less common exposures that add to the library of exposures and related respiratory conditions include 1) hypersensitivity pneumonitis (farmer's lung), 2) silo gas from non-airtight silos, 3) hydrogen sulfide and ammonia from decomposing livestock manure, 4) fumigant pesticides or biocides, 5) zoonotic infectious agents, and 5) the herbicide paraquat.

Agricultural Skin Diseases (Chapter 4)

Several reports suggest that skin conditions may be the most frequently reported type of agricultural illness. The most common category of skin conditions is contact dermatitis, which may occur as irritant or allergic contact dermatitis. The latter may occur as immediate or delayed allergic contact dermatitis. Contact dermatitis may also occur with sun exposure causing a chemical change to a substance on the skin, which then becomes an irritant or allergen. Some irritant or allergenic substances may be contracted from airborne exposures as well (airborne contact dermatitis).

There are several plants that may cause delayed allergic contact dermatitis, including those that contain the allergen erushiol (poison ivy, poison oak, poison sumac).

Sun and heat exposure are the second most common causes of skin conditions. Sunburn and miliarial rubra (prickly heat) are the two most common acute skin conditions caused by sun and heat. Chronic sun exposure causes wrinkling and thickening of the skin, precancerous lesions called actinic keratoses, and the skin cancers squamous cell carcinoma and basal cell carcinoma. Melanoma is thought to be caused by multiple sunburns at an earlier age in life, but may also be related to total impact of sunshine.

Ringworm (dermatophytosis) contracted from cattle is the most common infectious skin problem among farmers handling animals, especially dairy farmers.

There are numerous arachnids and insects (including mites, ticks, spiders, and stinging or biting insects—wasps, ants, mosquitoes), that may cause minor to severe irritation of the skin.

Cancer in Agricultural Populations (Chapter 5)

The farming population (primarily because they smoke less) benefits from lowered overall cancer because they have less lung cancer (one of the most common cancer fatalities) and fewer other smoking-related cancers. Overall colon and rectal cancer seems to be lower, and farm women also have less overall breast cancer. Besides not smok-

ing, there may be other protective factors in farming, but they are not clearly identified at this time. However, there are several cancers for which the farming population may be at increased risk, including lymphoma, leukemia, multiple myeloma, prostate, skin, and brain. Of various speculative risk factors for these cancers, only excessive sun (skin cancers), methylbromide (prostate cancer), and acetic acid herbicides (non-Hodgkin's lymphoma and soft-tissue sarcoma) exposures are relatively proven risk hazards.

Toxicology of Pesticides (Chapter 6)

Although the issue of pesticide exposure is often a dominant concern among the farming population, acute poisonings and fatalities are far less common than acute traumatic injuries or respiratory illnesses. Although some of the pesticides (especially the cholinesterase-inhibiting chemicals) used are very toxic, others are severe irritants or sensitizers (especially herbicides), and contact dermatitis is a common result of pesticide contact.

General Environmental Health Hazards in Agriculture (Chapter 7)

Adverse water quality from nitrate contamination is the most important general (nonwork) environmental health hazard for agricultural workers. High nitrates are converted to nitrites in the gastrointestinal tract, which can lead to the formation of methemoglobinemia, which does not carry oxygen efficiently. The condition is most critical to infants (blue baby). Lower levels of nitrates may be a carcinogenic risk, and nitrates in the presence of amino acids or the herbicide atrazine may form nitrosamines, which are known carcinogens.

There are other water, air, and solid waste problems in agriculture, but these are more directly related to environmental quality degradation and ecologic change, rather than direct individual worker health hazards.

Musculoskeletal Diseases in Agriculture (Chapter 8)

Low back pain and degenerative osteoarthritis of the hip and knee are common problems among the agricultural work force. These conditions are worsened by poor ergonomic working conditions, long working days, and heavy work loads. Furthermore, carpal tunnel syndrome is common among those working in meat and poultry processing. A number of other musculoskeletal disorders are related to physical work, and these kinds of ailments are the most common cause for farmers to contact the occupational health service or general practitioner/outpatient clinics. Managing these conditions includes modifying work practices with sound ergonomic practices.

Physical Factors Affecting Health (Chapter 9)

The work environment in production agriculture can be a hot or cold and noisy place. All of these physical elements create risk of injury to the farm worker.

HEAT

A great deal of agricultural work is in hot environments and/or in direct sunlight. As agricultural work requires a great deal of energy consumption, risk for heat-induced illnesses is common. Heat exhaustion may be a minor problem treatable by protection from the sun, periodic rest from a hot environment, and increasing fluid intake. Untreated heat exhaustion may lead to the more serious condition of heatstroke, which physiologically includes incapacitation of the body temperature regulatory mechanism. Body temperature may raise high enough to cause brain damage, and combined with dehydration and electrolyte imbalance may result in death if not treated as an emergency condition.

COLD

Work outdoors in extremely cold environments is often a requirement of agricultural work. Frostbite, which is the freezing of tissues, is a risk. Furthermore, cold environments exacerbate a condition called "white finger" or *Raynaud's phenomenon,* which may be a result of chronic high frequency vibration damage to the nerves and blood vessels in the hand (today often referred to as HAVS, Hand Arm Vibration Syndrome). When the hands become cold, the vessels of the affected hands "shut down" circulation, leading to painful symptoms and loss of refined hand movements.

NOISE-INDUCED HEARING LOSS

The agricultural work environment is very noisy, leading to the very common problem among farmers of noise-induced hearing loss. Excessive noise causes direct damage to the hair cells of the middle ear, which transmit the energy of sound to the brain. Once damaged, the cells will not repair themselves, and the loss is permanent. Loss of hearing increases the risk of injury to farmers and leads to social isolation.

VIBRATION-RELATED INJURIES

High-frequency vibration may impact the hands through handheld tools, chainsaws, and machinery operation. Long-term exposure to vibration can lead to Raynaud's phenomenon, which results in decreased blood supply to the hands with associated pain when exposed to the cold. Associated symptoms may be seen in the arm and shoulder as well. Low-frequency vibration can lead to subtle symptoms that might include back pain, nausea, and fatigue.

Mental and Social Health in Agriculture (Chapter 10)

Farming is an occupation that is increasingly filled with stress, mainly due to the diminishing profit margins. The culture of the agriculturalist is to persevere, rather than admit to needs and seek help. Mental health issues carry more of a stigma than in urban communities. The social structure of the rural community is changing, as the population becomes sparser and social structures and customs that enhance "neighboring" are changing. As old social support structures are declining and formal mental health services are rare, many farmers and family members suffer as chronic stress builds to depression. One of the most severe outcomes of this situation is suicide, as chronicled in the true account of a stressed farmer in a midwestern community in the U.S. (Brown 1989). Gunderson documented higher rates of suicide compared to the general population in the north central states of the U.S. (Gunderson and others 1993).

Acute Agricultural Injuries (Chapter 11)

Acute physical injuries are the primary occupational health concern in agriculture (as far as surveillance data allows), causing more fatalities and disabilities than any other agricultural illness category. Tractor-related events account for the majority of fatalities. Farm machinery in general along with animal-related injuries are the primary causes of acute injuries, each accounting for about equal numbers of injuries. However, machinery injuries are usually more serious than animal-related injuries.

In the Scandinavian countries, Baltic States, Scotland, and some other areas farming is combined with forestry. The risk of injuries is very high for farmers active in forestry who very often work without proper equipment and education.

Medical treatment of agricultural injuries is complicated by the extreme severity of tissue damage and even amputation that often results. Severe trauma, often combined with extensive contamination with soil and animal fecal material, increases the risk for wound infections with anaerobic organisms, and antibiotic-resistant organisms. Delayed finding, rescuing, and emergency transport of victims to appropriate medical facilities further complicate the prospects for good outcomes of these cases. Finally, rehabilitation of these victims often falls upon the rural primary care physician. Almost all injured farmers want to resume their farming activities, and it is up to the primary care provider in conjunction with organizations like Breaking New Ground (a rehabilitation program in the U.S.) to help rehabilitate injured farmers and return them to farming.

Human Health Hazards of Veterinary Pharmaceuticals (Chapter 12)

There are many products used for animal health or growth promotion that may cause illness in humans who may come in contact with these products. Accidental needle sticks are common among veterinarians and animal handlers, and they carry the consequence of unintended trauma, infections, and toxicity or inflammation. Antibiotics, immunization products, and hormones used in obstetrical procedures are common substances that may result in unintended illness in veterinarians or animal handlers. The largest concern with antibiotic use is the enhancement of resistant organisms and resulting resistant infections. Other concerns include severe toxic reactions (tilmicosin) and allergies. Accidental inoculation with immunization products may result in an infection of the product itself (live products), inflammation, or allergic response. Oxytocin and prostaglandins are two products that may cause abortion in women if they are accidentally inoculated.

Agricultural and Rural Zoonotic and Emerging Infectious Diseases (Chapter 13)

There are over 25 different infectious diseases that may be transmitted from animals or the environment that may produce occupational illnesses in agricultural workers. The specific diseases reviewed in this text include swine and avian influenza, leptospirosis, anthrax, bovine spongiform encephalitis, West Nile Valley virus, and Hantavirus, among others.

Prevention of Illness and Injury in Agricultural Populations (Chapter 14)

Although the degree of regulations that affect family farms varies among the developed countries, agriculture is relatively unregulated compared to other industries in regard to occupational health. Reasons include the fact that the majority of these people work as self-employed individuals on thousands of individual operations, which are often remotely scattered across the countryside. This fact creates political and practical difficulties in development and enforcement of regulations. The most common regulation is the requirement of rollover protective structures on farm tractors. However, this is not universally required in either North America or some other countries.

Most countries have regulations that apply to children and hired agricultural workers. However, some countries have a minimum number of workers per operation before the regulations take effect. For example, in the U.S., there must be at least 11 employees in an operation before federal funds can be used to inspect and enforce occupational safety and health regulations. The International Labor Organization has produced model guidelines for member countries to adopt for the protection of agricultural workers (ILO 2001). Details of regulations applicable to agricultural health and safety are seen in Chapter 14.

AGRICULTURAL HEALTH AND SAFETY ORGANIZATIONS

The past two decades have witnessed extensive growth in agricultural health and safety activities in the industrialized countries, in both governmental and nongovernmental organizations. The sum effort of these organizations has advanced the field significantly, creating a new discipline and changing the field to a public health concern from a previously low-profile interest of agricultural colleges. Chapter 14 on prevention discusses in detail these organizations and how they have advanced the field of occupational safety and health.

MEGATRENDS IN AGRICULTURE AND HEALTH—SAFETY IMPLICATIONS

Domestic and international economic and policy changes have caused major changes in the industry of agriculture over the years. These changes not only have an effect on the structure of agriculture, but also affect socioeconomic conditions and the health of the agricultural work force (Cleveland 1998). The following paragraphs trace some of the primary megatrends in agriculture, and the resulting effects on the agricultural work force (Martin 2001).

The roots of many of the megatrends in agriculture can be traced to the 1940s, as the effects of the industrial revolution started to reach production agriculture. Development of new technology for agriculture has been a progressive trend since that time, leading to tremendous growth in productivity (more commodities produced per person hours of labor). Another major force began in the 1970s when many industrialized countries began to view agricultural products as an important export commodity to trade for manufactured goods, which have been increasingly moving to China and the Pacific Rim countries. The 1990s was the decade of enhanced globalization, which has and is continuing to have major effects on the structure of agriculture and demographics of the agricultural work force. International trade agreements have enhanced globalization. Consolidation and vertical integration of industries including agriculture have been enhanced and are a result of the policies put in place in our new global economic system. The combined effects of globalization, consolidation, and vertical integration are having a major impact on the structure and demographics of agriculture. The following list is a combination of some of the major forces and resulting megatrends in agriculture (Doane Agricultural Services 2004; Frank and others 2004). Following each megatrend there is a comment for clarification to emphasize the effect on the agricultural work force:

1. **Decrease in farm numbers.** Farm numbers have been decreasing at around 1–5% per year in industrial countries over the past several decades. One of the reasons has been the difficulty making a living wage in agriculture as the cost of production has continued to rise and the price the farmer obtains for the commodity has remained stable, leaving a very low profit margin. This obviously leads to economic stress, which increases the risk for mental as well as physical health issues for those farmers remaining. Furthermore, as there is less incentive for new farmers to come into the industry, the family farm sector continues to consist of older producers. Geriatric health issues become blended with occupational health concerns in the agricultural work force.

2. **Continued increase in farm size.** As profit margins decrease, the primary way to gain a living wage is to produce more commodities. Thus in many instances, the typical farmer has had to expand his or her operation to remain in business. The effect of this is to increase the work load and stress on the individual operator, often using the same older equipment, which may not have the safety features of newer equipment. Increased work load, increased stress, older less-safe equipment all contribute to increased injury risk. The increase in farm size often is based on rented land, which increases insecurity and instability in the farm community because of competition and uncertainty of obtaining land contracts from year to year.

3. **Decrease in the number of individual independent owner-operator farm enterprises (e.g., family farm operations).** The primary agricultural sector losing operations is the mid-sized owner-operator family farm. The reason is that these farms have more difficulty raising the resources to expand to remain competitive, as compared to large corporate commercial farms, which have many advantages, including tax advantages, subsidies, and economies of scale (Lyson and Green 1999). Furthermore, they are losing access to their usual markets, which are being more controlled by large corporations. Increased costs for compliance with stricter environmental standards also may affect some smaller operations. The result is that the agricultural work force is moving from a self-employed owner-manager operation to a wage-earning farm worker base and a shift to an overall lower socioeconomic status farm work force (Mann 1992).

4. **Increasing numbers of wage-earning farm laborers, especially foreign-born migrant and seasonal workers moving from developing to industrialized countries.** Increasing numbers of workers from developing countries are leaving for work in the industrialized countries, and farm work is a common possibility for many migrant and seasonal workers. These workers are a vulnerable population for many reasons, as detailed in Chapter 2.

5. **Specialization of production (from diversified to mono-crop commodity operations).** Operations are more efficient and easier to manage if they are specialized. Furthermore, if the operation is tied into an industrial vertically integrated processing and marketing company, the product demand may be fixed. Mono-crop operations may be more likely to cause environmental pollution, and may be less sustainable. However, it is possible (because of increased resources and management) for these firms to develop highly effective environmental and occupational health programs.

6. **Industrialization of agriculture leading to consolidation and vertically integrated international food corporations.** Domestic and international economic policies have facilitated the rise of international vertically integrated food corporations (Gregory 2000; Viatte 2001). These food giants have tremendous political power and can certainly have an effect on policies that facilitate their operations (Moran and others 1996). This is in contrast to smaller family operations that find economic stress with little control over their situations, which exacerbates their personal stress.

7. **The subsidization of agriculture in industrialized countries, and the effect on developing countries.** Many countries are conflicted about the direction of agriculture enterprises. For historical, social, political, and domestic food security purposes, they would like to maintain a family farm structure (Jussaume 1998; Kim 1999). However, this requires a significant farm subsidy, which is rather significant in North America and the EU (Bruhnsma 2003). These subsidies create difficulties for developing countries, as they cannot compete on the world markets with subsidized products from the industrialized countries. Furthermore, these subsidies retard development of a domestic food industry in developing countries, as they often can import food cheaper than they can grow it. The international lenders to developing countries often require that their money funds potential export commodities, which may differ from commodities that are a domestic food supply, thereby inhibiting their domestic food security.

8. **The rise and international expansion of the fast food industry.** The fast food industry has enhanced the industrialization of agriculture. It demands very uniform and strict delivery times and amounts that individual family farms find difficult to supply. Thus, industrial farms are much more able to meet the demands of the fast food industry.

9. **Capital financing pressures.** Larger and industrial farms may find funding easily from venture capitalists or stockholders. Smaller and family operations do not have these sources of funds available, and therefore their borrowed money costs more.

10. **Decreased access to markets.** As industrial agriculture has become more vertically integrated, corporations have gained more control of the markets. Many small farmers have found it hard to sell product unless they are tied into a vertically integrated scheme as a contract grower.

11. **The rise of new players in the world market of agricultural commodities, including Brazil, Argentina, India, and China.** These countries have emerged as major agricultural exporting nations. As the industrialized countries have depended so much on export markets, these countries are lower-cost producers, making exports more difficult, which trickles down to lower prices for commodities to small farms, decreasing profit margins.

12. **The improved developing economies (e.g., China, India) and new increased demand for animal protein.** While these countries may export grains (e.g., rice), their improved economies have increased their appetite for and ability to purchase red meat and poultry (Bradford 1999; Van Der Zijpp 2000). This in

turn has increased the markets for livestock producers in the industrialized countries, where livestock production can expand to fill the new market demand (Meeker 1999).

13. **The increasing voice and power of the consumer and general public (e.g., public demand for products free of antibiotics, infectious organisms, antibiotic-resistant organisms, pesticides, growth promoters, genetically modified organisms (Buttel 1999), and food animals raised under humane circumstances).** Demand for more "naturally produced products" has created new niche markets for a variety of producers (Nielsen and Anderson 2001). Organic products are increasingly in demand by consumers (O'Hara 2001). Food animals grown under more "natural conditions" are increasingly in demand by consumers. This helps provide more opportunities for many small family farms. The negative part of this growth is that the equipment and methods are similar to farming several generations ago. The equipment may be old and without modern safety features. Also, the nature of the work may be very strenuous and conducted with poor ergonomic standards.

14. **Increased health consciousness and diet fads of industrialized countries (e.g., low-fat and low-carbohydrate diets).** In the 1970s, a dramatic shift began in the dietary demands of people in industrialized countries regarding the increased demand for low-fat foods. The agricultural industry responded by dramatically increasing the amount of chicken produced and changing the genetics of livestock toward leaner pork and beef. The 1990s saw the birth of the low-carbohydrate diets. This trend increased demand for animal protein of all types. Both of these trends favored larger and more industrialized farms, which could respond rapidly to these product demands.

15. **The rise of niche markets, including production of exotic crops, and community-sponsored agriculture.** Additional to the demand for organic products, there are new demands for products such as ginseng; meat from ostriches, emus, deer, and buffalo; and many other exotic products. Furthermore, there is a new demand for a closer connection between the grower and the consumer (Hinrichs 2000; La Trobe 2000). Farmers' markets and local farms are linking with local urban dwellers (community-sponsored agriculture), increasing opportunities for small farms to provide fresh farm products to their local community. Energy production is a developing sector also. Wood chips and oats are being used for fuel in Europe. Farmers in many areas of North America and Europe are building wind generators to power their own farms and to sell power commercially. Local cooperative grain-processing plants are providing value-added grain for ethanol and biodiesel fuel production.

16. **Increasing public scrutiny of agriculture regarding environmental contamination, consumer health, and occupational health.** The general public in most industrialized countries is no longer willing to give agriculture special considerations regarding the environment. Many people feel that agriculture should be held to the same standards as any other industry. New and more regulations are being developed that will create the need for farms to expend resources to meet these regulations. Although these regulations will focus more on large operations, they will have some economic impact on small farms.

17. **Emerging, zoonotic, and livestock health infectious diseases, including**
   a. **Bovine spongiform encephalopathy or mad cow disease**
   b. **Foot and mouth disease**
   c. **NIPAH and HENDRA viruses**
   d. ***Escherichia coli*** H157, *salmonella, campylobacter*
   e. **Hantaviruses**
   f. **Avian and swine influenza**

   These infectious diseases have significant occupational health concerns for farmers, their families, and the general public. They also have significant implications for markets. For example, Canada's and England's beef export markets were eliminated because of mad cow disease. The 2001 outbreak of foot and mouth disease caused the destruction of thousands of English farmers' cattle and sheep herds. The associated economic and mental stress among the farming population was profound.

SUMMARY

This chapter is meant to provide a broad overview of agricultural health and safety. Aided by this overview, the reader can approach the subsequent chapters with a background that provides broad connections to the field of agriculture and a perspective that enhances greater comprehension of the material.

REFERENCES

AAEM. 2005. Annals of Agricultural and Environmental Medicine Home Page.

Berry C. 1965. Organized research in agricultural health and safety. American Journal of Public Health 55(3):424–428.

——. 1971. Rural employment. American Journal of Public Health 61(12):2474–2476.

Blondell J. 1988. Urban-rural factors affecting cancer mortality in Kentucky 1950–1969. Cancer Detect Prevention 47:803–810.

Bradford G. 1999. Contributions of animal agriculture to meeting global human food demand. Livestock Production Science 59(2–3):95–112.

Brown B. 1989. Lone Tree: A True Story of Murder in America's Heartland. New York: Crown Publishing Group.

Bruhnsma J. 2003. World Agriculture: Towards 2015/2030. London: Earthsan Publications, Ltd.

Bureau US. 2005. U.S. Census Bureau. USA Statistics in Brief—Agriculture and Business. www.census.gov/statab/www/agbus.html. First accessed on March 2, 2005.

Burmeister L, Morgan D. 1982. Mortality in Iowa farmers and farm labors, 1971–1978. Journal of Occupational Medicine 24:898–900.

Buttel F. 1999. Agricultural biotechnology: Its recent evolution and implications for agrofood political economy. Sociological Research Online 4(3):u292–u309.

Campos H, Mata L, Siles X, Vives M, Ordovas J, Schaefer E. 1992. Prevalence of cardiovascular risk factors in rural and urban Costa Rica. Circulation 85:648–658.

Canada S. 2001. 2001 Agricultural Census.

Cleveland D. 1998. Balancing on a planet: Toward an agricultural anthropology for the twenty-first century. Human Ecology 26:323–340.

Davies J, Smith R, Freed V. 1978. Agromedical approach to pesticide management. Annual Review of Entomology (23):353–366.

Doane Agricultural Services. 2004. Megatrends in Agriculture.

Doll R. 1991. Urban and rural factors in the aetiology of cancer. International Journal of Cancer 47:803–810.

Donham K, Mutel C. 1982. Agricultural medicine: The missing component of the rural health movement. Journal of Family Practice 14(3):511–520.

Economics ABoAaR. 2005.

Elliott C. 1979. Agricultural medicine. Journal of the Royal Society of Medicine 72:949.

European Commission. 2004. Eurostat year book 2004: The statistical guide to Europe 1992–2002.

Figa-Talamanca I, Mearelli I, Valente P, Basherina S. 1993. Cancer mortality in a cohort of rural licensed pesticide users in the province of Rome. International Journal of Epidemiology 22:579–583.

Fleming M. 2004. Agricultural Health. A new field of occupational health nursing. American Association of Occupational Health Nursing 52(9):391–366.

Food and Agricultural Organization of the United Nations Economic and Social Department The Statistics Division. 1997. China Agricultural Census.

——. 2002. New Zealand Agricultural Census.

Frank A, McKnight R, Kirkhorn D, Gunderson P. 2004. Issues of agricultural safety and health. Annual Reviews of Public Health 25:225–245.

Gay J, Donham K, Leonard S. 1990. The Iowa Agricultural Health and Safety Service Program. American Journal of Industrial Hygiene 18:385–389.

Gomez M, Hwang S, Lin S, Stark A, May J, Hallman E. 2004. Prevalence and predictors of respiratory symptoms among New York farmers and farm residents. American Journal of Industrial Medicine 46:42–54.

Gregory P. 2000. Global change and food and forest production: Future scientific challenges. Agriculture Ecosystems & Environment 82(1–3):3–14.

Gunderson P, Donner D, Nashold R, Salkowicz L, Sperry S, Wittman B. 1993. The epidemiology of suicide among farm residents or workers in five north-central states, 1980–1988. Farm injuries: A public health approach. American Journal of Preventive Medicine 9:26–32.

Gwyther M, Jenkins M. 1998. Migrant farmworker children: Health issues, barriers to care and nursing innovations in health care delivery [Review]. Journal of Pediatric Health Care 12(2):60–66.

Hinrichs C. 2000. Embeddedness and local food systems: Notes on two types of direct agricultural market. Journal of Rural Studies 16(3):295–303.

Horvath E, Donham K. 1987. Agricultural Medicine. Zenz C, editor. Chicago: Year Book Medical Publisher.

IAAMR. 2005. International Association of Agricultural Medicine and Rural Health Home Page.

IAREH. 2005. Institute of Agricultural Rural and Environmental Health Home Page.

ILO. 2001. R192 Safety and Health in Agriculture Recommendation, 2001, Geneva.

ILO. 2004. Towards a fair deal for migrant workers in the global economy. Geneva: International Labour Office.

James P. 1994. Agromedicine: What's in a name? Journal of Agromedicine 1(3):81–87.

Jones S, Dann D, Coffey D. 2004. Strengthening the nursing curriculum. An interdisciplinary course addressing agricultural health and safety. American Association of Occupational Health Nursing Journal 52(9):397–400.

Jussaume R. 1998. Globalization, agriculture, and rural social change in Japan. Environment and Planning Agriculture 30(3):401–413.

Kim J. 1999. Sustainable agriculture development in Korea in a global economy. Journal of Sustainable Agriculture 13(3):73–84.

Knapp L. 1965. Agricultural injury prevention. Journal of Occupational Medicine 7(11):545–553.

La Trobe H. 2000. Localizing the global food system. International Journal of Sustainable Development and World Ecology 7(4):309–320.

Leigh J. 2001. Costs of occupational injuries in agriculture. Public Health Reports 116:233–248.

Lundvall A, Olson D. 2001. Agricultural health nurses: Job analysis of functions and competencies. American Association of Occupational Health Nursing Journal. 49(7):336–46.

Lyson T, Green J. 1999. The agricultural marketscape: A framework for sustaining agriculture and communities in the Northeast. Journal of Sustainable Agriculture 15(2–3):133–150.

Mann S. 1992. The survival and revival of non-wage labor in a global economy. Sociologicia Ruralis 32(2–3):231–247.

Marshfield Clinic Research Foundation. 2005. Marshfield Clinic Research Foundation Home Page.

Martin M. 2001. The future of the world food system. Outlook on Agriculture 30(1):11–19.

McCrane J. 1999. A case control study of the Health status of male farmers registered at Ash tree house surgery. Journal of the Royal Society of Health 119(1):32–35.

McCurdy S, Carroll D. 2000. Agricultural injury. American Journal of Industrial Medicine 38:463–480.

Meeker D. 1999. What are the livestock industries doing, and what do they need from us? Journal of Animal Science 77(2):361–366.

Mexico Agricultural Census. 1991. Mexico Agricultural Census 1991—Main Results. Available at http://www.fao.org/es/ESS/census/wcares/Mexico_1991.pdf.

Moran W, Blunden G, Workman M, Bradley A. 1996. Family farmers, real regulation, and the experience of food regimes. Journal of Rural Studies 12(3):245–258.

MUSC, Division of Public Health and Public Service, Department of Family Medicine. 2005. South Carolina Agromedicine Program.

Mutel C, Donham K. 1983. An Extended Role for the Rural Physician. In: Stone JL, editor. Medical Practice in Rural Communities. New York: Springer-Verlag. p 117–139.

NAC. 2005. The North American Agromedicine Consortium Home Page.

Nielsen C, Anderson K. 2001. Global market effects of alternative European responses to genetically modified organisms. Review of World Economics 137(2):320–346.

NIFS, Inc. 2005. National Institute for Farm Safety Home Page.

Nordic Statistical Secretariat. 1988. Occupational mortality in the Nordic countries 1971–1980. Copenhagen: Nordic Statistical Secretariat, 49.

Notkola V, Husman K, Laukkanen V. 1987. Mortality among male farmers in Finland during 1979–1983. Scandinavian Journal Work Environment Health 13:124–128.

NRHA. 2005. National Rural Health Association.

O'Hara S. 2001. Global food markets and their local alternatives: A socioecological economic perspective. Population and Environment 22(6):533–554.

Pickett W, Hartling R, Brison R, Guernsey J. 1999. Fatal work-related farm injuries in Canada, 1991–1995. Canadian Agricultural Injury Surveillance Program 160(13):1843–1848.

Pomrehn PR, Wallace RB, Burmeister LF. 1982. Ischemic heart disease mortality in Iowa farmers; the influence of life-style. Journal of the American Medical Association 248:1073–1076.

Popendorf W, Donham K. 1999. Agricultural Hygiene. Clayton G, Clayton F, editors. New York: John Wiley and Sons Inc.

Ramazzini B. 1713. Diseases of Workers. Wright W, translator. New York: Hafner Publishing Company. 550 p.

Rasmussen R, Cole G. 1976. The spectrum of agricultural medicine. Minnesota Medicine 59(8):536–539.

Rautiainen R, Lange J, Hodne C, Schneiders S, Donham K. 2004. Injuries in the Iowa Certified Safe Farm Study. Journal of Agricultural Safety and Health 10(1):51–63.

Rautiainen R, Reynolds S. 2002. Mortality and morbidity in agriculture in the United States. Journal of Agricultural Safety and Health 8(3):259–246.

Reed D. 2004. Caring for the families that feed the world. American Association of Occupational Health Nursing 52(9):361–362.

Reed D, Wachs J. 2004. The risky business of production agriculture. Health and safety for farm workers. American Association of Occupational Health Nursing 52(9):401–408.

Rywik S, Kupsc W. 1985. Coronary heart disease mortality trends and related factors in Poland. Cardiology 72;81–87.

Spengler S, Browning S, Reed S. 2004. Sleep deprivation and injuries in part-time Kentucky farmers. American Association of Occupational Health Nursing.

Stark A, Change H-G, Fitzgerald E, Ricardi K, Stone R. 1987. A retrospective cohort study of mortality among New York state farm bureau members. Architectural Environmental Health 42:204–212.

Stiernström E-L, Hölmberg S, Thelin A, Svardsudd K. 1998. Reported health status among farmers and nonfarmers in nine rural districts. Journal of Occupational Environmental Medicine 40(10):917–924.

Thelin A. 1991. Morbidity in Swedish Farmers, 1978–1983, according to national hospital records. Soc Sci Medicine 32:305–309.

Thelin A, Höglund S. 1994. Changes of occupation and retirement among Swedish farmers and farm workers in relation to those in other occupations. Social Science Medicine 38:147–151.

Thu K, Donham K, Yoder D, Ogilvie L. 1990. The farm family perception of occupational health: A multistate survey of knowledge, attitudes, behaviors and ideas. American Journal of Industrial Medicine 18:427–431.

Top F. 1962. Occupational health in agriculture. Modern Medicine:140–184.

University of Iowa. 2005. Ag Safety and Health Training at The University of Iowa.

University of Minnesota. 2005. Environmental Health Sciences, School of Public Health.

US Census Bureau. 2001. North American Industrial Classification System. 2001 Statistical Abstracts of the United States.

US Department of Labor. 2002–2003. Career Guide to Industries: US Government Printing Office.

USDA. 2004. Agricultural Statistics UASS.

Vagero D, Norell S. 1989. Mortality and social class in Sweden—Exploring a new epidemiological tool. Scandinavian Journal Social Medicine 17:49–58.

Van Der Zijpp A. 2000. Role of global animal agriculture in the 21st century. Asian-Australasian Journal of Animal Sciences 13:1–6.

Viatte G. 2001. Global economy in the new millennium: A dynamic context for food and agricultural policies. Outlook on Agriculture 30(1):21–25.

Wakatsuki T. 2003. Getting Among Farmers. Yoshimoto S, translator. Usuda, Japan: Saku Central Hospital. 313 p.

Walsh M. 2000. Farm accidents: Their causes and the development of a nurse led accident prevention strategy. Emergency Nurse 8(7):24–31.

Wheat J, Turner T, Weatherly L, Wiggins S. 2003. Agromedicine focus group: Cooperative extension agents and medical school instructors plan farm trips for medical students. Sournter Medical Journal 96(1):27–31.

WSI IM. 2004. Historia Instytutu Medycyny Wsi w Lublinie; Struktura Organizacyjna. Available at http://www.imw.lublin.pl/.

Zhai S, McGarvey S. 1992. Temporal changes and rural-urban differences in cardiovascular disease risk factors and mortality. Human Biology 64:807–819.

2
Special Risk Populations in Agricultural Communities

Kelley J. Donham and Anders Thelin

INTRODUCTION

Special risk groups in agriculture may sometimes be in the shadow of the mainstream of agricultural health and safety research and programming; however, they are important in production agriculture. Special risk populations discussed in this chapter may be considered minorities within the agricultural population, relative to the general working-age male owner-operator. The groups considered here have special risks associated with age, gender, low socioeconomic status, ethnicity, race, or religious beliefs. This chapter intends to highlight these populations and their special health issues. There are many additional special risk populations in agriculture not discussed here, including the Laps who live in the northern regions of Scandinavia crossing over into Finland and Russia. There are also the Romans or Gypsies who live across Central and Eastern Europe. There are new immigrant farmers from Japan and China on the west coast of the U.S. There are new immigrant farm workers in the U.S. from Somalia, Bosnia, and South East Asia (the Hmong) who inhabit the Midwest. Furthermore, there are African-American farmers scattered across the southern U.S. who have special risks. However, it is beyond the scope of this text to cover all of these groups. Therefore, we have chosen five special risk populations to profile in this chapter: 1) children, 2) women, 3) elderly, 4) migrant and seasonal workers, and 5) Anabaptist religious groups. The first four of these are common to all industrialized agricultural countries.

Women experience certain reproductive risks from agricultural exposures, including infertility and abortion. Youth may experience illnesses and injuries because of living and playing in or adjacent to hazardous work sites. Their mental, emotional, and physical skills may not be adequately developed to handle certain kinds of work safely. The elderly may have lost certain physical, mental, and emotional skills, which increases their risk for injury or illness. Furthermore, the elderly often end up operating the older machinery on the farm (more hazardous because it may lack modern safety features) that they are familiar with, and they may be conducting more hazardous tasks such as mowing ditches and waterways. Migrants often have cultural and language barriers they must overcome before they can safely perform their jobs. Furthermore, they may feel they are in a powerless position, unable to complain of hazards for fear of losing their jobs or being deported. The Old Order Anabaptists often use older farming methods and machines, which increases their risk of injury. In addition, their spiritual beliefs may delay their seeking traditional health care, including the lack of recommended immunizations. These are only a few of the risks to these special risk populations. This chapter elaborates on the demographics of these populations, the exposures, subsequent risks, and epidemiology of related illnesses and injuries. Finally we describe any special considerations for treatment and recommendations for prevention.

WOMEN

The authors recognize that some farm women may take offense to being classified as a special

Figure 2.1. A woman working with her swine. (This figure also is in the color section.)

risk population. The reason is that such categorization might appear to diminish their role in agriculture relative to men, or imply that they are in some way not as strong as men. Neither of these implications are accurate. Women are extremely important in production agriculture in a variety of roles—as principal owner-operator of a farming enterprise, as a member of a family farming unit, or as a farm worker (Figure 2.1). However, as in general medicine and health, there are women's health issues, different from men, and those are the points we intend to highlight here.

Women are clearly increasing their role in the ownership and management of agricultural operations in most industrialized countries (Dimich-Ward and others 2004). In Canada between 1951 and 1997, the percentage of women reported employed in agriculture grew from 4 percent to 25 percent. Similar demographic changes in females as primary operators are occurring in the U.S.

Females comprise 47% (including primary operators and those living on the farm, who likely contribute to the farm labor) of the total farm population (Dimich-Ward and others 2004).

Women are also extensively involved in European agriculture. Farm women in Finland contribute approximately 25% of the total work hours involved in farming (Institute AER 2000). Sixty-six percent of Polish women living in rural areas are associated with agriculture (Sawicka 2001). Many women are farm managers as the wife took over running the farm when the husband was killed or injured (66% of cases studied), or because husbands and sons work off-farm (30% of cases studied).

In the U.S., farm women in Wisconsin contribute on the average 20 hours/week to farm labor versus 50 hours/week for men (Nordstrom and others 1995). In a survey conducted by Reed and colleagues (1999) of rural farm households in Kentucky and Texas, about 50% of the 1600 women surveyed described themselves solely as homemakers, and yet 40% reported regularly working with farm animals and 30% reported regularly driving tractors, leading to potentially dangerous situations.

Many women have a combination of family, farm, and off-farm careers also, and as increasing numbers of men seek off-farm employment, some women find it necessary to increase their work on the farm, leading to increased risk of agricultural injury (Carruth and others 2001; Stueland 1997). Figure 2.2 (Hawk 1991, 1994) illustrates the percentage of off-farm employment in Iowa. This pattern is similar to that experienced by many farm families in industrialized countries. Sixty-three percent of women and 35% of men have additional off-farm work, either full or part time, see Figure 2.2.

Figure 2.2. The men of farm families are increasingly taking off-farm jobs, increasing the pressure on a woman to assume multiple roles on the farm and increasing her risk of injury. (This figure also is in the color section.)

Work Exposures/Hazards

As with many areas of health research, farm women have been studied less than men. However, the research available suggests that women do have significant exposure to agricultural work and thus related health and safety risks from farm exposures. Among injuries to women on the farm, livestock-related incidents are the most prevalent causes of acute injuries. High risk is found in working directly with large animals in enclosed spaces, and during feeding, cleaning, or milking. Injuries to women are most likely to occur in the lower extremities. In a 2-year case-control study in central Wisconsin, 40 acute injuries were studied in adult women. Most, or 55%, of the injuries occurred in a barn, and in 42.5% of the injuries a cow was the main agent of injury (Stueland 1997). Being crushed or struck by an animal is also the most common cause of injury for women ages 15–59. For women above 60, falls are the most common cause of nonmachinery injury and fatality. According to Dimich-Ward and others (2004), of machine-related fatalities, 48% are the result of tractor runovers. Tractors also are the most common cause of machine-related nonfatal injuries, causing 28% of the farming-related hospitalizations for females, and the most common means of death for adult women on the farm.

Women are highly concerned about pesticide exposures to them and their families (Alavanja and others 1994; Castorina and others 2003; McDuffie 1994; Thu and others 1990). Forty percent of the wives of agricultural producers in rural Iowa and North Carolina have assisted with mixing or application of pesticides (Gladen and others 1998). Furthermore, about 45% of farm wives in North Carolina performed at least one or more of the following activities that could expose them to pesticides: worked in the fields tilling the soil (16%), planted (55%), and hand-picked crops (54%). Other tasks noted that may cause either increased exposure to other potentially hazardous chemicals or injuries include applying manure fertilizer (27%), applying chemical fertilizer (26%), and driving combines (4%). Indirect exposures of women to pesticides include 1) drinking water from wells located at close proximity to areas where pesticides are stored, mixed, or applied; 2) storage of pesticides in the home; and 3) contamination of the home by pesticides brought in on boots or clothing. Safe methods for launder-

ing clothes contaminated with pesticides (see Chapter 6 for details) include washing them separately from other clothing, and if possible, using a separate machine and using three rinses. Unfortunately only about 6% of farm families practice this method (Gladen and others 1998).

Dust-related respiratory diseases are also a major concern for women on the farm (Molocznik and Zagorski 2000). A study of 10 selected family farms in Poland indicated that women spend a significant amount of their farm work time involved in dust-generating activities, including care of animals (205 hr/mo ave); sowing, planting, and harvesting of crops (17 hr/mo ave); and manual loading (30 hr/mo ave) of grain. These women were exposed to dust ranging from 3.5–9.3 mg/m^3. The exposure limit for grain dust in most industrialized countries is around 4.0 mg/m^3. Furthermore Donham and others (2000) have recommended that dust exposures in confined animal feeding operations (CAFOs) should be no higher than 2.5 mg/m^3. Although there has been little research on farm women to determine the specific prevalence of organic dust-related respiratory illnesses one study suggests that farmer's lung is highly prevalent among Finnish farm women (Terho and others 1987). The probable risk factor here is cow milking and exposure to moldy hay, among other organic dust exposures.

Women's work on farms is expanding in the area of swine production, as they are sought out for their skills in working with the birthing (farrowing) process. Carbon monoxide is a toxic gas frequently encountered in livestock buildings where fossil fuel heating units are used or where high pressure washers powered by internal combustion engines are used (Donham and others 1982; Fierro and others 2001). Although the risk of acute CO poisoning to the woman is small, the risk of poisoning to an unborn fetus is much higher. Because of the physiology of the human placenta, the effective CO levels experienced by the fetus may be twice as high as that experienced by the mother. Exposures to levels of 250–450 ppm CO (levels commonly found in swine barns with propane-powered radiant heaters, see Figure 2.3) are levels that may be associated with 20% carboxyhemoglobin in women and lowered birth weights and retarded mental development in their newborn infants. Higher levels (above 450 ppm)

Figure 2.3.   A propane-powered radiant heater, which increases the level of CO in swine barns and the risk of prenatal and birthing complications in women. (This figure also is in the color section.)

may be associated with acute abortion (Donham and others 1982).

Women working in livestock production are likely to administer hormones used to assist parturition in swine and cattle. Oxytocin is one commonly used product to assist uterine contractions and milk let-down. Prostaglandins are also commonly used hormones in livestock production to induce partition, terminate a pregnancy, or stimulate estrous. A needle stick and accidental inoculation of a pregnant woman with either of these products while vaccinating animals also can cause her to miscarry. Details of these needle stick hazards are described in Chapter 12.

Contracting any one of the zoonotic or environmental infections, including brucellosis, Q fever, or Listeria, also may cause abortions. Pregnant women working with cattle, sheep, and goats need to be especially aware of this risk (see Chapter 13 on zoonoses).

Pesticides pose a risk to all who regularly come into contact with them. (These general risks are described in Chapter 6.) This section highlights the variety of health problems that are particularly associated with women's health, including reproductive and fetal developmental outcomes, breast cancers, and other women's cancers.

Fecundity decreased 20% among women who were exposed to the herbicides dicamba, glyphosate, phenoxy acetic acids, and thiocarbamates (rate ratio 0.5–0.8 compared to expected) (Curtis and others 1999). Notably, when women on farms did not report having these exposures, fecundity ratios were higher (0.75–1.50). Mixing and applying herbicides and fungicides up to 2 years before attempting conception was shown to decrease fertility (Greenlee and others 2003). Spontaneous abortion rates have also been shown to be affected by pesticide exposure. An increase in risk for an early abortion (<12 weeks) was shown when preconception exposures occurred with phenoxy acetic acids, triazines, and other herbicides. Elevated risk of late abortion (12–19 weeks) was associated with preconception exposures to glyphosate, thiocarbamates, and a miscellaneous class of pesticides (Arbuckle and others 2001). The critical window of exposure to pesticides for spontaneous abortion appears to be during the fourth through sixth month of gestation (Arbuckle and others 2001).

Chapter 5 reviews in general cancers that are of concern to the farm population. Mortality for all cancers is lower in farm women as well as farm men. Risks in one study were 0.84 (CI 0.76–0.92) for all cancers; 0.32 (CI 0.20–0.50) for lung cancer, and 0.33 (CI 0.12–0.92) for bladder cancer, compared to nonfarm residents. The overall risk for breast cancer in farm women is also lower than expected. In fact, there seems to be a protective factor in that breast cancer risk declines with increased duration of farming. However, there is an elevated risk of breast cancer in those women who reported being present in fields during or shortly after pesticide application, and for those who reported not using protective clothing while applying pesticides (Duell and others 2000). Furthermore, risk is elevated in female farm residents for non-Hodgkin's lymphoma (RR = 1.52, CI 0.96–2.39) (Folsom and others 1996).

Other General Health Concerns

Depression is a common disorder among both men and women on farms, but it is more common (nearly 2–3 times) in farm women. This gender difference in depression among the farming population, appears to be similar among all industrialized countries (Carruth and Logan 2002). A major cause of depression for women is chronic stress associated with the economic, social, emotional, and physical well-being of all of the family members (Carruth and Logan 2002). Women commonly feel responsible for their family's health care, and difficult access to adequate health care services in many rural areas may add to a woman's stress level. In a survey by Hemard and others (1998) less than half of rural women were satisfied with the health care services available to them, and this dissatisfaction was a strong determinant of stress for them. Women may feel the strain of the financial burdens on the family, a major contributor to stress (Melberg 2003). Fatigue also increases stress. Fatigue can originate from the woman's combined workload of jobs on the farm, in the home, and (in many instances) off-farm employment. As mentioned above, in North America, nearly 60% of farm women work off of the farm either part-time or full-time. These multiple roles must balance or there is an increased risk for stress and depression. Farm pressures have a tendency to strain marital and family relationships, influencing the happiness of children, which allows for additional tension build-up and increases the likeliness of depressive symptoms and substance abuse (Carruth and Logan 2002). Symptoms of depression most likely to be reported are headaches, backaches, muscle pain, sleep problems, fatigue, and abdominal pain (Van Hook 1996). However, rural women are not likely to discuss depression with their primary health care providers (Elger and others 1995) due to lack of time, geographic isolation, the stigma about depression, and perceptions that primary providers are not interested (Coakes and Kelly 1997).

Prolonged depression can lead to more severe illness. Women with severe and persistent mental illness (SPMI) have more medical illnesses and earlier mortality than women in the general population (Dalmau and others 1997). They also can be very troubled over gender-related health concerns, such as unresolved grief over child loss, isolation from family, bodily changes, lack of sexual partners, and diminished sexuality. Few studies have been performed on investigating preventive health care for rural women with mental illness (Hauenstein 1996; Lyon and Parker 2003).

Misconceptions and lack of information about health care and farming risks can also be health concerns themselves. In a study of 102 rural women, most underestimated their risk of certain chronic illnesses (Fiandt and others 1999). Farm women generally perceive they have a lower risk for coronary heart disease, stroke, breast and colorectal cancer, osteoporosis, and depression than in actuality; this can lead farm women to not seek professional health care for themselves and their families, and can lead to physical decline and increased mental strain. Utilization of preventive health care services is lower in rural populations than in urban populations, including utilization of screenings such as mammograms (Carr and others 1996).

Recommendations for Prevention

Too often, safety education for the farmer is aimed at the men in the farming operations. Education should include women and the whole family and hired workers. This family focus not only helps the woman directly, but indirectly helps her entire family, as social support is important in promoting safe behavior. Furthermore, as women often take on a special caretaker role in the family, this information can provide them with the support and confidence to be the family safety and health advocate and to manage their concern, thus helping them feel more in control and thereby reducing stress. In the rural community, women on the farm need to be trained to perform their own agricultural work more safely, such as by safe handling of large animals and safe pesticide application. They also may be undereducated about the risks of farming in general, and knowledge of these risks might keep them and their families from dangerous exposures and situations. Women must be made aware of the seriousness of the illnesses and injuries so better preventive measures may be taken (Fiandt and others 1999). Examples of helpful information for them could be general preventive measures on the farm and the importance of cancer screenings and other preventive health care services. Knowledge of availability of health care services in the area for agricultural families is essential for utilization (Rosenman and others 1995). Assurance of adequate health services for farm women and their families may secondarily help reduce stress and depression.

Risk of women's exposures to agricultural chemicals such as pesticides can be significantly reduced if general preventive measures are taken, as detailed in Chapter 6 in this text. Women, men, and children who may wash clothes of those exposed to pesticides should wear protective gloves. These clothes should be washed separately and rinsed at least three times. When working in operations where there are respiratory exposures, appropriate respirators should be worn, as described in Chapter 3.

In addition to standard safety practices, pregnant women must take extra precautions to prevent accidental needle sticks with either oxytocin or prostaglandins. Veterinarians dispensing these drugs to farm operators should be sure to transmit appropriate safety precautions for women. Furthermore, pregnant women should not work in livestock or other buildings where heaters, internal combustion engines, or another source of carbon monoxide is present, unless work practices or an environmental assessment indicates the environment is free of CO.

YOUTH

Young people under the age of 19 commonly work on farms (Figure 2.4), either as part of a family operation, as part-time workers on others' farms, or as labor accompanying their parents who may be migrant or seasonal workers (Berman 2003). Furthermore, children living on farms may also be exposed every day to potential hazards living or playing on the farm, either because that is where they live, or because they are visiting a grandparent or other relative. In the U.S., there are around 1.3 million children the age of 18 and under, living and/or working on U.S. farms. Detailed statistics of youth working on farms is reviewed by the National Children's Center for Rural and Agricultural Health and Safety (Little and others 2003; NCCRAHS 2001). Most farms are still family businesses; the practice of children working on the farm is a part of the culture and tradition of farming and in some instances may be an economic necessity. Children typically do not have to be forced to work. Quite

Figure 2.4.   A child on a farm. (Source: *Medical Practice in Rural Communities,* Chapter 2, page 20, by Cornelia F. Mutel, Kelley J. Donham, Rural Health and Agricultural Medicine Training Program, Department of Preventive Medicine and Environmental Health, College of Medicine, The University of Iowa, Iowa City, IA 52242, U.S.A., © Springer-Verlag New York Inc., 1983. Image reprinted with kind permission of Springer Science and Business Media.)

the contrary, they typically want to work. It is something they can be proud of as their work creates appreciation and acceptance from the adult world which helps build their self esteem. Figure 2.5 lists positive attributes to children who work on farms.

The demographics of children working on farms have been reviewed in several studies (Parker and others 2002). In Kentucky, 82% of children on farms surveyed by Browning and others (2003) were involved in agricultural work with animals. Seventy percent reported having worked in the production of tobacco (Browning

Child Development Opportunities of Growing up on a Farm

➤ **Responsibility**
➤ **How to Work**
➤ **Independence**
➤ **Skills**
➤ **Understanding of life
  and food cycles**
➤ **Initiative**
➤ **Problem Solving**
➤ **Parental Connections**
➤ **Neighboring**

Figure 2.5.   The positive aspects of child involvement in agricultural work.

and others 2001). Seven percent of hired agricultural workers in crop production are between the ages of 15 and 17, according to the National Agricultural Workers Survey, and in 1996, an estimated 300,000 15–17-year-olds worked on farms in America (National Research Council 1998).

Work Exposures/Hazards

Children in rural areas, and specifically children on farms in the U.S., have a high mortality rate for unintentional injuries relative to other causes of death (Little and others 2003; Meyers and others 2001; Pryor and others 2002). Approximately 22,000 injuries and 100 unintentional injury deaths occur annually to children and adolescents under the age of 20 on U.S. farms (American Academy of Pediatrics 2001). Injuries to youth may be related to occupational or recreational exposures. Major sources of injury exposure for youth include tractors and other farm machinery, animals such as cattle and horses, mixing or applying pesticides, motor vehicles such as all-terrain vehicles and pickup trucks, nearby water (e.g., farm ponds), electrical and flash burns, and work from high places (Park and others 2003). Males are more often injured than females, by a ratio of 2:1 (Meiers and Baerg 2001).

Drowning has proven to be a legitimate concern for young farm residents and visitors (Adekoya 2003). An average of 32 farm youth drowning deaths occur annually; a slightly higher annual rate than the U.S. overall (2.3 deaths/ 100,000 farm youth resident years compared to 2.2 deaths/100,000) (Adekoya 2003). Drowning was found to be the most common cause of death for farm youth under 10 years of age in Australia (Mitchell and others 2001b). Drowning fatalities most commonly occur in dams, tanks, and in creeks or rivers.

Although children involved in farm work are more commonly boys (about a ratio of 3:1), girls also are involved (Marlenga and others 2001). Figure 2.6 indicates most boys begin working at the age of 10, and 67% of boys ages 15–18 operate tractors unsupervised (Hawk and others 1994). Girls more commonly start working at 15 (Hawk and others 1994). Numbers for farm-related fatalities of children in Iowa are shown in Figure 2.7.

According to several studies, injuries and fatalities involving animals as well as tractors are the most common among farm children of age groups under 19 years (Little and others 2003; Meiers and Baerg 2001; Pryor and others 2002). Boys tend to perform tractor-related chores more often than girls. Boys aged 16–18 have the highest risk of injury from tractors (Browning and

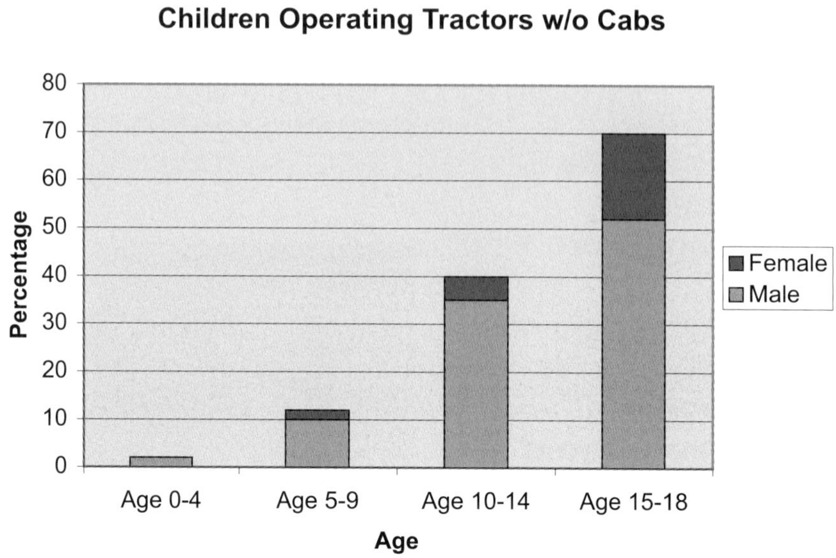

Children Operating Tractors w/o Cabs

Figure 2.6. Children, especially male children, are exposed to high levels of risk, sometimes at very early ages. (Hawk et al. 1994)

others 2003). Injuries involving tractors are most likely rollovers or runovers, and these incidents result in a very low (33%) survival rate (Meiers and Baerg 2001).

Girls are usually assigned more of the animal-related tasks (Marlenga and others 2001). One out of five youth injuries on farms is animal related, and of these injuries, 69% are work related (Hendricks and Adekoya 2001). Work-related injuries are mainly associated with cattle, and nonwork-related injuries involved mainly horses (Hendricks and Adekoya 2001).

Youth who work on family farms or are employed on nonfamily farms are both at high risk for injury (Chapman and others 2003; Gerberich and others 2001; Munshi and others 2002). Teen-age workers on a family operation are likely to work fewer hours per week but more seasons of the year, and perform a greater variety of (potentially hazardous) tasks compared to teens working on nonfamily-owned farms (Bonauto and others 2003).

Common serious injuries and those most often requiring hospitalization include orthopedic, neurological, thoracic, and abdominal injuries (Meiers and Baerg 2001). The majority of injuries on farms occur in the summer (44%), when children are out of school and most field work is ac-

complished. Injuries are also much more likely to occur on a Saturday (27%) or Sunday (28%) (Little and others 2003).

Pesticide exposure is a possible concern for children working, playing, or just living on farms. Some child activities that increase pesticide exposure are playing in farm fields, in dirt near fields, and in irrigation channels (Cooper and others 2001; Dunn and others 2003). Furthermore, exposure may occur by eating food while working, eating fruits and vegetables without washing, and applying pesticides. Children might not be provided with protective clothing for work activities where exposure may be possible. In fact, it is difficult to find personal protective equipment that will fit young people.

Pesticides can also be transported into homes on clothing and boots. Although pesticide sampling in homes indicates this possibility, urine tests indicate there is little absorption among children from this exposure (Fenske and others 2002).

Epidemiology of Illnesses and Effects of Exposures

There appears to be a weak association between pesticide exposures to parents and certain developmental disabilities and cancer in their children.

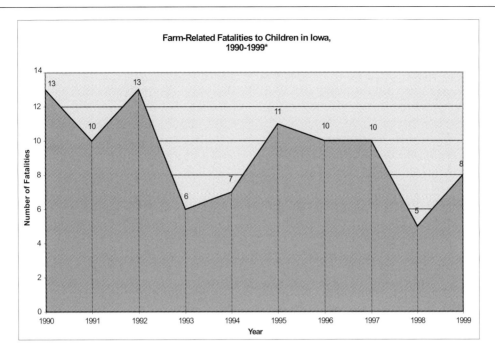

Figure 2.7. (The title and caption are part of the image)
(Source: Iowa Department of Public Health—SPRAINS, http://www.idph.state.ia.us/ems/sprains.asp)

Certain neurobehavioral developmental disorders and birth defects have shown weak association with pesticide application by parents (March of Dimes 2001). These defects were seen primarily with male children, born to applicators of the fumigant phosphine (OR = 2.48) and the herbicide glyphosate (OR = 3.6). Phosphine was also found to be associated with congenital cataracts, autism, and variably attention deficit disorders (Garry and others 2002).

One study (Efird and others 2003) suggested a risk of brain tumors in children born to women who had various farm exposures, including farm chemicals and farm animals. The odds ratios for these associations ranged from 1.5 to 2.3.

One other study revealed an association between childhood cancers (primarily lymphoma with an SIR of 2.18)and fathers who had pesticide exposure (Flower and others 2004). Associated risk factors included not wearing chemically resistant gloves and exposure to aldrin.

Another study of childhood cancers associated with pesticide exposure did not find an overall increase in childhood cancers, but did find a small increase in leukemia (Reynolds 2002). Specific

chemicals associated with the increased risk included the acaracide proparagite (used in orchard crops) and the organophosphates Azinphosmethyl and ziram.

The associations described above are not very strong and not very consistent, and there is no obvious plausible biological explanation for these findings. However, these findings do need further research, including animal studies to ascertain their significance.

Other General Health Concerns

Young children under 10 are generally at increased risk (two–10 times higher compared to the general population) for acquiring infections and particularly the animal fecal associated organisms *Campylobacter* and *Salmonella* (FSIS/CDC/FDA 1997; USDA 1998). Furthermore young children are at risk of developing infections of antibiotic-resistant organisms connected to the agricultural use of antimicrobials (Shea 2003).

Different from findings that show increased disease risk in farm-exposed children, being born and raised on a farm exposure seems to be protective for allergic sensitization and asthma (Horak and others 2002; Klintberg and others 2001). This protective factor is seen in farm children who lose their atopic status (Horak and others 2002; Portengen and others 2002). Farm life seems to have a lifelong protective effect against the development of allergy and other atopic-mediated allergic conditions (Leynaert and others 2001). Furthermore the protective factor may extend beyond the farm gate; one study has shown evidence for a general rural asthma protective factor (Chrischilles and others 2004). The hypothesis of protection relates to the control mechanisms of the immune systems and inflammation (Lauener and others 2002). It is hypothesized that early exposure down-regulates the usual mediators of inflammation and the cellular immune system. A more detailed discussion of this observation is seen in Chapter 3 of this text.

Contrary to these protective findings, a recent study has shown an increase of asthma in children who live on swine farms and especially those farms that used antibiotics in the feed (Merchant and others 2005). The discrepancy with previous findings is uncertain. The methods of asthma diagnosis were different which may explain the variant results.

In summary, there are several studies suggesting that farm exposures for children are protective for atopic allergic diseases. One should understand, however, that asthmatic symptoms in agriculture populations are often not of the usual atopic mechanisms and are inflammatory processes. Down-regulation of inflammatory reactions as well as allergic-mediated illnesses can also occur. Studies are inconclusive at present regarding the protective factors of farming to allergic or nonallergic asthma. Furthermore, the hypothetical mechanisms for these observations await further research.

Recommendations for Prevention

Protection of children in agriculture must include those on family farms, children of farm workers, and children visiting farms (e.g., a grandparent or other relative). Injuries that are not work related can be diminished by proper adult supervision of play, children's instruction of safety rules, and designation of specific play areas that are out of harm's way (Esser and others 2003; Hawk 1991, 1994).

Supervision of young children who are around operating farm machinery or farm animals also could keep them from life-threatening situations. Taking children (or anyone) as an extra rider on a tractor, is very temping and commonly practiced. However, it is done so with risk, as 25% of all tractor-related deaths are a result of being run over (many because of extra riders). Farm safety specialists are strongly against extra riders without a properly designed extra rider seat.

As farm families have multiple roles of on- and off-farm work, the need for rural child care becomes more important. Figure 2.8 indicates that full-time maternal employment implies child care is a necessity to reduce child exposure to farm hazards. However, part-time off-farm employment or no off-farm employment dramatically increases the exposure of children to farm hazards. It takes a devoted person or organization in the community to create a sustainable rural child care service. Solutions may include the involvement of retired persons or high school age children for in the home or in facilities out of the home.

For children of migrant and seasonal farm workers, housing is an extremely critical issue. Migrant housing has been notoriously poor, but new standards are now in force in many areas that help assure more safe housing for children.

Awareness of the issues and practice of safe behavior in the home is critical. One important organization facilitating in this area is Farm Safety 4 Just Kids (FSJK 2004) (http://www.fs4jk.org/). Based out of Earlham, Iowa (U.S.), this organization educates and advocates, and has produced numerous programs with special methods required (Liller 2001) to help families create and maintain a safe environment for farm children. This organization has developed and maintains a network of some 150 local chapters in North America, with activities in Europe and Australia. Local communities linking up with this organization can help obtain the support, advocacy, and information dissemination of parents and children to develop and maintain safe environments for farm children. Whereas FSJK is primarily aimed at families and children up to high school age, another curriculum for high school–age youth, has been developed called Agricultural Disability Awareness and Risk Education (AgDARE). Students participating in this program perform two types of simulations. In one simulation stu-

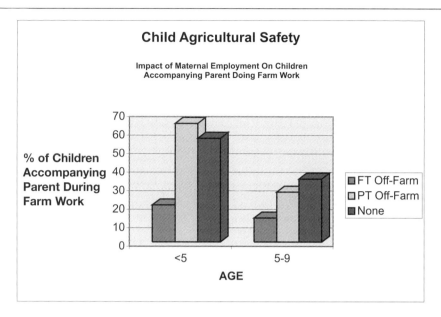

Figure 2.8. Children are more likely to accompany a parent working on a farm when the mother spends some or all of her time on the farm. This shows the need for off-farm supervision to decrease the risk of injury to children that is associated with accompanying a working parent. FT = full time; PT = part time. (Hawk et al. 1994)

dents make decisions about actions they would take in hazardous farm work situations. The others are physical simulations, which allow the student to assume a disability and perform simulated farm work. More information about this program can be found at its web address: http://www.mc.uky.edu/SCAHIP/projects/agdare-2.htm (Reed and others 2001).

Researchers evaluating the effectiveness of these programs have had difficulties in demonstrating long term effectiveness (DeRoo and Rautiainen 2000). However, there have been some positive results; students who complete programs such as those described above show some statistically significant positive changes toward safety on the farm (Baker and others 2001; Reed and others 2001). Parent involvement in educating children on safety measures is essential. Educating children requires that the parents must educate themselves about the dangers in agricultural work, and they must demonstrate that they value and practice safe behaviors. Furthermore, families planning to have children should also know of the preconception and prenatal risks involved in agriculture exposures (discussed above) and take measures to prevent and avoid unnecessary exposure.

The tradition of family participation in farm work is strong. Parents want to maintain this tradition (Thu and others 1990), and it is important for the best development of the child and family and for preparing the next generation of farmers. An important concept in helping to assure this work is safe for the child is assignment of tasks that match the developmental capabilities of the child and supervision of the children. Mason and Earle-Richardson (2002) studied a group of young farm workers over a 6-year period and found about half of the children injured were below the appropriate developmental ages recommended for those tasks. To assist farm parents and supervisors of young workers in assigning safe tasks and supervising those tasks, a group of farm safety specialists and developmental specialists developed the North American Guidelines for Children's Agricultural Tasks (NAGCAT) (Lee and Marlenga 1999; NCCRAHS 2005). These guidelines list appropriate developmental stages for safe conduct of a variety of farm tasks. Ability to perform specific tasks must be assessed for each individual child. Examples of some of the guidelines include 1) children must be 16 or older to drive an articulated tractor or to drive on a road, 2) children should be at least 16 to work in a CAFO unsupervised, and 3) children should be at least 16 to clean grain bins unsupervised. More guidelines for determining appropriate tasks can be found at www.nagcat.org.

Additional to voluntary educational-based programs, most countries have some regulations regarding child labor. For example, the U.S. Department of Labor developed the Hazardous Work Orders for Youth in Agriculture. These regulations are aimed at employers of youth off the family farm (they do not apply to the family farm). This act is reviewed in detail in Chapter 14 of this text.

ELDERLY

Farming is the occupation with the oldest workers in the United States (Occupational Health & Safety 2000) (Figure 2.9). (This applies primarily to owner operators of family farming, rather than hired farm workers.) In the year 2000, 68.5% of farmers were age 45 and older, quite a difference from the entire U.S. work force, where only 33.7% of employees are 45 and older. The farming population's average age is also increasing. From 1982–1997 the number of farmers aged over 64 years increased 24%, and the average age of a farmer in 1997 was 54.3 years (Olenchock and Young 1997). The longevity in farming is not as pronounced in many European countries, as the culture and social systems make it easier to retire and transfer the holdings compared to North America. The reasons for the longevity in farming in North America and many other countries are many, but primarily it is a fact that family farming is still a way of life, and not just a job. Farmers do not retire as others do because they find great satisfaction in the work. For many it is, in fact, their life. Many want to work until they die. As one older man said in a study of elderly farmers, "If I die in a farming accident, it would be an honorable way to die; better than being hooked up to tubes and machines in a hospital bed." The only thing that seems to make them stop and think about their own health risks is their grandchildren (Huneke and others 1998).

In some cases elderly farmers think they must work to augment their retirement resources. In some cases they may work because there is a child that they hope will farm in the future, or is in the process of taking over the farm, and they

Figure 2.9.   Farmers often keep working well past usual retirement age. The man on the far right in this photo is 91 years old and still works daily. (This figure also is in the color section.)

can phase out of the operation over a couple of decades, working as long as they feel able. They may work because there is not a young person available to take over the operation. Elderly farmers work 50–100% the amount of time worked by younger farmers, and they may work even when injured (Voaklander and others 1999). The risk for injury increases at age 55 and older, and continues to increase as they continue to work, many past the age of 85. Furthermore, risk for injury increases if the farmer has had a previous injury, probably due to the increased disability from the previous injury (secondary injury) (Browning and others 1998). It is clear that elderly farmers are the norm rather than the exception, and they have special health and safety risks. Self-perceived risk does not match up with actual risk, as most elderly farmers think there is a low risk (Tevis 1995). However, their perceived risk does increase with previous injury. Furthermore, even though risk for acute injury from machinery or animals is by far their highest occupational risk, elderly farmers perceive their greatest on-farm safety risks are from electrocutions and chemical exposure (Fiedler and others 1998).

Work Exposures/Hazards

Epidemiology studies of acute fatal injuries of elderly farmers reveals 65–84 years is the highest risk age group for fatal injuries (Horsburgh and others 2001). Data from Iowa show a similar ex-

**Farm-Related Fatal Injuries by Age in Iowa,
1990 - 1999***

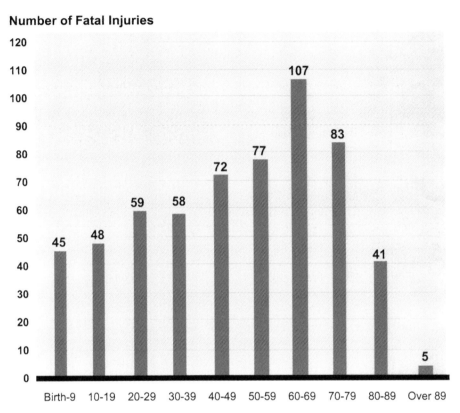

N = 595
*Recreation Injuries Excluded 1998 and 1999
Iowa Department of Public Health - SPRAINS

For the decade, the leading age group for fatalities was 60 - 69 years of age. Age groups 70-79 and 50-59 were the next highest.

Figure 2.10. In Iowa, the highest rate of fatal injury belongs to the age group of 60–79 years. (Source: Iowa Department of Public Health—SPRAINS, http://www.idph.state.ia.us/ems/sprains.asp)

perience (Figure 2.10). Tractor-related injuries are the most common cause of fatal injuries (Fiedler and others 1998; Horsburgh and others 2001; Janicak 2000; Mitchell and others 2001a, b). These injuries are related to tractor rollovers, being run over by a tractor, and falls from a moving tractor or equipment. Falls are also a common cause of fatal injuries (Sprince and others 2003), especially in the group of farmers 65 and older. Many of these falls are from tractors or farm equipment. Other causes of death may be moving

vehicles, other machinery, livestock, electrocution, or being struck by falling objects.

Epidemiology of nonfatal injuries in elderly farmers reveals a similar pattern to fatal injuries (Figure 2.11). Falls account for 18–21% of these injuries, compared to 5% for all other workers (NIOSH 2001; U.S. Dept. of Labor 2002). Most of these falls are from machinery or from heights in barns. Tractors are associated with only 11% of the nonfatal injuries. Animals are also an important cause of nonfatal injuries, with elevated

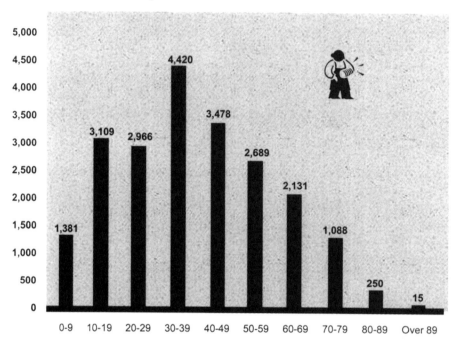

Farm-Related Non-Fatal Injuries by Age in Iowa, 1990 - 1999*

Number of Non - Fatal Injuries

N = 21,527
*Recreation Injuries Excluded 1998 and 1999
Iowa Department of Public Health - SPRAINS

For the decade, the leading age group for non-fatal injuries was 30 - 39 years of age except for 1999 when it was 40 - 49.

Figure 2.11.   Non-fatal agricultural injuries occur less frequently in the elderly age groups, perhaps because an older person is less likely to survive an injury.
(Source: Iowa Department of Public Health—SPRAINS, http://www.idph.state.ia.us/ems/sprains.asp)

risk on beef and dairy farms. Injuries from animals most often involve the arm, shoulder, hip, or leg (Xiang and others 1999). Associated co-morbidities which are risk factors for injuries in elderly farmers include the effects of normal aging (e.g., deterioration of hearing, vision, smell, and touch; loss of muscle mass, strength and dexterity). Arthritis and use of prescription medications may also increase risk (Browning and others 1998; Mitchell and others 2002; Pickett and others 1996; Sprince and others 2003).

Of farmers 55 years of age and above, 56% use prescription medications. Compared to a control group, the odds ratio for injuries among farmers using prescription medications was 2.8 (CI 1.0–7.7) (Xiang and others 1999). The reasons are likely impairments associated with co-morbidities or the actual effects of the medications, which could include sedation, impaired balance, and decreased reaction time.

Epidemiology studies reveal that arthritis is an important chronic injury issue in the elderly farmer: There are two basic types of arthritis, rheumatoid arthritis and osteoarthritis (OA). The former is caused by an altered immune process of the person and has a genetic predisposition. OA is caused by the normal aging process and perhaps is exacerbated by excessive and chronic wearing and tearing on the joint surfaces. The strenuous requirements of farm work may contribute to the development of OA. Specific risk factors are characterized by repeated lifting of heavy weights, kneeling, bending, squatting, working long hours, starting heavy work as a young person, and continuing work into years well past the usual age of retirement.

Degenerative osteoarthritis of the knee and hip are very common among agricultural populations in Europe as well as the U.S. (Axmacher and Lindberg 1993; Sandmark and others 2000; Homberg and others 2003). Risk factors related to farming are unclear, but there is some evidence that both male and female dairy farmers have higher risk for degenerative osteoarthritis of the knee, compared to other populations (Homberg and others 2003). The severity of the disease is a common cause for knee replacement surgery in this population (Sandmark and others 2000). Forestry and construction work are other occupations with high risk for degenerative osteoarthritis of the knee. Details of hip and knee osteoarthritis are seen in Chapter 8.

Degenerative osteoarthritis of the hip is also a common problem among farmers in Europe and the U.S. (Axmacher and Lindberg 1993). Risk for this condition is much higher than in urban comparison populations.

Other disability concerns of elderly farmers include a general greater loss of function relative to a comparison population of age-matched white-collar workers (Geroldi and others 1996). Elderly farmers had a mean of 2.0 functions lost, compared to .61 functions loss in age-matched white-collar workers. Examples of function losses were demonstrated in a study of Kentucky farmers age 55 and older, which revealed that 34.2% had hearing difficulty, 11.4% had vision difficulty, and 50.4% had arthritis (Browning and others 1998). Slower cognition problems in elderly farmers may impede safe judgment and proper self-care of co-morbidities (Sinclair and others 2000), and difficulty seeing things may impede reacting to them in time to prevent an injury.

Depression is an important problem among many elderly people, affecting 19% of Americans 65 years or older (19%). Older women (like all women) are twice as likely to report depressive symptoms as older men (NAMI 2005; Bartels and others 2003). Elderly Hispanics and blacks (males and females) are more likely to have depression than elderly whites. The causes of these changes include biological (body, hormone changes), environmental (social isolation, not feeling appreciated), psychological, and environmental (Turner 1996). As women generally tend to outlive their spouses, many farm women are left alone, increasing their isolation. Depression in elderly may have different manifestations than in younger people, characterized by and exacerbating usual aging changes, including memory problems, confusion, social withdrawal, inability to sleep, delusions, hallucinations. All of these factors can increase risk for injury when performing dangerous work tasks, as in agriculture (NIMI 2005). Depression in the elderly may also be complicated with co-morbidities, such as heart disease, stroke, diabetes, cancer, and Parkinson's disease. Health screenings are important to pick up conditions such as blood chemistry, cholesterol, and cancer, but just as with farmers in general, some elderly farmers are even more resistant to see a health care provider, as they perceive this as "weak" or may not want to realize or "give in" to sickness, and therefore may not make appointments until they feel quite ill (Engler 2002).

Elderly farmers are reluctant to adopt use of personal protective equipment (PPE) or retrofit equipment with safety and guarding structures. They may not understand the present protective value of wearing UV protective sunglasses and ear, eye, and respiratory protection because they think the damage is already done. However, they are very supportive that their children, grandchildren, and other younger people should use PPE and proper machine guarding.

Recommendations for Prevention

GENERAL RECOMMENDATIONS

Farm safety education for the elderly should include encouraging taking short breaks and not to work when ill, as fatigue and a cold or flu can increase their risk of an injury. As elderly farmers generally work alone, and may have a significant risk of delayed treatment from an injury, they have an increased risk for serious complications or death. There should be an emphasis on structuring work so that the elderly do not work alone or have a communication system in place (such as where and when the elderly person is working and time of expected return) or a portable communication device (cell phone, two-way radio) to request assistance if needed. An innovative program that seems to attract elderly farmers together with their grandchildren is the Generations Project (FSJK 2004). This project requires grandparents and their grandchildren working together, teaching by stories, and correcting hazards on the farm.

TRACTOR SAFETY PROGRAMS

Since tractor overturns are the number one cause of fatality in the elderly (North America), emphasis should be placed on installing rollover protective structures (ROPS) with seat belts on tractors, retiring tractors without ROPS, or strongly discouraging elderly from operating tractors without ROPS. Over 60% of elderly farmers commonly operate tractors lacking ROPS, seat belts, working lights, slow moving vehicle (SMV) emblem, and power take-off (PTO) shields (Tevis 1995). Only 26% of elderly farmers believe the benefits of ROPS outweigh the costs of installation (even though 88% believe they are effective). "In general, senior farmers rate tractor operation as a moderate to low-level risk—much like the perceived risk of driving in a car without a seatbelt." (Tevis 1995).

There have been several tractor safety programs that have shown some effectiveness in ROPS installation. One that is applicable to the elderly is the Tractor Risk Abatement and Control (TRAC-SAFE) at the University of Iowa (IREH 2005; NIOSH 2005). This community-based program includes in-depth education, involvement of local machinery dealers, and community incentives. This program also emphasizes reassigning tractors without ROPS away from use in hazardous tasks, such as operating on inclines; mowing ditches, banks, or waterways; and use with front-end loaders.

DECREASING RISKS OF FALLS

Falls are a common cause of acute injury among farmers in general, and they are even more of a risk for the elderly farmer. Falls on ice are common in northern climates. These can be reduced by assuring that places where water can collect and freeze are eliminated by drainage, rain gutters and downspouts are in proper working order, walkways are covered, and a container of sand is kept at sites that may be slippery during cold weather. Using traction treads on shoes and strap-on ice grips may be very beneficial. Other general fall prevention includes safe ladder usage and wearing a safety harness when working in high places. Install high-quality lighting, especially where there might be trip hazards.

PHYSICAL FITNESS

Although farming is hard work, the type of work may not be appropriate to maintain muscle tone, bone density, and cardiovascular fitness. Elderly farmers should be encouraged to participate in either home programs or group programs that encourage weight-bearing exercise and cardiovascular fitness.

ISSUES FOR CLINICAL SCREENING OF ELDERLY FARMERS

When conducting physical exams and screenings for elderly farmers, several issues should be considered:

1. Screen all medications, current and new, for side effects that may impair the patients physically, increasing their risk for injury in their work. Screen also for use of over-the-counter medications and home remedies.
2. Check vision and hearing regularly.
3. Examine skin for skin cancers, keratoses, etc.
4. Provide yearly influenza and pneumonia immunizations, as necessary.
5. Examine for arthritis and balance.
6. Consider the following screenings:
   a. Diabetes (blood sugar level)
   b. Hypertension (blood pressure)
   c. Osteoporosis (bone density)
   d. Prostate cancer (PSA for men)
   e. Breast cancer (mammograms for women)
7. Question and counsel on the following:
   a. Regular exercise to maintain strength and bone density
   b. Sleep disorders
   c. Depression and opportunities to manage it
   d. Wearing appropriate PPE for relevant work (e.g., UV protective glasses, hearing and respiratory protection)
   e. Proper posture to reduce risks of sprains and strains
   f. The realities of prebycusis and how to manage it

g. Prevention of hyper- and hypothermia
h. Symptoms of stroke and heart attack

Note that recommendations from general medicine practice may vary from country to country, and that some of these recommendations are more appropriate for general medical screening.

MIGRANT AND SEASONAL FARM WORKERS

Productivity in agriculture in many industrialized countries around the world has increased by the input of a large cadre of farm workers (Arcury and others 1998; Schenker 1996; Paltiel 1998) (Figure 2.12). This increase has been especially prominent in industrialized countries the past two decades because of the increased pressure for low labor costs in order to compete in the global market. This expansion in the use of hired farm labor has been most prominent in North America, where consolidation and industrialization of farming enterprises has resulted in larger but fewer farms and the need for hired farm labor to replace family labor. Also, growth in more labor intensive crops has increased farm labor demands, evidenced by a 66% increase in the U.S. annual production of fruits and vegetables (Villarejo and Baron 1999). In the U.S. migrant and seasonal workers are predominantly Latino (mainly from Mexico and Central America) though many are African-American, Haitian, Anglo, or Asian. Once only prominent in states such as California, Arizona, Texas, and Florida, hired farm labor has expanded tremendously into the whole of the Southeast and upper Midwest of the U.S., and Eastern Canada.

Hired farm workers are, as defined by the U.S. Census Bureau, "people employed to perform tasks on farms for the purpose of producing an agricultural commodity for sale." Farm workers may be divided into several groups: 1) migrant farm workers (those required to travel more than 75 miles and stay away from their home), and 2) seasonal farm workers (those performing agricultural work of a seasonal or temporary nature, who are not required to be away overnight from their home). The former two groups are referred to as migrant and seasonal farm workers (MSFW), 3) immigrant farm workers who have come to stay and either initiate their own operations or work in a more permanent job and location, and 4) indige-

Figure 2.12. Migrant worker (Source: Photograph by Tierra Nueva, www.peoplesseminary.org/images/013102.a.jpg). (This figure also is in the color section.)

nous full- or part-time hired farm workers. The groups of farm workers given most publicity and concern are the MSFW, which make up about 50% of all hired farm workers in the U.S., are considered minority groups in society, and are usually foreign-born (Culp and Umbarger 2004). These people may come from outside the country for agricultural work at certain times, depending on the crop cycle, and return to their home at intervals. It is difficult to estimate the number of MSFW in most countries. The U.S. Department of Labor's National Agricultural Workers Survey reports that there are approximately 1.4 million migrant workers in the U.S., and other estimates have reported that between 3 and 5 million people immigrate with their dependents for agricultural work (Hansen and Donohoe 2003). Most are male (80%), born outside of the U.S. in Mexico or Central America (69%), poor—with an average personal income of $2,000–10,000—and are married with children (Alderete and others 1999; Coughlin and Wilson 2002). The share of foreign-born workers has increased dramatically in the U.S. during the past several decades due to crises in the home countries of the migrants, such as economic difficulties in Mexico and the civil wars in El Salvador, Guatemala, and Nicaragua. About 33% of these people are not authorized to work in the U.S. (Villarejo and Baron 1999). Details of the demographics of this population are found in Table 2.1. Women and children are a part of this work force, mainly as members of family units that have settled out and stay in limited geographic areas (Farr and Wilson-Figueroa 1997).

Describing the geographic patterns of MSFW has used a "three streams" concept, although this pattern seems to be changing with changing availability of work. Generally MSFW move through the country from south to north: from South California to North California, Oregon, and Washington; from Texas and Arizona to Ohio, Michigan, Indiana, Illinois, and Iowa in the Midwest; and from Florida to Georgia, the Carolinas, and New England in the East (Figure 2.13) (Coughlin and Wilson 2002). Other migration patterns within these mainstreams are restricted circuit, point-to-point, and nomadic patterns. In the restricted circuit pattern, the MSFW travel throughout a season in a relatively small area to find work. MSFW who are on a point-to-point pattern follow the same route through the country to find work each season, usually with home bases in Florida, Texas, Mexico, Puerto Rico, or California. The nomadic farm worker travels from farm to farm and crop to crop, living away from home for a period of years (Migrant Clinicians Network 2005). Farm worker advocacy groups have brought a high level of attention to the potential problems of pesticide toxicity for farm workers, but in recent years, many other occupational problems previously overlooked are only now beginning to be realized.

Migrant and seasonal workers and immigrant farm workers are common in many other countries as well. For example, Brazil (perhaps the most rapidly developing industrialized agriculture in the Western Hemisphere), is utilizing migrant and seasonal labor. The United Kingdom utilizes migrant and seasonal farm workers from all over Europe. The horticultural industry alone employs about 1 million workers every year. Other farm labor jobs within the European community include fruit picking (e.g., strawberries, gooseberries) and vegetable picking (e.g., cauliflower) (Department for Work and Pensions 2005; PickingJobs.com 2005). French farmers employ migrant and seasonal workers to pick fruit such as grapes, blueberries, raspberries, and strawberries. Denmark uses migrant and seasonal labor to pick strawberries, among other crops, and Holland uses migrant and seasonal labor to pick fruit and flowers. The primary countries of origin for the bulk of these workers in the U.K. include India, Australia, and South Africa. Australia has developed a tradition of the "backpacking temporary" farm worker. These typically are students from Europe and North America traveling in Australia who stop to work for a few weeks to gain some extra money to continue their travel. They work at picking fresh fruits and vegetables. Australia fruit growers also hire "boat people" from surrounding Asian countries as seasonal labor (Farmindex.com 2005). There is a concern by the Australian government for the protection of these people regarding work exposures. On the other hand, the government is trying to manage the illegal immigration of these people into the country, causing the boat people to live in the bush, where living conditions are not necessarily safe and sanitary. New Zealand producers hire some seasonal workers to pick fruit, such as apricots, cherries, and apples (Farmindex.com 2005).

Work Exposures/Hazards

There are numerous occupational and environmental hazards for farm workers (Arcury and others 1998). Although detailed and accurate data are not readily available (Wilk and others 1999), the following paragraphs review the major risk concerns documented in the literature.

HOUSING

Adequate safe and sanitary housing has been a long standing problem for MSFW. Often the employer provides the housing. Although most countries have standards for worker housing, many labor camps still have marginal environmental conditions. Potable water, toilets, and bathing facilities are often a concern. MSFW who purchase

Table 2.1. Cultural, Legal, and Demographic Characteristics of Migrant Farm Workers by Percentage of the Total Migrant Farm Workers in the Three United States "Streams"

	Eastern (%)	Southwest and Midwest (%)	Pacific, Including California (%)
Birthplace			
United States	9.4	29.0	9.6
Puerto Rico	7.6	—	—
Mexico	73.9	70.2	87.3
Central America	8.0	0.6	0.2
Caribbean	0.9	—	0.2
Southeast Asia	0.1	—	2.2
Pacific Islands	—	0.2	0.2
Asia	—	—	0.1
Legal status			
U.S. citizen	18.2	33.6	13.5
Green card	14.0	37.5	41.7
Other work authorization	1.8	0.8	1.9
Unauthorized	66.0	28.0	42.9
Migrant type			
Seasonal	33.5	48.3	57.3
Follow-the-crop	31.4	24.7	10.7
Shuttle	35.1	27.0	32.0
Primary language			
English	7.3	17.9	6.7
Spanish	84.3	80.9	88.6
Creole	0.5	—	—
Cambodian	—	—	0.1
Tagalog/Ilocano	0.1	0.2	2.5
Mixtec	1.1	—	0.8
Other	6.7	1.0	1.3
Type of housing			
House	47.3	65.2	67.1
Flat or apartment	21.6	17.5	23.9
Room in hotel/motel	1.5	2.0	0.2
Room/bed in dormitory	4.0	2.6	1.1
Mobile home/trailer	22.3	10.5	5.8
Vehicle (car/camper)	—	—	0.2
Other	3.2	2.2	1.6
Demographic characteristics			
Gender			
Male	87.0	80.9	79.1
Female	13.0	19.1	20.9
Marital status			
Single	46.2	32.8	32.4
Married/living together	45.3	60.6	60.3
Separated/divorced/other	8.4	6.6	7.3
Age	30.6	34.6	34.0
Highest grade educated	6.0	7.5	6.6
Number of children	0.6	1.3	1.2
Years in U.S. farm work	6.6	11.0	11.0
Family income	$5,500 [5.2]	$10,250 [7.1]	$9,499 [6.8]

Source: 1998 National Agricultural Workers Survey, conducted by the U.S. Department of Labor (Ward and Alav 2004).

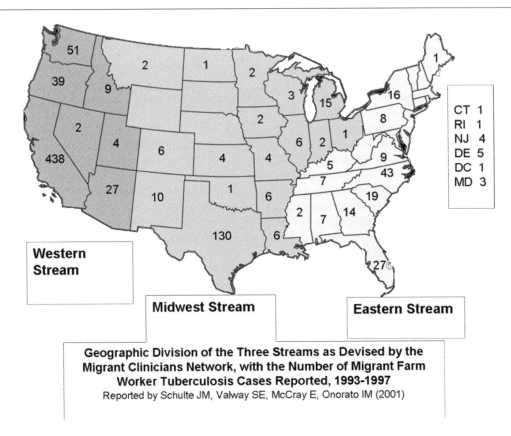

Geographic Division of the Three Streams as Devised by the Migrant Clinicians Network, with the Number of Migrant Farm Worker Tuberculosis Cases Reported, 1993-1997
Reported by Schulte JM, Valway SE, McCray E, Onorato IM (2001)

Figure 2.13.   MSFW cases of tuberculosis are of higher frequency in the homebase states of California, Texas, and Florida, but occur in all places where they may find employment. (Source: USDA website, http://www.usda.gov/wps/portal/usdahome)

housing on the available market may also be relegated to less than standard housing, as wages are not very high and they are trying to save as much money as possible to send home. Housing may be located close to the fields where pesticides are stored and applied, creating exposure risk through contamination of the local premises, air, and water (McCauley and others 2001). Environmental sampling of homes has revealed it is common to find low levels of the pesticides used on crops in the region within the housing units. Azinphos-methyl or AZM (McCauley and others 2001) is one chemical commonly found in farm worker housing in California.

Pesticide Exposure Through Work
MSFW may have exposures to pesticides through contact with fruit and vegetable crops that have been treated with pesticides (Mobed and Schen-

ker 1992), or through mixing and loading or applying pesticides. MSFW are not as likely to wear personal protective equipment and likely may not follow hygienic practices at home (showering, washing clothes frequently, etc.). Though U.S. farm operators are required to provide safe hand washing facilities in the fields, not all comply, increasing hazardous exposures (Villarejo and Baron 1999). Acute and chronic pesticide health effects are covered in Chapter 6.

Other common health problems of MSFW include eye problems, which are commonly reported (up to 40% prevalence) by MSFW. This includes physical trauma (corneal scratches or ulcerations) and conjunctivitis. Causes include punctures with foreign objects (from working through foliage and other plant materials), chemicals such as sulfur and propargite, and field dust, which also may cause eye irritation. Working in the bright sunlight creates risk of retinal and corneal exposure to ultraviolet light, which can lead to long-term damage. Wearing hats and safety/sunglasses would be helpful PPE. However, nearly 99% do not wear eye protection. Part of the reason for this may be the negative cultural issue associated with wearing sunglasses, as drug dealers in Mexico and Central America wear sunglasses, and MSFW do not want to be associated with them (Quandt and others 2001).

MUSCULOSKELETAL INJURIES

The labor intensive nature of MSFW employment results in heavy and chronic strain on the musculoskeletal system. Thirty-one percent of MSFW report muscle and joint strain (Earle-Richardson and others 2003). Approximately half of the musculoskeletal injuries experienced are back injuries (Bean and Isaacs 1995). Causes include heavy lifting; working in stressful postures, such as reaching above the head to pick with a heavy sack on the back or bending at the waist for long periods; working at an excessively fast pace; whole body vibration from operating or working on farm machines; and working in hot or cold climates (Mobed and Schenker 1992; Villarejo and Baron 1999).

SKIN CONDITIONS

A survey of North Carolina MSFW revealed 24% experienced skin conditions early in their seasonal farm work. However, a follow-up survey revealed a prevalence of 37% for skin conditions (Arcury and others 2003). Specific outbreaks of skin conditions have been noted in California related to harvest of grape, tomato, and citrus crops. Most of these conditions are irritant contact dermatitis. A number of different insecticides and herbicides are known causes of contact dermatitis (see Chapters 4 and 6). Elemental sulfa and the insecticide/fungicide propargite are known strong dermal irritants. Furthermore, there are plant components that can also cause dermatologic problems. Plants—including carrots, celery, and parsnips—contain the chemical furocoumarin, which can cause a severe contact dermatitis in combination with sun exposure (see Chapter 4).

RESPIRATORY ILLNESSES

MSFW have similar respiratory exposures to other farmers, but there are some specific risks that are of special concern for this population. MSFW smoking prevalence is similar to the general Hispanic population, which was 18.7% in 2000 (CDC 1992). This compares to around 10% prevalence for the general farming population. Smoking is a contributing factor to other occupational respiratory exposures. Pulmonary function screenings of MSFW show reduced forced vital lung capacities, which may be explained by exposures to agricultural organic and inorganic dusts. These changes are independent, but in the magnitude of that expected of cigarette smoking (Mobed and Schenker 1992). (There is no Hispanic normal comparison population for respiratory function, which may result in false low pulmonary function values. Therefore, to determine occupation-related acute obstructive pulmonary function deficits, it is best to compare before- and after-work measures.)

MSFW have a sixfold risk of having tuberculosis compared to other employed adults (CDC 1992; Schulte and others 2001). The risks include having been born in a country that has a high prevalence of tuberculosis. Commingling of tuberculosis-infected persons in migrant camps or in the workplace creates risk of infecting other workers. Additional risk factors include poverty, crowded living conditions, mobility, poor access to health care, and noncompliance with tuberculosis medications. Additional social factors in some MSFW, such as homelessness and alcohol abuse, increase the risk of tuberculosis in some workers (Poss 1998). Cases of tuberculosis by state, along with the three migrant streams, are shown in Figure 2.13.

Mental health conditions are also a concern in MSFW. There are many stressors MSFW have to endure that can lead to depression, including uprooting and separation from nuclear and extended family, separation from community and cultural origins, and adapting to host culture. Additional stressors include low and unpredictable income, language barriers, discriminatory treatment, inadequate housing, rigid work demands, and fear of being deported (Magana and Hovey 2003). The native Hispanic culture is a collectivist society, and people thrive on strong family connections, which include a strong support system (Alderete and others 1999). Many of the previous factors take them out the realm of self control (a strong stressor), and transfer control to their employer, their guest culture, and society. Several surveys of MSFW have revealed a prevalence of depressive symptoms from 20–40% (Alderete and others 1999), 28.9% (Vernon and Roberts 1982), and 39% (Magana and Hovey 2003). Vega and others (1998) found rates of various psychiatric disorders far lower in Mexico and in recent immigrants relative to immigrants with lengthy U.S. residence or those of Hispanic workers born in the U.S.

Children of MSFW are at risk because of the social and environmental conditions in which they live (Berman 2003; McCurdy and others 2002). Dental caries is an important problem in these children. Regular dental check-ups are rare. Lukes and Miller (2002) found that 51% of MSFW children had not sought oral health in the previous year, as they typically do not see a dentist unless there is pain. Nearly 50% of MSWF had symptoms of peridontal disease. Caries in MSFW children have followed a worldwide decrease, but this was primarily seen in permanent teeth, and not in deciduous teeth. School-age children of MSFW are still at high risk of dental caries (Chaffin and others 2003).

BARRIERS TO HEALTH CARE SERVICES

Language barriers still persist (Weathers and others 2003) even though most hospitals employ translators. There are also transportation barriers, as most MSFW do not have automobiles. Most MSFW do not have health insurance. The fact that many MSFW are not authorized to work in the U.S. keeps them from going to governmental health care agencies or sometimes from seeking any health care, for fear of being caught and deported (Villarejo and Baron 1999). Cultural barriers include the fact that Hispanics do not like to be examined by a person of the opposite gender. Furthermore, folk medicine beliefs are still strong in the Hispanic community, which might interfere with western health care treatment. Because of the migrant nature of this population, medical records may not be available. MSFW primarily live in the present, and the concept of preventive care is not relevant to their frame of reference. Lack of information and cultural barriers exacerbate this problem. Therefore, their medical care is primarily acute illness care. One example of the former is the finding that MSFW women underutilize mammograms and Pap tests because of limited awareness of benefits and cultural beliefs

(Coughlin and Wilson 2002). In spite of these barriers to health care, neonatal survivability and the birth weights of Hispanic farm worker babies are significantly greater than the general population (Ronda and Regidor 2003). The reasons for this observation are not known.

Prevention and Protection of Migrant and Seasonal Farm Workers

The Commission on Human Rights of the World Health Organization (WHO) held an International Convention on the Protection of Migrant Workers and Members of Their Families. This convention developed a document laying out a series of rights that should be afforded all migrant workers. It was adopted by the General Assembly of the WHO in December 1990, and has been ratified by 25 member states. The resolution went into force July 1, 2003. The primary points of the International Declaration of Migrant Workers Human Rights include the following human rights provisions, which are considered universal, indivisible, interconnected, and interdependent (OHCHR 2003):

- Assurance of a sufficient wage that contributes to an adequate standard of living
- Freedom from discrimination based on race, national or ethnic origin, sex, religion, or any other status, in all aspects of work, including in hiring, conditions of work, and promotion, and in access to housing, health care, and basic services
- Equality before the law and equal protection of the law, particularly in regard to human rights and labor legislation, regardless of a migrant's legal status
- Equal pay for equal work
- Freedom from forced labor
- Protection against arbitrary expulsion from the state of employment
- The right to return home if the migrant wishes
- The human right to a standard of living adequate for the health and well-being of the migrant worker and his or her family
- Safe working conditions and a clean and safe working environment
- Reasonable limitation of working hours, and rest and leisure
- Freedom of association and to join a trade union

- Freedom from sexual harassment in the workplace
- Protection during pregnancy from work proven to be harmful
- Protection for the child from economic exploitation and from any work that may be hazardous to his or her well-being and development
- Education of children of migrant workers
- Reunification of migrants and their families

If this Bill of Rights for migrants and seasonal workers is universally practiced, the health risks for this population should significantly decrease.

In the U.S., the Migrant and Seasonal Agricultural Worker Protection Act (MSPA) provides employment-related protections to migrant and seasonal agricultural workers. Culp and Umbarger (2004) provide an excellent review of the laws affecting the health and welfare of MSFW in the U.S. (U.S. Dept. of Labor 2005). Every nonexempt farm labor contractor, agricultural employer, and agricultural association who "employs" workers must

- Provide written disclosure of the terms and conditions of employment in the worker's language.
- Post information about worker protections at the worksite in the worker's language.
- Pay workers the wages owed when due and provide an itemized statement of earnings and deductions.
- Comply with the terms of any working arrangement made with the workers.
- Make and keep payroll records for 3 years for each employee.

Recommendations for Prevention

Lack of health and safety training has been cited as a major concern for MSFW. Several studies have shown that only 7% of farm operations who hire MSFW provide health and safety training for their workers, and less than 30% of MSFW have had adequate health and safety training for jobs they do (Arcury and others 2001; Bean and Isaacs 1995; Salazar and others 2004). One of the most important factors in improving this situation is to educate farm owners how to better teach occupational health and safety to MSFW. Culturally appropriate and translated materials are needed.

Explicit safety rules should be developed, and effectively communicated to the workers. Pesticide safety exposure has dominated the discourse on MSFW safety, to the detriment of developing more comprehensive prevention that includes eye protection, skin protection, musculoskeletal injuries, infectious diseases, and respiratory illnesses. Finally there needs to be a renewed emphasis to help assure that MSFW have safe and sanitary housing, water, and sanitary facilities.

The following are suggestions from the Migrant Health News line (DHS 2005) to assist rural health practitioners provide quality health services to MSFW:

1. To better understand this patient population, seek information and skills on cultural competency for MSFW in your area by reading or attending a workshop.
2. Be aware of the living and working situations of MSFW in your area by asking your patients and the MSFW employers questions. Use this information to provide care and advice to your patients (*Example:* be sure a patient has a refrigerator before prescribing medication that needs refrigeration or asking them to apply ice to an injury).
3. The first visit of a MSFW and family members may take extra time because of language and transportation barriers. Consider booking two time slots for them.
4. When communicating to the patient or interpreter, look directly at the patient and speak slowly but not loudly. Communicate in simple but organized sentences. (*Example:* You have _____ because of _____. You must do three things to get better: 1._____, 2._____, and 3. _____.)

More tips can be found at the Migrant Clinicians Network (2005).

ANABAPTISTS

A variety of new religious groups were spawned out of the protestant reformation of 1517 (Figure 2.14). One such group is the Anabaptists. The Anabaptist movement began in Switzerland following reforms begun by the ideas of Ulrich Zwingli (Schwieder and Schwieder 1975). The term *Anabaptist* is the Greek word for *rebaptized*, referring to the belief that adult instead of child

Figure 2.14.   Amish men working in their modified traditional ways (Source: iStockphoto.com, photograph by Diane Diederich, www.istockphoto. com). (This figure also is in the color section.)

baptism is essential to the Christian faith (Jones and Field 2002). Although there were at least 50 Anabaptist sects that evolved, with over 1 million members around the world today, there are four main sects discussed here: Amish, Mennonites, Brethren, and Hutterites. These people reside in communities scattered mainly in North America. The values that link these groups together are both religious and philosophical. Besides the religious practice of adult baptism, these groups value pacifism, separation of church and state, agricultural lifestyle, and hard work. Branches of these groups that adhere to their early beginnings (Old Order) are centered even more in agriculture, and in social and economic separation from the society around them. They also practice a similar plain clothing dress. However, most Mennonite and Brethren groups have moved much closer to mainstream protestant religious traditions and life-styles (Redekop 1989).

The large Anabaptist groups still existing today split apart by following different leaders. The Mennonites are originally followers of Menno Simons. The Amish, followers of Jakob Ammann,

split from the Mennonites in 1693. The Hutterites (from 1528 in Austria) are originally followers of Jakob Hutter (Hostetler 1974). The Brethren (from 1708 in Central Germany) are followers of Alexander Mack. As these groups were counter to state and religious doctrine at the time, they survived by fleeing from persecution, moving many times around Europe during 16th through the 18th centuries. During the 19th century, there were many voyages of the Anabaptists to North America, and later (Mennonites) to South America, where they were able to benefit from greater religious freedom and more available farmland (Agarwala and others 2001; Ober and others 1999).

There were many other groups of this same time period that split from the state church at the time and practiced a communal social structure of living (some 500 of these groups existed in the U.S., most not associated with the Anabaptists). Although most of these groups no longer function as a communal society (for example, the Amana Colonies in Iowa is now a private corporation), the Hutterites (an Anabaptist group) still practice a communally social form of life, and are growing in number. Nearly all of the research of Anabaptists has been reported on the more traditional, Old Order, Anabaptist communities, such as the Amish and conservative Mennonites; consequently the discussions of this text are centered around these conservative groups who define themselves partially by their value for being "nonworldly" and for their selective use of technology (the latter seen as a threat to their value of hard work). These groups primarily practice farming (with some allowance for other basic supporting occupations as carpentry, horse breeding and training, etc.). They practice limited use of modern conveniences, such as electricity, in the home and continue to use animal labor for transportation or field work (however, there is variance among groups, with some using cars and some using tractors with steel wheels). Most travel is accomplished by horse and buggy, but trips to town are infrequent since many of these communities are nearly self-sufficient.

The Amish are the largest of the Old Order Anabaptists with approximately 180,000 living in the United States and 20,000 living in the Province of Ontario, Canada (Hostetler 1995). There are 1,200 different settlements or church groups with 25–50 families each. Most reside in the central U.S., with the highest populations in Iowa, Missouri, Northern Indiana, Ohio, and Pennsylvania (Aaland and Hlaing 2004). The Amish in North America have been growing about 4% per year in number over the past two decades. The last Amish settlement in Europe dissolved in 1937. Only about 10% of the approximate 1 million Mennonites worldwide are Old Order. Approximately 31,000 live in settlements in the East portions of the U.S., with the largest population in Pennsylvania. Approximately 38,000 Hutterites live in 425 different communal colonies of about 90 members each in Canada and the western U.S. Maps of the locations of these groups in North America are seen in Figures 2.15 through 2.17.

Approximately 6,000 of the some 190,000 Brethren in North America are considered Old Order, or the old German Baptist Brethren. They have a similar life-style to the Old Order Amish

Figure 2.15.   Amish Congregations in North America in 2000. (Source [information source, but map has been redrawn]: Donald B. Kraybill and Carld Desportes Bowman. *On the backroad to heaven: Old Order Hutterites, Mennonites, Amish, and Bretheren.* Figure 4.2 on page 104, Baltimore, MD; London: © 2001, Johns Hopkins University Press)

Figure 2.16. Old Order Mennonite Congregations in North America in 1999. (Source [information source, but map has been redrawn]: Donald B. Kraybill and Carld Desportes Bowman. *On the backroad to heaven: Old Order Hutterites, Mennonites, Amish, and Bretheren.* Figure 3.3 on page 71, Baltimore, MD; London: © 2001, Johns Hopkins University Press)

Figure 2.17. Hutterite Colonies in North America in 2000. (Source: [information source, but map has been redrawn] Donald B. Kraybill and Carld Desportes Bowman. *On the backroad to heaven: Old Order Hutterites, Mennonites, Amish, and Bretheren.* Figure 2.2 on page 31, Baltimore, MD; London: © 2001, Johns Hopkins University Press)

and Old Order Mennonites, and live primarily in the Eastern U.S.

Work Exposures/Hazards

In Old Order Anabaptist communities such as the Amish, farming is central to their way of life. The Church and their culture are anchored to the farm, where everyone works as a family. The community act of farming pulls members together, and it is on the farm where families work, play, and worship (Fisher and others 2001). There seems to be an unwritten agreement among groups such as the Amish that requires a living be made from farming, rural, or semirural occupations (Hostetler 1987).

Agriculture is handled much as it was 100 years ago. Most groups shun modern technology, except when it is required by state health laws, such as in the pasteurization and refrigeration of milk (Rhodes and Hupcey 2000). They also may modify modern technology to make it consistent with their traditional less worldly ways, such as using electrical generators in barns for feed grinding, cooling milk for bulk storage, putting steel wheels on tractors (taking the rubber off wheels), and using horses to pull modified modern machinery (for example, putting a gasoline engine on a hay baler so it can run with independent power but still be pulled by horses—note an illustration of this usage in Figure 2.14). They believe in a sparing use of agricultural chemicals, preferring to farm and live as naturally as possible. One result of this type of farming is that older and/or modified or homemade machines are typically used—this equipment often lacks the more faultless functioning of newer equipment and is not equipped with safety devices. As a result, injury from this alternative farming style is a major concern (Rhodes and Hupcey 2000).

Injury statistics sometimes are not easy to report because these Old Order communities are closed and do not often keep good records. A study of Amish health records by Jones and Field (2002) revealed 39% of Amish farm–related fatalities were due to being run over by a vehicle. A high proportion of these injuries involve horse-drawn vehicles. These types of injuries are sometimes caused by human behavior, such as the operator not knowing where a child was, or caused by fractious animal behavior. The second most prevalent source of injury was direct animal injury. Comparing these populations to the general farming population, only 11% of Amish farming–related fatalities involved agricultural tractors (compared to 50% on non-Amish farms) (Jones and Field 2002).

Old Order Anabaptist children often begin participating in farm work activities by age 5 or 6 (Rhodes and Hupcey 2000), and since formal schooling ends with eighth grade, by age 14 boys are able to work full time on the farm and girls can work in the garden and in the home (Fisher and others 2001). Differences in these groups and non-Anabaptist family farms are that mothers and fathers are much less likely to work off the farm. Furthermore, conservative Anabaptists have their own schools in their communities, and children have many fewer extracurricular activities around their schooling, creating more available time for farm work. Available statistics indicate that these children are even more at risk of injury than are children on non-Anabaptist family farms (Hubler and Hupcey 2002). Amish children of age under 16 accounted for 64% of injuries reported in one study (Jones and Field 2002). Runovers by vehicles are the most frequent primary cause of fatal injuries for children (Jones and Field 2002) and can be a result of hazards such as children riding without proper restraints on vehicles and children being unseen by operators. The most common contributing factor of injury to children reported in a survey by Fisher and others (2001) was a farm animal (53%). From the available research, we can conclude that danger of injury and death exists, especially for children, around vehicles and animals, especially around animal-drawn vehicles.

The Amish use horse-drawn vehicles as their main method of transportation. Motorized vehicle crashes with horse-drawn buggies on roads are a major hazard to the Amish (Aaland and Hlaing 2004). In Ohio over 20 crashes between motorized vehicles and buggies occur annually, resulting in severe property damage, injury, and death (Eicher and others 1997). Injuries from these crashes are usually severe, as the buggy offers little protection and the victims are often thrown out upon impact. Prevention in most communities includes a slow-moving vehicle sign on the back of the buggy; widened shoulders, where possible, on heavily traveled roads; and lanterns and sometimes strobe lights on the back of the buggy to increase visibility to motorists. However, religious beliefs may deter these practices in some Amish communities because of the bright color or warning signs (Donnermeyer 1997).

Effects and Special Considerations for Treatments of Injuries

Old Order Anabaptist groups' religious and cultural beliefs have both positive and negative effects on their health. The belief in "worldly separation" puts mainstream health care in the category of secondary consideration. Lack of telephones can delay outside health care assistance in emergencies (Brewer and Bonalumi 1995). The common belief in predetermination or "God's will", and a delayed gratification in the after life, creates a resistance to modern medical technology and preventive measures (Greksa and Korbin 1997). They generally believe, however, that it is important to stay away from substances (tobacco, alcohol etc.) and to work hard. "Unnatural" or heroic interventions for a health condition may sometimes be unwanted (Huntington 1984).

Amish do use mainstream health care services, but only when necessary, and they often use them late in the course of disease, after other more "natural methods" have failed. They typically do not seek out the full course of recommended immunizations, although this varies among the different groups, and between individuals (Kulig and others 2002). They typically do not carry health insurance (although the tendency to do so is increasing in some groups), and they usually pay for health care services out of pocket. If there is a large medical expense (cancer, major surgery, etc.), the community often will pay the bill for the individual. Most often, when injuries are mild, such as strains and minor trauma, Old Order groups deliver their own home health care. The Amish women traditionally are the healers in the community and deliver non-mainstream medicine or complementary and alternative medicine. In one report by von Gruenigen and others (2001) 36% of 66 Amish women used at least one form of complementary and alternative medicine. The forms of this traditional healing style most often reported are diet and nutrition programs, herb therapies, and chiropractic medicine. These traditional therapies may have adverse effects, especially during pregnancy and in the elderly.

The remarkably strong beliefs of the Old Order groups also ground them in the value of faith and prayer in healing. The majority in a study by Gerdner and others (2002) reported that faith has a major role in curing and healing in their lives. The belief also exists that "the best is yet to come," implying that ascension into heaven occurs after death. This idea, in some instances, may keep them from readily seeking advanced health care (Jones and Field 2002; Schwartz 2002). The strong family support system of the Amish makes family involvement in health maintenance essential (Armer and Radina 2002). High levels of self-care in the Amish communities—including feelings of hope, satisfaction, control, and support—contribute to a tightly knit, nearly self-sufficient culture (Baldwin and others 2002). All of the support within the community greatly diminishes the need to seek outside health care. Their traditional diets are low in simple sugars but high in protein, fat, and complex carbohydrates. However, it is apparent today that their total caloric intake is still high, with increased simple sugars, and their caloric expenditures are decreasing. Today obesity is not uncommon in Old Order Anabaptist, though still less common than in the general population.

There has been an increasing trend in the Amish people's utilization of mainstream health care services in the recent years (Greksa and Korbin 1997), especially for pregnancy health care services (though home births are still preferred) (Donnermeyer and Friedrich 2002). One creative service in Ohio (Amish birthing center) has shown excellent value in the community with its service blend of Amish values with modern obstetrical standards (Campanella and others 1993). Health services for this community are best adopted if they balance quality health care with the Amish values of simplicity, low cost, closeness to community, accommodation to use of horse and buggy, and agrarian style (Kreps and Kreps 1997).

There has been an increasing trend of health insurance provided by non-Amish employers and use of major medical insurance administered by businessmen within the churches, which are called Amish Aid or Mennonite Aid (Greksa and Korbin 1997; Morton and others 2003).

A very important consideration for health care providers who serve Old Order Anabaptist groups is to strive to understand their culture and beliefs, accept them, and to not judge them. Failing in the latter (as with most any minority group) serves only to create mistrust and a reluctance to use mainline health services.

Other General Health Concerns

GENETIC DISORDERS

Some Anabaptist populations have been relatively closed over time (Sivakova and Pospisil 2001), creating concern for specific genetic illnesses in some of the more closed communities (Agarwala and others 2001). Infantile refsum disease (Bader and others 2000) is an autosomal recessive genetic disorder, more common in the Old Order Amish populations. One of a group of leukodystropy diseases that affects the myelin sheath covering of nerve fibers in the brain, the condition results in varying degrees of impaired vision, hearing loss, delayed development and neuromotor deficiencies. A yellow to orange stain of the teeth is common among affected females. Maple syrup urine disease (inability to metabolize certain amino acids, resulting in acidosis and mental retardation) and glutaric aciduria type 1 (inability to convert lysine to tryptophan, resulting in central neurological dysfunction and abnormal muscle control) are other conditions found in certain Anabaptist groups (Morton and others 2003).

In spite of what may appear on the surface as a common problem, current genetic disorders are clustered in few families at present, and overall genetic disease prevalence among these populations is relatively low.

Lifestyle may promote good health in some Old Order Anabaptist communities. Smoking, alcohol consumption, and other substance abuse are extremely low. Children of Hutterites in Saskatchewan, Canada, are observed to experience less asthma (2.4% versus 9.2% in a neighbor comparison group) (Anthonisen 2002). As the general life-style is family oriented and social in nature, a great deal of the elderly care is conducted by the family (often more mentally healthful to the elderly) (Andreoli and Miller 1998).

The percentage of overweight (>25% of mean body mass index [BMI]) in an Ontario, Canada, Amish community, was 25% (men) and 27% (women). This is nearly half those percentages of overweight persons in the general population of North America (51% in Canada, and 64% in the U.S.). Only 4% of this Amish community was considered obese (>30% of mean BMI), compared respectively to 15% and 31% for the U.S. and Canada (Bassett and others 2004). This can be explained by the increased caloric expenditures of Amish at 3100 Kcal/day (men) and 1850 Kcal/day (women), relative to the general population. Men and women took about 50% more steps per day than a Swiss comparison population and about 80% more than the recommended 10,000 steps per day recommended for cardiovascular fitness (Bassett and others 2004).

This degree of BMI fitness is not necessarily shared by all Old Order Anabaptist groups. Those people highly engaged in farming, particularly those still using horses for power, expend more energy than communities that use tractors, or have become less agrarian as the general population has grown around them (e.g., Northern Ohio and Eastern Pennsylvania).

Recommendations for Prevention

Safety and healthful practices, which are harmonious with their culture, need to be developed and implemented for this population. For example, research is needed to develop culturally relevant ways to signal slow-moving vehicles to vehicular traffic. Installing mirrors on buggies has been suggested, but it may not be culturally acceptable in some groups. Widening of roads with shoulders for buggies to operate on is needed in many areas (Eicher and others 1997).

Safety and health education programs and delivery methods need development for parents and children, and they should include the following topics: 1) age appropriate tasks, 2) the importance of child supervision, 3) sources of injuries and illnesses, 4) causes and effects of genetic disorders, 5) common sources of injuries, 6) home and farm preventive measures, and 6) adverse effects of traditional therapies. Furthermore, these programs must be developed with cultural relevancy and by a facilitator that has credibility within the Amish community. It is important that health professionals understand the cultural roots that affect health behavior of Anabaptist groups in order to establish effective health promotion (Brunt and Shields 1996). Conversely, Anabaptist groups need to seek out and be informed of health care providers in the community who understand and accept their beliefs, and who will work to provide therapy within their belief system.

REFERENCES

Aaland M, Hlaing T. 2004. Amish buggy injuries in the 21st century: A retrospective review from a rural level II trauma center. American Surgeon 70(3): 228–234; discussion 234.

Adekoya N. 2003. Trends in childhood drowning on U.S. farms, 1986–1997. J Rural Health 19(1):11–14.

Agarwala R, Schaffer A, Tomlin J. 2001. Towards a complete North American Anabaptist Genealogy II: Analysis of inbreeding. Human Biology 73(4): 533–545.

Alavanja M, Akland G, Baird D, Blair A, Bond A, Dosemeci M, Kamel F, Lewis R, Lubin J, Lynch C and others. 1994. Cancer and noncancer risk to women in agriculture and pest control—The agricultural health study. J Occup Med 36(11):1247–1250.

Alderete E, Vega W, Kolody B, Aguilar-Gaxiola S. 1999. Depressive symptomology: Prevalence and psychosocial risk factors among Mexican migrant farm workers in California. J Community Psychol 27(4):457–471.

American Academy of Pediatrics Co, Injury, Poison, Prevention, Committee on Community Health Services. 2001. Prevention of agricultural injuries among children and adolescents. Pediatrics 108(4): 1016–1019.

Andreoli E, Miller J. 1998. Aging in the Amish community. Nursing Connections 11(3):5–11.

Anthonisen N. 2002. Asthma in Hutterites. Canadian Respiratory Journal 9(5):289–290, 293–294.

Arbuckle T, Lin ZQ, Mery LS. 2001. An exploratory analysis of the effect of pesticide exposure on the risk of spontaneous abortion in an Ontario farm population. Environ Health Perspect 109(8):851–857.

Arcury T, Quandt SA, Mellen BG. 1998. Occupational and environmental health risks in farm labor. Hum Organ 57(3):331–334.

——. 2003. An exploratory analysis of occupational skin disease among Latino migrant and seasonal farmworkers in North Carolina. J Agric Saf Health 9(3):221–232.

Arcury T, Quandt SA, Rao P, Russell GB. 2001. Pesticide use and safety training in Mexico: The experience of farmworkers employed in North Carolina. Hum Organ 60(1):56–66.

Armer J, Radina M. 2002. Definition of health and health promotion behaviors among midwestern Old Order Amish families. Journal of Multicultural Nursing & Health 8(3):43–52.

Axmacher B, Lindberg H. 1993. Coxarthrosis in farmers. Clinical Orthopaedics and Related Research February(287):82–86.

Bader P, Dougherty S, Cangany N, Raymond G, Jackson C. 2000. Infantile refsum disease in four Amish sibs. American Journal of Medical Genetics 90(2):110–114.

Baker A, Esser N, Lee B. 2001. A qualitative assessment of children's farm safety day camp programs. J Agric Saf Health 7(2):89–99.

Baldwin C, Hibbeln J, Herr S, Lohner L, Core D. 2002. Self-care as defined by members of the Amish community utilizing the theory of modeling and role-modeling. Journal of Multicultural Nursing & Health 8(3):60–64.

Bartels S, Dums A, Oxman T, Schneider L, Arean P, Alexopoulos G, Jeste D. 2003. Evidence-based practices in geriatric mental health care: An overview of systematic reviews and meta-analyses. Psychiatr Clin North Am 26(4):971–990.

Bassett D, Schneider P, Huntington G. 2004. Physical activity in an Old Order Amish community. Med Sci Sports 36(1):79–85.

Bean T, Isaacs LK. 1995. An overview of the Ohio migrant farmworker needs assessment. J Agric Saf Health 1(4):261–272.

Berman S. 2003. Health care research on migrant farm worker children: Why has it not had a higher priority? [comment]. Pediatrics 111(5 Pt 1): 1106–1107.

Bonauto D, Keifer M, Rivara F, Alexander B. 2003. A community-based telephone survey of work and injuries in teenage agricultural workers. J Agric Saf Health 9(4):303–317.

Brewer J, Bonalumi N. 1995. Cultural diversity in the emergency department: Health care beliefs and practices among the Pennsylvania Amish. Journal of Emergency Nursing 21(6):494–497.

Browning S, Truszczynska H, Reed D, McKnight R. 1998. Agricultural injuries among older Kentucky farmers: The farm family health and hazard surveillance study. Am J Ind Med 33(4):341–353.

Browning S, Westneat SC, Donnelly C, Reed D. 2003. Agricultural tasks and injuries among Kentucky farm children: Results of the Farm Family Health and Hazard Surveillance Project. South Med J 96(12):1203–1212.

Browning S, Westneat SC, Szeluga R. 2001. Tractor driving among Kentucky farm youth: Results from the farm family health and hazard surveillance project. J Agric Saf Health 7(3):155–167.

Brunt J, Shields L. 1996. Preventive behaviours in the Hutterite community following a nurse-managed cholesterol screening program. Canadian Journal of Cardiovascular Nursing 7(2):6–11.

Campanella K, Korbin J, Acheson L. 1993. Pregnancy and childbirth among the Amish. Social Science & Medicine 36(3):333–342.

Carr W, Maldonado G, Leonard PR, Halberg JU, Church TR, Mandel JH, Dowd B, Mandel JS. 1996. Mammogram utilization among farm women. J Rural Health 12(4 Suppl S):278–290.

Carruth A, Logan CA. 2002. Depressive symptoms in farm women: Effects of health status and farming lifestyle characteristics, behaviors, and beliefs. J Community Health 27(3):213–228.

Carruth A, Skarke L, Moffett B, Prestholdt C. 2001. Women in agriculture: Risk and injury experiences on family farms. J Am Med Wom Assoc 56(1):15–18.

Castorina R, Bradman A, McKone TE, Barr DB, Harnly ME, Eskenazi B. 2003. Cumulative organophosphate pesticide exposure and risk assessment among pregnant women living in an agricultural community: A case study from the CHAMACOS cohort. Environ Health Perspect 111(13):1640–1648.

CDC (Centers for Disease Control). 1992. HIV Infection, Syphilis, and Tuberculosis Screening Among Migrant Farm Workers. Available from http://www.cdc.gov/mmwr/preview/mmwrhtml/000 17692.htm.

Chaffin J, Pai SC, Bagramian RA. 2003. Caries prevalence in northwest Michigan migrant children. J Dent Child (Chicago, Ill) 70(2):124–129.

Chapman L, Newenhouse AC, Meyer RH, Karsh BT, Taveira AD, Miquelon MG. 2003. Musculoskeletal discomfort, injuries, and tasks accomplished by children and adolescents in Wisconsin fresh market vegetable production. J Agric Saf Health 9(2):91–105.

Chrischilles E, Ahrens R, Kuehl A, Kelly K, Thorne P, Burmeister L, Merchant J. 2004. Asthma prevalence and morbidity among rural Iowa schoolchildren. J Allergy Clin Immunol 113(1):66–71.

Coakes S, Kelly GJ. 1997. Community competence and empowerment: Strategies for rural change in women's health service planning and delivery. Aust J Rural Health 5(1):26–30.

Cooper S, Darragh AR, Vernon SW, Stallones L, MacNaughton N, Robison T, Hanis C, Zahm SH. 2001. Ascertainment of pesticide exposures of migrant and seasonal farmworker children: Findings from focus groups. Am J Ind Med 40(5):531–537.

Coughlin S, Wilson KM. 2002. Breast and cervical cancer screening among migrant and seasonal farmworkers: A review. Cancer Detect Prev 26(3):203–209.

Culp K, Umbarger M. 2004. Seasonal and migrant agricultural workers: A neglected work force. AAOHN J 52(9):383–390.

Curtis K, Savitz DA, Weinberg CR, Arbuckle TE. 1999. The effect of pesticide exposure on time to pregnancy. Epidemiology 10(2):112–117.

Dalmau A, Bergman B, Brismar B. 1997. Somatic morbidity in schizophrenia—A case control study. Public Health 111(6):393–397.

Department for Work and Pensions. 2005. Statistics.

DeRoo L, Rautiainen R. 2000. A systematic review of farm safety interventions. Am J Prev Med 18(4S):51–62.

DHS Oregon. 2005. Health Issues Among Migrant/Seasonal Farm Workers.

Dimich-Ward H, Guernsey JR, Pickett W, Rennie D, Hartling L, Brison RJ. 2004. Gender differences in the occurrence of farm related injuries. J Occup Environ Med 61(1):52–56.

Donham K, Cumro D, Reynolds SJ, Merchant JA. 2000. Dose-response relationships between occupational aerosol exposures and cross-shift declines of lung function in poultry workers: Recommendations for exposure limits. J Occup Environ Med 42(3):260–269.

Donham K, Knapp L, Monson R, Gustafson K. 1982. Acute toxic exposure to gases from liquid manure. J Occup Med 24(2):142–145.

Donnermeyer J. 1997. Amish society: An overview. Journal of Multicultural Nursing & Health 3(2):6–12.

Donnermeyer J, Friedrich L. 2002. Amish society: An overview reconsidered. Journal of Multicultural Nursing & Health 8(3):6–14.

Duell E, Millikan RC, Savitz DA, Newman B, Smith JC, Schell MJ, Sandler DP. 2000. A population based case-control study of farming and breast cancer in North Carolina. Epidemiology 11(5):523–531.

Dunn A, Burns C, Sattler B. 2003. Environmental health of children. Journal of Pediatric Health Care 17(5):223–231.

Earle-Richardson G, Jenkins PL, Slingerland DT, Mason C, Miles M, May JJ. 2003. Occupational injury and illness among migrant and seasonal farmworkers in New York State and Pennsylvania, 1997–1999: Pilot study of a new surveillance method. Am J Ind Med 44(1):37–45.

Efird J, Holly EA, Preston-Martin S, Mueller BA, Lubin F, Filippini G, Peris-Bonet R, McCredie M, Cordier S, Arslan A, Bracci PM. 2003. Farm-related exposures and childhood brain tumours in seven countries: Results from the SEARCH International Brain Tumour Study. Paediatr Perinat Epidemiol 17(2):201–211.

Eicher C, Bean T, Buccalo S. 1997. Amish buggy highway safety in Ohio. Journal of Multicultural Nursing & Health 3(2):19–24.

Elger U, Wonneberger E, Lasch V, Fuhr D, Heinzel W. 1995. [Stresses and health risks of farm women]. Soz Praventivmed 40(3):146–156.

Engler J. 2002. The Medicaid bind. Fiscally burdened states need short-term help, long-term fix for program. Mod Healthc 32(30):35.

Esser N, Hieberger S, Lee B. 2003. Creating Safe Play

Areas on Farms, Second Edition. Marshfield, WI: Marshfield Clinic.

Farmindex.com. 2005. Farm Index.

Farr K, Wilson-Figueroa M. 1997. Talking about health and health care: Experiences and perspectives of Latina women in a farmworking community. Women & Health 25(2):23.

Fenske R, Lu C, Barr D, Needham L. 2002. Children's exposure to chlorpyrifos and parathion in an agricultural community in central Washington State. [see comment]. Environ Health Perspect 110(5): 549–553.

Fiandt K, Pullen CH, Walker SN. 1999. Actual and perceived risk for chronic illness in rural older women. Clin Excell Nurse Pract 3(2):105–115.

Fiedler D, Von Essen S, Morgan D, Grisso R, Mueller K, Eberle C. 1998. Causes of fatalities in older farmers vs. perception of risk. J Agromed 5(3):13–22.

Fierro M, O'Rourke MK, Burgess JL. 2001. Adverse health effects of exposure to ambient carbon monoxide: University of Arizona, College of Public Health. Available at http://www.airinfonow.org/pdf/CARBON%MONOXID2.PDF.

Fisher K, Hupcey J, Rhodes D. 2001. Childhood farm injuries in Old-Order Amish families. Journal of Pediatric Nursing 16(2):97–101.

Flower K, Hoppin JA, Lynch CF, Blair A, Knott C, Shore DL, Sandler DP. 2004. Cancer risk and parental pesticide application in children of agricultural health study participants. Environ Health Perspect 112(5):631–635.

Folsom A, Zhang SM, Sellers TA, Zheng W, Kushi LH, Cerhan JR. 1996. Cancer incidence among women living on farms—Finding from the Iowa women's health study. J Occup Environ Med 38(11): 1171–1176.

FSIS/CDC/FDA. 1997. FSIS/CDC/FDA Sentinel Site Study: The Establishment and Implementation of an Active Surveillance System for Bacterial Foodborne Diseases in the United States.

FSJK. 2004. Farm Safety 4 Just Kids.

Garry V, Harkins ME, Erickson LL, Long-Simpson LK, Holland SE, Burroughs BL. 2002. Birth defects, season of conception, and sex of children born to pesticide applicators living in the Red River Valley of Minnesota, U.S.. Environ Health Perspect 110(Suppl 3):441–449.

Gerberich S, Gibson RW, French LR, Renier CM, Lee TY, Carr WP, Shutske J. 2001. Injuries among children and youth in farm households: Regional Rural Injury Study-I. Inj Prev 7(2):117–122.

Gerdner L, Tripp-Reimer T, Sorofman B. 2002. Health beliefs and practices: The Iowa Old Order Amish. Journal of Multicultural Nursing & Health 8(3):65–71.

Geroldi C, Frisoni G, Rozzini R, Trabucchi M. 1996. Disability and principal lifetime occupation in the elderly. International Journal of Technology & Aging 43(4):317–324.

Gladen B, Sandler DP, Zahm SH, Kamel F, Rowland AS, Alavanja MCR. 1998. Exposure opportunities of families of farmer pesticide applicators. Am J Ind Med 34(6):581–587.

Greenlee A, Arbuckle TE, Chyou PH. 2003. Risk factors for female infertility in an agricultural region.

Greksa L, Korbin J. 1997. Influence of changing occupational patterns on the use of commercial health insurance by the old order Amish. Journal of Multicultural Nursing & Health 3(2):13–18.

Hansen E, Donohoe M. 2003. Health issues of migrant and seasonal farmworkers. J Health Care Poor Underserved 14(2):153–164.

Hauenstein E. 1996. Testing innovative nursing care: Home intervention with depressed rural women. Issues Ment Health Nurs 17(1):33–50.

Hawk C, Donham KJ, Gay J. 1994. Pediatric exposure to agricultural machinery: Implications for primary prevention. J Agromed 1(1):57–73.

Hawk C, Gay J, Donham KJ. 1991. Rural Youth Disability Prevention Project survey: Results from 169 Iowa farm families. J Rural Health:170–179.

Hemard J, Monroe PA, Atkinson ES, Blalock LB. 1998. Rural women's satisfaction and stress as family health care gatekeepers. Women & Health 28(2): 55–77.

Hendricks K, Adekoya N. 2001. Non-fatal animal related injuries to youth occurring on farms in the United States, 1998. Inj Prev 7(4):307–311.

Homberg S, Thelin A, Steiernstrom E, Svardsudd K. 2003. The impact of physical work exposure on musculoskeletal symptoms among farmers and rural non-farmers. Ann Agric Environ Med 10(2): 179–184.

Horak FJ, Studnicka M, Gartner C, Veiter A, Tauber E, Urbanek R, Frischer T. 2002. Parental farming protects children against atopy: Longitudinal evidence involving skin prick tests. Clin Exp Allergy 32(8): 1155–1159.

Horsburgh S, Feyer A, Langley J. 2001. Fatal work related injuries in agricultural production and services to agriculture sectors of New Zealand, 1985–1994. Occupational and Environmental Medicine 58(8): 489–495.

Hostetler J. 1974. Hutterite Society. Baltimore: Johns Hopkins University Press.

——. 1987. A new look at the Old Order. The Rural Sociologist 7(4):278–292.

——. 1995. The Amish. Scottdsale, PA: Herald Press.

Hubler C, Hupcey J. 2002. Incidence and nature of farm-related injuries among Pennsylvania Amish children: Implications for education. Journal of Emergency Nursing 28(4):284–288.

Huneke J, Von Essen S, Grisso R. 1998. Innovative approaches to farm safety and health for youth, senior farmers and health care providers. J Agromed 5(2):99–106.

Huntington G. 1984. Cultural interaction during time of crisis: Boundary maintenance and Amish boundary definition. In: Enninger W, editor. Internal and External Perspectives on Amish and Mennonite Life. Essen: Unipress. p 92–118.

Institute AER. 2000. Results from book-keeping farms Fiscal years 1996 and 1997 (In Finnish).

IREH, University of Iowa. 2005. Agricultural Safety and Health Training Program. Iowa City, IA.

Janicak C. 2000. Occupational fatalities to workers age 65 and older involving tractors in the crops production agriculture industry. Journal of Safety Research 31(3):143–148.

Jones P, Field W. 2002. Farm safety issues in Old Order Anabaptist communities: Unique aspects and innovative intervention strategies. J Agric Saf Health 8(1):67–81.

Klintberg B, Berglund N, Lilja G, Wickman M, van Hage-Hamsten M. 2001. Fewer allergic respiratory disorders among farmers' children in a closed birth cohort from Sweden. Eur Respir J 17(6):1151–1157.

Kreps G, Kreps M. 1997. Amishizing "medical care." Journal of Multicultural Nursing & Health 3(2):44–47.

Kulig J, Meyer C, Hill S, Handley C, Lichtenberger S, Myck S. 2002. Refusals and delay of immunization within southwest Alberta. Understanding alternative beliefs and religious perspectives. Canadian Journal of Public Health Revue Canadienne de Sante Publique 93(2):109–112.

Lauener R, Birchler T, Adamski J, Braun-Fahrlander C, Bufe A, Herz U, von Mutius E, Nowak D, Riedler J, Waser M, Sennhauser FH, Alex study group. 2002. Expression of CD14 and Toll-like receptor 2 in farmers' and non-farmers' children. Lancet 360(9331):465–466.

Lee B, Marlenga B. 1999. North American Guidelines for Children's Agricultural Tasks. Marshfield, WI: Marshfield Clinic.

Leynaert B, Neukirch C, Jarvis D, Chinn S, Burney P, Neukirch F, European Community Respiratory Health, Survey. 2001. Does living on a farm during childhood protect against asthma, allergic rhinitis, and atopy in adulthood? Am J Respir Crit Care Med 164(10 Pt 1):1829–1834.

Liller K. 2001. Teaching agricultural health and safety to elementary school students. J Sch Health 71(10):495–496.

Little D, Vermillion JM, Dikis EJ, Little RJ, Custer MD, Cooney DR. 2003. Life on the farm—Children at risk. J Pediatr Surg 38(5):804–807.

Lukes S, Miller FY. 2002. Oral health issues among migrant farmworkers. J Dent Hyg 76(2):134–140.

Lyon D, Parker B. 2003. Gender-related concerns of rural women with severe and persistent mental illnesses. Arch Psychiatr Nurs 17(1):27–32.

Magana C, Hovey JD. 2003. Psychosocial stressors associated with Mexican migrant farmworkers in the midwest United States. J Immigrant Health 5(2):75–86.

March of Dimes. 2001. March of Dimes: Infant health statistics. On an Average Day in the United States. Available at http://www.modimes.ord/HealthLibrary2/InfantHealthStatistics/avg-day2001.htm.

Marlenga B, Pickett W, Berg R. 2001. Agricultural work activities reported for children and youth on 498 North American farms. J Agric Saf Health 7(4):241–252.

Mason C, Earle-Richardson G. 2002. New York State child agricultural injuries: How often is maturity a potential contributing factor? Am J Ind Med Suppl(2):36–42.

McCauley L, Lasarev MR, Higgins G, Rothlein J, Muniz J, Ebbert C, Phillips J. 2001. Work characteristics and pesticide exposures among migrant agricultural families: A community-based research approach. Environ Health Perspect 109(5):533–538.

McCurdy S, Samuels SJ, Carroll DJ, Beaumont JJ, Morrin LA. 2002. Injury risks in children of California migrant Hispanic farm worker families. Am J Ind Med 42(2):124–133.

McDuffie H. 1994. Women at work—Agriculture and pesticides. J Occup Med 36(11):1240–1246.

Meiers S, Baerg J. 2001. Farm accidents in children: Eleven years of experience. J Pediatr Surg 36(5):726–729.

Melberg K. 2003. Farming, stress and psychological well-being: The case of Norwegian farm spouses. Sociologia Ruralis 43(1):56–76.

Merchant JA, Naleway AL, Svendsen ER, Kelly KM, Burmeister LF, Stromquist AM, Taylor CD, Thorne PS, Reynolds SJ, Sanderson WT, and others. 2005. Asthma and farm exposures in a cohort of rural Iowa children. Environ Health Perspect 113(3):350–356.

Meyers D, Wjst M, Ober C. 2001. Description of three data sets: Collaborative Study on the Genetics of Asthma (CSGA), the German Affected-Sib-Pair Study, and the Hutterites of South Dakota. Genetic Epidemiology 21(Suppl 1):S4–8.

Migrant Clinicians Network. 2005. Migration Patterns.

Mitchell B, Hsueh W, King T, Pollin T, Sorkin J, Agarwala R, Schaffer A, Shuldiner A. 2001a. Heritability of life span in the Old Order Amish.

American Journal of Medical Genetics 102(4): 346–352.

Mitchell R, Franklin R, Driscoll T, Fragar L. 2001b. Farm-related fatalities involving children in Australia, 1989–1992. Aust N Z J Public Health 25(4):307–314.

——. 2002. Farm-related fatal injury of young and older adults in Australia, 1989–1992. Aust J Rural Health 10(4):209–219.

Mobed K, Schenker MB. 1992. Occupational-health problems among migrant and seasonal farmworkers. Western Journal of Medicine 157(3):367–373.

Molocznik A, Zagorski J. 2000. Exposure of female farmers to dust on family farms. Ann Agric Environ Med 7(1):43–50.

Morton D, Morton C, Strauss K, Robinson D, Puffenberger E, Hendrickson C, Kelley R. 2003. Pediatric medicine and the genetic disorders of the Amish and Mennonite people of Pennsylvania. American Journal of Medical Genetics 121C(1): 5–17.

Munshi K, Parker DL, Bannerman-Thompson H, Merchant D. 2002. Causes, nature, and outcomes of work-related injuries to adolescents working at farm and non-farm jobs in rural Minnesota. Am J Ind Med 42(2):142–149.

NAMI. 2005. About Mental Illness, Depression in Older Persons.

National Research Council IoM. 1998. Protecting youth at work: Health, safety, and development of working children and adolescents in the United States. Washington, DC: National Academy Press.

NCCRAHS. 2001. Fact Sheet, Childhood Agricultural Injuries. Marshfield, WI: NIOSH, National Institute for Occupational Safety and Health.

——. 2005. North American Guidelines for Children's Agricultural Tasks.

NIMI. 2005. Depression.

NIOSH (National Institute of Occupational Safety and Health). 2001. Worker Health Chartbook 2000, Nonfatal Injury. Available at http://www.cdc.gov/niosh/pdfs/2002-119.pdf.

NIOSH. 2005. TRAC-SAFE.

Nordstrom D, Layde PM, Olson PM, and others. 1995. Incidence of farm-work-related acute injury in a defined population. Am J Ind Med 28:551–564.

Ober T, Hyslop T, Hauk W. 1999. Inbreeding effects on infertility in humans: Evidence for reproductive compensation. American Journal of Human Genetics 64:225–231.

Occupational Health & Safety. 2000. Occupation with Oldest Workers: Farmers. Occupational Health & Safety 69(10):44.

OHCHR. 2003. International Convention on the Protection of the Rights of All Migrant Workers and Members of Their Families.

Olenchock S, Young N. 1997. Changes in agriculture bring potential for new health and safety risks. Wis Med J 96(8):10–11.

Paltiel FL. 1998. Workers and health. In: Stellman JM, Warshaw LJ editors. Encyclopaedia of Occupational Health and Safety Vol I. Geneva: International Labour Office.

Park H, Reynolds SJ, Kelly KM, Stromquist AM, Burmeister LF, Zwerling C, Merchant JA. 2003. Characterization of agricultural tasks performed by youth in the Keokuk County Rural Health Study. Appl Occup Environ Hyg 18(6):418–429.

Parker D, Merchant D, Munshi K. 2002. Adolescent work patterns and work-related injury incidence in rural Minnesota. Am J Ind Med 42(2):134–141.

Pickett W, Chipman M, Brison R, Holness D. 1996. Medications as risk factors for farm injury. Accident; Analysis and Prevention 28(4):453–462.

PickingJobs.com. 2005. Jobs List. Available from http://pickingjobs.com/Find_a_job/find_a_job.htm.

Portengen L, Sigsgaard T, Omland O, Hjort C, Heederik D, Doekes G. 2002. Low prevalence of atopy in young Danish farmers and farming students born and raised on a farm. Clin Exp Allergy 32(2): 247–253.

Poss J. 1998. The meanings of tuberculosis for Mexican migrant farmworkers in the United States. Soc Sci Med 47(2):195–202.

Pryor S, Caruth AK, McCoy CA. 2002. Children's injuries in agriculture related events: The effect of supervision on the injury experience. Issues Compr Pediatr Nurs 25(3):189–205.

Quandt S, Elmore RC, Arcury TA, Norton D. 2001. Eye symptoms and use of eye protection among seasonal and migrant farmworkers. Southern Medical Journal 94(6):603–607.

Redekop C. 1989. Mennonite Society. Baltimore, MD: Johns Hopkins University Press.

Reed D, Kidd PS, Westneat S, Rayens MK. 2001. Agricultural Disability Awareness and Risk Education (AgDARE) for high school students. Inj Prev 7(Suppl 1):i59–63.

Reed D, Westneat SC, Browning SR, and others. 1999. The hidden work of the farm homemaker. J Agric Saf Health 5:317–327.

Reynolds P, Von Behren J, Gunier RB, Goldberg DE, Hertz A, Harnly ME. 2002. Childhood cancer and agricultural pesticide use: An ecologic study in California. Environ Health Perspect 110(3):319–324.

Rhodes D, Hupcey J. 2000. The perception of farm safety and prevention issues among the Old Order Amish in Lancaster County, Pennsylvania. J Agric Saf Health 6(3):203–213.

Ronda E, Regidor E. 2003. Higher birth weight and lower prevalence of low birth weight in children of agricultural workers than in those of workers in other occupations. J Occup Environ Med 45(1): 34–40.

Rosenman K, Gardiner J, Swanson GM, Mullan P, Zhu ZW. 1995. U.S. farm women's participation in breast cancer screening practices. Cancer 75(1):47–53.

Salazar M, Napolitano M, Scherer JA, McCauley LA. 2004. Hispanic adolescent farmworkers' perceptions associated with pesticide exposure. West J Nurs Res 26(2):146–166.

Sandmark H, Hogstedt C, Vingard E. 2000. Primary osteoarthrosis of the knee in men and women as a result of lifelong physical load from work. Scandanavian Journal of Work, Environment & Health 26(1):20–25.

Sawicka J. 2001. The role of rural women in agriculture and rural development in Poland. Electronic Journal of Polish Agricultural Universities 4(2).

Schenker M. 1996. Preventive medicine and health promotion are overdue in the agricultural workplace. Journal of Public Health Policy 17(3):275–305.

Schulte J, Valway SE, McCray E, Onorato IM. 2001. Tuberculosis cases reported among migrant farm workers in the United States, 1993–1997. J Health Care Poor Underserved 12(3):311–322.

Schwartz K. 2002. Breast cancer and health care beliefs, values, and practices of Amish women: Wayne State University.

Schwieder E, Schwieder D. 1975. A Peculiar People. Ames, IA: Iowa State University Press.

Shea K. 2003. Antibiotic resistance: What is the impact of agricultural uses of antibiotics on children's health? Pediatrics 112(1 Suppl S):253–258.

Sinclair A, Bayer A, Girling A, Woodhouse K. 2000. Older adults, diabetes mellitus and visual acuity: A community-based case-control study. Age Ageing 29(4):335–339.

Sivakova D, Pospisil M. 2001. Dermatoglyphic analysis of Habans (Hutterites) from Slovakia. Anthropologischer Anzeiger 59(4):355–363.

Sprince N, Zwerling C, Lynch C, Whitten P, Thu K, Gillette P, Burmeister L, Alavanja M. 2003. Risk factors for falls among Iowa farmers: A case-control study nested in the Agricultural Health Study. Am J Ind Med 44(3):265–272.

Stueland D, Lee BC, Nordstrom DL, Layde PM, Wittman LM, Gunderson PD. 1997. Case-control study of agricultural injuries to women in central Wisconsin. Women & Health 25(4):91–103.

Terho E, Husman K, Vohlonen I. 1987. Prevalence and incidence of chronic bronchitis and farmer's lung with respect to age, atopy, and smoking. European Journal of Respiratory Diseases 71(Suppl 152):19–28.

Tevis C. 1995. Why veteran farmers fall victim to ag injuries. Successful Farming 93(3):24.

Thu K, Donham K, Yoder D, Ogilvie L. 1990. The farm family perception of occupational health: A multi-state survey of knowledge, attitudes, behaviors, and ideas. Am J Ind Med 18(4):427–431.

Turner M. 1996. Female farm operators: Attitudes about social support in their retirement years. J Women Aging 8(3–4):113.

U.S. Dept. of Labor. 2002. Injuries, Illnesses, and Fatalities. Available at http://stats.bls.gov/iif/home.htm#tables.

———. 2005. Compliance Assistance—The Migrant and Seasonal Agricultural Worker Protection Act (MSPA).

USDA. 1998. U.S. Department of Agriculture Food Safety and Inspection Service. Report to Congress. Food Net: An Active Surveillance System for Bacterial Foodborne Diseases in the United States. April 1998. Available at www.fsis.usda.gov/ophs/rpcong98/rpcong98.htm.

Van Hook M. 1996. Challenges to identifying and treating women with depression in rural primary care. Soc Work Health Care 23(3):73–92.

Vega W, Kolody B, Aguilar-Gaxiola S, Alderete E, Catalano R, Caraveo-Anduaga J. 1998. Lifetime prevalence of DSM-III-R psychiatric disorders among urban and rural Mexican Americans in California. Arch Gen Psychiatry 55(9):771–778.

Vernon S, Roberts R. 1982. Use of the SADS-RDC in a tri-ethnic community survey. Arch Gen Psychiatry 39(1):47–52.

Villarejo D, Baron SL. 1999. The occupational health status of hired farm workers. Occup Med 14(3): 613–635.

Voaklander D, Hartling L, Pickett W, Dimich-Ward H, Brison R. 1999. Work-related mortality among older farmers in Canada. Can Fam Physician 45:2903–2910.

von Gruenigen V, Showalter A, Gil K, Frasure H, Hopkins M, Jenison E. 2001. Complementary and alternative medicine use in the Amish. Complementary Therapies in Medicine 9(4):232–233.

Ward LS, Atav S. 2004. Migrant farmworkers. In: Glasgow N, Morton LW, Johnson N, editors. Critical Issues in Rural Health. Ames, IA: Blackwell Publishing Professional, 169–181.

Weathers A, Minkovitz C, O'Campo P, Diener-West M. 2003. Health services use by children of migratory agricultural workers: Exploring the role of need for care. [see comment]. Pediatrics 111(5 Pt 1): 956–963.

Wilk V, Holden R, Bauer S, Brock S, Hendrikson E, Lombardi G, Keifer M, Monahan P, O'Malley M, Sandoval S, Torres E, Villarejo D. 1999. New directions in the surveillance of hired farm worker health and occupational safety. NIOSH, 1999 Jan:1–41.

Xiang H, Stallones L, Chiu Y. 1999. Nonfatal agricultural injuries among Colorado older male farmers. Journal of Aging and Health 11(1):65–78.

3
Agricultural Respiratory Diseases

Kelley J. Donham and Anders Thelin

SECTION 1: OVERVIEW

Introduction

Knowledge of the occupational source of common respiratory diseases is important for purposes of diagnosis, control, prevention, and workers' compensation issues. Because repeated attacks of illness or prevention of chronic impairment can be controlled only through preventing further exposure, each illness should be traced to its source. The emphasis on prevention explains inclusion of material on measurement of dusts and gases, and proper selection and use of personal respirator protection.

Sources of Respiratory Hazards in Agriculture

Perhaps more than any other occupational group, agricultural workers are exposed to a tremendous variety of agents potentially harmful to the respiratory system. Depending on the type of agricultural operation, workers are likely to inhale substances on a daily basis that originate from the soil; the animals and plants raised on farms; animal wastes and their breakdown products; animal feeds; products of decay or fermentation of stored plant materials; applied pesticides, fertilizers, and their residues; and exhaust fumes from operation of farm machinery. Table 3.1 summarizes some of the primary substances and sources of hazardous aerosol exposures in agriculture that may lead to respiratory disease (Kern and others 2001).

Farmers are commonly exposed to low concentrations of these substances when performing numerous daily chores. Periodic high concentration exposures of certain harmful dusts or gases are typical of many agricultural activities, including moving or transporting grain, uncapping silos, breaking open bales of spoiled hay, mixing feed or feeding animals, cleaning moldy grain out of storage structures, entering recently filled non-airtight silos, working in a livestock confinement unit or near its associated manure storage area, applying pesticides or anhydrous ammonia fertilizer, and working with diseased animals. Exposure to any substance is likely to be more intense, and its effects more damaging, if it occurs within an enclosed structure. Many of these exposures are most typical in winter months, when livestock structures are more tightly closed to conserve heat.

Since some of these activities are performed by agricultural workers other than farmers or ranchers, exposure hazards extend well beyond the farm or ranch to a variety of related occupational groups, including grain elevator workers, animal slaughterhouse workers, truckers, professional pesticide applicators, and veterinarians. Farm family members (including spouses and children) who help part-time with farm operations are also exposed.

The respiratory system has a limited number of responses to inhaled substances. The particular response is modulated by the concentration, type of inhaled matter, and biological variation among individuals. Details of the respiratory system responses are provided later in this chapter.

The great majority of agricultural dusts are organic particles. Inorganic particles such as mineral particles in soil, which primarily produce a nuisance effect (excepting silica and asbestos),

Table 3.1. Hazardous Agricultural Aerosol Exposures: Substances and Their Sources

Animals	Plants	Insects	Microbes & Metabolites	Pesticides	Infectious Agents	Feed Additives	Gases & Fumes
Dander	Grain dust	Feces of mites, roaches	Bacteria & endotoxins	Residues on crops	Rickettsia (ornithosis, Q Fever)	Antibiotics	Anhydrous ammonia fertilizer
Broken bits of hair	Plant particles	Insect parts	Fungi glucans & mycotoxins	Fumigants: (methyl bromide, phosphine, formaldehyde, carbon tetrachloride)	Bacteria (tularemia, anthrax)		Ammonia from animal wastes
Dried fecal material	Tannins				Fungi (histoplasmos, blastomycosis)		Hydrogen sulfide from animal waste
	Glucans				Virus (avian, swine influenza, Hantavirus)		Nitrogen oxides from Silo gas
							Welding fumes
							Diesel exhaust

may be linked to chronic bronchitis and mixed dust pneumoconiosis seen in some California agricultural workers (Schenker 2000). Organic particles are often biologically active and capable of producing irritating, allergic, toxic, inflammatory, or infectious conditions. Responses may be acute or chronic, resolving completely or resulting in permanent impairment and disability, and sometimes ending in death. Agricultural dusts are common exposures in nearly all types of production agriculture. The greatest mass of agricultural particles is above respirable size, affecting primarily the airways. However they range in diameter from under 1μ to over 50μ and thus may be deposited throughout the airways and into the lung tissues proper. Agricultural dust exposures are often mixed exposures (e.g., dust and endotoxin), increasing their biological effects. Unlike most industrial settings where one or a few known dust components produce a specific identifiable respiratory response, components of farm dusts are never completely known. These components vary with the season, type of farming operation, and type of chore being performed. Thus an individual may be experiencing more than one respiratory exposure, which may be additive or synergistic and cumulative, as has been proven recently with dust and ammonia exposures in livestock buildings (Donham and others 2002).

In addition to dust, gases, pesticides, and fertilizers, there are several infectious diseases common to animals and man (zoonoses) that cause respiratory disease in agricultural workers. Histoplasmosis, swine influenza, and ornithosis are just three of at least eight zoonoses affecting the respiratory tract of agricultural workers.

Respiratory Responses to Inhaled Agricultural Substances

An individual's respiratory response to this complex mixture of dusts, gases, and infectious agents will depend on many factors: specific composition; irritant and antigenic properties of inhaled substances; size and shape of dust particles; location of deposition in the respiratory tract; concentration; duration of exposure; and the individual's relative genetic susceptibility to organic dust diseases, immunologic status, smoking history, and occupational or environmental exposures off the farm. For all these reasons, respiratory responses to dusts and gases in agricultural settings are complex.

Nevertheless, research in recent years has documented clusters of symptoms and specific conditions that clearly make up a well-recognized group of agricultural respiratory diseases. Respiratory responses to agricultural dusts and gases are listed in Table 3.2 and described in the following paragraphs. Specific exposure circumstances and resulting disease entities are outlined in Sections 2–8 of this chapter. Details of diagnosis and treatment of these conditions are also explained in these seven sections.

SECTION 2: AGRICULTURAL STRUCTURES AND RESPIRATORY HAZARDS

Introduction

Frequently, agricultural respiratory problems are associated with work in agricultural structures because of the confined environment, resulting in concentrated exposures for workers. Therefore, an understanding of the design, function, and products stored in these structures is important to obtain an accurate occupational history, diagnosis, and prevention strategy. Table 3.3 lists various agricultural structures (Figure 3.1), related exposures, and resultant disease conditions. The following sections describe common agricultural structures, their functions, and related work activities that are associated with common respiratory health problems.

Feed Grain, Silage, and Other Commodity Storage Structures

More than any other structures, feed grain, silage, and other commodity storage structures and associated activities are closely linked to respiratory problems.

GRAIN BINS

Grain bins are used to dry and store shelled corn, oats, beans, rye, barley, wheat, sorghum, and other small grains. Typically, grain bins are cylindrical galvanized steel structures placed on a cement slab. The floor is concrete or (if a drying bin) steel with very small slots to allow ventilation from underneath (a drying floor). Grain is put into the bin by augering it up and into an opening in the roof. Grain is augured out from the floor surface or through a hole in the middle of the bin floor, with gravity or a "sweep" auger keeping the grain flowing into the hole. Some

Table 3.2. Agricultural Respiratory Conditions

1) Conditions Arising Primarily from Organic Dust Exposure
    a. Mucous membrane irritation (MMI)
        • Sinusitis
        • Rhinitis
        • Pharyngitis
        • Laryngitis
        • Tracheitis
    b. Bronchitis
        • Acute/subacute bronchitis (symptoms of dry cough)
        • Chronic bronchitis (cough with phlegm lasting more than 2 years)
    c. Nonallergenic Asthma-like Condition (NALC, also called hyperresponsive airways, or reactive airways disease)
    d. Monday morning disease (MMD)
    e. Organic Dust Toxic Syndrome (ODTS, also called mycotoxicosis, silo unloader's disease, or toxic alveolitis)
    f. Allergic-mediated illnesses
        • Extrinsic asthma (atopic-related disease)
        • Allergic rhinitis
        • Hypersensitivity pneumonitis (HP), also called farmer's lung (FL) or extrinsic allergic alveolitis) may result in
            – Acute HP (a delayed influenza-like illness clinically similar to ODTS)
            – Chronic HP (interstitial fibrosis)
2) Conditions Arising Primarily from Inhaled Agricultural Gases
    a. Hydrogen sulfide
        • Respiratory depression/respiratory arrest
        • Pulmonary edema (acute or delayed)
    b. Anhydrous ammonia
        • Laryngeal edema
        • Pharyngitis
        • Tracheitis
        • Bronchitis
        • Pulmonary edema (acute)
        • Bronchiectasis (delayed)
    c. Silo gas (nitrogen oxides)
        • Pharyngitis
        • Tracheitis
        • Bronchitis
        • Pulmonary edema (acute or delayed)
        • Bronchiolitis obliterans (delayed)
    d. Fumigant pesticides
        • Pharyngitis
        • Tracheitis
        • Bronchitis
        • Pulmonary edema
    e. Ingested paraquat
        • Progressive malignant pulmonary fibrosis
3) Respiratory Response Conditions to Infectious Agents
    a. Pneumonitis/pneumonia

Figure 3.1. Agricultural respiratory conditions often result from work inside various agricultural structures where aerosols are concentrated, increasing hazardous exposures.

bins may not have an integral auger and a portable auger or grain vacuum must be used. Augers carry the grain to a truck, wagon, automated feeding system, or other machine (mixer/grinder) for feed preparation.

Bins cannot be emptied completely with an auger. To empty the last several inches of grain, the grain must be hand shoveled or vacuumed. (*Note:* there is also an injury hazard in this process because a sweep auger is often left running during the process, which may entangle the operator's leg or legs.) During this operation, large quantities of aerosolized grain dusts, microorganisms and their by-products, and other particles may be inhaled, creating risk of organic dust-related respiratory diseases (discussed later in this chapter). Often the residual grain left in the bottom has become spoiled because moisture collects there, substantially increasing the microbial and microbial by-product (endotoxins and glucans) contents of the dust. Therefore, the risk of respiratory disease is increased. Grain truckers also may inhale these dusts while helping a farmer empty the bin or while leveling grain that is filling the truck. Prevention includes drying the grain down to at least 15% moisture prior to storage and bin maintenance to prevent water leakage and condensation. Grain vacuums may be used in place of shovels and will help reduce, but not eliminate, dust. Therefore, particulate (dust) respirators should be used during these tasks (see

Section 9 of this Chapter for more detail on respirator selection and use).

When bins have grain in them and it is being removed from the bottom, a rapid downward-flowing cone-shaped column of grain develops from the surface to the bottom of the bin. This grain movement can bury a person in the bin in less than 60 seconds, causing the person to suffocate. Once trapped in the flowing grain, even a strong person cannot remove him- or herself. Removal by other persons is difficult because of the pressure, weight, and friction of the grain. Details of grain engulfment are seen in Chapter 11, acute injuries.

Farmers and grain elevator employees often apply fumigants to grains in long-term storage. These highly toxic and irritant pesticides, if inhaled, can cause **respiratory tract irritation, laryngeal edema, bronchospasm, pulmonary edema, respiratory depression, and sudden death.** The details of this hazard are covered in Section 6.

GRAIN ELEVATORS AND FEED MILLS
Grain elevator and feed mill enterprises are commonly found in grain and livestock production areas. Elevator operators either store grain for producers or they purchase grain from producers (farmers) that they dry, store, and grind for feed or which they sell and transport to larger terminals. Farms and ranches that raise or feed out

Table 3.3. Agricultural Structures and Respiratory Problems

Agricultural Structures	Activities Resulting in Respiratory Exposures	Major Resulting Respiratory Problems	Causation of Respiratory Problems	For Detailed Discussion See:
Feed Grain, Silage, and Grain Storage and Handling Buildings				
Corn cribs	Moving and shelling corn out of crib; cleaning out crib	Asthmatic attack Nonallergenic asthma-Like Condition (NALC) Bronchitis ODTS (Organic Dust Toxic Syndrome) Hypersensitivity pneumonitis (rare)	Grain dust, including bacterial and fungal spores and by-products (e.g., endotoxin, glucans)	Section 3 (agricultural dusts) Section 4 (confinement animal feeding operations)
Grain bins	a) Moving grain out of storage, cleaning out moldy residual grain	Asthmatic attack ODTS Bronchitis Acute inflammatory response Airways obstruction Increased airways reactivity Hypersensitivity pneumonitis (rare)	Grain, dust, including bacterial and fungal spores and by-products (e.g., methyl bromide)	Section 3 (grain dusts)
	b) Entering bin that is being emptied (entrapment)	Suffocation	Grain entrapment	Section 2 (structures)
	c) Grain fumigation; entering a fumigated structure too early	Irritation Laryngeal edema Bronchospasm Pulmonary edema Respiratory depression Death	Fumigants (e.g., methyl bromide)	Section 6 (pesticides)

Table 3.3. Agricultural Structures and Respiratory Problems (continued)

Agricultural Structures	Activities Resulting in Respiratory Exposures	Major Resulting Respiratory Problems	Causation of Respiratory Problems	For Detailed Discussion See:
Grain elevators and feed mills	a) Loading or unloading grain, cleaning grain bins, grinding and mixing feed	ODTS NALC Bronchitis Chronic obstructive pulmonary disease Occupational asthma Hypersensitivity pneumonitis (rare)	Grain dust, including bacterial and fungal spores and by-products (e.g., endotoxin, glucans)	Section 3 (grain dusts)
	b) Improper application of fumigants; entering a recently fumigated structure	Irritation Laryngeal edema Bronchospasm Pulmonary edema Respiratory depression Death	Fumigants (e.g., methyl bromide)	Section 6 (pesticides)
Airtight silos	Entering silo filled with silage or high-moisture grain	Asphyxiation	Anoxia	Section 2 (structures)
Upright, non-airtight silos	a) Entering silo within 2 weeks of filling with silage	Sudden death, pulmonary edema, delayed reaction with bronchiolitis obliterans (silo filler's disease)	Oxides of nitrogen (silo gas)	Section 5 (oxides of nitrogen)
	b) Unloading silage, throwing off top layers of moldy silage	ODTS Hypersensitivity pneumonitis	Bacteria, fungi, and their by-products	Section 3 (spoiled grains)

(continued)

Table 3.3. Agricultural Structures and Respiratory Problems (continued)

Agricultural Structures	Activities Resulting in Respiratory Exposures	Major Resulting Respiratory Problems	Causation of Respiratory Problems	For Detailed Discussion See:
Fruit and root storage buildings	Operating machinery or working in buildings with improperly working heaters (CO poisoning)	Carbon monoxide poisoning	CO	Section 2
Confinement Animal Feeding Operations and Processing				
Confinement houses: swine and poultry	Working inside building	ODTS Mucous membrane irritation (MMI) NALC Bronchitis Chronic obstructive pulmonary disease (COPD)	Organic dusts from animals, their feed and wastes), and gases (e.g., endotoxin, glucans, NH_3, H_2S <500 ppm)	Section 4 (livestock confinement)
Confinement houses and manure storage pits (swine, sheep, veal calf, dairy or beef cattle, with liquid manure system)	Entering building during manure pit agitation; entering manure pit	Pulmonary edema, possibly respiratory arrest and death	H_2S (levels >250 ppm)	Section 4 (livestock confinement)
Confinement houses: poultry	a) Administering aerosol vaccines; working inside building during disease outbreak	Newcastle disease	Infection with Newcastle disease virus	Section 7 (infectious diseases)
	b) Working with diseased animals	Ornithosis (turkey only) (rare)	Infection with *Chlamydia psittaci*	Section 7 (infectious diseases)

72

Table 3.3. Agricultural Structures and Respiratory Problems (continued)

Agricultural Structures	Activities Resulting in Respiratory Exposures	Major Resulting Respiratory Problems	Causation of Respiratory Problems	For Detailed Discussion See:
Sheep and dairy cattle housing	Cleaning buildings, especially those where animals are born; assisting during birth process (Q fever)	Q fever	Infection with *Coxiella burnetii*	Section 7 (infectious diseases)
Conventional chicken coops (or other structures where wild birds have roosted)	a) Cleaning or razing houses not in use for several years	Histoplasmosis Ornithosis	Infection with *Histoplasma capsulatum*	Section 7 (infectious diseases)
	b) Working with infected poultry	Ornithosis (rare)	Infection with Newcastle disease virus	Section 7 (infectious diseases)
Turkey processing plants	Slaughter operations	Ornithosis	Infection with *Chlamydia psittaci*	Section 7 (infectious diseases)
Cattle, sheep, or goat slaughterhouses	Slaughter operations	Q fever (rare) Brucellosis	Infection with *Coxiella burnetii* or *Brucella* spp.	Section 7 (infectious diseases)
Equipment and Supply Buildings on Farms Machine shops, garages, machine storage buildings	a) Running gasoline- or diesel-fuel–powered engines when buildings are not adequately ventilated; heating buildings with heaters that are not working properly (CO_2)	Carbon monoxide poisoning	CO	Section 2 (structures)

Table 3.3. Agricultural Structures and Respiratory Problems (continued)

Agricultural Structures	Activities Resulting in Respiratory Exposures	Major Resulting Respiratory Problems	Causation of Respiratory Problems	For Detailed Discussion See:
	b) Welding galvanized iron (metal fume fever)	Metal fume fever	Zinc oxide	Section 2 (structures)
BARNS Barns	Moving or feeding hay, grain; grinding and mixing feed (organic dust); caring for animals	ODTS Allergic rhinitis NALC Bronchitis Chronic obstructive pulmonary disease Occupational asthma Hypersensitivity pneumonitis Q fever Histoplasmosis	Hay dust Grain dust Bacteri, fungi, and their by-products in moldy fodder Dusts from animals, their feed and wastes Infection with *Coxiella burnetii* Infection with *Histoplasma capsulatum*	Section 2 (spoiled hay, grain) Section 3 (grain dusts) Section 4 (livestock confinement) Section 7 (infectious diseases) Section 7 (infectious diseases)

large numbers of livestock may also have extensive grain handling facilities, including a system of bins, tanks, and hoppers for grain and feed. Both elevators and feed mills have a series of bins and tanks for grain and feed storage. Similar to farm storage structures, they are tightly closed to prevent infestation and spoilage. Unless grain is being stored as part of a government program, it is continuously being moved in and out of storage. Elevators and mills also have large wooden or steel buildings for storage of bagged feed, seed, pesticides, and other products.

Since grain elevator and feed mill workers are exposed to grain dusts regularly over an extended time period, chronic as well as acute responses may be elicited (Buchan and others 2002; Dosman and Cotton 1980). Clouds of dust are especially prominent whenever moving or grinding the grain. Acute and chronic responses to these organic dusts are described in Section 3 of this chapter.

Similar to on-farm operations, the same acute conditions may result as described in the discussion of grain bins. Fumigants are routinely applied to grain stored in elevators. Most fumigants are extremely toxic or lethal when inhaled. Respiratory responses to fumigants are reviewed in Section 6 of this chapter.

OXYGEN-LIMITING SILOS

Called *oxygen-limiting structures* by their manufacturers (also called *airtight silos*), these structures are made of glass fused to steel or concrete-stave construction with epoxy lining (Figure 3.2). They can be tightly closed to limit the entrance of air. Internal air pressure is regulated by valves and by a large vinyl bag that displaces air, collapsing as the silo is filled.

Plant materials placed in the silo continue aerobic cellular respiration using the residual oxygen until a low-oxygen, high-carbon dioxide atmosphere develops, in which metabolic processes are halted and plant material is preserved. Silage or high-moisture corn is blown or augured into the silo at the top and removed from the bottom with an auger. Because these bottom unloaders often need repair, top unloaders are now available for conversion. Silage consists of the whole, chopped plant (usually corn, but may also be alfalfa, clover, mixed grasses and legumes, sorghum, oats, wheat, or milo) harvested at a relatively high moisture content compared to conventionally stored hay or grain. Silage is fed to ruminant animals, mainly to dairy and beef cows, although alfalfa haylage may also be fed to sheep. Corn silage is harvested during late summer or early fall and typically fed out in winter (when pastures are depleted), but may be fed out all year long. High-moisture corn is fed to feeder cattle or swine and may be used year round.

This atmosphere is **oxygen deficient** and there is a high potential for **asphyxiation** for anyone who enters. **As long as feedstuff remains in the structure, airtight silos should not be entered under *any* circumstances, except by a trained repair technician familiar with the use of a self-contained breathing apparatus (SCBA).**

NON-AIRTIGHT SILOS

These conventional (concrete stave, brick, metal, or wooden) silos are used to store the same kinds

Figure 3.2. Oxygen-limiting silos (the three on the left), and non–oxygen-limiting silos (the two on the right) present different occupational exposures to farm workers.

of chopped plant material as are airtight silos and occasionally are used to store grain (see Figure 3.2). Use of non-airtight silos predates the newer airtight ones, and there is a much greater diversity of design among them.

When freshly chopped plant material is placed in these silos, natural processes can result in formation of oxides of nitrogen (NO_x) from nitrates in the plant material. Anyone who enters a silo to level silage when these gases are present may inhale enough to cause **sudden death, acute or delayed pulmonary edema,** or **latent bronchiolitis obliterans,** reactions commonly called **silo filler's disease.** The danger period extends for 2 weeks after filling. (Details of silo filler's disease are seen in Section 5 of this chapter.) After the silo is filled, usually a sheet of plastic is placed over the top and weighted down with an additional 6–12 inches of silage. The purpose is to help preserve the silage underneath. However, the top few inches of silage (regardless of whether plastic has been placed over the top) will spoil. This material must be discarded before the "good" material underneath can be fed. Unloading this material leads to massive exposure of bioaerosols, and resultant organic dust respiratory diseases. Organic Dust Toxic Syndrome (ODTS) (also called *silo unloader's disease* or *mycotoxicosis*) has been associated with this exposure. Detailed descriptions of ODTS and other organic dust respiratory conditions are presented in Section 3.

FRUIT AND ROOT STORAGE STRUCTURES

Located in orchard and root crop growing regions are special buildings used for storage of potatoes, apples, and bananas (Figure 3.3). These buildings are usually large enough for a forklift truck, tractor and wagon, or large straight truck to enter. These storage structures are mentioned here to assist the health professional to differentiate respiratory and nonrespiratory conditions in this type of agricultural structures exposure, which involves carbon dioxide asphyxiation or carbon monoxide intoxication. There are no known other toxic or inflammatory exposures in these environments.

LIVESTOCK AND POULTRY HOUSING AND PROCESSING

Confined animal feeding operations (CAFOs) differ from conventional housing in that large numbers of animals are housed in tightly constructed buildings, for most or all of their lives. The structure must include a ventilation system for control of heat and humidity, a system for watering the livestock, a system for feeding the livestock and a system for handling animal wastes. Extensive management and maintenance are needed to insure that all systems are working pro-

Figure 3.3. Potato and other fruit storage (e.g., apple storage) facilities may have high carbon dioxide concentrations to preserve the product, and may result in asphyxiation hazard to workers.

perly. Additional work includes monitoring the animals, and assisting in birthing and treating the newborn pigs. As the ventilation system is designed to control only humidity and heat, environmental control of dusts and gases generated by the animals, their feed, and their wastes is often excessive, especially in cold weather. Thus, workers in confinement buildings, especially those housing swine or poultry, may experience any or a combination of a complex set of acute and chronic respiratory reactions to these dusts and gases. Organic dust diseases are described in Section 4 of this chapter.

Workers in swine confinement buildings may be exposed to the **swine influenza** virus, which typically causes subclinical or a mild, transient illness in humans (see Section 7). Poultry confinement workers may be exposed to the **Newcastle disease** virus, particularly when spraying live attenuated vaccines (aerosol immunization) onto poultry flocks. Veterinarians working with infected birds or performing a diagnostic postmortem are at risk of exposure to ornithosis or avian influenza (World Health Organization 2005).

SHEEP AND DAIRY CATTLE HOUSING

Dairy barns or other conventional buildings sheltering cattle or sheep may be used to confine animals while they give birth. Farmers typically assist with the birth or clean out the straw bedding after a birth, and thus may come in contact with the rickettsia *Coxiella burnetii,* which cause **Q fever**.

CONVENTIONAL CHICKEN COOPS

Conventional chicken coops are not used much today, having been largely replaced by large confinement structures (however, the trend for increasing organic corn niche agriculture may result in renewed use of these facilities). Old unused buildings may harbor the fungus *Histoplasma capsulatum*, which grows in dried avian feces or soil contaminated by same. Inhalation of fungal spores released into the air when these houses are cleaned or torn down causes **histoplasmosis**.

POULTRY AND MEAT PROCESSING PLANTS

Workers in plants where turkeys are slaughtered and processed, especially workers eviscerating birds, may inhale the chlamydial organism that causes **ornithosis** (*Chlamydia Psittasi*), an infectious disease ranging from influenza-like infection to acute fulminating pneumonia. The risk is especially high in those plants processing the breeding birds from range-reared flocks because these birds are at higher risk of contracting ornithosis. Workers in cattle, sheep, and goat slaughterhouse operations could contract **Q fever** from infected animals.

Equipment and Supply Buildings on Farms

MACHINE SHOPS, GARAGES, MACHINE SHEDS

Since farmers are occupied from spring through fall with the cycle of crop production, they use the winter season to do the bulk of their machinery repair and maintenance work. This usually means working inside closed garages, machinery shops, machine storage buildings, or any other available structure. Operating gasoline or diesel fuel engines indoors can result in **carbon monoxide poisoning** when the buildings are closed to prevent heat loss and when heaters are not working properly. Adequate ventilation of buildings is necessary at all times. Farmers do much of their own repair and maintenance work. Welding is one of those tasks. Farm shops may not be adequately ventilated for this purpose. Welding galvanized steel (which contains zinc) produces zinc oxide fumes which, when inhaled, cause the syndrome **metal fume fever** (Zimmer 2002). Symptoms begin 4–12 hours following exposure (Figure 3.4). The exposed person first notices a sweet or metallic taste in the mouth, followed by throat dryness or irritation. Later symptoms include cough and shortness of breath, general malaise, weakness, fatigue, and muscle and joint pains. Leukocyte count and serum LDH may be elevated. Fever and shaking chills then develop, followed by profuse sweating. Resolution occurs in 24–48 hours. Metal fume fever can be prevented through adequate ventilation indoors, completing welding operations outdoors, or use of a dust and fume respirator.

BARNS

Barns are multipurpose farm structures. They may be used to store straw and hay, to mix or grind feed, as farm shops, or (especially in winter) to house livestock. Although many farms now have specialized storage structures for agricultural chemicals, many barns are used to store pes-

Figure 3.4. Welding is commonly practiced on many farms and may create a risk for metal fume fever, skin burns, and retinal inflammation.

ticides, treated seeds for planting, and various chemicals. Chemical dust residue from open or broken containers may be present. Respiratory or toxic problems associated with any of these activities may therefore occur in barns as well as in other more specialized structures.

Dairy barns are the most common type of barn in use. Any activity associated with milking and caring for dairy cattle can be done in the typical multistoried dairy barn. Dust can be aerosolized from the hay and grain often stored here and from animals housed and fed in the barn. If grain or hay has spoiled, shoveling the grain or breaking open moldy bales releases clouds of dusts containing bacteria, fungal spores, and endotoxins. On the barn's ground floor there may be cow stanchions or a milking parlor; free stalls; holding and sorting pens; an area with feeding bunks; box stalls for calving, calves, or other cow isolation; and areas for storing, mixing, and grinding feed. In some dairy operations, free stalls or "loafing sheds" are used where cattle can go to feed and be out of the elements. Bedding for the cattle may be blown into these areas with a bale shredder. This device takes a whole bale of straw, chops it up and blows it into the stall area. This creates a massive organic dust aerosol and risk of acute or chronic organic dust respiratory disease for the operator. Sawdust or wood chips may also be used for bedding and may present respiratory hazards similar to hay or straw. Sand may also be used for bedding, and it is free of respiratory hazards.

The complex mixture of organic dusts inhaled during these activities can result in the complex of organic dust diseases described in Section 3 of this chapter.

A number of infectious diseases could be transmitted to humans from animals housed in barns. If sheep or cattle give birth there, **Q fever** could be contracted while assisting in the birth or from contact with infected placentas, reproductive discharges, or contaminated bedding. **Bovine TB** would have been a health hazard in past years, but it is rare today because this illness has been nearly eradicated in the developed world. Old barns where chickens or wild birds have roosted are ideal sites for proliferation of the fungus causing **histoplasmosis**.

Newer steel dairy buildings contain the same types of areas as older barns, except that they are single-storied buildings with hay or straw storage on the ground floor. Persons can experience the same respiratory responses listed previously in these newer buildings.

SECTION 3: AGRICULTURAL DUSTS

Introduction

As previously mentioned, agricultural dusts are by far the most common respiratory hazard for agricultural workers. Agricultural dusts are characterized as complex mixtures of organic materials (see Table 3.1). The main toxic effect is inflammation and irritation (allergic responses are less common). Often agricultural dusts occur as mixed exposures with gases (such as hydrogen sulfide or ammonia) that may be additive or synergistic with dust in regard to health effects. There is a well-recognized cluster of symptoms and conditions that result from short- and long-term agricultural dust exposures, including the following conditions (see Table 3.2):

1. Bronchitis (acute and chronic)
2. Mucous membrane irritation (MMI), an inflammatory condition of mucous membranes impacted by organic dust; may include rhinitis, sinusitis, pharyngitis, and conjunctivitis
3. Monday morning syndrome (MMS) or byssinosis
4. Nonallergic asthma-like condition (NALC), also referred to as *nonallergic occupational asthma,* or *reactive airways disease*
5. Allergic asthma (atopic individuals)
6. Organic Dust Toxic Syndrome (ODTS), also called *toxic alveolitis, silo unloader's disease,* or *mycotoxicosis*; a delayed onset influenza-like condition
7. Hypersensitivity pneumonitis (HP) or farmer's lung (FL) in the agricultural setting

General Mechanisms of Agricultural Dust Toxicity

Environmental exposure to agricultural dusts can result in two pathogenetic pathways: 1) classic allergenic disease (rare among farmers) leading to 1 and 7 in the list above, and 2) respiratory tract inflammation (common among farmers) leading to 1, 2, 3, 4, and 6 above (Rylander 2004). Toll-like receptors are components of the innate immune system of mammals, which can be activated upon exposure to certain inhaled substances within organic dusts, such as endotoxin, beta glucans, and allergens among others. Activation of these receptors results in a "turn on" of the immune system that affects the atopic process and/or initiates an inflammatory cascade, an initial neutrophilic response followed by a mononuclear cell recruitment to the airways and lungs, resulting in tissue damage and respiratory disease (Hoffmann and others 2004). Genetic variation (polymorphisms) among individuals of these Toll-like receptors results in individual response variations among exposed individuals, which may exacerbate or attenuate the inflammatory response (Kline and others 2004). Some characteristics of the specific dust may increase its toxic effect (e.g., hog barn dust increases lymphocyte adhesion to respiratory epithelial cells) (Mathisen and others 2004).

Age of first exposure is also a variable in response to organic dust. Several studies have shown that childhood exposure to agricultural dusts (livestock mainly) offer a protective effect for atopic asthma in later childhood and in adult life (the hygiene hypothesis) (Begany 2003). The mechanisms of these responses are under study, but probably have to do with down-regulation of the cellular immune response (Radon and others 2004). Somewhat contrary to the previous findings, Merchant and co-workers (2005) found an increase in either doctor diagnosis of asthma or one of several self-reported symptoms of asthma, relative to swine production on their home farms. The latter studies did not reveal any objective finding relative to swine exposures. Therefore, the meaning of these findings are unclear at this time.

Additional to individual biological responses, different circumstances of agricultural dust exposures (relative to time and concentration) result in different syndromes of exposure responses. There are **periodic, acute, massive, moldy (PAMM)** exposures to agricultural dusts. These dusts are usually highly contaminated with bacteria and fungi and their by-products. Secondly, there are chronic exposures to **lower level concentrations (CLLC)** of less highly contaminated dusts. The first situation may result in ODTS and/or FL. The second circumstance results in bronchitis, NALC, MMS, MMI, and chronic forms of ODTS or FL. The relative range of concentrations of these two exposure conditions are on an order of magnitude (Table 3.4). For example, CLLC exposures to dust, endotoxin, and microbes may range from $0.5–10.0$ mg/m^3, $50–500$ EU, and $10^4–10^6$ organisms/m^3, respectively (Donham 1986, 1991). Comparatively, PAMM exposures to agricultural dust endotoxin and microbes range respectively from $10–100$ mg/m^3, $500–2500$ EU, and $10^6–10^9$ microbes/m^3 (May and others 1989).

Examples of situations that generate PAMM exposures include opening moldy bales of hay, removing the top layer of silage from a silo (silo unloading), shoveling grain or cleaning out the residual (often spoiled) grain from storage structures, power washing a swine or poultry building,

Table 3.4. Comparisons of Tasks and Resulting Exposures to Agricultural Dusts, Relative to Time and Concentration

PERIODIC, ACUTE, MASSIVE AND MOLDY (PAMM) EXPOSURES TO AGRICULTURAL DUST	
Specific Tasks or Structure	Exposure Characteristics
Silo Unloading	3–4 times/year
Moving Moldy Grain	30 minutes to 2 hour duration
Emptying Grain Bins	Total dust 10–50 mg/m^3
Moving and Sorting Pigs in a Confinement Building	Total viable microbial count 10^7–10^{10} microbes/m^3
Power Washing the Inside of a Swine Confinement Building	Endotoxin concentration 500–1000 EU
Loading or Caging Chickens or Turkeys	Micrograms concentration 10^6–10^9 organisms/m^3
Moving or Loading Wood Chips	

CHRONIC LOWER LEVEL CONCENTRATION (CLLC) EXPOSURES TO AGRICULTURAL DUST	
Specific Tasks or Structure	Exposure Characteristics
Swine Confinement Buildings	Daily exposures
Poultry Confinement Buildings	2–8 hours/day duration
Dairy Barn	Total dust 1–9 mg/m^3
Work in a Grain Elevator or Feed Mill	10^3–10^5 microbes/m^3
	Endotoxin concentration 100–400 EU/m^3

moving and sorting or load-out of hogs or poultry, and shredding bales of hay or straw for bedding of dairy cows. The agents are found not only in moldy feed products but also grow in wood chips (which are commonly used as heating fuel in rural Scandinavia) (Swedish National Board of Occupational Safety and Health 1994) or sawdust (used for bedding for dairy cattle and horses).

Common examples of CLLC exposures occur in work sites such as swine and poultry confinement buildings, dairy barns, and feed preparation and grain handling sites. The differences in the types and concentrations of exposure incidents are shown in Table 3.4.

PAMM exposures can induce several acute respiratory reactions: ODTS, FL, NALC, allergic asthma, and acute bronchitis. At least two of these respiratory conditions (FL and possibly ODTS) may also present with a chronic response. CLLC exposures induce acute or chronic bronchitis, NALC, MMS, and MMI. Individual susceptibility variables include smoking status, atopic status (allergic asthma), previous cellular sensitization

to certain inhaled microbes (FL), and genetic sensitivity to endotoxin (bronchitis, NALC, and ODTS) (Schwartz 2002). Table 3.5 summarizes the symptoms and different diseases resulting from PAMM exposures to the organic components of agricultural dust.

The following paragraphs describe, in more detail, PAMM and CLLC exposures and the resulting health effects.

PAMM, Exposures to Agricultural Dusts: Causative Agents, Pathogenesis, and Treatment of FL and ODTS

Animal feed (hay, silage, and grain) that is put into storage with a high-moisture content favors the growth of bacteria and fungi. These microorganisms produce spores and by-products (mycotoxins, endotoxins and ß,1–3 glucan, among others) of respirable size (less than 10 microns in diameter) which, when released into the air, can be inhaled in large quantities and induce inflammatory effects within the airways or alveoli (Lacey and Lacey 1964; Rylander 1987).

Table 3.5. Symptoms and Conditions Resulting from Two Types of Exposure Conditions to Agricultural Dusts

Periodic, Acute, Massive and Moldy (PAMM) Exposures	Conditions
Delayed Response—2–6 hours following exposure Cough Chest Tightness Malaise Headache Myalgia Arthralgia Fever	Organic Dust Toxic Syndrome (ODTS)—toxic alveolitis and/or Hypersensitivity Pneumonitis (HP)

Chronic Lower Level Concentration (CLLC) Exposures	Conditions
Cough—intermittent, associated with exposure Phlegm Production—intermittent, associated with work exposure	Acute Bronchitis
Cough and Phlegm Production—occurring more than 3 weeks out of the year, for longer than 2 years	Chronic Bronchitis
Chest Tightness—wheezing upon exposure	Nonallergenic asthma-Like Condition (NALC)
Sore Throat Nasal Irritation	Mucous Membrane Irritation (MMI)
Eye Irritation Stuffy Nose	Chronic Sinusitis (one manifestation of MMI)
Difficult Nasal Breathing Complaints of Plugged Up or Persistent Cold	

The most common constituents of agricultural dust are microbes and associated products, which generally proliferate in moist, warm environments (Donham and Popendorf 1985; Lacey and Lacey 1964). Hay, silage, and grain are stored with different optimal moisture contents; microorganism growth resulting from heat and excess moisture may be beneficial or damaging. More specifically, hay baled while too moist may deteriorate into an unsuitable animal feed since it heats spontaneously and proceeds through a natural succession of fungal and bacterial populations, including the thermophilic species *Micropolyspora faeni* and several species of *Thermoactinomyces* (Lacey and Lacey 1964). When silage with high moisture content is stored, it undergoes anaerobic ensilage processes characterized by formation of high levels of short-chain fatty acids, which preserve the plant material. When stored in a non-airtight silo (e.g., traditional concrete-stave silo), the uppermost layer of silage spoils. When whole grain (corn, oats) is stored in a bin before being adequately dried, it will spoil (similar to the top layers of silage) and harbor large quantities of microorganisms, but throughout the entire bin. The species of fungi and bacteria that dominate will change with the type of plant material in storage, moisture, heat, and oxygen-level conditions, but typically include the thermophilic actinomycetes and a variety of fungi.

Who Is Exposed to These Dusts, and When?

Almost any farmer, but especially those working with livestock or grain, is likely to be exposed to dust containing high levels of bacteria and fungi

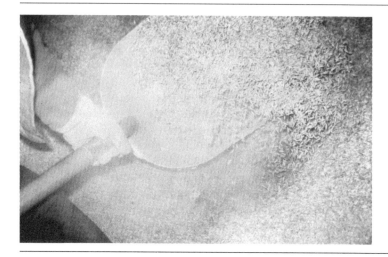

Figure 3.5. Shoveling or other disturbance of moldy grain can create a risk for organic dust toxic syndrome and other respiratory conditions.

and their by-products. This agricultural dust will be released whenever a farmer moves or otherwise works with moldy grain, hay, or silage (Figure 3.5). Occupational exposures that occur in a confined space are most hazardous. However, cases of asthma have been documented in people living downwind from grain elevators where grain dust is released into the general environment (Pont and others 1997).

What Are Typical Exposure Concentrations Resulting in Respiratory Diseases?

A limited amount of dose-response data is available for agricultural dusts, and only for cotton dust, swine, and poultry confinement exposures. These dose-response data typically apply only to CLLC exposures. Decreased pulmonary function has been demonstrated in persons exposed to cotton dust levels with endotoxin in excess of 100 EU/m^3. These levels of endotoxin may also produce ODTS symptoms. Research in swine and poultry confinement workers has shown the following concentrations significantly increase the risk for pulmonary function decline and respiratory symptoms: dust (2.5 mg/m^3), endotoxin (100 EU/m^3), and microbes (10^5/m^3). PAMM exposures, such as a disturbance of spoiled plant material, can produce spore clouds of very high concentration. For example, loosening bales of hay in a confined space, such as in a barn, has produced clouds of 1.6 × 10^9 organisms/m^3 of thermophilic microbes. A person doing light work in this setting may retain 7.5 × 10^5 spores/minute in

the lungs. Concentrations of 4 × 10^9 viable spores/m^3 have been documented in silo openings (May and others 1989). Endotoxin is found to reach levels in silos of 8.8–87.3 ng/mg of dust during unloading of silos (May and others 1989). Respirable histamine levels have been recorded as high as 10 nmol/cu.m after chopping bedding (Siegel and others 1992).

Acute ODTS and acute cases of FL are the two principal conditions resulting from PAAM agricultural dust exposures. They present acutely in a similar way: febrile illness of variable severity with chills, cough, dyspnea (more prominent in FL), myalgia, malaise, headache, fever, and muscle aches and pain, which appear 4–6 hours after exposure. However, the course epidemiology and outcomes of these diseases present in dissimilar ways, as described in the following paragraphs.

Furthermore, there are some historical and geographical variations in terminology of these diseases. The generic name of FL is *hypersensitivity pneumonitis (HP)*. The British prefer to call this *extrinsic allergic alveolitis. Farmer's lung disease (FL)* is the conventional expression within the United States for HP in affected farmers (Do Pico 1986; Emanual and Kryda 1983; Emanual and others 1989; Schenker 1998; Von Essen and others 1999). There are many other common names used for HP disease (listed in Table 3.6), according to the environment where exposure occurred. ODTS has previously been called *atypical farmer's lung, silo unloader's disease, inhalation fever, toxic alveolitis,* and *pulmonary mycotoxico-*

Table 3.6. Agents and Common Names for Hypersensitivity Pneumonitis (HP) Among Agricultural Workers and Food Product Processors

Exposure	Agent[1]	Common Name or Condition
Moldy hay or silage	*Thermophilic actinomycetes* *Micropolyspora faeni* Others	Farmer's Lung
Mushroom production/mushroom compost	*Thermoactinomycetes vulgaris* Others	Mushroom Worker's Lung
Moldy sugar cane plant residue— postprocessing	*T. viridis* *T. sacharii* Others	Bagassosis
Fungi		
Moldy maple bark	*Cryptostroma corticale* Others	Maple Bark Stripper's Disease
Moldy malt	*Aspergillus clavatus* Others	Malt Worker's Lung
Moldy dust	*Penicillium frequentans* Others	Suberosis
Surface mold on cheese	*P. caseii* Others	Cheese Worker's Lung
Moldy wood chips	*Altervaria spp.* Others	Wood Worker's Lung People using wood chips for heating fuel
Moldy redwood dust	*Pullularia spp.* Others	Sequiosis
Paprika dust	*Mucor spp.* Others	Paprika Splitter's Lung
Arthropods		
Infested wheat	*Sitophilus granarius* Others	Wheat Weevil Disease
Fresh avian droppings	Avian proteins	Bird Breeder's Lung

[1]There likely are multiple agents of hypersensitivity pneumonitis for each specific condition. Specific agents listed here have been associated with the condition, but are probably just one of several agents involved.

sis (Emanual and others 1989; Pratt and May 1984). However, today *ODTS* remains the most common term used.

Farmer's Lung (FL)

FL is a delayed allergic response at the alveolar level, with a variable systemic presentation depending on host factors and specific circumstances of exposure. Although several alternative pathways to FL disease have been proposed (Olenchock and others 1992; Pesci and others 1990; Rylander 1987), the commonly accepted pathogenesis of FL includes formation of IgG an-

tibodies to inhaled antigens found in agricultural dusts (Emanual and Kryda 1983; Jones 1982; Schatz and Patterson 1983; Schwartz and others 1995; Seaton 1994). Inhalable, thermophilic organisms of several actinomycetes species and *Micropolyspora faeni* are commonly incriminated allergens and often included in the FL battery for serologic testing. Once an individual becomes sensitized and develops circulating antibodies, subsequent exposures to these antigens may result in antigen-antibody complexes at the alveolar level. Secondarily, these antigen-antibody complexes are then recognized by sensitized macrophages as foreign and, thus, engulfed stimulating a type IV (delayed) immunologic response in the interstitium. Subsequently, even low dose exposures may result in macrophage infiltration. Histologic appearances of lung tissue reveal granulomas with giant cells in the interstitium. Pathogenesis may include macrophage transformation into fibrous tissue and a progressive interstitial fibrosis. Illness covers a continuum from acute reversible to chronic debilitating disease. Symptoms of acute illness are observed 4–8 hours following massive exposure to organic dusts. Acute symptoms are very similar to acute ODTS or influenza but usually more severe, with chest x-ray evidence of an infiltrate, lowered PO_2, and restrictive pulmonary function decrements. The acute illness subsides in 2–5 days, and initial respiratory impairment usually resolves completely within weeks to a few months, if exposure ceases.

However, multiple acute attacks or prolonged (years) low-level exposure can lead to irreversible, progressive interstitial lung fibrosis that can decrease total lung capacity and diffusion capacity (Braun and others 1979). This chronic form of FL is characterized by a progressive illness with weight loss, fatigue, and insidious onset of cough and dyspnea. Progression is similar to that of any chronic interstitial pulmonary fibrosis, with end-stage disease characterized by respiratory insufficiency or *cor pulmonale*. Some patients may also have emphysematous lung changes.

FARMER'S LUNG—EPIDEMIOLOGY
The prevalence of FL varies from study to study (Gruchow and others 1981; Guernsey 1989; Madsen and others 1976; Marx and others 1990). Because the symptoms of acute FL and ODTS may be nearly identical, many of the FL studies, particularly those prior to the mid1980s, probably included many ODTS cases. Prevalence in the U.S. has been considered to be below 5% of the farming population. For example, 3.9% of one surveyed group of Wyoming farmers and dairy producers gave a history typical of FL. A population-based prospective survey of over 1,500 Wisconsin farmers which began in 1976, revealed a case prevalence rate of 4.2 per 1000 farmers (0.42%) (Jones 1982). However, 10% of this cohort had circulating antibodies to at least one FL antigen. The 1984 follow-up study revealed a

prevalence of clinical cases at 0.9% and sero-positive rate to *M. faeni* at 2.09% (Marx and others 1990). The highest prevalence was among dairy farmers with the largest farms and largest herds. Sero-positivity to one of the usual farm-related FL antigens is poorly correlated to FL disease. Sero-positivity indicates only antigen exposure. Prevalence appears higher in England, Scotland, and Finland (Notkola and others 1992).

FL is most typical of those who tend barn-enclosed dairy cattle. In some populations (e.g., Finland), this task is often performed by women (Notkola and others 1992). Examples of dust-disturbing activities attributable to HP disease include breaking open bales of hay for feed or bedding, generally in winter or spring of the North Temperate Zone.

Disease in children is extremely rare. Two cases in Denmark (ages 10 weeks and 3 years) led researchers to causal exposures to decentralization of grain silos and dryers, and farmers' tendency to store overly moist grain in silos adjacent to the residence (Thorshauge and others 1989). Why many farmers with FL antibodies fail to develop clinical disease is unknown. Interestingly, it has been observed that there is an unexpected, significantly lower prevalence of *M. faeni* antibodies and lower disease prevalence among tobacco smokers (Kusaka and others 1989; Warren 1977). The reason for this protective factor is thought to be that smoking down-regulates the cellular immune system (Blanchet and others 2004).

DIAGNOSING FARMER'S LUNG
Clinical presentation of FL is difficult to distinguish from ODTS. In fact, it is likely that most PAMM exposures resulting in acute symptoms are likely ODTS, and that a small percentage of those cases progress or develop simultaneously with FL. FL is highly variable; no single symptom is diagnostic. Acute FL should be suspected in any farmer with acute influenza-like pneumonitis or active interstitial lung disease. In most circumstances, the following combination of factors is necessary for a tentative diagnosis of acute and subacute FL disease: 1) the cluster of symptoms including cough, fever, malaise and dyspnea, and basal crepitant rales; 2) a history of exposure to decayed plant material dusts. Supporting evidence includes positive serology to

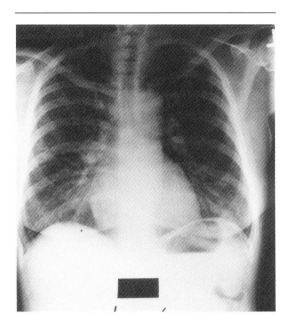

Figure 3.6. This lung-field radiograph shows a micronodular infiltrate in the lower lung field, commonly seen in acute cases of Farmer's Lung.

any of a panel of 15 or so fungal or thermophilic actinomycetes antigens (assuming offending antigen is included in a farmer's lung panel); 3) a chest radiograph revealing bilateral, micronodular lung infiltrates in the lower two-thirds of peripheral lung field (Figure 3.6); and 4) abnormal pulmonary function tests of acutely ill patients, including decreased lung volumes with a restrictive pattern, small airways obstruction, and decreased carbon monoxide diffusing capacity. Patients may be cyanotic or hypoxemic. Neutrophils dominate broncheoalveolar lavage findings in the first few hours following exposure. This picture changes some hours following onset of illness by a mononuclear cell infiltrate with a predominance of lymphocytes. This lymphocytosis has a high T-cell to B-cell ratio (as compared to peripheral blood) (Pesci and others 1990). Also notable is the decrease in T4 to T8 cell ratios; such increased numbers of suppressor cells are opposite to the finding in sarcoidosis. Chronic cases of FL show a spectrum of abnormalities, including lymphocytosis in bronchoalveolar lavage (BAL) fluids (also evident in subacute cases), pneumonitis, fibrosis, and hyperexpansion or honeycombing of lungs (Braun and others 1979).

Although the presentation, history of exposure, and x-ray and laboratory results may be sufficient to diagnose acute cases, diagnosis of subacute or chronic cases is more difficult. Continuous low-level exposure may be difficult to identify and onset of disease is insidious. Although FL patients usually demonstrate a positive serology, 10% or more of the farming population may possess FL-antibodies and only a small number of these experience clinical illness. Also, care must be taken to use an appropriate battery of FL antigens. Both chest radiographs and pulmonary function tests may be highly variable, and in some cases either or both may be normal. However, pulmonary function is likely to show a restrictive pattern. Chest radiograph may show nodular thickening of the alveolar interstitium. High-resolution computerized tomography will more readily demonstrate the latter lesions (Wouters and others 2004).

Lung biopsy and lung lavage are not normally required or advised, but may be useful in an exceptional case when a specific diagnosis is needed (e.g., for worker's compensation), or with a difficult differential diagnosis. Lung biopsy in HP reveals a characteristic mononuclear cell infiltrate or granulomatous interstitial pneumonitis, possibly with giant cells (Figure 3.7). Gross thickening of the alveolar capillary membranes results from mononuclear infiltration into interstitial tissues, resulting in obliteration of the alveoli. Mononuclear cells often form noncaseating granulomata that may occlude bronchioles. Multinucleated Langerhan's giant cells and foreign body type cells that may be birefringent or nonrefringent are common in areas of inflammation. Spores of the causative molds usually are not recognized in tissues.

Bronchoprovocation by FL antigens has been used as a definitive diagnostic test, but can involve significant risk of exacerbating the disease in the patient and is not recommended today.

DISTINGUISHING CHRONIC FARMER'S LUNG IN A
DIFFERENTIAL DIAGNOSIS
Chronic FL can be misdiagnosed as depression, chronic bronchitis, or any chronic interstitial lung disease. Pulmonary sarcoidosis may prove an especially difficult differential because of histopathologic and other similarities to FL. Occupational history and other history of exposure to spoiled plant material are critical in establishing a diagno-

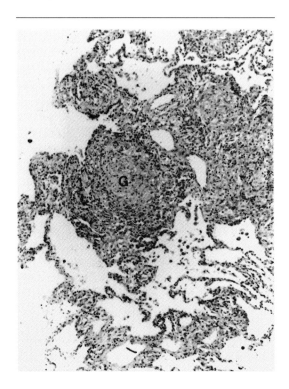

Figure 3.7. This biopsy of a patient with acute Farmer's Lung shows the characteristic mononuclear cell infiltrate. The letter *G* indicates a forming granuloma. Multinucleated giant cells may also be seen.

sis of chronic FL. BAL or bronchoprovocation may be helpful in exceptional cases, but slightly increased numbers of lymphocytes in a BAL sample may be found in asymptomatic farmers and thus is not diagnostic by itself. Acute FL can be misdiagnosed as ODTS (previously mentioned ODTS may often occur simultaneously with FL), influenza, a bad cold, infectious pneumonia, or atopic asthma. Differentiation of FL from most cases of asthma can be based on the presence of wheezing and the absence of rales in asthma, and chest x-ray changes (infiltrate in lower lung fields) in FL. Patients with FL will have the typical pattern of restrictive pulmonary function: reductions of both FEV_1 and FVC, often yielding a normal FEV_1/FVC ratio, with decreased compliance and diffusing capacity.

Organic Dust Toxic Syndrome (ODTS)

ODTS is a nonallergic, inflammatory reaction of the airways and the alveoli with systemic in-

fluenza-like symptoms due to high levels of organic dust and endotoxin exposure (Clark and others 1983). Clinically, acute cases present very much like acute FL, with cough, fever and chills, headache, fatigue, myalgia, and anorexia occurring 4–8 hours following exposure to high levels of organic dusts. Severity varies from a mild, influenza-like illness to profound illness with dyspnea, but rarely death (Von Essen and others 1990, 1999). The peripheral blood neutrophil count will be elevated but with no eosinophilia. Symptoms subside in 2–5 days, and permanent lung damage does not appear to occur. Multiple exposures simply produce repeated acute illness. However, subsequent cases of ODTS occur with lower exposures and more severe symptoms. Furthermore, bronchitis may remain for several weeks following an acute case. An acute ODTS case may occur simultaneously with acute FL, or it may be the sensitizing incident for chronic FL.

ODTS is a disease of toxic-mediated inflammation, not an allergic-mediated disease. The initial response of neutrophils results in chemotactic factor release, attracting macrophages to the affected tissues. The macrophages in turn release a cascade of mediators including INF-alpha and tumor necrosis factor, which result in alveolitis and generalized febrile and influenza-like symptoms. There are also anecdotal reports of a chronic form of ODTS. This poorly documented condition is characterized by chronic fatigue and a persistent pulmonary infiltrate in workers with CLLC exposures to agricultural dusts (such as livestock confinement workers) (Larsson and others 1992).

ORGANIC DUST TOXIC SYNDROME—
EPIDEMIOLOGY
ODTS may occur at any time of the year, though some exposures may be more common in late summer and early fall, during processes where vast amounts of respirable organic dust are released in an enclosed workspace. Typical work situations include "silo uncapping" and grain bin cleaning, both of which entail strenuous physical activity within an enclosed space. (Other exposure risks for ODTS are seen in Table 3.3.)

Silo uncapping involves removing moldy silage from a conventional non-airtight upright silo in preparation for mechanical unloading and feeding to cattle. The silage had been stored for

Table 3.7. Epidemiologic and Clinical Differentiations of Organic Dust Toxic Syndrome (ODTS) and Farmer's Lung (FL)

ODTS	FL
Often "mini" epidemics—all persons exposed to an acute episodic massive exposure affected	NOT every person affected by exposure must first be immunologically sensitized.
Acute symptoms, delayed response to exposure (2–6 hours)—cough, chest tightness, malaise, headache, myalgia, arthralgia, fever—lasting 24–72 hours	Acute symptoms, delayed response to exposure (2–6 hours)—cough, chest tightness, malaise, headache, myalgia, arthralgia, fever—lasting 24–72 hours. HP typically has more severe acute symptoms than ODTS.
Elevated WBC, with left shift	Elevated WBC, mononuclear cells may be relatively elevated.
Usually normal chest film	Usually finely nodular infiltrate more evident in lower lobes and mediastinum.
Pulmonary function shows obstruction	Pulmonary function shows restriction.
PO_2 usually normal	PO_2 is below normal values.
BAL—elevated white cells with neutrophil predominant	BAL—elevated white cells, mononuclear cells may predominate.
Biopsy—acute inflammation of alveolitis, bronchitis	Biopsy—more chronic inflammation of alveoli, with mononuclear cells, possible granulomas and/or giant cells.

weeks or months in the silo beneath a "cap" of silage a foot or so in depth, atop a plastic sheet which covers the silage underneath. Low oxygen conditions beneath this cap impede fungal growth, but the exposed, upper silage layer becomes grossly contaminated with microorganisms. Before dispensing the silage for winter's feed, someone must climb into the silo, pitch off this molded silage cap, and lower the mechanical unloader into place. High concentrations of microorganisms and their by-products can be inhaled during this task, possibly exacerbated by the greater respiratory load.

ODTS resulting from exposure to spoiled plant material has been recognized as a response distinctive from FL only as recently as 1984 (Do Pico 1986; Pladsen 1984). It is far more common than FL or other illnesses associated with feed storage. A 5-year prospective study of nearly 1000 Iowa swine farmers revealed a self-reported prevalence rate of approximately 30% (Donham and others 1990). In a study of New York dairy farmers, 14 of 26 feed-related episodes of respiratory illness were identified as ODTS (Pratt and May 1984). Ten percent of Finnish farmers are reported to have experienced symptoms indicative of ODTS.

DISTINGUISHING FARMER'S LUNG AND ODTS

A number of features differentiate acute FL and ODTS. These are summarized on Table 3.7 and discussed below (Pratt and May 1984; Rask-Andersen 1989; Von Essen and others 1999). FL occurs in only a small subset of any exposed population. Although predisposing factors must exist, these have not yet been defined. Among sensitized farmers, even CLLC exposures to aerosolized mold and bacteria can elicit an (often silent) progression of the disease. ODTS, in contrast, can affect a high percentage of exposed individuals following PAMM exposures. Thus, cases often are clustered, with all individuals in certain PAMM exposures affected simultaneously. Patients may or may not have serum precipitins to FL antigens. As previously mentioned, ODTS may also occur in a chronic form, manifesting as chronic fatigue, myalgia, shortness of breath, and persistent pulmonary infiltrates. This may be seen with CLLC exposure situations.

A number of laboratory tests can help distinguish FL from ODTS. Chest radiographs of FL patients characteristically reveal a finely nodular density in the lower lung fields, while chest radiographs of ODTS patients characteristically are clear (although occasionally may have an abnormal infiltrate). Blood gas measurements often show decreased PO_2 for FL, but are usually normal for ODTS. Immunoserology to antigens in a farmer's lung battery in FL patients is usually positive, while ODTS patients may or may not have these antibodies. Pulmonary function tests, although usually showing marked restriction with FL, commonly show mild or no change or obstruction with ODTS. BAL in FL more typically yields fluids with a predominance in lymphocytes; in ODTS, BAL fluids are typically dominated by neutrophils ODTS (Von Essen and others 1999).

DISTINGUISHING FARMER'S LUNG AND ORGANIC DUST TOXIC SYNDROME FROM NITROUS OXIDE POISONING AND METAL FUME FEVER

Because acute FL and ODTS may result from exposure within a silo, either may be confused as silo filler's disease resulting from exposure to nitrous oxides (see Section 5) (Pladsen 1984). For example, the dyspnea often present for several weeks after acute silo gas exposure may be confused with FL. However, silo gas exposure can be traced to silos filled within the previous 2 weeks, exemplifying the need for an accurate exposure history. In general, acute exposure to agricultural dusts occurs when unloading the silo occurs at least 1 or 2 months after it has been filled. The primary physical finding with silo gas exposure is pulmonary edema, not generalized, and febrile influenza-like symptoms as in FL or ODTS.

Exposure to welding fumes may also create a syndrome similar to ODTS. Welding, particularly with galvanized metal (contains zinc) produces aerosols that can create a syndrome similar to ODTS (Zimmer 2002). A good exposure history will help differentiate these conditions.

TREATING FARMER'S LUNG AND ORGANIC DUST TOXIC SYNDROME

Since both acute ODTS and acute FL are self-limiting, with severe symptoms resolving in 2–5 days and complete resolution occurring within 10–60 days, a health care provider's help is often not solicited by afflicted persons. The treatment is the same for patients with acute FL or ODTS. Treatment is primarily supportive and symptomatic (Rask-Andersen 1989). As the patient may be hypoxic with a respiratory alkalosis, supportive therapy includes oxygen via nasal canula and rehydration with a balanced electrolyte solution. Nonsteroidal and/or a burst and taper of corticosteroids may diminish the acute symptoms and shorten the duration of illness (although there are no research data to back up this practice). Bronchodilators may be used if bronchoconstriction is a predominant symptom. Desensitization is not effective; antibiotics and antihistamines are ineffective.

The primary importance of a proper diagnosis of FL is that a small concentration of antigen can provoke illness in highly sensitive individuals, and continued exposure can lead to permanent impairment. Avoidance of spoiled plant material is imperative in these individuals. Early diagnosis and avoidance are most important in preventing irreversible lung damage of FL.

Chronic Lower-Level Concentration (CLLC) Exposures to Agricultural Dusts

Handling and moving grain at a local elevator, working at a grain terminal, working in a livestock or poultry confinement facility or dairy barn, grain harvesting and cotton ginning, and further processing of cotton or flax are all examples of CLLC exposures to agricultural dusts. CLLC exposures are those occurring daily in an agricultural workplace, but at lower levels in a relative context to the PAMM exposures previously described. The chronic illnesses associated with CLLC exposures include bronchitis, NALC, MMS, MMI, and atopic asthma. Details of exposure circumstances, symptoms, clinical signs, diagnosis, treatment, and prevention are described in the following paragraphs. Grain dust exposures and confinement animal feeding operations (CAFOs) will be used as examples and discussed in detail in the following section of this chapter.

GRAIN DUSTS EXPOSURES IN AGRICULTURE

Dusts from grain consist of a complex mixture of organic and inorganic particles from sources as diverse as leaves, soil, and insect parts. The mixture varies with the type of grain, where it is grown, growing conditions and methods of harvest, storage, and processing. A significant fraction of grain dust is composed of biologically active substances, which are inhalable (less than 10 microns in diameter). Dusts of certain grains such as durum wheat, barley, and milo are more irritating than others (Warren 1983). Adverse health effects of grain dust also increase as moisture content and spoilage of grain increases.

The bulk of particles are from fruits of grasses such as wheat, and legumes such as soybeans, or oil seeds such as rapeseed. Bits of leaves and stems also may be present. Nonplant contaminants are numerous. Animal material (bits of insects, rodents, or birds, or their excreta), mites, chemical residues (pesticides used to grow or later treat the grain), and inorganic matter (soil including silica particles) all may be intermixed in small quantities. A variety of fungi and bacteria, their spores, and their by-products (endotoxin and glucans) also pose a respiratory hazard. Species of microorganisms vary with regional climate and change from harvest through storage. In North America, major fungi are *Penicillium* and *Aspergillus* species; thermophilic actinomycetes increase in wet and overheated grain. Many of the components of grain dust are capable of affecting the respiratory tract individually; together, they produce a heterogeneous array of biological effects, as outlined in Section 1.

WHO IS EXPOSED TO GRAIN DUSTS, AND WHEN?
Anyone involved in production, storage, transportation, or processing of grain can suffer the effects of chronic exposure to grain dusts. Exposure starts with farmers and farm workers who grow, harvest, sometimes store, and then transport grain to local storage facilities. Exposure extends far beyond the farm to workers in feed mills, grain elevators, and grain transportation industries. These latter workers are more routinely exposed to grain dust and suffer from respiratory responses more commonly and severely than do farmers. Small grain elevator agents and workers clean, fumigate, and store grain and may grind and mix it into animal feed. Other workers load and transport it by truck or rail to larger elevators, where it is handled by additional workers, and eventually to terminal elevators at harbors or milling. Grain is then prepared by dockworkers and longshoremen for transport abroad, or is processed into products such as feed, flour, alcohol, sweeteners, and breakfast cereal. This section covers health effects of grain dust inhalation among this great variety of agricultural workers who grow, transport, and store grain. Occupational asthma from exposure to processed grains (e.g., baking flour) is well documented among bakers and food han-

dlers, but will not be discussed here (Becklake 1980; Salvaggio and others 1986).

Exposure to grain dusts can occur at any stage of the above process. Clouds of grain dusts are most evident whenever grain is moved, and especially heavy exposures among any grain handlers occur during loading and unloading grain. Workers in grain storage elevators have the highest total number of dust exposures during performance of housekeeping and maintenance chores and working in the towers and transfer galleries (Farant and Moore 1978).

HOW COMMON IS EXPOSURE TO GRAIN DUSTS?
An estimated half-million workers (excluding farmers) are involved in storage, transportation, and processing of grain in North America. The grain industry is a major source of community pollution: 27 pounds of dust are emitted for every ton of grain handled, resulting in 1.7 million tons of grain dust produced per year (Vandergrift and others 1971). The concentration inside workplaces varies widely. Measurements inside elevators have ranged from 0.18 to 781 mg/m^3 of total dust, with the respirable range extending up to 76.3 mg/m^3 (Becklake 1980; Farant and Moore 1978; Viet and others 2001). Airborne concentrations of fungal spores often exceed one million spores per cubic meter. Although prevalence of respiratory responses to grain dust varies from study to study, the presence of cough and phlegm (indicators of bronchitis) and airway obstruction are consistently high; about twice that of unexposed populations (Becklake 1980; Do Pico 1982). Prevalence rates are thought to be higher than those documented, since studies are completed on "survivor populations." The variation in prevalence of disease probably stems from the heterogeneous nature of grain dust, as well as from variations in study populations and research techniques.

Respiratory Health Effects of Grain Dust Inhalation

Respiratory health effects of grain dusts are similar to other organic dust exposures. The exposures and resulting respiratory conditions described with grain dust exposures may be applied to most of the CLLC exposures (Dosman and Cotton 1980). As with other agricultural dust exposures, disease mechanisms are inflammatory, immunological, toxicological, physical responses, or a combination thereof. Since dust particles range from less than 5 microns to greater than 20 microns, responses can occur in the upper airways, large and small bronchi, and the alveoli. Responses are exacerbated in those who smoke, and by length of employment and specific jobs performed while working as a grain handler. Certain grains such as sorghum and durum wheat seem to create a higher risk for respiratory symptoms (Schenker and others 1998). The set of acute and chronic health effects of grain dust are similar to other organic agricultural dust exposures (Flaherty 1982).

Bronchitis and Airways Obstruction

The most common condition caused by grain dust is bronchitis. A significant portion of grain workers will develop chronic bronchitis (cough with phlegm for at least 3 months of the year, for more than 2 years). The severity of bronchitis is variable, but total disability is unusual unless the person also smokes or has other cardiorespiratory disease. Chest tightness and wheezing may accompany bronchitis (see NALC, which follows).

The prevalence of bronchitis among grain producing farmers has been reported as high as 15% of nonsmokers and up to 25% of smokers (Bar-Sela and others 1984; Donham and Leininger 1984; Donham and others 1984b; Dosman and Cotton 1980).

LUNG FUNCTION

After adjusting for age, loss of lung function is greater among grain handlers than expected values for a comparable worker group not involved in handling grain. Initially, airflow obstruction (measured by FEV_1) or decreased midexpiratory flow rates (FEF_{25-75}) is observed during the work period, but this is reversible and improves when the grain handler is not at work. With repeated exposure, this obstruction may become permanent. Workers who have an acute response to grain dust (i.e., decline in lung function over the course of a work shift) are more likely to have accelerated baseline declines in lung function over time (Chan-Yeung and others 1983; Eduard and others 2001).

TOBACCO SMOKE INVOLVEMENT

While bronchitis and airways obstruction are common among nonsmoking grain handlers, the conditions are more common among grain handlers who smoke cigarettes. Asthma symptoms among grain handlers who smoke has been reported to be nearly five times (50% prevalence) greater relative to a comparison nonsmoking population (11% prevalence) (Chan-Yeung and others 1983; Donham and others 1990). Cigarette smoke and grain dusts are additive and possibly synergistic, accelerating changes in the peripheral airways causing more severe, frequent, and earlier onset of respiratory symptoms.

Nonallergic Asthma-like Condition (NALC)

NALC (nonatopic, or nonallergic asthma) is much more common among grain workers than allergic or atopic asthma among farmers (Eduard and others 2004). With farmers working in CLLC environments, NALC commonly occurs in conjunction with chronic bronchitis (Donham 1993; Von Essen and others 1999). The complaints appear as episodic chest tightness and wheezing, following exposure to dusty work environments. The symptoms do not appear immediately upon exposure, but usually after 1–2 hours within CLLC work environments, such as intensive swine or poultry production facilities. The following observations indicate that the mechanism of NALC is chronic inflammation, not allergy in the classical sense: 1) few production workers are atopic by history or skin test; 2) development of NALC is not related to atopic status (Chan-Yeung and others 1983; Do Pico 1986); 3) grain triggers histamine and leukotriene release, indicating the inflammation mechanism of the nonallergic asthmatic response of NALC (Rylander and others 1990); 4) work shift declines in lung function have been positively associated with increases in peripheral blood leukocyte count. Most individuals with these symptoms have increased sensitivity to methacholine challenge (Rylander and others 1990). It is important to note that many of these workers may develop asthma symptoms following nonspecific exposures such as cold air or exercise. Although NALC may develop in workers within days or weeks of first exposure (Dosman and others 2004), more commonly it is seen in workers with greater than 6 years exposure, and with exposure concentrations above 2.5 mg/m^3 for 2 or more hours per day (Do Pico 1982, 1986; Donham and others 2000, 2002).

Monday Morning Syndrome (MMS)

Symptoms similar to a mild case of organic dust toxic syndrome (chills, fever, flushed face, myalgia, malaise, headache, cough, wheezing, and shortness of breath) may be seen in recently hired workers or on Monday or a return to work following vacation. (Becklake 1980). Symptoms typically commence after work and may last several hours or a few days. Symptoms wane over the workweek, only to recur in the future following a subsequent absence from work. This response is thought to be a down-regulation of macrophages from inhalation of endotoxins over time. MMS was first recognized in the cotton processing industry and coined *byssinosis*. However, MMS can be seen in many different CLLC agricultural dust exposures, such as grain handling (then called *grain fever*) and work in livestock buildings (Donham and others 1984b; Thelin and others 1984).

Respiratory Symptoms During Grain Harvest

Grain producers exhibit a condition that is probably akin to MMS, particularly prevalent during grain harvest. Even though most combines have air filtering cabs, particulate matter inside combines is still high in many instances, averaging around 1 mg/m^3 with excursions to 15 mg/m^3. The same study revealed endotoxin commonly higher than 100 EU/m^3, the suggested limit for chronic exposure (Viet and others 2001). During the first days of operating a grain harvester, symptoms develop including throat congestion and an unproductive cough with dyspnea in the evening. The farmer may be awakened in a state of breathlessness with wheezing and bouts of coughing. Sixty percent of exposed workers had a cross-shift drop in peak flow (Viet and others 2001). Analysis of bronchoalveolar lavage samples reveals acute airways inflammation. The mechanism is likely the same as MMS, which is up regulation of "inexperienced" microphages by endotoxin and/or ß,1–3 glucan. The symptoms typically disappear at harvest's end.

Atopic Asthma

"Classical" atopic or allergic asthma is an IgE antibody response to a specific antigen, as compared to the nonspecific inflammation mechanism of NALC. Atopic asthma is much less common than NALC among agricultural workers generally, especially those exposed to CLL environments. Symptoms of atopic asthma and NALC are similar: bronchoconstriction with cough, wheezing, and dyspnea. Symptoms usually immediately follow exposure, but may be delayed several hours following exposure, may have a dual response (immediate and delayed), or recur on successive nights following exposure. The delayed response is associated with an IgG antibody response (Rylander and others 1990). The asthmatic reaction may be caused by any number of grain dust components. Studies of grain farmers implicate storage mites as the allergen most responsible for sensitization and atopic asthma development. Atopy evaluations, however, do not dependably identify sensitization in cross-sectional studies of grain workers and likely underestimate the risk of atopic asthma because of self-selection out of the environment (healthy worker effect).

The results of a 1990 study of grain workers and an unexposed comparison population demonstrated very similar prevalence rates of asthma (2.4% and 2.7%, respectively) (Chan-Yeung and others 1983). However, the prevalence of NALC is more common (5–20%) in grain handlers. The term *grain asthma* is commonly used in the industry. However, this term has been applied to workers with asthmatic symptoms regardless of the mechanism.

Mucous Membrane Irritation (MMI)

MMI presents with one or more of the following symptoms: a sore/scratchy throat, irritation of the eyes and nose, and noninfectious sinusitis. Often workers complain of a "persistent cold," stuffy head, headache, and sometimes "popping" ears (Von Essen and others 1999). These latter symptoms of MMI are associated with chronic noninfectious sinusitis. The constituents of agricultural dust can cause an inflammatory response in all mucous membranes.

Other Health Risks Associated with Grain Dusts

Grains stored in elevators and transported long distances are commonly fumigated with pesticides such as phosphine and methyl bromide. Fumigants can be lethal upon inhalation and exposure may occur during the fumigation process or during the first few days following fumigation, before the residual dissipates. Respiratory effects of fumigants are discussed in Section 6.

In addition to reactions of the respiratory system, grain dust can produce dermatitis, either a contact dermatitis, or from grain mites. Dust explosions in grain handling facilities have killed many grain elevator employees in the past, and continue to pose a hazard, even though federal regulations have resulted in much cleaner and less dusty operations. Any time the dust levels rise above 50 grams/m^3 air (minimum explosive concentration), explosions are possible (Noyes 2002).

Diagnosis of Agricultural Dust Respiratory Diseases

Although there are a group of common symptoms and conditions that could be called a complex or agricultural dust syndrome, each patient's reaction varies, because of the varying mixtures of causative agents and varying individual responses. Diagnosis depends on a thorough occupational history documenting type and time of exposure and correlating these to onset of symptoms. Workers in grain elevators, feed mills, and grain transportation industries may, on a regular basis, experience any one or a combination of the responses listed in Table 3.2. Farmers may experience similar conditions as grain handlers. However, their complex of symptoms varies depending on the type of agriculture they conduct. Those farmers with CLLC exposures (e.g., livestock confinement) have a relative greater frequency and severity of symptoms. Grain producers may have a higher prevalence of ODTS relative to chronic symptoms because of a higher probability of exposure, such as uncapping a silo or cleaning out a grain bin (Figure 3.8).

Chronic obstructive pulmonary disease (COPD) in farmers can develop as a likely result of multiple environmental exposures, such as grain dusts and cigarette smoking (Monso and others 2004; Sigurdarson and others 2004a). Diagnosis of COPD is dependent upon history of chronic cough with phlegm and wheezing (and baseline obstructive lung function test result). The definition of COPD is as follows: the presence of a persistent cough with mucus production that continues for at least three-month periods for at least two consecutive years. The importance of monitoring grain industry and

Figure 3.8.   Cleaning out a grain bin is one of the most common tasks associated with organic dust toxic syndrome.

agricultural production workers for early onset of COPD cannot be overemphasized. When workers are symptomatic, work shift spirometry or peak flow testing should be performed annually to determine whether obstruction is increasing over time, which is predictive of baseline or permanent declines in PFT (Schwartz and others 1990). Work shift declines may be the only objective data seen in symptomatic workers. Workers with NALC symptoms will likely be hyperresponsive to methacholine or histamine inhalation challenge. Furthermore, air trapping may be evident.

Diagnostic tests for occupational asthma include history of chest tightness and wheezing following work exposure and a decline in FEV_1 or peak flow rates (PFR) before and after work. A negative skin test, IgE antibody negative, and negative family history would help rule out allergic asthma, pointing to NALC as the diagnosis. Additionally, bronchoalveolar lavage or induced sputum tests revealing a neutrophilic response would point to NALC. The presence of eosinophils in the same samples would suggest that an allergic basis is involved. Skin prick tests or RAST testing to specific allergens would also be helpful to determine an allergic basis (Wouters and others 2004).

Diagnosis of ODTS (also called *grain fever* in the industry) is common among grain industry workers and depends on presentation with appropriate symptoms and signs following a known exposure to grain dusts. Affected individuals may have normal pulmonary functions and usually have a clear chest radiograph. Chest auscultation is usually normal, as is PO_2 or chest x-ray. PFT usually will result in an obstructive pattern. Lavage and peripheral blood will show a neutrophilic response. Diffusion capacity of CO is often reduced (Wouters and others 2004).

These authors think it is important to consider apparent differences in basic pulmonary function of farmers relative to comparison populations. We have seen through experience that farmers commonly have higher baseline volumes and flows compared to standard industrial comparison populations. This is probably because of the heavy work they do and healthy worker effect. Pulmonary function of farmers with mild lung impairments might show close to 100% of predicted value, whereas if there were a farmer comparison group, that person might show 70% or 80% of predicted. The results are that one might not detect low-level lung disease in farmers with baseline testing. Cross–work-shift testing is ideal, as each person serves as his or her own control. Furthermore, forced expiratory flows (FEF $_{25-75}$) are commonly quite low in farming populations, even in asymptomatic individuals. This suggests that small airway obstruction is common.

Diagnosis of FL and distinctions of acute hypersensitivity pneumonitis and ODTS are described in depth in Section 2.

Treatment of Agricultural Dust-Related Disease

The major thrust of any treatment program among both grain industry workers and farmers should be to control exposure sources through applied industrial hygiene principles. Exposure control is critical for preventing repeated episodes of ODTS and FL and preventing progression of obstructive disease, FL, or MMI. Grain industry workers with severe NALC, progressive COPD, or FL should be advised to relocate to low- or no-exposure areas. Farmers with these illnesses often can alter work practices to prevent or significantly reduce exposure. Use of personal protective equipment, med-

ical surveillance, and other preventive steps are discussed in the following section.

There is no specific treatment for ODTS or acute FL. Symptomatic treatment for either condition was discussed earlier in this chapter. Medication for patients with occupational asthma is similar to that for other asthmatics (i.e., bronchodilators, corticosteroids), but should be combined with exposure reduction. Desensitization is not helpful. Smoking cessation is especially critical to patients with bronchitis, hyperreactive airways disease, and chronic obstructive pulmonary disease. Bronchodilators (with or without antiinflammatory agents) may be given temporarily as indicated. However, patients should never be kept on these for long periods of time without managing exposures.

Prevention of Agricultural Dust-Induced Illness

Prevention of excessive exposure to agricultural dusts is advisable for all farmers; it is imperative for persons sensitized to components of agricultural dust in their environment that cause them allergic asthma or FL. Prevention should take a multifaceted industrial hygiene approach. Some dairy farmers with a history of FL have successfully managed their illness by wearing a respirator regularly and by delegating to other workers those jobs with the potential of exposure to mold (Muller-Wening and Repp 1989). Other farmers with predisposing conditions have undertaken more dramatic steps. Strategies include eliminating the agent of disease (by installing airtight silos) or avoiding exposure to the aerosolized particles (by using a respirator, installing a completely mechanizing cattle-feeding system, or installing a more effective ventilation system in the barn). Although very expensive, these latter measures have allowed sensitized farmers to stay working on the farm.´

The generic paradigm for prevention from an industrial hygiene standpoint includes the following: 1) emission control or keeping dust out of the workplace environment, 2) removing dust from the air once it is there, 3) protecting the individual with personal protective equipment (e.g., a respirator), 4) modifying the work to decrease exposure or assigning or choosing jobs that decrease exposure for sensitive individuals, 5) educating the workers, 6) medical monitoring, and 7) smoking cessation.

EMISSION CONTROL

As high microbial concentration in agricultural dusts increases its health hazard, preventing microbial growth in feedstuff is an important prevention strategy. Capping silage with a plastic sheet held in place by rocks or a heavy chain (rather than additional plant material) reduces the mold and dust in the top layers of silage. Switching to glass-lined, airtight silos (realizing that these silos pose the health risk of asphyxiation) also will reduce mold growth, but may be economically impossible for many farmers. Grain and hay should be dried to less than 15% moisture before storage. There are many instances when mold growth cannot be prevented. For prevention of aerosolization of moldy plant material when silo caps are removed, one minimal-cost option is to wet down the top layer of silage prior to disturbing the material. Personal protective respiratory equipment must be considered an important adjunct in prevention when engineering and management methods are not successful. (Section 9 of this Chapter provides guidelines for selection and fit of respirators.)

The use of the new plastic silage bags significantly reduces worker exposure, because it is an anaerobic process, (reducing much of the microbial growth). Furthermore, these plastic bags eliminate the process of uncapping the silo, and the process is done outdoors, eliminating the confined space problem of dust and silo gas concentration. Pit silos also are outside, and are thus another way to reduce emission exposure to the workers. However, it should be noted that the quality of the feedstuff may be reduced and overall spoilage may be increased with pit storage. Furthermore, pit silos add the risk of silage collapsing on the person loading out of the pit with skid steer loaders or front-end loaders.

Emission control of all agricultural dust is important even though it may not have a high microbial concentration. Controls on the farm now include structural improvements such as combines or tractor cabs with filtered air. In newer grain elevators, totally enclosed conveyor belts, spraying the grain with a vegetable oil, dust collectors, and effective ventilation systems have greatly reduced dust levels. These improvements are lacking on older farms and in older or smaller rural elevators. Retrofitting older facilities with newer dust control technology, while desirable, is both costly and difficult. In addition to these emission controls, farmers and other grain handlers should be taught the hazards of grain dust inhalation. They should also be taught techniques for decreasing these hazards based on good practice techniques in growing, harvesting, and storing the grain, and on good housekeeping and work practices in elevators. Grain placed in storage should always be top quality, with insect and animal contamination kept to a minimum. Fumigation will help reduce the latter. When grain dust levels are above 4 mg/m^3 (a common grain dust standard in many countries), or when workers are especially sensitive, personal protective equipment (a certified dust mask, see Section 9) should be used.

REMOVING DUST FROM THE AIR

Properly functioning ventilation systems in animal buildings may prove beneficial, albeit costly, in reducing dust level concentrations. There are dust filtration systems, cyclone separators, oil misting systems, and ionization systems that can help remove dust from the air. Many of these systems are applied in grain elevators, but on-farm usage is not common because of the expense and extra management required.

USE OF RESPIRATORS

There are several factors that must be considered when recommending use of respirators. Respirators are hot, and difficult to breathe through when working. It is difficult to talk or be understood when wearing them. Because of these reasons, the individual must be highly motivated to wear a respirator. Also, for a respirator to be effective, the correct respirator for the job must be selected and fitted for the individual. Convenient supply and knowledgeable resource persons are rare in rural areas. Although respirators have been shown to be helpful, they are not completely protective as acute pulmonary function changes, and biological markers of inflammation are still present in workers who wear respirators in swine barns (Dosman and others 2000; Palmberg and others 2004). Because of all these difficulties with respirator use, they should be considered only one component of a control program.

REASSIGN JOBS TO PROTECT VULNERABLE INDIVIDUALS

Persons with a history of FL, asthma, or bronchitis should ordinarily take extra precaution to prevent agricultural dust exposure. For example, un-capping a silo or performing other tasks that have a high probability of exposure to massive quantities of agricultural dusts should be assigned to a nonsensitized person. Workers who develop airflow obstruction or significant respiratory symptoms should transfer to low-dust jobs.

CHANGING PROCESS TO REDUCE EXPOSURE
As a preferred work practice, the farmer should be kept separate from the moving, grinding, and mixing of feedstuff, especially when the process is in an enclosed space. Mechanical handling of dusty silage or hay in a manner that keeps the farmer separate from the dust is a desirable work practice. The widespread acceptance of large round bales that are transported by a tractor (instead of small, square, hand-carried bales) has decreased exposure to dust from moldy hay.

EDUCATION OF OWNER/OPERATORS AND WORKERS
Education of the owner/operator and workers is important in control of respiratory diseases. Although farmers may be quite knowledgeable of the causes of acute traumatic injuries on the farm, they are not very educated on the source and causes of illnesses on the farm. Education must be incorporated with other means of prevention. Education must be frequent and relevant. Health care practitioners in rural areas can have an important role in health education through one-on-one education to the patient or by facilitating community education programs.

MEDICAL MONITORING
Medical monitoring of agricultural workers should be considered in two circumstances: 1) persons with evidence of FL disease, and 2) persons working in CLL exposure situations. Monitoring of patients with confirmed FL should include periodic testing of pulmonary function and chest radiographs, as well as physical examination and history regarding dyspnea following exposure and on exertion. Measurement of blood gases and exercise tolerance may be useful in assessing degree of impairment. If modification of work behaviors and environmental control measures do not prevent the progression of symptoms and clinical signs of FL, further protection from exposure is indicated, including industrial hygiene control and being fitted with an appropriate respirator. Recommending that a farmer leave the farm to prevent permanent impairment should be done judiciously, as this is not a social or economic option for most farmers. Furthermore, studies in Wisconsin and Quebec have shown little difference in long-term outcome for patients who leave farming (Leblanc and others 1986). Medical surveillance is recommended in many CLL agricultural dust exposures (such as livestock confinement work or grain handling). New workers on these operations should have a preemployment examination that includes an occupational and medical history, physical examination, and spirometry. Any potential worker who has a history of respiratory symptoms or disease, atopy, or pulmonary function abnormalities, or who smokes cigarettes, should ideally be placed in a job where less exposure to dusts will occur. Annual medical workups, which include lung function tests (preferably as quickly before and after work exposure as possible), will allow detection of developing airflow obstruction while it is still reversible. Inquiries should be made into the cause of absenteeism and complaint of respiratory or other grain dust–related symptoms.

Medical evaluation also should address nonrespiratory effects of grain dust exposure, such as skin rash and eye irritation.

SMOKING CESSATION
Because of the proven additive adverse effect of smoking and agricultural dust exposure, smoking cessation programs must be recommended for anyone in regular contact with agricultural dust. This is one of the most important prevention actions that a health care practitioner can recommend to his or her farm patients.

SECTION 4: CONFINED ANIMAL FEEDING OPERATIONS (CAFOS) AND THE RESPIRATORY DISEASE HAZARDS
Introduction
The U.S. Environmental Protection Agency defines a Confined Animal Feeding Operation (CAFO) as a facility that concentrates livestock or poultry production in enclosed areas where grass or other plants will not grow, and is a potential surface water pollutant. CAFOs began with poultry production in the late 1950s, and swine CAFOs began to appear in the late 1960s. More

recently, sheep, beef cattle, dairy cattle, and veal calves may be housed in confinement buildings, although less commonly than swine and poultry (Donham and Gustafson 1982). CAFOs may be open feedlots or totally enclosed buildings. The latter are of more concern from an occupational health standpoint. CAFOs are production systems that include facilities for ventilation, heating, feed preparation, and delivery, and for disposal of animal wastes (Donham 1991). This system of production began to proliferate rapidly in the last half of the 1980s (Donham 1993).

Toxic gases emitted from manure are a health concern in many of these buildings. Manure is handled in one of two ways: it either drops through a slatted floor into a deep (6–8 feet) pit beneath the house where it remains until the manure slurry is pumped out to be distributed on fields (usually twice a year), or it is frequently removed from shallow pits (2–4 feet) under the building through any of several mechanisms to a solid storage structure or earthen lagoon outside the building. Outside storage was typical of most systems built in the 1980s, but a large number of buildings with deep storage pits directly below the house were built prior to 1990. Since 1995, there has been a return to more under-building storage.

What Toxic Dusts and Gases Are Found in Confinement Houses?

CAFO dust is a complex mixture of potentially hazardous agents that is generated primarily from the animals (hair and dander), dried feces, and feed (see Table 3.8) (Donham and Gustafson 1982; Donham and others 1985a; Nilsson 1982).

Additional to the dust, gases are generated inside the building, which arise from decomposition of animal urine and feces (ammonia, hydrogen sulfide, and methane among others) (Donham and Gustafson 1982; Donham and Popendorf 1985; Donham and others 1995), fossil fuel-burning heaters (carbon dioxide and carbon monoxide), or by the animals' respiration (carbon dioxide). These dusts and gases often accumulate to concentrations that may be acutely hazardous to human and animal health (Donham and Gustafson 1982) or may add to workers' chronic hazardous exposures.

The mixture and concentrations of dusts and gases inside CAFOs vary depending on numerous

Table 3.8. Potentially Hazardous Agents in Dusts from Livestock Buildings[a]

Feed particles[b]
Swine proteins (urine, dander, serum)
Swine feces[c]
Mold
Pollen
Grain mites, insect parts
Mineral ash
Gram-negative bacteria
Microbial by-products
Endotoxin
(1–3)-ß-D-glucan
Microbial proteases
Mycotoxins
Histamine
Ammonia adsorbed to particles
Infectious agents
Plant parts and by-products
    Tannins
    Plicatic acid

[a](Donham 1989).
[b]Grain dust, antibiotics, growth promotants.
[c]Gut, microbial flora, gut epithelium, undigested feed.

factors including management practices; ventilation and other engineering controls; the age, number and type of animals in the building; and the design and management of the feeding and waste handling systems.

Dust and gas concentrations and composition vary over time relative to season of the year and age of the animals. There are different types of CAFOs for different animal species (e.g., swine, poultry, dairy or beef cattle, and sheep). This section will focus on swine CAFOs primarily and poultry secondarily, as these operations have been most extensively studied and have been most commonly reported as potential health risks for workers (Bar-Sela and others 1984; Donham 1986; Jones 1982; Donham and others 1989; Lenhart 1984; Marrie 1989; Olenchock and others 1982; Thelin and others 1984).

Dust particles in CAFOs contain approximately 25% protein and range in size from less than 2μ to 50μ in diameter (Donham and others 1985a, b). One-third of the particles are within the respirable size range (less than 10μ in diameter) (Donham and others 1985a; Nilsson 1982). Fecal material particles are quite small ($\leq 10\mu$) relative to other dust components and consist of high concentrations of gut-flora bacteria and exfoliated gut epithelium. This component of the dust constitutes a major burden to small airways and alveoli. The larger particles are mainly of feed grain origin and primarily impact the upper airways. Also present are animal dander, broken bits of hair, bacteria, endotoxins, pollen grains, insect parts, and fungal spores (Donham 1986; Donham and others 1985a). In recent years, researchers have focused on the microbial by-products contained in this dust as the primary hazardous substances. Endotoxin and (1–3)-ß-D-glucan, respectively originate from the cell wall of gram-negative bacteria, and certain yeasts, molds, and bacteria. They are known toxins and inflammatory mediators. The dust absorbs NH_3 and possibly other toxic or irritating gases adding to the potential hazards of the inhaled particles (Do Pico 1986; Donham and Gustafson 1982; Donham and others 1982b; Sigurdarson and others 2004b). A recent study has shown that the mixed exposure to dust and ammonia in CAFOs has a synergistic toxic effect on the airways as measured by cross-shift pulmonary function decline in workers. Dust combined with ammonia results in 2–4 times the extent of cross-shift decline as compared to a single exposure of dust or ammonia (Donham and others 2002).

At least 40 different gases are generated in anaerobically degenerating manure. Hydrogen sulfide (H_2S), ammonia (NH_3), carbon dioxide (CO_2), methane (CH_4), and carbon monoxide (CO) are common hazardous gases present (Donham and others 1988, 1982a, b; Donham and Gustafson 1982; Donham and Popendorf 1985). These gases may escape into the work environment when pits are under the building. Furthermore, 40% of the ammonia measured in-building is released by bacterial action on urine and feces on the confinement house floors (Donham and Gustafson 1982). Carbon monoxide and CO_2 are generally not produced in hazardous concentrations from the manure pit. They are produced in much higher concentrations by fossil-fueled heating systems in winter, as well as by the animals' respiration (CO_2 only) (Donham and others 1982b). These latter gases are usually hazardous only when the heaters or ventilation system mal-

functions. Methane is not a respiratory hazard in these buildings. However, methane may be an occasional fire or explosive hazard in some buildings.

Who Is Exposed to These Dusts and Gases, and When?

The risk of chronic respiratory health effects are related to the length of time the person has worked in these buildings. Those who have worked more than 2 hours daily and for 6 or more years are at greatest risk (Donham and others 1977, 2000; Donham and Gustafson 1982). Owners and managers, hired hands, and family members of traditional family-owned CAFOs may work in the houses anywhere from a few hours a week to 8 or more hours daily. However, as livestock production has become more specialized, workers may spend 40 or more hours per week in the building.

During this time, workers are preparing feed, feeding animals, cleaning the buildings, sorting and moving animals from one pen or building to another, performing routine vaccinations and treatments, breeding sows, tending to birthing sows and little pigs, or other management and maintenance procedures. Women commonly work in these facilities, particularly in the farrowing operations. Farm family children may often be exposed because they may be helping out or may be accompanying parents due to lack of convenient childcare options. Veterinarians who provide services for these farm operations are exposed. Additionally, larger corporate-style farms employ service technicians who work in maintenance or animal health in these facilities.

Dust and gas concentrations increase in winter when the houses are tightly closed and ventilation rates reduced to conserve heat (Donham and others 1977). Dust concentrations increase when animals are being moved, handled, and fed, or when buildings are being cleaned by high-pressure spray washing or sweeping (Nilsson 1982). Ventilation systems are designed to control heat and humidity in the building and often will not reduce dust or gas levels adequately to insure a healthful environment for humans. During the cold seasons, should the ventilation systems fail for several hours, CO_2 from animal respiration, combined with CO_2 and CO from heaters and manure pits can rise to asphyxiating levels in a matter of hours. In the warm seasons, the greater risk to animals from ventilation failure is heat stress from high temperature and humidity. Although massive animal losses have been attributed to these latter situations, they would not create an acute human health threat, as they are not so acute as to prevent workers from leaving the building in a safe time.

Hydrogen sulfide may pose an acute hazard when the liquid manure slurry is agitated, an operation commonly performed to suspend solids so that pits can be emptied by pumping (Donham and others 1982b; Osbern and Crapo 1981). However, agitation may occur in many ways, such as draining the pits by gravity flow (an optional design in some buildings) or wash water running into the pits from above. During agitation, H_2S can be released rapidly, soaring from usual ambient levels of less than 5 ppm to lethal levels of over 500 ppm within seconds (Donham and others 1982b, 1988). Generally, the greater the agitation, the more rapid and greater amount of H_2S is released. Animals and workers have died or become seriously ill in swine CAFOs when H_2S has risen from agitated manure in pits under the building. Several workers have died when entering a pit during or soon after the emptying process to repair pumping equipment or clean out solids (Donham and others 1982b). Persons attempting to rescue these workers also have died. Workers may be exposed to high H_2S levels when they enter the pit to retrieve animals or tools that have fallen in, or to repair ventilation systems or cracks in the cement. Hydrogen sulfide exposure is most hazardous when the manure pits are located beneath the houses. However, an acutely toxic environment may result from outside storage facilities if gases backflow into a building, due to inadequate gas traps or other design fault, or if a worker enters a separate confined-space storage facility.

CAFOs in North America are concentrated in North Carolina in the East, most states of the Midwest, and the Far West including Oklahoma, Texas, Colorado, and Utah. CAFOs are also found in the Mideast and prairie provinces of Canada, Northern European countries, Australia, and Brazil. Poultry CAFOs (which include turkey, broiler, and egg production) are concentrated in the East-Central, Southeast, Midwest, and Far West of the U.S. Poultry confinements are also found in Europe, Australia, and Brazil. Other types of CAFOs (beef, dairy, veal) are not nearly as common as swine and poultry and are located in regions where principal feedstuff (corn and wheat) are grown.

How Commonly Does Excessive Exposure Occur?

In the United States, an estimated 700,000 persons work in livestock and poultry confinement operations (Donham 1990). This number includes owner-operators, hired farm workers, spouses, children, veterinarians, and service technicians. Included in the hired farm workers in the U.S. are minority populations, including Hispanic, Asian, and Bosnian, among others.

The largest group of CAFO-exposed workers with the most frequent and severe health problems is swine CAFO workers (Donham 1990; Donham and others 1989; Radon and others 2002). Here, typical dust concentrations are 2–6 milligrams per cubic meter (Donham and others 1985a). Buildings with 10–15 milligrams of dust per cubic meter may be seen during cold weather or when moving or sorting the pigs. Concentrations in this range are high enough to create an unclear view across a 50-foot room. Concentrations of dust, endotoxin as well as H_2S, CO_2, and CO may exceed safe levels. Furthermore, research has shown safe dust and gas concentrations in CAFOs are considerably lower than levels set for industrial standards. Table 3.2 compares recommended maximum exposure concentrations from current research to levels set by the Occupational Safety and Health Act (OSHA) and The American Conference of Governmental Industrial Hygienists (ACGIH). The more toxic nature of this dust, relative to others, is thought to occur because of the high degree of its biological activity, its inflammatory nature, and the additive and synergistic actions of the mixed dust and gas exposures. Nearly 60% of swine confinement workers who have worked for 6 or more years experience one or more respiratory symptoms (Clark and others 1983; Donham and others 1984a, 1989; Thedell and others 1980). Prevalence of respiratory symptoms among workers in nonconfinement swine workers is generally less than half of that reported by swine confinement workers (Donham 1990).

Respiratory Effects of Inhaling Confinement House Dusts and Gases

Human health effects of work in swine CAFOs was first described in veterinarians by Donham and others in 1977. Since that time, numerous studies by many different authors in various countries around the world have been published regarding the health of CAFO workers. Even with improvements in the engineering of these buildings over the subsequent 30 years, veterinarians as well as others still commonly experience the common complex of agricultural dust respiratory conditions (Andersen and others 2004). Inhalation of confinement house dusts and gases produces a complex set of respiratory responses, and is an example of a specific type of CLLC exposure to an agricultural (organic) dust (Donham and Zejda 1992; Mustajabegovic and others 2001; Thelin and others 1984). CAFO workers experience the same type of symptoms as grain handlers, including acute and chronic bronchitis, NALC, MMI, and MMS. An individual's specific response depends on characteristics of the inhaled bioaerosol (such as particulate size, endotoxin, ammonia, and total inhaled mass) and on the individual's susceptibility, which is moderated by coexisting factors (including atopic status, relative genetic sensitivity to endotoxin, length of exposure, and smoking history). Inflammation and direct epithelial damage are the primary pathological processes involved (Eckert 1997). However, toxic or allergic processes may also be involved, alone or in combination, resulting in one or perhaps several conditions simultaneously (Donham 1991; Donham and others 1989; Prior and others 2001). Since dusts include both inhalable (<10μ) and noninhalable-sized (>10–50μ) particles, lung tissues, lower airways, and upper airways may all be affected. Although endotoxin appears to be a primary hazardous substance in this dust, responses depend on additional agents in the bioaerosol (Heederik and others 1991). General inflammatory mechanisms involve recruitment of neutrophils into the airways and lung tissue, and stimulation of macrophages and epithelial cells to liberate a host of mediator substances that continue the inflammation cascade. However, the specific mechanisms involved often cannot be precisely defined, and conditions are best described symptomatically.

Table 3.9. Acute Symptoms of Swine Confinement Workers

Symptom	Prevalence
Cough	67%
Sputum or phlegm	56%
Scratchy throat	54%
Runny nose	45%
Burning or watering eyes	39%
Headaches	37%
Tightness of chest	36%
Shortness of breath	30%
Wheezing	27%
Muscle aches and pains	25%

(Donham and others 1977; Donham 1993).

The prevalence of acute symptoms of swine CAFO workers is listed in Table 3.9. Acute, delayed, and chronic conditions of swine confinement workers with associated history, diagnostic aids, and prognosis are described in the following paragraphs and outlined in Table 3.10. The most common respiratory symptoms (cough, sputum production, chest tightness, shortness of breath, wheezing) are manifestations of airways disease, composed of **bronchitis** (dry cough or cough with phlegm), which is often associated with **increased airways reactivity or NALC** (Donham and others 1989). Evidence suggests that those exposed become increasingly reactive to the confinement environment with increasing exposure (greater than 2 hours per day and 6 years' work experience) (Donham and Gustafson 1982; Donham and others 1989). In general, the symptoms are more frequent and severe among smokers (Donham and Gustafson 1982; Markowitz and others 1985; Marmion and others 1990) and in those working in larger swine operations (related to longer hours working inside CAFO buildings) or working in buildings with high levels of dusts and gases (Donham and others 1985, 2000; Reynolds and others 1996). Health effects are also greater among those with preexisting respiratory problems (hay fever, bronchitis) and among those with heart disease or allergies (Donham and Gustafson 1982). Chest tightness, coughing, nasal, and eye irritation symptoms have been experienced in some persons within 30 minutes of entering these houses for the first time. However, (especially in persons who have had prior expo-

Table 3.10. Occupational Respiratory Conditions Associated with Swine Livestock Confinement Diagnosis, Treatment, and Control

	Symptoms/History	Work Exposure	Diagnostic Aids	Treatment/Control	Prognosis
Bronchitis	• Cough, with sputum production, possibly tightness of chest. • Very frequently seen among swine confinement workers; somewhat less often in poultry workers. • Smoking associated with increased frequency and more severe symptoms. • Symptoms continue for 2 or more years classified as chronic bronchitis.	• Usually occurs in those who work in swine confinement for 2 or more hours per day. More frequent and severe for those who have worked 6 or more years in confinement. Generally occurs in buildings with poor environment: dusty (appears hazy and dust accumulates on horizontal surfaces), poor ventilation, often older building (built before 1985). Nursery buildings and those with manure pits under slatted floors may be biggest offenders. Usually worst during cold weather.	• Symptoms and history usually sufficient for diagnosis. • PFT may show decreased flow rates. • Skin tests or other immunological tests not indicated.	• Protection from environment most important action. • Medications usually not indicated. • Broncodilators and antiinflammatories may provide temporary relief of symptoms but should not be used long-term. • Improved ventilation crucial. • Employ management procedures to limit dust generation (i.e. frequent cleaning). • Install dust and gas control technology. • Establish a respirator program. • Abstain from smoking.	• Most improve if environmental exposure is controlled through engineering, management, or use of respirator. Cessation of smoking also crucial. • Temporary removal from the environment or use of a respiratory may help until other measures can be taken. • Progressive obstructive disease occurs. • Usually not necessary to quit working.
Non-allergic Asthma-like Condition (NALC)	• Chest tightness, mild dyspnea, some restriction and obstruction during breathing.	• Following a work exposure (usually at least a 2-hour exposure).	• PFT following a work shift shows decreased flow rates, primarily FEV_1 and FEV_{25-75}.	• Identical to bronchitis (above).	• Identical to bronchitis (above).

Table 3.10. Occupational Respiratory Conditions (*continued*)

	Symptoms/History	Work Exposure	Diagnostic Aids	Treatment/Control	Prognosis
	• Often accompanied by bronchitis. • Very common in exposed workers. • History similar to bronchitis, but often with a nonproductive cough.	• Identical to bronchitis (above).	• Respiratory challenge with methacholine or histamine shows decreased PFT flow rates.		• Same as for any asthmatic.
Allergic Occupational Asthma	• Wheezing within minutes (immediate asthma) or for up to 24 hours (delayed asthma) following exposure. • Only seen in small percentage of workers (less than 5%).	• Among atopics or those who already have asthma from another source, often occurs with first exposure. • With other workers, a period of sensitization is required, which may vary from a few months to several years. • Extent of exposure not as important (environment may be relatively clean, and a person may spend very small amount of time in building).	• Same as asthma from any other source: obstructive air flow patterns following exposure; skin test often positive to one or more of feed grains, hog dander, hog hair, various molds, dusts; associated with atopic status and increased airways reactivity. Reversible with bronchodilators.	• Medication and treatment same as for any asthmatic. • Attempts to control exposures by environmental control and respirators may or may not be helpful. • Desensitization usually not applicable because of multiple antigens and irritant gases.	• Depending on degree of sensitivity, may be almost impossible to protect these people from their environment. • This may be one condition for which patient must quit working in confinement house. • Increased airway reactivity and asthma may continue past employment.

(*continued*)

Table 3.10. Occupational Respiratory Conditions (*continued*)

	Symptoms/History	Work Exposure	Diagnostic Aids	Treatment/Control	Prognosis
Organic Dust Toxic Syndrome (ODTS)	• Fever, muscle aches, chest tightness, cough, malaise. • Symptoms develop 4–6 hours following exposure. • Self-limited symptoms usually resolve in 24–72 hours. • Recurrent episodes common. • Seen in 25–35% of the swine farming population.	• Usually condition associated with work in a totally enclosed building. • Usually occurs following working in situations involving heavy or high dust exposure or concentrations (e.g., 4–6 hours of very dusty work, such as handling or sorting hogs, loading crates of poultry, power-washing building (interior).	• Elevated white blood cell count, usually neutrophilia. • PFT will show decreased FEV_1 and diminished flow rates. • $pO2$ may be decreased. • Bronchoalveolar lavage usually shows PMN response. • X-ray may show scattered patchy infiltrates. • Lung biopsy may show inflammatory polymorphonuclear cell infiltrates.	• Symptomatic treatment in acute stages may include oxygen, IV fluids to correct acid-base in balance and dehydration. • Steroidal and/or non-steroidal antiinflammatories may be used to control fever and myalgia. • Most cases do not seek medical attention; often confused with influenza or Farmer's lung.	• Usual recovery period is 3–4 days, but patient may feel tired and have shortness of breath for several weeks. • Subsequent attacks may occur in future following heavy exposure.
H_2S Intoxication	• Sudden and immediate onset of nausea, dizziness, possibly sudden collapse, respiratory distress, apnea. • May lead to sudden death or patient may recover if removed from environment, often with dyspnea,	• Almost always occurs with agitation of a liquid manure pit while emptying it. • Respiratory effects will occur within seconds of encountering high concentration of H_2S.	• If patient survives: –X-ray often shows pulmonary edema. –possibly presence of sulfhemoglobin and sulfide in blood. • If deceased: –autopsy shows pulmonary edema, froth in trachea,	• Avoidance. • Remove exposed person from environment (without exposing others) and resuscitate. May have to ventilate. • Seek medical care, watch for and control pulmonary edema.	• If patient survives initial exposure, will probably recover usually with minimal loss of lung function. • Recovery period may be from days to 2–3 years, depending on severity of exposure.

Table 3.10. Occupational Respiratory Conditions (*continued*)

	Symptoms/History	Work Exposure	Diagnostic Aids	Treatment/Control	Prognosis
	hemoptysis, and pulmonary edema, following intensive exposure.		possibly greenish tinge to *viscera*. –blood contains sulfide and sulfhemoglobin.		
Mucous Membrane Irritation (MMI) and Sinusitis	• Conjunctivitis, rhinitis, pharyngitis. • Stuffy nose, constant cold, headache, ears popping.	• Identical to bronchitis and NALC above.	• Symptoms and history sufficient.	• See bronchitis (above).	• Good, depending on environmental control and appropriate respirator use.

sure) 2 or more hours exposure are required to develop symptoms. These symptoms can diminish 24–72 hours after leaving the unit, although they can persist for several days or weeks or even months among workers exposed for several years. In some cases, symptoms may be continuous (even if the affected individual leaves the facility), particularly if chronic bronchitis and/or reactive airways ensue. Only a small percentage (less than 2%) of these cases are thought to be specific allergic-mediated (IgE) illnesses such as **asthma** in atopic individuals (classical type 1 reactions), while the remaining proportion appears to be chronic inflammatory reactions (Bar-Sela and others 1984; Pederson and others 1990; Prior and others 2001; Rylander 1987).

ODTS does occur in CAFO workers. Cases may occur in workers who have had 4–6 hours of a PAMM exposure such as handling, moving, sorting or feeding animals, or washing with a high-pressure sprayer. Symptoms are the same as ODTS caused by other PAMM organic dust exposures. Symptoms are delayed 2–6 hours following exposure and include fever, malaise, muscle aches and pains, headache, cough, and tightness of chest. This episodic problem is experienced occasionally by about 30% of workers in confinement buildings (Do Pico 1986; Donham and others 1990). CAFO ODTS is thought to be caused mainly by inhaled endotoxin from aerosolized gram-negative bacteria as is the cause in other organic dust exposures (Rylander 1987).

Symptoms of irritation of the eyes, nose, sinuses, and pharynx (MMI) are common. This is thought to be an inflammatory process by virtue of the inflammatory substances contacting the mucous membranes, similar to MMI described with other organic dust exposures. However, swine CAFO MMI appears to involve the sinuses more frequently than other organic dust exposures. Swine CAFO patients commonly complain of a "persistent cold." Symptoms include stuffy nose, headache, and "popping ears." These patients typically have **chronic noninfectious sinusitis**, produced by long-term inhalation of inflammatory aerosols within the CAFOs.

Chronic lower airways effects manifest as **chronic bronchitis** with or without **obstruction**, and are experienced by 25% of all swine CAFO workers. This is the most commonly defined health problem of this occupational group, and is

suffered two to three times more frequently compared to farmers who work in conventional swine housing units or in agricultural operations other than swine or poultry production (Donham 1990). Symptoms include chronic cough, with excess production of phlegm and sometimes chronic wheezing and chest tightness. Smokers experience a greater prevalence and severity of chronic bronchitis than do nonsmokers. Most workers removed from the confinement house environment become asymptomatic (in the absence of smoking) within a few months, but bronchitic symptoms in some workers can persist for years.

Although irreversible airways obstruction has not been a general finding in confinement house workers, there is objective evidence that long-term lung damage may be occurring. Pulmonary function studies show evidence of air trapping in the lungs. Lavage studies of bronchial fluids show a persistent leukocytosis, and sputum studies show persistent inflammatory cells and epithelial cells (Djuricic and others 2001; Schwartz and others 1990). Baseline pulmonary function studies (FVC, FEV_1) of healthy confinement workers usually do not differ significantly from those of workers in conventional swine buildings (Donham and others 1989; Donham and others 1990). However, flow rates at 25–75% of lung volume (FEF_{25-75}) are significantly lower. Furthermore, workshift declines in FEV_1 and flow rate values are seen in most confinement house workers following 2–4 hours of exposure (Donham and others 1984b). These work shift declines are predictive of future baseline declines (Schwartz and others 1995). Another longitudinal study has shown a decline in PFT with increasing evidence of obstruction over the years in a cohort of CAFO workers (Eckert 1997). These findings suggest that **chronic obstructive pulmonary disease** may occur among these workers in future years (Schwartz and others 1995). Although end-stage lung damage in CAFO workers has not yet been systematically studied, the authors have experienced many anecdotal case studies where workers have quit because of health reasons. One study of owner/operators revealed a dropout rate of 10% over a 6-year period for respiratory health reasons (Holness and others 1987). Experimental animal studies have shown that long-term CAFO exposures create a risk for pneumonia, pleuritis, and bronchitis (Donham 1991; Donham and Leininger 1984).

Although dust exposure is the most common hazardous exposure in CAFOs, the most dramatic acute response results from exposure to hydrogen sulfide (H_2S). At moderately high concentrations (100–400 ppm), the irritating properties of H_2S produce rhinitis, cough, dyspnea, tracheobronchitis, and possibly pulmonary edema; at higher concentrations (400–1500 ppm), H_2S will cause sudden collapse, respiratory paralysis, pulmonary edema, and death. In addition to its irritant properties, H_2S is a general cellular toxin which works by disrupting the cellular metabolic system and has a predilection for the central nervous system. At least 25 deaths of confinement workers in the U.S. have been reported from this exposure up through the early 1990s (Donham and Gustafson 1982). Often multiple deaths occur during exposure events, as would-be rescuers become victims.

Although respiratory exposures are extremely common among CAFO workers, there are several other occupational hazards that should be considered. There are certain infectious agents involving the respiratory tract that humans may contract from animals in CAFOs. These include, but are not limited to, swine influenza, ornithosis, and Q fever (Holness and others 1987). These diseases are covered in detail in Section 7 of this chapter. Injuries to CAFO workers from animals, pinch points in gates and pens, cuts, and needle sticks are common. Furthermore, high noise levels in these facilities can lead to noise-induced hearing loss. All of these hazards are discussed in subsequent chapters of this text.

Diagnosis

Use of diagnostic aids to identify these conditions is of secondary importance to a detailed occupational and clinical history. It is important to recognize that a patient's response to confinement dusts and gases is variable, and that one or more conditions may be occurring simultaneously (e.g., chronic bronchitis, occupational asthma, and sinusitis). Question a patient in detail about chief complaints, including questions on how long symptoms have been present and the time relationship of symptoms to work exposure. Work exposure for more than 2 hours per day and 6 or more years of total exposure are related to increased frequency and severity of symptoms. Patients may complain of having a continual cold or feeling tired most of the time, or say they think

they are allergic to the dust (but are negative atopic status). Improvement in symptoms over a vacation period with greater than normal symptoms on return to work (Monday morning syndrome or MMS) is an indicator of a work-related condition. Question the patient on the specific jobs he/she does and on the environmental air quality in the building. Moving and sorting animals and power washing the building inside are high-exposure tasks leading to increased exposure and possible episodes of ODTS. Assess the patient medically for proper selection, fit, and use of respirators. Take an in-depth personal and family medical history, including questions on allergies, asthma, heart conditions, and hobbies or personal habits (such as smoking) that might complicate the work exposures. Ask how many hours per day or week the patient works in confinement buildings, and how long the patient has held this job. Does the patient recognize the air quality in the building as particularly bad? Have environmental assessments been conducted? If so, maximum concentrations should be under the current research recommendations seen in Table 3.2. Pulmonary function tests (PFT) may be useful. Lowered volume and flow rates (FEV_1 and FEF) over the work period of 5–30% are common in people having symptoms. Less commonly, decreases of 5% or more in volumes (FVC) over the work period may be seen. However, baseline PFT values may be normal. A decreased tolerance to methacholine challenge is common. Dermal prick tests for suspected feed or swine allergens are usually negative.

Without a proper environmental history, the health care provider may fail to relate the patient's symptoms to a CAFO atmosphere, resulting in misdiagnosis and treatment of CAFO-related respiratory conditions as allergic responses.

Table 3.10 summarizes the primary respiratory conditions associated with swine confinement dusts and gases. Conditions provoked within other types (animal species) of confinement buildings may be similar, but typically less severe and less common.

Treatment

Medically, little can be prescribed to cure respiratory conditions of CAFO workers, excluding treatment of some of the acute illnesses (ODTS, asthma, pulmonary edema from H_2S intoxica-

Table 3.11. Comparison of OSHA and ACGIH TLVs to Recommended Exposure Limits to Toxic Dusts and Gases Based on Current Research

Toxic Substance	Current Research Recommendations for CAFOs	Typical Findings In CAFOs	ACGIH[1]	OSHA[2]
Total Dust	2.5 mg/m^3	3–6 mg/m^3	4 mg/m^3	15 mg/m^3
Respirable Dust	0.23 mg/m^3	0.5–1.5 mg/m^3	—	—
Ammonia	7 ppm	5–15 ppm	25 ppm	25 ppm
Hydrogen Sulfide	—	0.5–5 ppm	10 ppm	10 ppm
Carbon Dioxide	1,500 ppm	1,000–4,000 ppm	5,000 ppm	5,000 ppm
Endotoxin	100 EU	50–1,000 EU	—	—

[1](ACGIH 1985).

[2]Occupational Safety and Health Association, Permissible Exposure Limits, http://www.osha.gov/pls/oshaweb/owadisp.show_document?p_table=STANDARDS&p_id=9992.

tion). Bronchitis and NALC may respond temporarily to enteral or inhalant-administered bronchodilators and/or corticosteroids. However, these treatments address symptoms, not the underlying causes of the problem. Details of these treatments, specific control measures, and the prognosis for these illnesses are listed in Table 3.10.

Medical treatment must be accompanied by reducing exposures to dust and gas by management and engineering controls, appropriate use (selection and fitting) of respirators, and/or temporary removal from the work site. In almost all cases, with appropriate use of these modalities, workers can return and be kept working safely on the job. In order to reduce dusts and gases, a patient may need to contact a consulting veterinarian or agricultural engineer who has knowledge of environmental control. The local veterinarian or the Cooperative Extension Service agricultural engineer should be able to recommend an appropriate expert. Monitoring air quality in these buildings is essential to assurance of a healthful work environment. Minimum assessment includes ammonia and total dust (mass) two times yearly, one of which should be in cold weather conditions. Contaminant concentration should be below those levels listed in Table 3.11 under current research findings.

Regarding CAFO workers who smoke, the health care provider must direct these patients to smoking cessation. Health care providers must be cognizant of the patient's emotional wellness in addition to the patient's physical problems. A rec-

ommendation to quit raising animals is difficult for many farmers as there may be no other reasonable occupational choice other than to continue working in the CAFO. Such a recommendation may produce extreme mental stress since quitting farming is leaving a life-style as well as a job (especially in family owner-operator businesses). Recommendations to leave farming are often unnecessary and should be given only after the cause and prognosis of illness have been determined and other avenues of controlling harmful exposures have been fully explored. Some guidelines the health care provider might want to consider in a recommendation include the following:

1. If the patient has severe symptoms, a 2–3–week "vacation" from the work environment might be indicated. Reducing work periods to 2 hours or less per day may help. In the interim, the patient can be fitted with an appropriate respirator.
2. A temporary course of inhaled steroid and/or bronchodilators may be tried.
3. Arrangements can be made to assess the work environment and establish controls for dust to less than 2.5 mg/m^3 and ammonia to 7 ppm or less (see Section 8 for specific recommendations).
4. Monitor the patient's symptoms and pulmonary function at least annually.

Farmers are becoming increasingly aware of confinement-associated respiratory conditions.

A health care provider can explain potential long-term respiratory conditions but also instill confidence regarding maintenance of the farmer's health status and assist in protecting the farmer from health problems of the work environment. Annually monitoring the patient's respiratory status may be reassuring to many patients and may encourage behavior changes in the patient to institute environmental control measures and comply with proper selection and use of respirators.

Prevention

Health hazards associated with confinement houses must be addressed through improvements in the environment through 1) decreased generation of dusts and gases by improved management procedures or engineering controls; 2) removal of contaminants once in the air, e.g., ventilation; and 3) proper protection of the individual with respirator use. A prevention model for confinement house problems based on education and industrial hygiene consultation has demonstrated its effectiveness (Donham and others 1990). Some examples of management practices to reduce the sources of dusts and gases include 1) delivering feed by extension spouts into covered feeders, rather than letting feed fall freely several feet from automatic delivery systems into open feeders; 2) using extra fat or oil in the feed reduces dust; 3) sprinkling or misting the environment with vegetable-based oil and regularly (every 3–4 weeks) washing buildings with power sprayers (operators must use respiratory protection during this procedure) to keep them as clean as possible; 4) using flooring which is more self-cleaning (e.g., plastic-coated wire mesh); and 5) assuring that heating units are clean, vented, and functioning properly. Details of control measures are published elsewhere (Donham 1991). Effectiveness of control techniques can be assessed by measuring dust and gas concentrations. These buildings should be routinely monitored (previously mentioned) to assure air contaminants are within healthful limits.

Because it is economically impossible to completely eliminate the formation of dusts and gases in CAFOs, techniques for removing contaminants from the air of confinement houses are critically important. Ventilation will help reduce dusts and gases to healthful levels. Ventilation systems must be properly designed and maintained, and ventilation rates adjusted to include consideration of air quality. Operators often keep these rates low in winter because of concerns for conserving heat, causing dust and gas concentrations to rise. A number of engineering techniques (e.g., use of heat exchangers which allow increased ventilation while capturing some waste heat) have been tried with varying degrees of success (Donham 1993).

Anyone working in a swine or poultry confinement house should be advised, at a minimum, to wear a NIOSH-approved two-strapped dust mask, even if the concentrations of dusts and gases are below recommended limits. Persons exposed to houses with high dust or gas concentrations, or persons with respiratory conditions, may need to use a more sophisticated respirator, such as a half-mask cartridge respirator or powered air-supplying respirator (e.g., air helmet).

Preventing exposure to high concentrations of H_2S from manure pits requires stringent controls. General safety measures include constructing manure pits outside of the confinement building, constructing openings so that lids or other objects cannot fall into the pit (requiring a worker to enter the pit for retrieval), and erecting safety guards and warning signs around open pits. Whenever a pit that is under a confinement house is being agitated, people should stay out of the building, ventilation of the house should be maximized, and animals should be removed or observed from outside the building.

Even when not being agitated, manure pits can seldom be entered safely. If entrance is imperative, adequate protection is assured only when a self-contained breathing apparatus is worn by an individual trained in its use. All operators should understand that high concentrations of H_2S cannot be smelled and that H_2S above 1000 ppm produces unconsciousness and respiratory arrest in only one to three breaths. A variety of H_2S gas alarms give an accurate indication of hazard.

Poor air quality in the confinement house has also been shown to be associated with health problems and lowered productivity in the swine (Donham 1991). Advising this economic fact to a swine producer may be the most expedient way to create environmental improvement that would help the person, the animals, and the economics of the operation.

SECTION 5: OXIDES OF NITROGEN ("SILO GAS")

Introduction

Oxides of nitrogen (NO_x or silo gas) are produced inside non-airtight silos and silage pits or bunkers by microbial action on nitrates in stored plant materials. NO_x gases are strong irritants, and if inhaled, can cause a range of injury to the respiratory system (**silo filler's disease**), including bronchial irritation, pulmonary edema to bronchiolitis obliterans (delayed), or sudden death. If the exposed individual survives the acute episode, he or she may develop advanced diffuse pulmonary fibrosis. The oxides causing silo filler's disease begin to form shortly after (within hours) the silage is stored in the silo or bunker and may persist in the silo for up to 2 weeks. Workers may be exposed when they enter a confined space where the gas collects, such as when entering a silo, the silo chute, or an adjacent feed room where the silo chute opens. In cases where silage is stored in pits or silage bags, there is a low risk for occupational exposure hazards as these are not confined spaces. Furthermore, silage bags limit the amount of air exposed to the stored silage and therefore, NO_x formation should be minimal. Silo gas is mainly a health hazard associated with non-airtight silos, as oxygen is required to form the NO_x from the nitrogen in the plant material.

Most commonly, workers inhale low concentrations of NO_x and develop delayed minor transient respiratory symptoms. However, high concentrations of NO_x can kill within minutes. Workers may inhale moderate concentrations for extended time periods without developing immediate symptoms or detecting danger. However severe reactions such as **pulmonary edema** can be delayed up to 30 hours. Another type of condition (bronchiolitis obliterans) can occur 2–6 weeks following exposure. Determination of exposure to oxides of nitrogen followed by proper treatment and monitoring of exposed patients are imperative to prevent death or serious complications. Preventing exposure to silo gas is paramount to the prevention of this illness.

Oxides of Nitrogen on the Farm

Silage is chopped corn, alfalfa, oats, grass, or other plant material that is stored in bulk containers where it undergoes a fermentation process that preserves it. Silage commonly constitutes the major portion of the feed ration for ruminant livestock, including beef and dairy cattle; alfalfa silage may also be fed to sheep. Oxides of nitrogen are produced as a by-product of the fermentation process (Fleetham and others 1978; Horvath and others 1978) through microbial oxidation of nitrogen compounds in green chopped plant material. Although silo gas may form in other storage structures such as silage pits, it is primarily a health hazard in the conventional non-airtight silo (see Figure 3.2) where the gas can accumulate to toxic concentrations in the confined space.

Production of silo gas commences within 4 hours after silo filling has begun.

Four oxides of nitrogen (collectively referred to as NO_x) may be found in silo gas on the farm; nitric oxide (NO), nitrogen dioxide (NO_2), nitrogen tetroxide (N_2O_4), and nitrous acid (HNO_2). Of these compounds, **NO_2 and N_2O_4 are medically significant due to their strong irritant properties** (Fleetham and others 1978).

Concentrations of NO_x reach a maximum in 1–3 days, after which production continues at a decreasing rate for a week to 10 days (Douglas and others 1989). Both NO_x and carbon dioxide, which are produced simultaneously, lie at or near the silage surface and in depressions in the silage, where they replace oxygen. Because NO_x **gases are heavier than air,** they will flow down the silo chute like water when a chute door, at the level of the silage, has been left open. NO_x may concentrate at the base of a silo, in an adjoining feed room, or within the chute.

Higher NO_x production is associated with certain crops (corn), crops fertilized heavily with nitrates, and certain weather and growing conditions (prolonged drought with rain just before ensiling, cloudy weather, damage to leaves or roots, harvesting after a frost) (Douglas and others 1989; Fleetham and others 1978; Horvath and others 1978).

When Are Farmers Exposed to NO_x?

Exposure to NO_x may occur during the first 2 weeks after a silo has been filled (typically late summer or early fall). (Haylage and oatlage may be harvested in summer, corn silage in late summer or early autumn.) Examples of areas of gas accumulation include a closed silo, in the chute

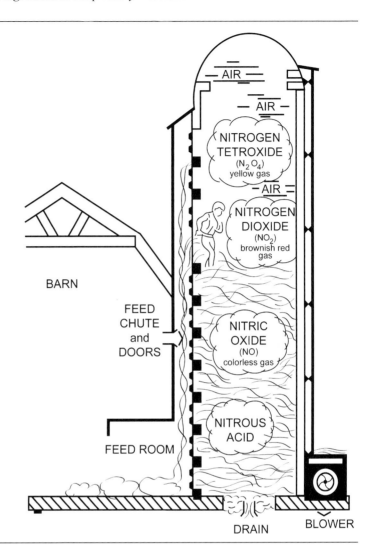

Figure 3.9. This diagram shows the formation of nitrogen oxides in a non-airtight silo, shortly after it is filled. (Source: United States National Library of Medicine, National Institutes of Health, http://www.nlm.nih.gov/)

(when the worker opens a chute door and a cloud of NO_x rolls past), at the base of a silo chute (when a silo door at the level of the silage has been left open), or in the feed room or barn next to the silo (when gases have flowed through open doors into these structures) (Figure 3.9) (Centers for Disease Control 1982; Douglas and others 1989). Typically, a worker climbs the chute and enters the silo to level silage to enhance storage capacity, and to cap the silage by putting a plastic sheet over the material and then adding an extra layer of 6–12 inches of silage on top of the plastic. This is done to prevent excess spoilage. Additionally, workers may enter a silo to feed silage out to animals, to prepare a silo unloading machine for use, or check on the level of silage.

Either task is potentially fatal when attempted soon after filling or when a silo has not been ventilated adequately.

Because NO_x is not highly irritating to the upper respiratory tract, farm workers may work in atmospheres with low to moderate NO_x concentrations for hours, without detecting danger. However these gases can be highly damaging to the alveoli. The high carbon dioxide and low oxygen atmosphere in the environment induces deep breathing, which speeds penetration of NO_x into the alveoli where the gas causes damage. When NO_x concentrations are high, a worker may become too weak to get out of the silo without assistance. The danger is increased when movement of a worker releases gases trapped in the silage, or when a worker enters or falls into a cavity in the silage. Gases concentrated in adjacent buildings or at the base of the silo may be inhaled by by-standers, children at play, or livestock.

The danger from NO_x is greatest **during the 2 weeks following filling, which is in summer or early fall.**

How Is NO_x Detected?

Nitrogen dioxide can be detected by its bleach-like odor and yellow to reddish-brown color that may be detected as a cloud, or by staining of nearby objects (the silage surface, silo wall, base of the chute or other structures). Other oxides of nitrogen may be less readily detected. Dead birds or insects may lie at the base of the chute or near the silo. Livestock near the silo may be sick. A health care provider establishing a diagnosis should question a patient about the presence of any of these warning signs.

Farm workers should understand that NO_x may be present in any recently filled conventional **non-airtight silo**, even if not smelled or seen.

How Common Are Farmer Exposures to NO_x?

Although medically diagnosed cases are not commonly reported, the true scope of exposure to NO_x is thought to be underestimated. A study in New York State estimated an annual incidence rate of 5/100,000 (Zwemer and others 1992). An unpublished survey in the late 1960s revealed that 4.2% of Wisconsin operators had developed symptoms of NO_x inhalation when working in or near recently filled silos (Horvath and others 1978; Maurer 1985). The severity of the hazard rests partially in the high case fatality rate. A total of 29% of the cases cited in medical literature have been fatal (Horvath and others 1978).

Respiratory Effects of Oxides of Nitrogen

Reactions to NO_x depend on the concentration of gas inhaled and the length of exposure. Nitrite formation in the alveoli from NO_2 inhalation may lead to methemoglobinemia, which has been seen in over 40% of cases in one report (Fleetham and others 1978). However, the primary health effect is one of direct tissue damage. Relatively mild exposure to NO_x produces ocular irritation and transient respiratory tract symptoms manifested as cough, possibly with dyspnea, and fatigue. Generalized symptoms may also be present, including

nausea, cyanosis, vomiting, vertigo, or somnolence. Symptoms may be severe enough to induce workers to leave the silo. However, when reactions to NO_x are minimal, workers may stay in the silo and increase their probability of a more severe reaction. Mild exposures may result in symptoms that persist 1–2 weeks. Chest films, pulmonary function tests, and blood gases may be normal in these instances and recovery is usually complete. Mild responses are more common than the more serious reactions described below.

Very high concentrations of NO_x induce immediate distress, resulting in collapse and death within minutes. The mechanism of this reaction is not completely understood. Death may be due to airway spasm or laryngospasm, reflex respiratory arrest, or simple asphyxiation due to low ambient oxygen concentrations (Douglas and others 1989; Epler 1989; Horvath and others 1978). Persons who collapse in silos and are rescued immediately may survive only to experience one of two possible delayed respiratory responses described in the following paragraphs.

At **less than acutely toxic** concentrations, NO_x may induce delayed **pulmonary edema (normally within 30 hours following exposure), bronchiolitis obliterans (within days to weeks), or both.** At the time of exposure, patients may have no or minimal symptoms. However, a slowly evolving and progressive inflammation of the lungs results in massive pulmonary edema most commonly from 6–12 hours later. Death from fulminant pulmonary edema may occur within hours. However, when patients receive appropriate care in time, they often recover within days or weeks.

In a small percentage of cases, recovery from this first phase of illness may appear to be complete, including clearing of chest films, only to be followed, 2–4 weeks later, by a **relapse characterized by bronchiolitis obliterans.** This fibrocellular obliteration of the bronchioles may also be the initial clinical manifestation. Relapse may lead to death or to slow recovery over a period of weeks or months, with varying degrees of diffuse pulmonary fibrosis and small airways obstruction (Stepanek and others 1998).

Severely exposed individuals may experience persistent pulmonary dysfunction of variable severity due to a diffuse pulmonary fibrosis.

The reactions to NO_x result from its hydrolysis

in the lungs to nitrous and nitric acid when the gases dissolve in mucosal or alveolar fluids. The delayed and insidious effects of NO_x reflect their relatively low solubility and result in slower conversion to acid. These factors also explain the gases' major involvement with lower rather than upper airways. The acid affects peripheral airways, bronchiole and alveoli, causing extensive tissue damage and formation of methemoglobin. Persons dying from fulminant pulmonary edema display intraalveolar edema and exudation and thickening of alveolar walls with lymphocytic cellular infiltrates.

Diagnosis

Patients in the acute stages of silo filler's disease may present with moderate to severe respiratory distress, hypotension, and severe hemoconcentration. Leukocytosis, methemoglobinemia, and metabolic acidosis may also be present, (Douglas and others 1989; Epler 1989; Fleetham and others 1978; Horvath and others 1978). Pulmonary function tests show a reduced vital capacity, increased airways resistance, and impaired CO_2 diffusion.

Because the initial illness may be mild, patients may present to a health care provider for the first time during a relapse, 2–6 weeks after exposure to NO_x. At this time, cough, tachypnea, dyspnea, fever, tachycardia, cyanosis, or other symptoms of respiratory distress are due to bronchiolitis obliterans. Small, discrete nodules, with or without confluence, will be evident on the chest radiograph.

Silo filler's disease may be confused with a number of illnesses including hypersensitivity pneumonitis or organic dust toxic syndrome (ODTS, see Section 2). However the exposure circumstances for the latter two conditions are entirely different.

The chest radiograph of bronchiolitis obliterans may resemble miliary TB among other diseases. An accurate occupational history and negative sputum smears for acid-fast bacillus will help avoid confusion. (However, it should be noted that military TB patients are often sputum negative.) A detailed medical and occupational history is crucial to correct diagnosis. In addition to noting exposure to a recently filled silo, most commonly in late summer or early fall, a patient may recall seeing signs of NO_x (i.e., yellow-

brown stains, dead birds) near the silo or experiencing the transient symptoms described previously. Since exposure to silo gas may have occurred from hours to weeks prior to onset of severe respiratory symptoms, patients may not associate their symptoms with exposure to silo gas.

Prompt diagnosis and treatment of patients with acute symptoms is vital to prevent possible death and, in the case of initial illness, to lessen the probability of relapse.

Treatment

Any symptomatic patient who has been exposed to NO_x **should be monitored closely by a health care provider for 48 hours because of the possibility for sudden development of acute pulmonary edema and acute respiratory distress, as in acute respiratory distress syndrome** (Douglas and others 1989; Horvath and others 1978; Stepanek and others 1998). Typically, these exposed patients should be hospitalized. In certain cases, patients could remain at home but should be warned to report immediately to a health care provider upon development of respiratory distress. Persons developing pulmonary edema or respiratory embarrassment must be placed on steroids (20–120 mg Prednisone/day, tapered over time) for 6–8 weeks, to decrease the probability of bronchiolitis obliterans (Epler 1989; Fleetham and others 1978; Muller-Wening and Repp 1989). Persons presenting for the first time with bronchiolitis obliterans also should receive steroid treatment.

Patients may require intensive supportive therapy, including oxygen, bronchodilators or assisted ventilation. Blood should be checked for methemoglobin and treated with methylene blue if indicated (Douglas and others 1989; Horvath and others 1978). Antibiotics may be required for secondary respiratory infections.

Any symptomatic patient should be monitored for 6 weeks following an exposure for possible development of bronchiolitis obliterans.

Prevention of Silo Filler's Disease

Farmers must thoroughly understand the hazards associated with recently filled silos. Once filled, no one should enter the silo **for at least 2 weeks**. If entry is imperative during filling, the blower (silo filler) should be run for 30 minutes prior to entering the silo and kept running while anyone is inside. All silo doors down to the level of the silage should be opened, and all roof sections should be in place to assure proper ventilation. Doors between the silo room and barn should be kept closed. Children and animals should be kept away from the silo and adjacent feed room during filling and for 2 weeks afterward. A few days before the silo is entered for the first time, the filler opening should be opened from the ground (not from the chute) with a rope. The blower should be operated for at least one-half hour prior to entrance in the manner described above. Detector tubes (also called *colorimetric tubes*) that measure the concentration of NO_x are reasonably priced and reliable if used properly. These tubes could be placed at the bottom of the silo chute. Detailed procedures for measuring toxic gases are included in Section 9 of this Chapter.

Because of small silo door openings and the deadly nature of NO_x, rescue of people down inside a silo is extremely difficult and hazardous. Unless testing has been done to assure the absence of toxic levels of NO_x, safe entrance into the silo within 2 weeks of filling can be accomplished safely only with a self-contained breathing apparatus.

If the slightest throat irritation or coughing occurs while working with fresh silage, a worker should exit the silo immediately. In any case of symptomatic exposure to silo gas, the worker should report to a health care provider immediately and be monitored for either immediate or delayed reactions.

SECTION 6: APPLIED AGRICULTURAL CHEMICALS

Contact with certain pesticides and the chemical fertilizer anhydrous ammonia can have adverse respiratory affects to exposed agricultural workers. The pulmonary effects of pesticides will be briefly reviewed here as they are covered in more depth in Chapter 6, "Health Effects of Agricultural Pesticides." Certain fumigants, cholinesterase-inhibiting insecticides, and the herbicide paraquat, can threaten life through their acute effects on the respiratory tract. Fumigants, commonly used in grain storage facilities, agricultural commodity storage and transport, and as soil sterilant treatment, can cause laryngeal edema, bronchospasm, and/or pulmonary edema. Cholinesterase-inhibiting insecticides, the pre-

dominate class of agricultural insecticides used today, not only can cause systemic poisoning, but they also cause respiratory depression, bronchoconstriction, and bronchorrhea. Paraquat, a broad-spectrum herbicide, can (particularly when ingested) lead to a delayed malignant proliferation of connective tissue (fibrosis) in the lungs and death through asphyxiation. The respiratory health effects of these chemicals are summarized in Table 3.12. Fumigants are biocides of low molecular weight, which have enormous powers of diffusion and penetration. Fumigants are highly irritating and toxic chemicals. This chemically diverse group of pesticides is classified in Table 3.12. The current major agricultural fumigants include **phosphine**, **methyl bromide**, **vikane**, **sulfur dioxide**, **acrolein**, and **chloropicrin**. Prompt and specific diagnosis and medical management is crucial if victims are to survive exposures to the previously described agents.

Although the fertilizer anhydrous ammonia is often a cause of skin or eye burns, it may cause direct injury to the respiratory tract. Anhydrous ammonia is injected into soil in many regions in spring or fall. It is the most commonly used nitrogen fertilizer today in many industrialized countries. The primary crop for its use is corn. Because it is stored in liquid form, it has a very low temperature ($-40°F$) and is under very high pressure. It has an extraordinary affinity for water, and a high pH with a strong corrosive power. Anhydrous ammonia is capable of causing severe tissue damage. When it is inhaled (because of its hydroscopic nature), the ammonia dissolves in the mucous fluids of upper airways, and becomes a severe irritant that may induce laryngospasm. Tissue damage from light to moderate exposures usually is limited to the upper airways. However, massive exposure forcing inhalation deep into the lungs produces severe inflammation at all levels of the respiratory tract. Pulmonary edema and/or bronchiectasis may follow—either acutely or within 48 hours. A large ammonia release in a confined space could produce fatal results or produce severe disabling airway disease. Chronic sequelae of massive ammonia exposure can cause chronic bronchitis, obstructive pulmonary disease, bronchial hyperactivity, or bronchiolitis obliterans. However, anhydrous ammonia is a more common cause of skin and eye damage in a farm setting.

Table 3.12. Respiratory Effects of Fumigants, Cholinesterase Inhibitors, and Paraquat

Pesticide Type	Representative Commercial Products	Use	Possible Effects on the Respiratory Tract

Fumigants	Halocarbons:
	  Methyl bromide
	  Vikane
	  Chloropicrin
	  Carbon tetrachloride
	  Chloroform
	Sulfur and Phosphorus
	  Phosphine
	  Carbon disulfide
	  Sulfur dioxide
	Oxides and Aldehydes
	  Ethylene oxide
	  Acrolein
		Used to kill insects, microorganisms, weeds, rodents in grain, agriculture, industry, homes	• Irritation
			• Laryngeal edema
			• Bronchospasm
			• Pulmonary edema
			• Respiratory depression

Cholinesterase Inhibitors (Organophosphates and Carbamates)	Parathion (Alleron®, Paramar®, Phoskil®), phorate (Agrimet®, Thimet®), fonofos (Dyfonate®); aldicarb (Temik®), carbofuran (Furadan®)	Insecticides, with widespread agricultural, industrial, home use	• Bronchoconstriction • Bronchorrhea • Pulmonary edema • Respiratory depression and death (due to inactivation of acetylcholinesterase)
Paraquat	Paraquat (Crisquat, Esgram; mixtures: Weedol®, Gramonol®, Pathclear®, Gramoxone®)	Herbicide, defoliant, widely used in agriculture and elsewhere	• Upper respiratory tract irritation • Following ingestion, pulmonary fibrosis leading to death

Anhydrous ammonia also causes topical injury by freezing and desiccating tissues and inducing severe alkali burns. If sprayed into the eyes, it penetrates the cornea and elevates pH to damaging levels in a few seconds, causing full or partial blindness. Eye exposure demands immediate and extended flushing with water; non-vented goggles should always be worn whenever working around anhydrous ammonia.

Anhydrous ammonia is used as a refrigerant in large cooling facilities, for example, in meat and poultry processing plants. Many of these plants are located in small towns in rural areas. Refrigerant leaks from these systems have resulted in severe occupational injuries and deaths.

Many small rural communities often have anhydrous storage facilities. Damaged or malfunctioning equipment at these sites has resulted in massive community exposures.

In recent years, the illegal manufacturing of methamphetamine has caused a new group of persons to be exposed. Anhydrous ammonia is an important component to illicit methamphetamine manufacturing. Farmers' or dealers' supply tanks of anhydrous have been targets of anhydrous theft. The thieves are at risk of getting exposed to anhydrous, causing eye, skin, or pulmonary damage.

Prevention of Agricultural Chemical-Induced Respiratory Problems

Prevention of poisonings and injuries depends on safe usage and storage practices, good personal hygiene, monitoring of health status of frequently exposed workers, compliance with governmental regulations, and intelligent use (proper selection, fitting, and maintenance) of personal protective equipment (Davies 1990). General pesticide protection methods are covered in Chapter 6. This chapter will deal only with protection from fumigants.

Avoidance also is the only effective technique for preventing fumigant poisonings: Fumigants must not be released until structures to be treated have been vacated; guards and warning signs should be posted around treated buildings. Slow-release pellets and more efficiently enclosed application systems assist in this practice. Farmers can avoid fumigants by eliminating the need for their use (through storing good quality, dry grain in clean, rodent- and insect-proof storage areas, which are monitored and aerated when necessary). If fumigation of farm-stored grain becomes necessary, it is advisable to hire a professional fumigator.

As fumigants are so highly penetrable, only a self-contained breathing apparatus should be considered for use in this environment. Even then, some fumigants may penetrate the rubber seal around the face. Precise recommendations are included with the package label of each fumigant.

SECTION 7: ZOONOTIC INFECTIONS CAUSING RESPIRATORY DISEASE

About 25 of the over 250 known zoonoses are considered to be a significant occupational threat to agricultural workers in the western industrialized countries. Of these, approximately eight may primarily affect the respiratory tract. These eight zoonoses produce systemic illness with major respiratory involvement, or the lungs are the primary site of entrance. These diseases include bovine tuberculosis, histoplasmosis, hydatidosis, Newcastle disease, ornithosis, Q fever, swine and avian influenza, and tularemia.

Farm operators, persons employed in agricultural operations, and family members of these sectors are at increased risk of contracting certain infectious diseases because of regular contact with domestic livestock, poultry, and the rural environment. These "zoonotic diseases" can be transmitted directly to humans from live animals, their carcasses or by-products (wool, bone, hides), indirectly through the environment where certain of these organisms survive, or by arthropod vectors such as ticks and mosquitos. Thus farmers and ranchers, seasonal farm workers, slaughterhouse workers, animal transporters, persons who process animal products, and veterinarians all constitute high-risk groups for certain

zoonoses. These eight zoonoses are summarized in Table 13.1 and are covered in detail along with other zoonoses in Chapter 13.

SECTION 8: GENERAL PREVENTIVE MEASURES

Although there were some specific preventive recommendations in the previous sections according to specific exposures, the following section provides a summary and overview of general preventive measures for agricultural respiratory health.

Medical Surveillance of the Agricultural Work Force

Preemployment medical screening or at least counseling may be appropriate in some instances, especially in swine or poultry confinement facilities or grain handling facilities where chronic organic dust exposure is common. People are more likely to have respiratory illnesses from these exposures if they 1) smoke cigarettes, 2) are atopic, 3) have a concurrent respiratory or cardiovascular condition (Donham 1990), or 4) have a genetic predisposition to adverse response to inflammatory substances in organic dusts (Kline and others 2004). (Detection of genetic susceptibility to organic dust is only in research stages at this time.)

Obviously, preemployment evaluations are not applicable to the many self-employed farmers and family members. Routine physical exams and evaluations of acute conditions should be conducted with the objective of preventing chronic pulmonary disease. Health care professionals need to be aware of the variety of agricultural processes resulting in occupational exposures and respiratory illness. Too frequently, occupational respiratory illnesses among farmers are not recognized, and farmers return to a work setting that further induces or aggravates the illness. A detailed occupational history should establish characteristics of the work environment and common activities that could lead to respiratory illness. Smoking history, second jobs in an industrial setting, and other environmental exposures that could contribute to respiratory illness should be noted. The frequency and temporal sequence of respiratory symptoms and signs in relation to exposure are particularly important in assessing asthma, NALC, FL, bronchitis, and ODTS.

Where the occupational history suggests an infectious disease hazard of zoonotic origin (e.g., histoplasmosis or ornithosis), blood should be drawn, and serum should be frozen for possible subsequent serological testing if a febrile illness is presented. Changes in levels of specific circulating antibodies enhance a rapid diagnosis of difficult-to-diagnose zoonotic infections.

If the occupational history indicates frequent organic dust exposures (e.g., working in livestock confinement, grain handling, etc.), an assessment of respiratory symptoms (see Table 3.13) and a baseline spirometry should be conducted. If the patient is symptomatic with dyspnea or low

Table 3.13. Agricultural Respiratory Disease Occupational History

Pertinent questions that suggest an agricultural-occupational relationship to one or more of the following: A. One or more of the components of Mucous Membrane Irritation (MMI), including sinusitis, rhinitis, pharyngitis, laryngitis, and tracheitis; B. Subacute or chronic bronchitis; C. Asthma-Like Condition; or D. Monday Morning Syndrome.

1) Do you have regular dust exposure (<2 hours/day and <6 years)? Such as:
   a. Working with livestock inside buildings? Such as:
      i. Dairy barns
      ii. Confined animal feeding operations (CAFOs)
         1. Swine CAFO
         2. Poultry CAFO (chickens or turkeys)
   b. Working with grain handling? Such as:
      i. Moving and storing grain, such as in a grain elevator
      ii. Grinding and mixing animal feeds

Pertinent questions that suggest an agricultural-occupational relationship to one or more of the following: A. Acute bronchitis; or B. A generalized febrile condition with symptoms of organic dust toxic syndrome, hypersensitivity pneumonitis, or pneumonia.

1) Two to six hours before these symptoms began, were you involved in any of the following activities that may have resulted in a heavy exposure to an agricultural dust?
   a. Unload a non-airtight silo (remove the moldy layers from the top of the silage)
   b. Move or handle grain that may have been moldy
   c. Clean out a grain bin
   d. Break open/handle straw or hay bales that were spoiled
   e. Use a bale shredder to prepare animal bedding
   f. Move, sort, or load swine or poultry in a confinement building
   g. Use a leaf blower to clean dust from surfaces in a swine or poultry building
   h. Use a high pressure washer to clean in a swine or poultry building
   i. Move or handle wood chips or sawdust that may have been spoiled
2) Had you been using a welder, and were any of the metals you were working with galvanized?
3) Had you been working in or around bird roosts (pigeons or starlings, etc.), an old poultry house, hayloft or silo in the past 10 days?
4) Had you been working with sick poultry in a confined space or in turkey processing in the past 10 days?
5) Had you been administering aerosolized poultry vaccines (particularly Newcastle virus) in the past 10 days?

Pertinent questions that suggest an agricultural-occupational relationship to one or more of the following: Difficult breathing related to pulmonary edema or laryngeal edema.

1) Had you been working with or around liquid manure storage or moving or pumping liquid manure (within the past 48 hours)?
2) Had you been working with or around fresh (within 10 days of cutting) silage, particularly in a non-airtight silo?
3) Had you been working with or around grain or pest fumigants such as methyl bromide or phosphine in the past 48 hours?

forced vital capacity (FVC), a baseline chest x-ray would be indicated. These procedures are valuable to monitor, over time, for developing occupational obstructive disease (asthma, bronchitis) or restrictive disease (HP). If the patient has allergy symptoms upon work exposure, family history and skin testing would be recommended to rule out an atopic disease. A spirometer is a valuable diagnostic tool for any rural clinic. Spirometry measurements should be conducted according to accepted standards (American Thoracic Society 1990). Typically, farmers have lowered flow rates (FEF 25–75) relative to comparison populations, suggesting that small airways obstruction is common, signaling risk of developing obstructive disease. However, flow rates are variable and therefore not considered highly reliable from an individual clinical case. One of the authors (KJD) has observed that farmers generally have larger lung capacities than standard comparison populations (Donham and others 1990). For example, 85% of predicted FVC or FEV_1 for a farmer may be similar to 75% or 80% of predicted if there was a relevant comparison farm population. Furthermore, baseline pulmonary functions tests (PFTs) are of limited value as they may not reveal acute obstruction from occupational exposures. Before and after a work exposure can be very helpful in diagnosing occupational NALC. A farmer may have a 90% or 100% of predicted FEV1 on baseline, but after 2 or more hours of exposure in their swine confinement building or other organic dust exposure, they may drop 10–20% in FEV1, revealing evidence of NALC and impending obstructive disease. Diffusion capacity measurements, conducted at a pulmonary function laboratory, would be important if chronic FL is suspected.

Following the initial examination, follow-up exams should be performed as dictated by symptoms, exposures, and lung function measurements. More detailed workups may be indicated for those with a history of bronchitis, asthma, frequent respiratory infections, or flu-like illnesses (Chaudemanche and others 2003). Routine monitoring should be conducted on those who have chronic occupational exposures to organic dusts, including grain elevator workers, dairy farmers, swine or poultry confinement workers, and smokers. In addition, routine examinations provide an excellent opportunity to provide education on occupational exposures and on smoking cessation. The respiratory evaluation may also be psychologically beneficial for agricultural workers who are concerned about their respiratory exposures.

Agricultural workers with chronic respiratory illness will require routine follow-up at a frequency determined by their lung function and continued exposure. Prevention of further exposure to causative agents and elimination of smoking are crucial for control of most chronic respiratory diseases (Hoppin and others 2003). Techniques for preventing exposure to specific agents are described in the following sections in this chapter. Every effort should be made to allow the patient to continue to work safely. It is a rare situation when a patient cannot continue working in farming if dedicated efforts are made to limit exposures, including education, engineering controls, work practice changes, and use of respirators. Exceptions might occur with severely atopic individuals, or those with progressive pulmonary fibrosis subsequent to FL. If control measures and medication do not control the respiratory response or prevent further deterioration of respiratory status, a job change may be necessary.

Approaches to Prevention

Because inhaled agricultural dusts and gases commonly produce debilitating acute respiratory illness, mitigation of hazardous exposures is crucial. As farmers are reluctant to take measures to protect their own health (for a variety of reasons), health care providers can play an important role in preventing respiratory disease by educating themselves and their patients in causes and prevention of occupationally induced respiratory illness. Agricultural environments may be assessed for presence of and concentrations of potentially harmful agents, to help protect workers and to assist in diagnosis of illness. Measurement of dusts and gases is described in Section 8. Agricultural workers must guard their health in four ways: by preventing **formation** of harmful substances in the work environment, by preventing harmful substances **from entering the air**, by **removing** them once they have become aerosolized, and by **preventing inhalation** of harmful substances. Engineering controls and management of the work process are preferred techniques for avoiding respiratory exposures.

Preventing formation of substances hazardous to the respiratory system is the optimal solution. Examples include proper storage of grain, hay, and silage to help prevent spoilage of these plant materials and eliminating growth of bacteria, fungi, and by-products that are responsible for causing hypersensitivity pneumonitis and ODTS. Livestock can be managed to limit transmission of respiratory infections to humans.

If formation of harmful agents cannot be avoided, attempts should be made to prevent them from becoming aerosolized or to rid or dilute aerosolized toxins within the structure. Ducting systems in grain elevators have been quite effective in reducing grain dust aerosolization. Building design and ventilation of livestock confinement operations are crucial for keeping dust and gas levels minimal. Non-airtight silos recently filled with silage must be well ventilated to rid them of NO_x. Tractor cabs with air filtration systems reduce inhalation of field dusts.

Often, the source cannot be eliminated completely, so steps must be taken to avoid their inhalation. Manure pits should be entered with extreme caution because of the potential presence of deadly concentrations of H_2S. Airtight silos must be entered with extreme caution because of the asphyxiation hazard (correct procedures for entering confined spaces are discussed later). Persons with documented episodes of FL should avoid the environments and activities that result in exposure to offending allergens. Field workers can avoid inhaling pesticides by vacating recently sprayed fields or orchards for the recommended reentry time.

Use of personal protective equipment should be considered an ancillary or temporary measure to engineering and management prevention procedures. Workers using respirators must use well-fitting and well-maintained respirators that protect against all hazardous agents present. Respirators may be imperative for highly sensitive individuals or in certain agricultural tasks, such as using a power sprayer to wash the inside of a swine building (Figure 3.10). Persons entering confinement operations or moving grain are wise to wear mechanical filter respirators as a matter of course. Use of respirators is discussed in Section 9. Specific preventive measures to the major respiratory exposures are discussed in Sections 2–7.

Figure 3.10. Operating a high pressure washer inside a livestock building produces a risk of respiratory exposures to toxic mists and gases.

SECTION 9: MEASUREMENT OF AGRICULTURAL DUSTS AND GASES

Introduction

Risk to the producer could be reduced by having a knowledge of type and concentrations of exposures. Environmental assessments can be conducted with informed use of gas or dust measuring devices. These assessments can also be used to monitor effectiveness of engineering and management controls.

Health care providers and veterinarians can play a crucial role by actually conducting assessment or encouraging patients/clients to use gas and dust measuring devices to assess risk in the workplace. Environmental assessment may be psychologically supportive as well as medically beneficial in some patients. Health care providers also can encourage local retailers to stock and supply commonly used devices. Agricultural operations are sufficiently large to have their own health and safety office. Smaller family farms may have to rely on community resources to provide environmental assessment, such as government extension services, insurance companies, or agricultural health clinics.

Measurement Devices for Gases and Dusts

Almost all of the many measuring devices for gases and aerosolized dusts were developed for use in industries other than agriculture and therefore some may not always be directly applicable. The instruments include small personal air samplers, handheld devices, and large permanently installed systems that measure several substances at once.

GAS-MEASURING DEVICES

The measuring devices that are best suited to most farm needs are those that give an immediate reading of the gas or dust concentration. Also, those devices that give an accumulated measurement over time (dosimeters) are useful. The following characteristics should be considered when choosing a gas measuring instrument:

- *Accuracy and precision*: How well does it measure what it is supposed to measure?
- *Specificity*: Does it measure just the substance of concern or also register other gases, giving a false high reading?
- *Sensitivity*: Can it measure accurately in the range of concentrations in the environment of interest?
- *Cost*: Is the instrument and any refills or replacements needed affordable?
- *Ease of operation*: Is the maintenance, service, training, and expertise required to use and maintain the device reasonable?
- *Calibration*: Can it be calibrated locally or does it have to be returned to the factory?
- *Durability* and *life expectancy*: Will the instrument withstand the use an agricultural settings and do components have a long shelf life?
- *Technical support*: Are technical services available?

The following paragraphs describe the basic classes of equipment available.

DETECTOR OR COLORIMETRIC TUBES

These instruments give a direct reading of the concentration of the gas being measured. Active dosimeter tubes require an air pump to pull air over the chemical matrix in the tube. A hand-operated bellows pump or a piston-type pump is fitted with a detector or colorimetric tube containing a chemical that changes color when in contact with the substance to be measured. Tubes for many types of gases are available. To operate, the appropriate detector tube is placed in the pump and pumped for a specified number of strokes. The length of the color change in the tube is proportional to the concentration of the gas.

An extension hose can be placed between the pump and the tube so that the air inside a structure can be measured from the outside, reducing risk to the operator. This basic pump costs about $150 (U.S.). Dosimeter tubes (about $3 each) are sold separately. Detector or colorimetric tubes adequately fulfill the farmer's requirements for accuracy and ease of operation and maintenance. Shelf life of the tubes is 2–3 years.

There are also passive diffusion colorimetric tubes available. These tubes have no pumps but rely on passive diffusion to affect the colorimetric reaction in the tube. Passive diffusing tubes give an average or integrated concentration over longer work exposure, such as 8 hours. Colorimetric tubes are accurate within the range of ±20%.

SENSING BADGES AND TAPES

Sensing badges and tapes are available that give an indication of gas concentrations by means of a color change. These devices are essentially dosimeters although they are less quantitative and they are unsuitable for uses where an immediate reading may be required, such as when pumping out a manure storage pit.

SOLID-STATE DETECTORS

These electronic instruments have sensors for detection of various gases. Typically they are battery-operated, but most can also be plugged into a standard electrical outlet. Gas concentrations are indicated by a digital readout, lights, and/or by alarms that indicate when gas concentrations reach hazardous concentrations. These devices just described are available for $1,000–3000 (U.S.). Most of these instruments have not been tested for use in agricultural operations. These meters require periodic calibration, which may require a return to the factory with a substantial cost. Sensors on the solid-state detectors must be replaced every 1–3 years.

DUST-MEASURING DEVICES

There are three types of methods used to measure dust in the air:

1. Gravimetric methods
2. Photometers
3. Particle counters

Gravimetric methods are most reliable, and are used for regulatory standards (e.g., OSHA-permissible exposures). A pump pulls a known quantity of air over a filter. The preweighed filter is weighed again after dust is collected, allowing computation of milligrams of dust in a cubic meter of air. The gravimetric methods require rather sophisticated laboratory support and the expense per sample is quite high. Results may take several days to be returned. Therefore, gravimetric methods have significant barriers for the farm owner-operator, but may be used by a professional health and safety consultant.

Photometers are electronic instruments that measure dust by the amount of light that is blocked or scattered between the light source and the photocell. There is a computer chip that uses an algorithm to estimate the dust concentration in mg/m^3, based on calibration with a "standard" dust. However, agricultural dusts have different physical qualities than the standard calibration dusts, and readings are not necessarily accurate for agricultural dusts. Their accuracy is reliable only within a limited range of dust concentrations and partial size. Research is taking place to provide greater accuracy for agricultural use. These photometer instruments do have the advantage of giving both immediate readout and cumulative measures. Their cost is from $2,000–3,000, and can be very useful in agricultural settings.

Particle counters are similar to photometers, but they record each particle of dust as it blocks light or a radiation emission. They do not give a mass measure, and thus, their readings are not relevant to exposure. These instruments can be useful to measure the size of particles and the effect (before and after) of a dust control method. However, the reading will not be related to a health effect and, therefore, has relatively little use in agriculture.

Bioaerosol assessments have been conducted to assess microbes, endotoxin, and glucans quantitatively. However, these procedures have primarily been used in research. They are expensive and difficult to adapt to agriculture at this time (Thorne and others 2004).

Using Gas-Measuring Devices on the Farm

The American Conference of Governmental Industrial Hygienists (ACGIH) reviews research on health effects of various concentrations of toxic gases, and then recommends the maximum concentration of each gas to which a worker should be exposed for a certain length of time called a *threshold limit value (TLV)*. The TLV for ammonia is 25 ppm. This means that a worker should be exposed to no more than 25 ppm of ammonia averaged over an 8-hour workday (ACGIH 1985).

These TLVs (see Table 3.11) are useful in industrial settings where the atmosphere and the workers can be monitored. These levels are determined when there is only one toxic substance. However, many agricultural operations, such as CAFOs, have **mixed exposure environments**. The rule is that substances affecting the same tissues should be considered additive in their effect (unless known otherwise). The formula to determine maximum exposures is as follows:

$$\frac{C_1}{TLV_1} + \frac{C_2}{TLV_2} \frac{n}{n} \text{ must be } \leq 1.$$

(C_1, C_2 up to C_n are concentrations of the various toxic substances in the work environment.)

For example, if a farm has exposures of both ammonia and hydrogen sulfide at half their TLV values, that would constitute a maximum exposure. Table 3.11 also lists maximum recommended levels of exposure for livestock confinement environments.

Farmers may need to measure gas concentrations in the following acutely hazardous situations.

FILLING A NON-AIRTIGHT SILO
Nitrogen dioxide (NO_2) can be measured with detector tubes or with one of the more expensive solid-state detectors. Electronic instruments with direct readouts are most useful. See Section 5, Oxides of Nitrogen ("Silo Gas").

PUMPING OUT A MANURE STORAGE PIT UNDER A CONFINEMENT BUILDING
Hydrogen sulfide can quickly reach harmful levels in manure pits and buildings over pits during pumping. Instruments with direct readouts must be used because colorimetric methods do not react fast enough. See Section 4: Livestock Confinement Dusts and Gases.

Monitoring Ammonia Levels in Swine and Poultry Confinement Buildings

Since ammonia levels do not generally change rapidly in confinement buildings and concentrations are not acutely hazardous, either direct readout instruments or dosimeters can be used.

Working in a Tightly Closed Building

Carbon monoxide levels may become hazardous in tightly closed buildings with poorly adjusted fossil fuel-fired heaters. Hazards may occur when exhaust gas from indirect-fired heaters may be drawn back down the exhaust stack by exhaust fans elsewhere in the building, or when internal combustion engines are operated inside the building. Direct readout detection devices or detector tubes may be used in these environments.

Oxygen-Deficient Atmospheres

Farmers may want to enter closed storage structures such as airtight silos, underground manure storage pits, or other structures where fermentation processes have depleted the oxygen supply. Unless there is absolute assurance that no toxic or oxygen-deficient atmospheres are present, these structures should not be entered without a self-contained breathing apparatus or a supplied-air respirator. Farmers and ranchers typically do not have this equipment or the training to use it but they will occasionally enter closed storage areas without proper protection. Before entering they can measure the oxygen concentration with detector tubes or solid-state sensors. By far, the safest way to handle emergencies in airtight silos is to call the serviceperson from the company who manufactures the silo. These service people are trained and equipped to handle the problem. They are trained in self-contained breathing apparatus (SCBA) use.

Alternate Ways of Measuring (or Estimating) When Instruments Are Cost-Prohibitive

If your farm patients are unable or unwilling to purchase measuring devices, some of the following options might be useful. Co-ops and farm supply stores should be encouraged to stock basic items such as detector tubes. These businesses also could purchase some air-sampling equipment, maintain the device for customers on a rental basis, and provide them with instructions for use.

Other sources of measuring equipment include veterinarians who often own measuring equipment for use in their practices. Farmers could inquire about using this equipment or hiring the veterinarian to take measurements. The extension service may be a resource for measurement in some areas. Also, farmers' health services such as Lantbrukshälsan in Norway or the AgriSafe Network of Clinics in the Upper Midwest of the U.S. often are equipped to conduct environmental assessments as well as clinical screenings. A private consulting agricultural engineer or industrial hygienist could be hired to take measurements. In many instances, hiring consultants may be the most practical and efficient way to obtain this service. However, with proper training and motivation a farmer may buy and use the instruments him/herself with good results. The policies of Occupational Safety and Health offices vary from country to country. Some state occupational administrations have consultation programs that offer free services. Agricultural extension services are becoming better equipped to meet these needs.

In many cases farmers can avoid danger by knowing what dusts or gases might be found in certain structures and being especially alert to any physical sense of their presence. People who have been overcome by silo gas often mention that they noticed eye or nose irritation while they still had a chance to escape, but decided to ignore that warning. Farmers who regularly store grain on the farm often can assess the condition of the grain by smelling the air that is being blown through the grain. They may notice that dusts from moldy grain (which are more dangerous than dusts from grain in good condition) have a different odor than dusts from grain that has no spoilage.

Encourage your farm clients to learn to pay attention to their senses and trust their experience. When farmers begin to take seriously any signs of danger they will do much to avoid respiratory hazards.

Use of smoke tubes can help farmers indirectly assess problems of dust or gas build-up in animal confinement buildings. Smoke is released in the building, and its movement (or lack of movement) enables farmers to identify inadequacies in the ventilation system. Smoke tubes are sold by the same outlets that sell gas detector tubes. Table 3.14 lists international companies where measurement equipment may be obtained.

Table 3.14. Sources of Information and Equipment on Air Sampling

Sources	Website Contact Information
Air Sampling Instruments, ACGIH 2001	http://www.acgih.org/home.htm
Am Conf Gov Ind Hygienists	http://www.acgih.org/store/ProductDetail.cfm?id=479
Drager International	http://www.draegersafety.com/ST/internet/CS/en/ index.jsp
Mine Safety Appliance	http://www.msanet.com/
Environmental Supply	http://www.envisupply.com/
Fisher Scientific Company	http://www.fisherscientific.com/
SKC	http://www.skcinc.com/
Casella CEL Inc.	http://www.casellacel.com/
Google Directory	http://directory.google.com/Top/Science/Environment/ Environmental_Monitoring/Products_for_ Sampling_and_Monitoring/Air/
Gray wolf sensing solutions	http://www.wolfsense.com/

SECTION 10: PERSONAL RESPIRATORY PROTECTIVE EQUIPMENT

Introduction

Although agricultural workers are commonly exposed to a vast array of agents that threaten their respiratory health, environmental control of these hazards and/or use of respirators is not as common as it should be. Although nearly 25% of swine workers have symptoms of respiratory illness, the few available statistics suggest that the minority (about 10%) of hog farmers wear a dust mask (Ferguson and others 1989). Observation suggests that workers in feed mills, grain elevators, and grain terminals, and farm chemical and feed supply companies only occasionally use respirators. Why is it that the use of respirators in agriculture is so uncommon, even though hazardous respiratory exposures are so common? The following are just some of the reasons:

1. They are hot and uncomfortable to wear.
2. Filtering respirators are hard to breathe through, especially when one is working hard and the filter becomes plugged with dust.
3. There are no regulations requiring respirator use in most agricultural settings.
4. There is lack of knowledge about exposures and protection.
5. There is lack of a safety culture in agriculture.
6. There is lack of accessibility, consultation on their use, selection, and fit testing.
7. There is lack of knowledge on the part of health care providers to counsel and recommend respirator use.

Fortunately, even with these negative factors, there appears to be a growing interest in personal protective equipment among agricultural workers in recent years. This probably resulted from greater awareness and concern about pesticide hazards, beginning in the late 1960s. The Worker Protection Standard of 1992 (U.S.) has further enhanced interest in this subject. More recently (1990s), there has been a much greater awareness of organic dust hazards in agriculture. This heightened awareness is reflected in the increasing number of articles, advertisements, agricultural extension service publications, and other literature focusing on respiratory hazards and respirators. Unfortunately, most farmers who try to purchase respirators often discover that distributors and knowledgeable personnel to help select and fit appropriate respirators is lacking in most rural areas, and only a limited supply is available locally.

As a result, farmers may increasingly turn to rural health care providers or their veterinarian to ask why they should wear respirators, in what situations, and what specific equipment should be used. Health care providers and other health care professionals can be effective if they are knowledgeable about respiratory use and encourage proper and safe use of respirators. Health care providers should educate their farming patients on the importance of selection, fit, and care of respirators in various agricultural environments. Selection and use of respirators is a complex process requiring evaluation of several variables, including the patient's medical condition.

Types of Respirators

The National Institute for Occupational Safety and Health (NIOSH) (and analogous agencies in other countries) test and certify respirators (Nickens 1991). Any respirator selected for use should have a NIOSH (or equivalent for the respective country) approval number (TC-*xxxx*) noted on the packaging and mask. Devices are tested under laboratory conditions and approved for protection against specific substances in specific concentrations. (These tests reflect industrial exposures.) **However, these same respirators will work adequately in agriculture, with appropriate selection and fit.** Approval is given for a class of chemicals or agents, for example, organic vapors.

NIOSH certification evaluation includes a laboratory evaluation of the respirator, an evaluation of the manufacturer's quality control (QC) plan, audit testing of certified respirators, and investigations of problems with certified respirators. NIOSH annually publishes a list of all certified respirators and cartridges.

All respirators have limits to the amount and duration of protection they can provide. Respirator users should thoroughly read, understand, and follow the manufacturer's instructions included with the respirator. Health care providers and veterinarians should impress on their farm patients that proper use is as essential to proper respirator functioning as it is to any of their other farm equipment.

Respirators are divided into two categories: 1) air-supplying respirators, and 2) air-purifying respirators. The former filter contaminants from the air inhaled by the wearer. **They DO NOT provide oxygen.** The latter provide clean, uncontaminated air from either compressed air in a tank (self-contained breathing apparatus or SCBA) or from a hose opening to a clean environment from which air is pumped.

AIR-SUPPLYING RESPIRATORS

Most farmers and ranchers do not have a routine need for air-supplying respirators. Air-supplying respirators are expensive and require extensive training and maintenance to ensure that they are used safely. However, they may be used in some large operations where safety support services are available.

AIR-PURIFYING RESPIRATORS

Air-purifying respirators are issued with four basic precautions that should always be kept in mind. Since they do not supply air, **they cannot be used in oxygen-deficient atmospheres** (less than 19.5% oxygen, such as in an airtight silo). **They cannot be used in any acutely toxic or asphyxiating environment that is immediately dangerous to life and health (IDLH)** (such as a freshly filled silo or manure pit where gases could quickly incapacitate a victim), **or when a contaminant has poor warning qualities**. **They should not be used in abrasive blasting (insufficient protection).** Use of odor as a warning of potential hazardous substances is a very poor practice. For example, H_2S deadens the olfactory senses at concentrations that are severely hazardous (>500 ppm). Because the effectiveness of this type of respirator depends on inhaled air passing through and not around the filter, **the respirator must have a tight seal around the face**. Therefore, people with beards, facial scars, unusual facial size or shape, etc., may not be able to wear a conventional air-purifying respirator effectively because of the difficulty of establishing a tight seal around the face. However, it should be known that some manufacturers make masks in small, medium, and large sizes to help allow for human physical size variations. Because the wearer of an air-purifying respirator must inhale air through a filter, all such respirators (except powered air-supplying respirators) offer breathing resistance. This is an important medical, as well as psychological, consideration and will be discussed later in this chapter.

Air-purifying respirators come in three basic face pieces: quarter-masks, half-masks, and full-face masks.

Face masks are made of flexible molded rubber, silicon rubber, vinyl, or plastic. They are attached to the head by two or more rubber or woven straps. Some of the powered air-purifying respirators have loose-fitting head coverings, helmets, or face shields (see discussion of A_5, below). The NIOSH-approved full-face mask has eye shields that have met impact and penetration standards. It is more expensive than the quarter-mask and half-mask. Additional to classifying respirators by type of face piece, air-purifying devices can be further divided into the following (see Table 3.15):

A_1: disposable mechanical filter respirators
A_2: nondisposable mechanical filter respirators
A_3: chemical cartridge respirators
A_4: powered air-purifying respirators
A_5: gas masks

The majority of agricultural applications can be filled by one of the respirators in categories A_1–A_4 above.

A_1, disposable mechanical filter respirators, are often called *dust masks* or *particulate respirators*. They consist basically of a shaped piece of filter material held onto a wearer's head by one or two straps. **(One-strap models often sold in stores frequented by farmers/ranchers are not recommended, as they cannot provide a good seal around the face.)** These respirators protect against airborne particles by trapping them mechanically in the filter medium as the worker inhales. Breathing resistance increases with use, and the respirator is ready to be discarded when that resistance becomes too great, or when absorbed moisture causes the filter to lose its "shape." Several kinds of mechanical filters are approved by NIOSH. The filter packaging lists the substances, particle size, and concentration for which the filter is approved. There are filters that are designed to protect against dusts and mists; dusts, fumes, and mists; or dusts, fumes, mists, and radionuclides. Those approved for dust and mists are appropriate for most agricultural usage.

Another feature of some disposable respirators is that they contain an exhalation valve. This helps prevent buildup of moisture in the mask material (helping to maintain its rigidity and shape), helps prevent the wearer's eyeglasses from fogging up in cold weather, reduces breathing resistance, and assists in maintaining a seal to the face. Also, some masks are now available that have a plastic mesh framework, which provides a longer-lasting rigidity and shape to the mask.

A_2, nondisposable mechanical filter respirators, remove particles in the same way as do the disposable filter respirators. However, filters in these nondisposable respirators are replaceable, and the face pieces are made of a rubber or plastic material that lasts longer than A_1 and generally affords a tighter seal to the face. Filters must be replaced when breathing resistance becomes evident. High efficiency filters offering greater protection are also available.

Nondisposable mechanical filter respirators require routine cleaning and maintenance to assure proper function.

A_3, chemical cartridge respirators, can be either half-mask or full-face piece. They protect against specific gases and vapors, or against narrowly defined classes of gases or vapors, through the use of replaceable screw-on cartridges containing an absorbent chemical. All NIOSH-approved cartridges are color coded according to a code established by the American National Standards Institute (ANSI) (e.g., ammonia = green, organic vapors = orange) (ANSI 1969, 1980, and 1992; NIOSH 1987). Most also have particulate prefilters (dust caps) that may be attached on top of the chemical cartridge and may be changed independently of the chemical cartridge. This is advantageous since the particulate filter often needs changing before the cartridge.

Most chemical cartridge respirator manufacturers also make a separate particulate filter cartridge so they can function as a mechanical particulate filtering respirator only, when chemical filtering is not necessary.

Manufacturers have developed a new variation of the chemical cartridge respirator called "maintenance-free." They are designed to be discarded after the chemical cartridge is spent. However, they must be checked for defects and fit before each use. They also have replaceable particulate prefilters that attach to the cartridge. The entire mask is ANSI color coded.

In contrast to mechanical filters, which become more efficient with use, chemical cartridges become less efficient with use. After the sorbent is completely utilized, gas or vapor will pass through the cartridge. This is known as **breakthrough** and the cartridge must be changed immediately. Breakthrough may be detected by the wearer upon noticing odor, taste, dizziness, or irritation. Air-purifying respirators are not recommended for use against chemicals with poor warning properties (e.g., no odor).

Table 3.15. Comparison of Air-Purifying Respirators

Types of Respirators	Used for Dusts	Used for Gases	Fit	Amount of Maintenance	Breathing Resistance	Remarks
A_1 Disposable	Yes	No	• Good fit hard to obtain • Adjustable and two-strap models available	Low	High (unless has exhalation valve)	• Easy to see over • Lightweight • Clog with moisture from wearer's breath in humid air • Cause glasses to fog over in cold weather
A_2 Non-disposable	Yes	No	• moderately good for half-mask models	Moderate (unless disposable)	Moderate	• Reusable mask with replaceable dust filter
A_3 Chemical Cartridge	Yes	Yes	• Good for full face piece models	Moderate	Moderate	• Some "disposables" available that are discarded when cartridge is spent • Selecting the correct type of particulate or gas cartridge is mandatory • Particulate prefilters available
A_4 Gas Masks	Yes	Yes	• Good for full face piece models	Moderate	Moderate	• Larger-capacity canisters rather than cartridges • Correct type of canister must be selected
A_5 Powered Air-Purifying	Yes	Yes	• Good	Moderate to high	None	• Add noticeable weight • Not recommended for protection against gases

A practical air-purifying respirator for general farm or ranch use is a half-face chemical cartridge, mechanical filter, or chemical cartridge that will accept a mechanical filter cap. This would increase the respirator's flexibility in varied farm situations. Users should be aware that different brands of masks and/or cartridges should not be mixed, and will usually not have compatible fittings.

The limitations of chemical cartridge respirators are similar to the limitations of mechanical filters:

1. They cannot be used in an O_2 deficient atmosphere.
2. It is important to select the proper respirator for the specific environmental exposure condition, e.g., do not use an organic vapors cartridge in an ammonia gas environment.
3. Color coding is not a reliable way of ensuring that the proper cartridge has been chosen. Always check the TC# of the cartridge.
4. They should not be used in IDLH atmospheres or with chemicals that cannot be detected by odor at the Threshold Limit Value or the Permissible Exposure Limit, whichever is lowest.

A$_4$, powered air-purifying respirators (PAPR), are equipped with a battery-powered motor that blows filtered air into a headgear with a face piece. Face pieces may be full-face or half-mask as in traditional respirators, but, more commonly, the headgear consists of a hard helmet with a clear plastic face shield. Many combinations of helmets and face shields are available; there are also some nonrigid head covers. In contrast to mechanical filter respirators, there is no breathing resistance, making them more suitable for individuals who may have a concurrent respiratory condition, such as COPD or asthma. Other advantages include a constant flow over the wearer's head and face that assists in cooling the wearer. Since powered air-purifying respirators provide constant positive pressure, they do not require a positive face seal to assure effectiveness. Thus, they can protect workers with beards, sideburns, odd-shaped faces, or facial scars, which other air-purifying respirators cannot do.

Most units are powered by a battery pack strapped to the wearer's waist or back. Batteries are rechargeable. An option for some units is a 12V or 24V DC adaptor, allowing the motor to be run by a vehicle battery for use in tractor cabs or combines.

Some smaller, lighter-weight units have mechanical filters only, but chemical cartridge filters are becoming more commonly available for these units. In most units, a replaceable filter and prefilter, as well as fan and motor assembly, are in the helmet itself. Some units with mechanical filters have fan, motor, filters, and batteries attached at the wearer's waist and blow purified air through a hose into the headgear.

Still other models must be attached to a stationary air supply, either in a workplace or in a vehicle. A long hose from the motor/filter assembly to the headpiece allows some worker movement.

There are many applications in the agricultural workplace where this type of respirator has advantages over other types of respirators. They are especially useful for workers with cardiorespiratory health limitations, those workers where a good facial seal is difficult, and in hot moist working environments. The authors (KJD and AT) have anecdotally noted that workers are more compliant in the use of this type of respirator; perhaps because of their belief in the effectiveness of the technology, and/or comfort in its use.

The limitations of powered air-purifying respirators are the same as for mechanical filter respirators.

Additional limitations are that peripheral vision may be distorted by the curved face shield. The face shield may fog up when going out into cold air from a warm building. In environments with odorous particulates (e.g., swine buildings), constant off-gassing of odors from dust trapped in the filters may be forced into the helmet. Finally, the cost ($400–600) may be a concern in many situations. Tables 3.16, 3.17, and 3.18 briefly compare types and brands of air-purifying respirators, relative to cost, size, and protection factor. The protection factor (PF) is a relative measure of how much the particular mask can cut down on exposure (e.g., a PF of 5 means exposure is reduced five times.)

A$_5$, gas masks, also are for use against gases and vapors. Although they have limited application in production agriculture, they are included in the discussion here for sake of completeness. They are often put in a separate category from the chemical cartridge respirators because gas masks

Table 3.16. Assigned Protective Factors (APF) for Selected Types of Respirators

APF	Protection from Particulates
5	Single-use or quarter-mask respirator
10	Any air-purifying half-mask respirator, including disposable equipped with any type of particulate filter except single use
25	Any powered air-purifying respirator (PAPR) with a particulate filter
50	Air-purifying full-face respirator equipped with a high-efficiency filter. Any PAPR with a tight-fitting face piece and a high-efficiency filter

APF	Protection from Gases and Vapors
10	Any air-purifying half-mask respirator (including disposable) equipped with appropriate gas/vapor cartridges
25	Any PAPR with loose-fitting hood or helmet with appropriate cartridges
50	Any air-purifying respirator equipped with a tight-fitting face piece and appropriate cartridges

have canisters. The canisters contain a much larger volume of sorbent than do cartridges. They may be used either in higher gas or vapor concentrations or for longer periods of time than chemical cartridges before breakthrough occurs.

Gas masks have a full-face piece (which adds eye protection). Canisters are color-coded with the same ANSI code used for chemical cartridges. Like all air-purifying devices, gas masks are not for use in oxygen-deficient or IDLH atmospheres.

AIR-SUPPLYING RESPIRATORS
These respirators, unlike air-purifying devices, provide clean, uncontaminated air from a source outside the wearer's environment. They are the only type of respiratory protection designed for use in oxygen-deficient atmospheres and some IDLH atmospheres. Most family farmers and ranchers would have little use for such a respirator: In most cases, it would be hard to justify the cost of such equipment, which would be intended only for emergency or rescue use. In addition, the

air-supplying respirators require careful maintenance and use, necessitating training for the wearer. However, large farming and food processing operators and structural fumigators may use them, especially where there are confined space hazards.

Briefly, air-supplying respirators are of two basic types:

B_1: supplied air respirators
B_2: self-contained breathing apparatus (SCBA)

B_1, supplied air respirators, are either air line or hose mask types. Air line devices use a stationary source of compressed air, which is delivered through a high-pressure hose. These respirators have half-mask or full-face pieces, helmets, hoods, or a complete suit. Air line respirators supply air on demand (requiring the user to create a slight negative pressure through inhalation), by pressure demand (always at a slight positive pressure within the face piece), or by continuous flow (the unit constantly feeds air into helmets, hoods, or suits). The highest degree of protection is offered by the pressure demand and continuous flow types, both of which are positive pressure systems.

They can be worn for relatively long, continuous periods of time. However, they cannot be worn in IDLH atmospheres because the wearer is too dependent on the air supply hose and source: If something were to happen to either, the wearer might not have enough escape time.

Another type of supplied air respirator is the hose mask. These deliver air from an uncontaminated source to a face piece through a connecting hose. They do not use compressed air or pressure regulation devices; they use fresh air open to the environment that the person pulls in with his or her own breathing or air that is forced by a fan or air pump. Those with a powered air supply may be used in IDLH atmospheres.

B_2, self-contained breathing apparatus, allows the user to carry up to 4 hours of air. There are closed circuit SCBAs in which air is rebreathed (with carbon dioxide scrubbed and oxygen added), and open circuit SCBAs in which exhaled air is exhausted into the atmosphere. There are also combination SCBA and supplied air respirators. They may be used in IDLH atmospheres. Although there are problems in recommending

Table 3.17.   Some Manufacturers of Personal Respiratory Protection Equipment

Company Name	Contact Information
Protect Respirators	107 East Alexander Street PO Drawer 339 Buchanan, MI 49107 616-695-9663
3M	Occupational Health and Safety Products Division 3M Center, Bldg 220-7W St Paul, MN 55144 800-328-1667
Moldex	Safety Products Division 4671 Leahy Street Culver City, CA 90230 800-421-0668
MSA	600 Penn Center Boulevard Pittsburgh, PA 15235 800-MSA-2222
National Draeger, Inc.	101 Technology Drive Pittsburgh, PA 15230 800-922-5518
North Safety Equipment	A Division of Siebe North, Inc. 2000 Plainfield Pike Cranston, RI 02920 800-581-0444/401-943-4400
Racal Airstream, Inc.	7305 Executive Way Frederick, MD 21701 800-682-9500
Scott Health Safety Line	800-247-7257
Survivair	3001 South Susan Street Santa Ana, CA 92704-5018 800-821-7236
U.S. Service Co (CESCO)	Kansas City, KS 800-821-5218 / 913-599-5555
Wilson Safety Products	PO Box 622 Reading, PA 19603 800-345-4112 / 610-376-6161

them for the typical family farm operation (as described above), they are standard equipment for firemen and emergency rescue personnel. In agricultural settings, firemen and rescue personnel are usually volunteers. There are several situations where rescue attempts may have to be made in oxygen-deficient or IDLH atmospheres. Therefore, it is incumbent that these rural services have properly trained and equipped personnel, which includes SCBA equipment and training.

Guidelines for Respiratory Selection and Usage

Proper and effective use of personal respiratory protective equipment necessitates that hazard, respirator, and wearer be properly matched (Olenchock and others 1992). It is possible that more harm may come to a person using improper respiratory protection than if that person were using no protection. For example, farmer Smith (who was told to wear a respirator by his doctor)

Table 3.18. Air-Purifying Respirators

Respirator[1]	Cost	Fit/Size	Remarks
A₁ Disposable Mechanical Filter Respirators			
3M 8710 Dust/Mist	The cost of respirators in this category varies from about $.30 to $9.00 each, according to the available features.		
3M 9920 Dust/Fume/Mist Respirator		Adjustable straps	Exhalation valve. For protection against metal welding fumes.
AO R1050 Dust Demon®			
AO R1070 Disposable		Adjustable straps	Inhalation/exhalation valve.
North 7170 Disposable Dust/Mist Respirator			
U.S. Safety (CESCO) Softseal-D			
Moldex 2200 Dust and Mist Respirator			
Gerson 1710 Dust and Mist Respirator			
A₂ Mechanical Filter Respirators			
North 7100V Reusable Dust/Mist Respirator	The cost of respirators in this category varies from about $12 to $20 each. The cost of filters for these respirators starts at about $2.50 for a package of 5 filters.	Quarter-mask/One size	Formable aluminum face piece.
MSA Dustfoe 66 Respirator		Quarter-mask/One size	Rubber face piece body.
MSA Dustfoe 77 Respirator		Quarter-mask/One size	
MSA Dustfoe 88 Respirator		Quarter-mask/One size	
AO R2090N Dust/Mist Respirator		Quarter-mask/One size	
AO Welding Fume Respirator		Half-mask/S, M, L	Designed for wear under most welding helmets.
Pulmosan Dust Respirator		Quarter-mask/S, M, L; Half-mask/S-M, M-L	
A₃ Chemical Cartridge Respirators			
3M Easi-Air	The costs of all respirators in this category except the disposables vary from about $15 to $65 for half-mask models, to about $75 to $125 for full face piece models.	Half-mask/S-M, M-L	Unless otherwise noted, all respirators in this group have 1) interchangeable dual cartridges and mechanical filters, and 2) replaceable particulate pre-filters.
MSA Confo II Respirator		Half-mask/S-M, M-L	
MSA Back-Mounted Respirator	Replacement cartridges cost from about $3.75 to $6.00 each.	Half-mask/S, M, L	Filters/cartridges suspended on wearer's back via two breathing tubes.

Table 3.18. Air-Purifying Respirators (*continued*)

Respirator[1]		Cost	Fit/Size	Remarks
MSA	Belt-Mounted Respirator		Half-mask/S, M, L	Filters/cartridges mounted on wearer's belt, attached to mask via breathing tube.
MSA	Ultra-Twin Respirator		Full face piece/S, M, L	
U.S. Safety (CESCO)	Series 150 (CESCO 95)		Half-mask/One size	
U.S. Safety (CESCO)	Series 151 (CESCO 96)		Full face piece/One size	
Willson	1200 Series Respirators		Half-mask/S, M	
Willson	1600 Series Respirators		Full face piece/One size	The North and 3M have several gas/vapor respirators. Replaceable prefilters are available for dust/mist, spray paint, pesticides. Both brands are single cartridge.
Willson	1700 Series Respirators		Full face piece/One size	
North	100 Series Disposable	The North and 3M models are "disposable" after the cartridge is spent. From $8.50 upward.	Half-mask/S, M	
3M	Maintenance-free Respirators		Half-mask/One size	
AO	Quantifit Series Respirators		Half-mask/S, M, L; Full face piece/One size	
AO	Commander Respirator		Half-mask/S, M, L	
Survivair	Blue 1 Air Purifying Respirators		Full face piece/One size	
Scott	Model 64 Respirator		Half-mask/S, M, L	
Scott	Model 65 Respirator		Full face piece/S, L	
HSC	Model 1482 Respirator		Half-mask/S, M, L	
Pulmosan	C-200 Series Respirators		Quarter-mask/S, M; Half-mask/S-M, M-L	Single cartridge.
North	7700 Series Respirators		Half-mask/S, M, L	Cradle strap suspension (with strap over top of head) improves fit.
North	7600 Series Respirators		Full face piece/One size	

(*continued*)

129

Table 3.18. Air-Purifying Respirators (*continued*)

	Respirator[1]	Cost	Fit/Size	Remarks
North	75 BP Series Respirators		Half-mask/M, L	Designed for wear under most welding helmets. Filters/cartridges suspended on wearer's back via two breathing tubes.
A₄ Gas Masks				
MSA	Type N Gas Mask Super Size Gas Mask Industrial Size Gas Mask Chin Type Gas Mask		Full face piece/S, M, L	Choice of two face pieces for all gas masks. All except chin type may be front or back mounted.
Scott	Model 63 "Chin style" Gas Mask		Full face piece/S, L	Choice of two face pieces.
A₅ Powered Air-Purifying Respirators				
Racal	Breathe-Easy Systems	The cost of respirators in this category varies from about $155 to over $500.	Full face piece, helmet/visor PVC hood, or crown/visor	Gas/vapor and dust/mist. Battery pack and turbo unit with filters and blower mounted at waist.
Racal	Airstream Systems		Helmet/visor, helmet/welding visor, or crown/visor	Dust/mist only. Three models with filter in helmet, three models with high efficiency filter in waist module.
Racal	Dustmaster System		Disposable hood/acetate visor, or helmet/rigid visor	
3M	Airhat Systems		Helmet/face shield	Dust/mist only. Filter and pre-filter or high efficiency filter(s) in helmet with blower. Battery pack at waist.
3M	Powered Air Purifiers		Hood, welding helmet, abrasive blasting helmet, helmet/visor, or hardhat assembly	

Table 3.18. Air-Purifying Respirators (*continued*)

	Respirator[1]	Cost	Fit/Size	Remarks
MSA	Powered Air-Purifying Respirator		Half-mask, full face piece, or welder's face piece	Dust/mist only. Filters and batteries in blower assembly worn at waist.
Neoterik	Breezer Powered Respirators		Hoods, face shield, or hardhat assembly	Dust/mist only.
Neoterik	Puriflo Powered Respirators		Half-mask or full face piece	Dust/mist or gas/vapor according to cartridge used.

[1]This is only a partial listing of all the available respirators and includes models, and comparable models, on which research is currently being done to determine their effectiveness in agricultural usage. Manufacturers and vendors may direct suggestions for product additions or deletions to the American Lung Association of Iowa, 1025 Ashworth Road, Suite 410, West Des Moines, IA 50265-6600.

The mention of manufacturers' names and products is solely for informational purposes and is not intended as endorsement over other comparable products, either mentioned or not mentioned, by the American Lung Association of Iowa, The University of Iowa, or the Iowa State University Cooperative Extension Service.

might continue to work in his swine confinement building, ignoring respiratory disease symptoms, because he assumes that his respirator is protecting him (even though it is not). In industry, it is not as simple as telling the employee to wear a respirator. There are legally enforced mechanisms employed to assure proper and safe respirator use. The following guidelines are those of the U.S., OSHA. Other industrialized countries have similar recommendations.

When there is a known respiratory hazard (e.g., assessment reveals air contaminants that exceed the legal permissible exposure or an industrial hygienist or other physician recommends the workers wear respirators) the usual legal requirements are that the business must have a **respirator protection program** (NIOSH 1978, 1987; Parkes 1982; U.S. Dept of Health, Education, and Welfare 1978; Vanchuk 1978). Under a respirator program, respirators must be worn all the time in the hazardous environment. NIOSH has published minimal requirements for a respirator protection program as follows:

1. Written standard operating procedure must be in place.
2. Knowledgeable health professionals must select the proper respirators for the job.
3. There must be an effective employee training program.
4. There must be an effective respirator inspection and maintenance program in place.
5. There must be an effective respirator cleaning and sanitizing program in place.
6. There must be proper facilities for storing respirators that protect them from contamination with dust or other environmental contaminants.
7. There must be an effective fit testing to assure respirators properly function when worn.
8. There must be medical fitness testing conducted by a knowledgeable health care professional.
9. There must be a program review/periodic program evaluation to assure effectiveness and compliance with the program.

Because of the obvious differences between industrial and agricultural work places, procedures designed for industrial use are not directly applicable to the typical diversified family farm or ranch setting; however, by U.S. OSHA law, if there are 10 or more employees, the industrial standards must be used (the numbers of workers required to activate legal requirements of the occupational health laws varies from country to country). Since there may be no legal requirement for a full respiratory protection program for many small farms, we propose a program modified from industry and adapted for use by small agricultural operations. To insure the safety and benefit from respirator usage, the following steps are recommended as practical farm operation guidelines to adopt in absence of an industrial standard respirator protection program:

1. Determine the source and amount of risk present.
2. Eliminate or reduce hazard by engineering and management strategies and then reassess risk.
3. Determine the worker's medical suitability to use respiratory protective equipment.
4. Select the proper respirator for the exposure and the person.
5. Establish recommendations in what environment and circumstances respirators should be worn.
6. Establish respirator care procedures.
7. Evaluate effectiveness.

DETERMINE THE SOURCE AND AMOUNT OF RISK PRESENT

The agricultural workplace has a different character than the industrial workplace. In industries, it is usually economically feasible to buy air-sampling equipment and assess the environment for hazardous exposures. Through assessment, occupational health personnel know exactly what offending substances are hazards in the work environment. The degree of protection required can then be determined by comparing monitored environments with recommended and government-established exposure limits (ACGIH 1985; Parkes 1982).

In agriculture, people work in a variety of environments and are exposed to a wide range of potentially harmful substances. It would be impractical for the average farm operator to purchase and properly use all the different equipment necessary to monitor the air in the various work environments on his or her farm. However, some more specific work environments (such as CAFOs) may warrant regular monitoring. Section 8, "Measurement of Agricultural Dusts and Gases," describes air quality measurement instruments potentially useful to the farmer.

Using Table 3.19 as a guide, farmers and health professionals should note all potentially hazardous substances that the farmer comes into contact with over a year. This type of list will pinpoint areas of respiratory health concerns. Often, farm patients will need protection against more than one class of hazard or substance. They may want to purchase the most protection possible while minimizing cost and maximizing comfort. Or, they might want to have a supply of disposable dust masks on hand as well as a respirator with interchangeable cartridges and filters. Each individ-

Table 3.19. Agricultural Respiratory Hazard Exposure

Class and Constituents of Hazard	Jobs Resulting in Exposure	A_1	A_2	A_3	A_4	A_5	B
A. Dusts/Aerosols							
1. Primarily inorganic dust							
a. Field dust/road dust [soil, sand, rock, small amounts of mold spores, bacteria, and other organic material]	• Soil tillage operations (plowing, disking, harrowing, etc.)	1	2	—	—	3	—
	• Driving farm equipment on or working near dirt or gravel roads.	1	2	—	—	3	—
	• Harvesting operations (combining soybeans, sorghum, or other grains; mechanical harvesting of potatoes, tomatoes, and other food crops; manual harvesting of grapes)	1	2	—	—	3	—
b. Silica sand	• Sand blasting	3	2	1	—	2	—
2. Primarily organic dust							
a. Grain dust	• Working at grain elevators or feed mills	2	2	—	—	3	—
	• Transporting (trucking) and storage of grain	1	2	—	—	3	—
	• On-the-farm handling, transporting, and storage of grain. Grinding and mixing feed, and feeding livestock	1	2	—	—	3	—
b. Dusts from swine operations [grain dust, manure dust, bacteria, mold spores, bacterial toxins (endotoxins), swine dander, insects, insect parts, ammonia adsorbed to dust]	• Working in confinement or other swine housing	2	1	—	—	3	—
	• Moving, sorting, trucking swine	1	2	—	—	3	—
	• Working in livestock barns	2	1	—	—	3	—
c. Dusts from poultry operations [grain dust, manure, feather dust, bacteria, mold spores, endotoxins, insects, insect parts, ammonia adsorbed to dust]	• Working in confinement poultry housing	2	1	—	—	3	—
	• Loading, sorting, unloading birds	2	1	—	—	3	—
	• Handling and treating birds	1	2	—	—	3	—
	• Poultry processing-unloading and live hand operations	2	1	—	—	3	—
	• Cleaning out old chicken houses	2	1	—	—	3	—

(continued)

Table 3.19. Agricultural Respiratory Hazard Exposure (*continued*)

Class and Constituents of Hazard	Jobs Resulting in Exposure	A_1	A_2	A_3	A_4	A_5	B
d. Moldy corn or other grains	• Moving spoiled grain out of storage	1	2	—	—	3	—
	• Cleaning out moldy grain from storage bins	3	1	—	—	2	—
e. Moldy silage	• Opening up non-airtight silos, throwing off the spoiled top layer	3	1	—	—	2	—
f. Moldy hay	• Moving, handling or feeding moldy hay, either loose hay or bales that have to be broken up (usually only important when done indoors)	2	1	—	—	3	—
B. Low Levels of Irritative Gases and Vapors							
1. Ammonia	• Working in poultry and livestock housing (primarily chickens, turkeys, swine, veal)	—	—	1	2	3	—
	• Working with anhydrous ammonia (see also C.6.)	—	—	1	2	3	—
2. Hydrogen sulfide	• Working in and around liquid manure storage or handling liquid manure from livestock confinement structures	—	—	1	2	3	—
	• Working inside livestock confinement structures which have liquid manure storage under a slatted floor	—	—	1	2	3	—
C. High Levels of Toxic Gases or Oxygen-deficient Environments							
1. Hydrogen sulfide	• Working inside livestock confinement buildings with storage pit under the building while the pit is being agitated or emptied	—	—	—	—	—	1
	• Entering a liquid manure storage pit anytime	—	—	—	—	—	1
2. Oxides of nitrogen	• Entering a silo (or chute) which has been filled with fresh silage within the previous 2 weeks	—	—	—	—	—	1
3. Oxygen-deficient environments	• Entering an airtight silo which has been filled with silage, haylage, or high moisture grain	—	—	—	—	—	1
	• Entering a non-airtight silo or grain bin which has recently been filled with high moisture grain	—	—	—	—	—	1
4. Carbon monoxide	• Working in an enclosed, poorly ventilated building having engine exhausts, open flames, improperly ventilated indirect-fired heaters, or improperly adjusted direct-fired or catalytic heaters	—	—	—	2	—	1

Table 3.19. Agricultural Respiratory Hazard Exposure (*continued*)

Class and Constituents of Hazard	Jobs Resulting in Exposure	Respirator Category[1]/Ranking[2]					
		A_1	A_2	A_3	A_4	A_5	B
5. Welding [metal fumes—zinc, cadmium, iron oxide, manganese; and gases—nitrogen dioxide, ozone, fluorides]	• Welding, especially in poorly ventilated areas, and especially on galvanized metals	—	—	1	2	—	3
6. Anhydrous ammonia	• Working with anhydrous ammonia in enclosed spaces	—	—	3	2	—	3
D. Pesticides							
1. Insecticides [such as organophosphates and some carbamates]	• Mixing and applying insecticides	—	—	1	2	3	—
	• Working in sprayed field before proper re-entry time	—	—	—	—	—	—
2. Fumigants [such as methyl bromide, chloropicrin, etc.]	• Putting fumigants on stored grain	—	—	1	2	3	—
3. Herbicides [paraquat]	• Working with concentrate						

N.B. For all pesticides follow the recommendations for all protective equipment

[1]AIR-PURIFYING respirators are divided into five categories:

A_1 Disposable mechanical filter respirators
A_2 Mechanical filter respirators
A_3 Chemical cartridge respirators
A_4 Gas masks
A_5 Powered air-purifying respirators

AIR-SUPPLYING respirators (B) should be used only by trained people. Local volunteer fire departments may be able to suggest where a farmer or rancher could get such equipment and training. It is not recommended that farmers or ranchers attempt to use such equipment without proper training.

[2]Rankings are based on a subjective priority ranking by the authors and content consultants: 1–3 is highest to lowest priority of acceptable respirators for use in the specified situation. The ranking is based on a combination of effectiveness of protection, comfort, and freedom from interference while performing required tasks, and cost.

ual will have different needs, based on his/her farming operation.

Once hazards have been documented, they should be eliminated or reduced to the extent possible, and appropriate respirators should be selected.

ELIMINATE OR REDUCE HAZARD BY ENGINEERING AND MANAGEMENT STRATEGIES, AND THEN REASSESS RISK

As mentioned previously in the text, hazard source reduction is the preferred way to deal with a potential agricultural respiratory hazard. Farmers and supervisors of agricultural workers may think that wearing a respirator would be the easy way, perhaps the only way, to protect their respiratory health. They might not think about controlling the work environment by engineering or management interventions, or that it would be too expensive, or make their work harder or more time-consuming. These are real issues, but respirators should never be considered as the first or only control measure for several reasons, including

1. Proper respirator selection, fitting, and maintenance are essential but may be neglected if

the wearer is not properly motivated or there is no technical assistance.

2. Wearing any nonpowered filter respirator increases the work of breathing. They can be hot, uncomfortable, and inconvenient.

3. Wearing a respirator may increase the risk of injury because of interference with vision, talking, or movement.

4. There are time and money costs to purchase and maintain respirators, train the users, and assure their proper selection, fit, and use.

5. At least two studies have shown that respirators are not completely effective; workers in swine buildings wearing respirators still had declines in pulmonary function over workshift and biomarkers of inflammation (Dosman and others 2000; Palmberg and others 2004).

Workers depending solely on respirators may not be protected if they are using ineffective equipment or if they fail to wear effective equipment when needed.

In addition, **elimination of the hazard is much safer both for family members or visitors and for the farmer/rancher: As long as a hazard exists, people unprotected by respiratory equipment may be accidentally exposed.**

For these reasons, ranchers and farmers must be encouraged to consider effective hazard source control. Each of the other sections in this series gives detailed suggestions for control of specific hazards discussed in those sections. In some cases, elimination of a hazard may also increase production and profit. For instance, adjusting ventilation in a swine house might produce healthier, faster-gaining animals. Extension agents consulting veterinarians, among others, can assist farmers in eliminating potential hazards.

Initial environmental assessments should be conducted, followed by reassessment of respiratory hazards after reduction efforts are completed. However, the reality is that some agricultural respiratory hazards cannot be sufficiently controlled or avoided within the time or finances available. In such cases, personal respiratory protection should be employed.

DETERMINE THE PATIENT'S MEDICAL SUITABILITY
TO USE RESPIRATORY PROTECTIVE EQUIPMENT
Some people, for medical or psychological reasons (a tendency for claustrophobia) should not use respirators, or a proper respirator should be selected to fit their medical need.

Mentioned previously, respirators increase the total work of breathing, the average and peak ventilation rates, and impose burdens on the pulmonary and cardiovascular systems. Some respirators impair sight, speech, or movement. Some add enough weight to be burdensome. With these respirator limitations in mind, prospective respirator wearers should be screened for the following (Harber 1984):

- Respiratory impairment, such as emphysema, asthma, chronic obstructive lung disease, restrictive lung disease, or breathing control disorders
- Cardiovascular impairment
- Poor eyesight
- Poor hearing
- Anemia
- Epilepsy, diabetes, or any other condition which may cause a person to rip off a respirator during an attack or seizure (this is of concern mainly in highly contaminated, toxic, or oxygen-deficient atmospheres)
- Hernia
- Lack of fingers or hands, or lack of full use of them (respirators such as a gas mask, supplied-air respirator, and self-contained breathing apparatus require manipulation during use that may give these persons difficulty)
- Deep facial scars or skin creases, facial hair, hollow temples, very prominent cheekbones, abnormally receding chin, full or partial dentures, or lack of teeth (these conditions may prevent a good seal between face piece and face)

Other information you should consider when evaluating medical suitability includes tasks to be performed while wearing the respirator, length of time the equipment will be worn, visual and audio requirements associated with the task, estimation of the energy requirements of the task, and the substances to which the wearer will be exposed.

Psychological conditions may also limit use of respirators. People who have had claustrophobic tendencies or those who do not understand the need for respirator use may not be reliable respirator users. This is especially true when the problem of discomfort arises. Some people are discouraged after trying a respirator that is uncomfortable. But by informed shopping around, your patient should be able to find a style of respirator that gives adequate protection with reasonable comfort and cost.

In summary, there are eight points recommended for medical evaluation on respirator use (NIOSH 1978):

1. A health care provider should determine fitness to wear a respirator by considering the worker's health, type of respirator, and the conditions of respirator use.
2. A medical history and at least a limited physical examination are recommended.
3. Although chest X-ray and/or spirometry may be medically indicated in some fitness determinations, these are not necessarily routinely performed.
4. The recommended periodicity of medical fitness determination varies according to several factors but could be as infrequent as every 5 years.
5. The respirator wearer should be observed during a trial period to evaluate potential physiological problems.
6. Health care providers may want to consider exercise stress tests with electrocardiographic monitoring when cardiovascular risk factors are present or when stressful workload conditions are expected, and when heavy respirators are used.
7. General medical or psychological work limitations are often exacerbated with respirator use.
8. In operations where there are several workers, the supervisor should consider the fitness of individuals and assign jobs accordingly.

SELECT PROPER RESPIRATOR FOR THE EXPOSURE AND THE PERSON

Table 3.19 lists agricultural respiratory hazards and agricultural tasks exposing workers to these hazards. Matched to each task or hazard are specific types of respiratory protective equipment.

Only classes of respirators are suggested, not specific respirators. Farm patients will need to base their choices on availability as well as medical suitability and specific exposure hazards. The use of air-supplying respirators by untrained farmers or ranchers is not recommended.

Proper respirator fit is important. Respirator manufacturers usually provide fitting information for their product. Some respirators come in two or three sizes, some in only one size. Some general fitting considerations include

- Beards and bushy sideburns prevent proper face piece sealing.
- Gum and tobacco chewing cause excess facial movement, which could break the seal.
- Prescription glasses and goggles may interfere with respirator fit.

Fit of nondisposable respirators can be evaluated by quantitative and qualitative tests.

Quantitative tests are more accurate but require expensive equipment and trained operators. You and your farm patient must rely on qualitative tests to determine the best fit.

Qualitative tests include the negative pressure test and the positive pressure tests, both of which are used for nondisposable respirators with tight-fitting face pieces (White 1978). For the negative pressure test, the wearer closes off the inlet of the canister, cartridge(s), or filter(s) by covering with the palm(s). The wearer then inhales so that the face piece collapses slightly. If after 10 seconds the face piece remains collapsed and no inward leakage is noticed, there is probably a tight seal. A drawback of this test is that wearer handling of the face piece may reposition the face piece and cause an improper seal.

Regarding the positive pressure qualitative test, the exhalation value is closed off and the wearer gently exhales into the face piece. If a slight positive pressure is built up, without outward leakage, the fit is considered to be satisfactory. This test should not be used if removal and replacement of the exhalation valve cover is required. Respirators should be tested for fit before each use.

Tests can be performed to determine the effectiveness of respirators (protection factors) quantitatively. Assigned protection factors (APFs) sometimes referred to in the literature as *respirator protection factors* are based on quantitative fit testing of the respirator under controlled laboratory conditions, based on the ratio of the substance concentration inside the respirator face piece as compared to the concentration outside the respirator. The APF is the minimum respirator protection provided by a properly fitted and functioning respirator on trained users. The APF can be used to predict whether certain respirators can provide adequate protection in specific environments (maximum specified use concentration). This is determined by multiplying the exposure limit for the contaminant in question by the APF assigned to that respirator.

Approximate APFs for the various classes of respirators for protection against particulate exposures and gas/vapor exposures are seen in Table 3.16.

There is currently no data available to demonstrate that the results of a quantitative fit test are sufficiently indicative of the protection that a given respirator provides in the workplace. Studies have shown that fit factors do not always correlate with worker protection factors provided by PAPRs and negative pressure half-face respirators (Popendorf 1995).

It must not be assumed that the actual level of protection that would be achieved in the workplace, or workplace protection factor (WPF), is generally less than the APF.

However, one study of WPF for various types of respirators in CAFO workers resulted in a favorable comparison to the APF. The following basic results were found (Pratt and May 1984):

Disposable respirators	WPF = 13
Quarter-faced masks	WPF = 22
Half-faced masks	WPF = 19
Powered air supply respirators	WPF = 30

A final but most important criterion for the effectiveness of any respirator is that *it must be worn*. One of the best predictors that a respirator will be worn is to assure its acceptability. Acceptability to the wearer is determined by 1) comfort, 2) ability to communicate verbally, 3) ability to see without hindrance, 4) no mechanical interference with work tasks, 5) cost, and 6) breathing resistance. A study of farm worker evaluation of respirator acceptability revealed a variation relative to the particular type of work (Pratt and May 1984). Poultry workers preferred a type of PAPR called a powered air helmet. Grain handlers preferred the half-mask mechanical filter. Swine producers were split between the quarter-mask and half-mask mechanical filter respirators. Disposable respirators were preferred by only 20% of workers in each setting. Among all groups, the powered air helmet received highest marks for breathing ease, communication ease, skin comfort, and inside mask temperature and humidity. Disposables were rated best for weight and convenience. Because of variation among workers preferences, the quarter-mask or half-mask appeared to be the best compromise without individual trial and selection.

ESTABLISH RESPIRATOR CARE PROCEDURES

Some respirators require considerable maintenance. Others, such as disposable ones, require very little. One should be sure that the farm patient/client is aware of (and willing to do) the maintenance required before purchase. Manufacturers' care instructions are provided with most respirators and should be followed.

Maintenance and inspection for defects is necessary for all respirators before each use. Encourage patients to check for stretched-out or torn headbands; bent, broken, or missing hardware; aging or damaged exhaust and intake valves; cuts, tears, or holes; melting or stiffening of face piece; and cracks or damaged threads in the filter/cartridge housings. Wearers should also check that a respirator disassembled for cleaning has been properly reassembled. Cleaning and disinfecting are required for all but disposable respirators. Manufacturer's instructions should be followed.

Proper storage means that all respirators should be protected from excess heat, cold, and humidity, and stored in an uncontaminated area (e.g., in plastic zip-lock bags).

One other wearer responsibility is to know when a disposable respirator or cartridge, canister, or filter needs replacing.

For dust/mist filters and respirators:

- Replace when first noticing that breathing becomes noticeably harder (wearer may need to try a clean filter periodically for comparison).
- Replace routinely after a standard period based on experience.

For gas/vapor respirators:

- Replace canister or cartridge upon initial odor breakthrough.
- Replace routinely after a standard period based on experience.

You and your farm patient/client can note any change in the patient's respiratory health status. How does the patient feel? If there were symptoms of respiratory irritation or disease before respirator use was begun, have the symptoms abated? Are any changes noted in pulmonary function tests?

Table 3.17 lists some of the major respirator manufacturers and their contact information. Any of these manufacturers will provide detailed information about any of their specified products.

ANSI (American National Standards Institute). 1969, 1980, and 1992. American National Standard for Respiratory Protection.

ACGIH (American Conference of Governmental Industrial Hygienists). 1985. Threshold Limit Values for chemical substances in the work environment adopted by ACGIH for 1985–1986 Cincinnati, Ohio.

American Thoracic Society. 1990. Standardization of spirometry—Statement of the American Thoracic Society. Am Rev Respir Dis 119:831–838.

Andersen C, Von Essen S, Smith L, Spencer J, Jolie R, Donham K. 2004. Respiratory symptoms and airway obstruction in swine veterinarians; a persistent problem. Am J Ind Med 46:386–392.

Bar-Sela S, Teichtahl H, Lutsky I. 1984. Occupational asthma in poultry workers. J Allergy Clin Immunol 73:271–275.

Becklake M. 1980. Grain dust and health: State of the art. In: Dosman J, Cotton DJ, editors. Occupational Pulmonary Disease, Focus on Grain Dust and Health. New York: Academic Press Inc. p 194–196.

Begany T. 2003. Hygiene hypothesis gains support in the United States and Europe. Respiratory Reviews.Com.

Blanchet M, Israel-Assayeg E, Cormier Y. 2004. Inhibitory effect of nicotine on experimental hypersensitivity pneumonitis in vivo and in vitro. Am J Respir Crit Care Med 169(8):903–909.

Braun S, Do Pico G, Tsiatis A, Horvath E, Dickie H, Rankin J. 1979. Farmer's lung disease: Long-term clinical and physiologic outcome. Am Rev Respir Dis 119:185.

Buchan R, Rijal P, Sandfort D, Keefer T. 2002. Evaluation of airborne dust and endotoxin in corn storage and processing facilities in Colorado. Int J Occup Environ Health 15(1):57–64.

Centers for Disease Control. 1982. Silo filler's disease in rural New York. MMWR 31:389–391.

Chan-Yeung M, Tabona M, Enarson D, MacLean L, Dorken E, Schulzer M. 1983. Factors affecting longitudinal decline in lung function among grain elevator workers. Am Rev Respir Dis Suppl 127:154.

Chaudemanche H, Monnet E, Westeel V, Pernet D, Dubiez A, Perrin C, Laplante J-J, Depierre A, Dalphin J-C. 2003. Respiratory status in dairy farmers in France; cross sectional and longitudinal analyses. Occup Environ Med 60:858–863.

Clark C, Rylander R, Larsson L. 1983. Airborne bacteria, endotoxin and fungi in dust in poultry and swine confinement buildings. Am Ind Hyg Assoc J 44:537–541.

Davies S. 1990. Histoplasmosis: Update 1989. Semin Respir Infect 5(2):93–104.

Djuricic S, Zlatkovic M, Babic D, Gligorijevic D, Dlamenac P. 2001. Sputumcytopathological finding in pig farmers. Pathol Res Pract 197(3):145–155.

Do Pico G. 1982. Epidemiologic basis for dose-response criteria. Agricultural Respiratory Hazards. Annals of the American Conference of Governmental Industrial Hygienists. p 189–196.

——. 1986. Workgroup report on diseases. Am J Ind Med 10:261–266.

Donham K. 1986. Hazardous agents in agricultural dusts and methods of evaluation. Am J Ind Med 10:205–220.

——. 1989. Relationships of air quality and productivity in intensive swine housing. Agri-practice 10:16.

——. 1990. Health effects from work in swine confinement buildings. Am J Ind Med 17:17–25.

——. 1991. Association of environmental contaminants with disease and productivity in swine. Am J Vet Res 52:1723–1730.

——. 1993. Respiratory disease hazards to workers in livestock and poultry confinement structures. Semin Respir Infect 14(1):49–59.

Donham K, Carsons T, Adrian B. 1982a. Carboxyhemoglobin values in swine relative to carbon monoxide exposure: Guidelines to monitor for animal and human health hazards in swine buildings. Am J Vet Res 5:813–816.

Donham K, Cumro D, Reynolds S. 2002. Synergistic effects of dust and ammonia on the occupational health effects of poultry production workers. J Agromed 8(2):57–76.

Donham K, Cumro D, Reynolds S, Merchant J. 2000. Dose-response relationships between occupational aerosol exposures and cross-shift declines of lung function in poultry workers: Recommendations for exposure limits. J Occup Environ Med 42(3):260–269.

Donham K, Gustafson K. 1982. Human occupational hazards from swine confinement. Annals of the American Conference of Governmental Industrial Hygienists 2:137–142.

Donham K, Haglind P, Peterson Y, Rylander R, Belin L. 1989. Environmental and health studies of workers in Swedish swine buildings. Br J Ind Med 46:31–37.

Donham K, Knapp L, Monson R, Gustafson K. 1982b. Acute toxic exposure to gases from liquid manure. J Occup Med 24:142–145.

Donham K, Leininger J. 1984. The use of laboratory animals to study potential chronic lung disease in swine confinement workers. Am J Vet Res 45:926–931.

Donham K, Leistikow B, Merchant J, Leonard S. 1990. Assessment of U.S. poultry worker respiratory risk. Am J Ind Med 17:73–74.

Donham K, Merchant J, Lassise D, Popendorf W, Burmeister L. 1990. Preventing respiratory disease in swine confinement workers: Intervention through applied epidemiology, education, and consultation. Am J Ind Med 18:241–262.

Donham K, Popendorf W. 1985. Ambient levels of selected gases inside swine confinement buildings. Am Ind Hyg Assoc J 46:658–661.

Donham K, Reynolds S, Whitten P, Merchant J, Burmeister L, Popendorf W. 1995. Respiratory dysfunction in swine production facility workers: Dose-response relationships of environmental exposures and pulmonary function. Am J Ind Med 27:405–418.

Donham K, Rubino M, Thedell T, Kammermeyer J. 1977. Potential health hazards of workers in swine confinement buildings. J Occup Med 19:383–387.

Donham K, Scallon L, Popendorf W, Treuhaft M, Roberts R. 1985a. Characterization of dusts collected from swine confinement buildings. Am Ind Hyg Assoc J 46:658–661.

Donham K, Yeggy J, Dague R. 1985b. Chemical and physical parameters of liquid manure from swine confinement facilities: Health implications for workers, swine, and environment. Agricultural Wastes 14:97–113.

——. 1988. Production rates of toxic gases from liquid swine manure: Health implications for workers and animals in swine buildings. Biol Wastes 24:161–173.

Donham K, Zavala D, Merchant J. 1984a. Acute effects of the work environment on pulmonary functions of swine confinement workers. Am J Ind Med 5:367–376.

——. 1984b. Respiratory symptoms and lung function among workers in swine confinement buildings: A cross-sectional epidemiological study. Arch Environ Health 39:96–100.

Donham K, Zejda J. 1992. Lung dysfunction in animal confinement workers—Chairman's report to the Scientific Committee of the Third International Symposium: Issues in health, safety, and agriculture, held in Saskatoon, Saskatchewan, Canada. Pol J Occup Med Environ Health 5(3):277–279.

Dosman J, Cotton D. 1980. Occupational Pulmonary Disease, Focus on Grain Dust and Health. New York: Academic Press Inc. 615 p.

Dosman J, Lawson B, Kirychuk S, Cormier Y, Biem J, Koehncke N. 2004. Occupational asthma in newly employed workers in intensive swine confinement facilities. Eur Respir J 24:698–702.

Dosman J, Senthilselvan A, Kirychuk S, Lemay S, Barber E, Willson P, Cormier Y, Hurst T. 2000. Positive human health effects of wearing a respirator in a swine barn. Chest 118:852–860.

Douglas W, Hepper NG, Colby TV. 1989. Silo-filler's disease. Mayo Clin Proc 64:291–304.

Eckert J. 1997. Epidemiology of Echinococcus multilocularis and E. granulosis in Central Europe. Parasitologia 39(4):337–344.

Eduard W, Douwes J, Mehl R, Heederick D, Melbostad E. 2001. Short term exposure to airborne microbial agents during farm work: Exposure response relations with eye and respiratory symptoms. Occup Environ Med 58(2):113–118.

Eduard W, Omenaas E, Bakke P, Douwes J, Heederik D. 2004. Atopic and non-atopic asthma in a farming and a general population. Am J Ind Med 46:396–399.

Emanual D, Kryda M. 1983. Farmer's lung disease. Clinical Review of Allergy 1:509–532.

Emanual D, Marx JJ Jr, Ault B, Roberts RC, Kryda MJ, Treuhaft MW. 1989. Organic dust toxic syndrome (Pulmonary Mycotoxicosis)—A review of the experience in central Wisconsin. In: Dosman JA, Cockcroft DW, editors. Principles of Health and Safety in Agriculture. Boca Raton, FL: CRC Press, Inc. p 72–75.

Epler G. 1989. Silo-filler's disease: A new perspective. Mayo Clin Proc 64:368–370.

Farant J, Moore C. 1978. Dust exposures in the Canadian grain industry. Am Ind Hyg Assoc J 39:177–194.

Ferguson K, Gjerde C, Mutel C, Donham K, Hradek C, Johansen K, Merchant J. 1989. An educational intervention program for prevention of occupational illness in agricultural workers. J Rural Health 5:33–37.

Flaherty D. 1982. Mechanisms of host response to grain dust. Annals of the American Conference of Governmental Industrial Hygienists 2:197–205.

Fleetham J, Munt P, Tunnicliffe B. 1978. Silo-filler's disease. CMA Journal 119:482–483.

Gruchow H, Hoffmann R, Marx JJ, Emanuel D, Rimm A. 1981. Precipitating antibodies to farmer's lung antigens in a Wisconsin farming population. Am Rev Respir Dis 124:411–415.

Guernsey JR, Morgan DP, Marx JJ, Horvath EP, Pierce WE, Merchant JA. 1989. Respiratory disease risk relative to farmer's lung disease antibody status. In: Dosman JA, Cockcroft DW, editors. Principles of Health and Safety in Agriculture. Boca Raton, FL: CRC Press, Inc. p 81–84.

Harber P. 1984. Medical evaluation for respirator use. J Occup Med 26:496–501.

Heederik D, Brouwer R, Biersteker K, Boleij J. 1991. Relationship of airborne endotoxin and bacteria levels in pig farms with the lung function and respiratory symptoms of farmers. Int Arch Occup Environ Health 62:595–601.

Hoffmann H, Iversen M, Takai H, Sigsgaard T, Omland O, Dahl R. 2004. Exposure to work-related levels of swine dust up-regulates CD106 on human alveolar macrophages. Am J Ind Med 46:378–380.

Holness D, O'Blenis E, Sass-Kortsak A, Deliger C, Nethercott J. 1987. Respiratory effects and dust exposures in hog confinement farming. Am J Ind Med 11:571–580.

Hoppin J, Umbach D, London S, Alavanja M, Sandler D. 2003. Animal production and wheeze in the Agricultural Health Study: interactions with atopy, asthma, and smoking. Occup Environ Med 60:63.

Horvath E, Do Pico G, Barbee R, Dickie H. 1978. Nitrogen dioxide-induced pulmonary disease, five new causes and a review of the literature. J Occup Med 20:103–110.

Jones A. 1982. Farmer's lung: An overview and prospectus. Annals of the American Conference of Governmental Industrial Hygienists 2:171–179.

Kern J, Mustajabegovic J, Schachter E, Zuskin E, Vrcic-Keglevic M, Ebling Z, Senta A. 2001. Respiratory findings in farm workers. J Occup Environ Med 43(10):905–913.

Kline J, Doekes G, Bonlokke J, Hoffman H, Von Essen S, Zhai R. 2004. Working group report 3: Sensitivity to organic dusts—Atopy and gene polymorphisms. Am J Ind Med 46:416–418.

Kusaka H, Homma Y, Ogasawara H, Munakata M, Tanimura K, Ukita H, Denzumi N, Kawakami Y. 1989. Five-year follow-up of *Micropolyspora faeni* antibody in smoking and nonsmoking farmers. Am Rev Respir Dis 140:695–699.

Lacey J, Lacey M. 1964. Spore concentrations in the air of farm buildings. Transactions of the British Mycological Society 47:547.

Larsson K, Eklund A, Malmberg P, Belin L. 1992. Alternations in bronchoalveolar lavage fluid but not in lung function and bronchial responsiveness in swine confinement workers. Chest 102:767–774.

Leblanc P, Belanger J, Laviolette M, Cormier Y. 1986. Relationship among antigen contact, alveolitis, and clinical status in farmer's lung disease. Arch of Int Med 146(1):153–157.

Lenhart S. 1984. Sources of respiratory insult in the poultry processing industry. Am J Ind Med 6:89–96.

Madsen D, Klock L, Wenzel F, Robbins J, Schmidt C. 1976. The prevalence of farmer's lung in an agricultural population. Am Rev Respir Dis 113:171–174.

Markowitz L, Hynes N, de la Cruz P, Campos E, Barbaree J, Plikaytis B, Mosier D, Kaufmann A. 1985. Tick-borne Tularemia: An outbreak of Lymphadenopathy in children. JAMA Nov 22/29:2922–2925.

Marmion BP, Ormsbee RA, Kyrkow M, Wright J, Worswick DA, Izzo AA, Esterman A, Feery B, Shapiro RA. 1990. Vaccine prophylaxis of abattoir-associated Q fever: Eight years' experience in Australian abattoirs. Epidemiol Infect 2:275–287.

Marrie T. 1989. Q Fever Pneumonia. Semin Respir Infect March 1989:47–55.

Marx J, Guernsey J, Emanuel D, Merchant J, Morgan D, Kryda M. 1990. Cohort studies of immunologic lung disease among Wisconsin dairy farmers. Am J Ind Med 18(3):263–268.

Mathisen T, Von Essen S, Wyatt T, Romberger D. 2004. Hog barn dust extract augments lymphocyte adhesion to human airway epithelial cells. J Appl Physiol 96:1738–1744.

Maurer W. 1985. Silo-filler's disease: A historical perspective and report of a case. Wi Med J 84:13–16.

May JJ, Pratt DS, Stallones L, Morey PR, Olenchock SA, Deep IW, Bennett GA. 1989. A study of dust generated during silo opening and its physiologic effects on workers. In: Dosman JA , Cockcroft DW, editors. Principles of Health and Safety in Agriculture. Boca Raton, FL: CRC Press, Inc. p 76–77.

Merchant J, Nleway A, Svendsen E, Kelly K, Burmeister L, Stromquist A, Taylor C, Thorne P, Reynolds S, Sanderson W, and others. 2005. Asthma and Farm Exposure in a Cohort of Rural Iowa Children. Environmental Health Perspectives 113: 350–356.

Monso E, Riu E, Radon K, Magarolas R, Danuser B, Iversen M, Morera J, Nowak D. 2004. Chronic obstructive pulmonary disease in never-smoking animal farmers working inside confinement buildings. Am J Ind Med 46(357–362).

Muller-Wening D, Repp H. 1989. Investigation on the protective value of breathing masks in farmer's lung using an inhalation provocation test. Chest 95:100–105.

Mustajabegovic J, Zuskin E, Schachter E, Kern J, Vrcic-Keglevic M, Vitale K, Ebling Z. 2001. Respiratory findings in livestock farm workers. J Occup Environ Med 43(6):576–584.

Nickens H. 1991. The health status of minority populations in the United States. West J Med July 1991: 27–32.

Nilsson C. 1982. Dust investigations in pig houses. Report 25. Department of Farm Buildings, Swedish University of Agricultural Sciences, Lund, Sweden.

NIOSH. 1978. Respiratory Protection: An Employer's Manual, U. S. Dept of Health, Education, and Welfare (NIOSH). Atlanta, Centers for Disease

Control: Superintendent of Documents, Washington, D.C. 20402.

NIOSH. 1987. Guide to Industrial Respiratory Protection: NIOSH Publication. p 87–116.

Notkola V, Husman K, Susitaival P, Taattola K. 1992. Morbidity and risk factors of Finnish farmers. Scandinavian Journal of Work and Environmental Health 18(2):51–54.

Noyes R. 2002. Preventing Grain Dust Explosions. Oklahoma State University Extension. Report nr CR-1737. p 1–4.

Olenchock S, Lenhart S, Mull J. 1982. Occupational exposure to airborne endotoxins during poultry processing. J Tox & Env Hlth 9:339–349.

Olenchock S, Murphy S, Mull J, Lewis D. 1992. Endotoxin and complement activation in an analysis of environmental dusts from a horse barn. Scandinavian Journal of Work and Environmental Health 18(2):58–59.

Osbern L, Crapo R. 1981. Dung Lung: A report of toxic exposure to liquid manure. Ann Intern Med 95:312–314.

Palmberg L, Larsson B-M, Sundblad B-M, Larsson K. 2004. Partial protection by respirators on airways responses following exposure in a swine house. Am J Ind Med 46:363–370.

Parkes W. 1982. Occupational Lung Disorders. London: Butterworth and Co. Ltd.

Pederson B, Iversen M, Dahl R. 1990. Bronchoalveolar lavage of pig farmers. Am J Ind Med 17:118–119.

Pesci A, Bertorelli G, Dall'Aglio P, Neri G, Olivieri D. 1990. Evidence in bronchoalveolar lavage for third type immune reactions in hypersensitivity pneumonitis. Eur Respir J 3(3):359–361.

Pladsen T. 1984. Silo emptiers' diseases. Minn Med May 1984:265–269.

Pont F, Gispert X, Canete C, Pinto E, Dot D, Monteis J. 1997. An epidemic of asthma caused by soybean in L'Hospitalet de Llobregat (Barcelona) (in Spanish). Arch Bronconeumol 33(9):453–456.

Popendorf W. 1995. Respirator protection and acceptability among agricultural workers. Appl Occup Environ Hyg 10(7):595–605.

Pratt D, May J. 1984. Feed associated respiratory illnesses in farmers. Arch Environ Health 39:43–48.

Prior C, Falk M, Frank A. 2001. Longitudinal changes of sensitization to farming-related antigens among young farmers. Respiration 68(1):46–50.

Radon K, Danuser B, Iversen M, Monso E, Weber C, Hartung J, Donham K, Palmgren E, Nowak D. 2002. Air contaminants in different European farming environments. Ann Agric Environ Med 9(1):41–48.

Radon K, Ehrenstein V, Praml G, Nowak D. 2004. Childhood visits to animal buildings and atopic diseases in adulthood: An age-dependent relationship. Am J Ind Med 46:349–356.

Rask-Andersen A. 1989. Organic dust toxic syndrome among farmers. Br J Ind Med 46:233–238.

Reynolds S, Donham K, Whitten P, Merchant J, Burmeister L, Popendorf W. 1996. Longitudinal evaluation of dose-response relationships for environmental exposures and pulmonary function in swine production workers. Am J Ind Med 29(33–40).

Rylander R. 1987. Role of endotoxins in the pathogenesis of respiratory disorders. Eur J Respir Dis Suppl 154:136–144.

——. 2004. Organic dusts and disease: A continuous research challenge. Am J Ind Med 46:323–326.

Rylander R, Essle N, Donham K. 1990. Bronchial hyperreactivities among pig and dairy farmers. Am J Ind Med 17:66–69.

Salvaggio JE, Taylor G, Weill H. 1986. Occupational asthma and rhinitis. In: Merchant JA, editor. Occupational respiratory diseases. Cincinnati, OH: U.S. Department of Health and Human Services, Public Health Service, Centers for Disease Control, National Institute for Occupational Safety and Health, DHHS (NIOSH) Publication No. 86-102, p 461–477.

Schatz M, Patterson R. 1983. Hypersensitivity pneumonitis—General considerations. Clinical Review of Allergy 1:451–467.

Schenker M. 2000. Exposures and health effects from inorganic agricultural dusts. Environ Health Perspect 108(suppl. 4):661–664.

Schenker, M.B., editor. American Thoracic Society. 1998. Respiratory health hazards in agriculture. Am J Respir Crit Care Med 158:S1–S76.

Schwartz D. 2002. The genetics of innate immunity. Chest 121:625–685.

Schwartz D, Donham K, Olenchock S, Popendorf W, Van Fossen D, Burmeister L, Merchant J. 1995. Determinants of longitudinal changes in spirometric function among swine confinement operators and farmers. Am J Respir Crit Care Med 151:47–53.

Schwartz DA, Donham K, Popendorf W, Lassise D, Hunninghake G, Merchant J. 1990. Are work shift changes in lung function predictive of underlying lung disease? Am Rev Respir Dis 131:A 593.

Seaton A. 1994. The breathless farm worker. Br Med J 288:1940–1941.

Senthilselvan A, Dosman J, Kirychuk S, Barber E, Rhodes C, Zhang Y. 1997. Accelerated lung function decline in swine confinement workers. Chest 111:1733–1741.

Siegel P, Shahan T, Sorenson W. 1992. Analysis of environmental histamine from agricultural dust. Scandinavian Journal of Work and Environmental Health 18(2):60–62.

Sigurdarson S, Donham K, Kline J. 2004a. Acute toxic pneumonitis complicating chronic obstructive pulmonary disease (COPD) in a farmer. Am J Ind Med 46(4):393–395.

Sigurdarson S, O'Shaughnessy P, Watt J, Kline J. 2004b. Experimental human exposure to inhaled grain dust and ammonia: Towards a model of concentrated animal feeding operations. Am J Ind Med 46:345–348.

Stepanek J, Capizzi S, Edell E. 1998. Case in point. Silo filler's lung. Hosp Pract 33:70.

Swedish National Board of Occupational Safety and Health. 1994. Organic Dust in Agriculture.

Thedell T, Mull J, Olenchock S. 1980. A brief report of gram-negative bacterial endotoxin levels in airborne and settled dusts in animal confinement buildings. Am J Ind Med 1:3–7.

Thelin A, Tegler O, Rylander R. 1984. Lung reactions during poultry handling related to dust and bacterial endotoxin levels. Eur J Respir Dis 65:266–291.

Thorne P, Duchaine C, Douwes J, Eduard W, Gorny R, Jacobs R, Reponen T, Schierl R, Szponar B. 2004. Working group report 4: Exposure Assessment for biological agents. Am J Ind Med 46:419–422.

Thorshauge H, Fallesen I, Ostergaard P. 1989. Farmer's lung in infants and small children. Allergy 44:152–155.

U.S. Dept of Health, Education, and Welfare (NIOSH). 1978. Respiratory Protection: An Employer's Manual. Atlanta, Centers for Disease Control: Superintendent of Documents, Washington, D.C. 20402.

Vanchuk J. 1978. What steps must be taken to establish a respiratory protection program? Occup Health Saf July/August 1978:26–29.

Vandergrift A, Shannon L, Sallee E. 1971. Particulate air pollution in the United States. J Air Pollut Control Assoc 21:324.

Viet S, Buchan R, Stallones L. 2001. Acute respiratory effects and endotoxin exposure during wheat harvest in northeastern Colorado. Appl Occup Environ Hyg 16(6):685–697.

Von Essen S, Fryzek J, Nowzkowski B, Wampler M. 1999. Respiratory symptoms and farming practices in farmers associated with an acute febrile illness after organic dust exposure. Chest 116(5):1452–1458.

Von Essen S, Robbins R, Thompson A, Rennard S. 1990. Organic dust toxic syndrome: An acute febrile reaction to organic dust exposure distinct from hypersensitivity pneumonitis. Journal of Clinical Toxicology 28(4):389–420.

Warren C. 1977a. Extrinsic Allergic Alveolitis: A disease commoner in non-smokers. Thorax 32(5):567–569.

——. 1983. Health and safety in the grain industry. In: Rom WN, editor. Environmental and Occupational Medicine. Boston: Little, Brown and Co. p 221–232.

White J. 1978. Respirator fitting: The key to protecting workers. Occup Health Saf July/August 1978:22–25.

World Health Organization. 2005. Death in a Veterinarian from Avian Flu.

Wouters I, Sigsgaard T, Gora A, Nowak D, Palmberg L, Sundblad B-M, Tutluoglu B. 2004. Working group report 1: Tools for the diagnosis of organic dusts-induced disease. Am J Ind Med 46:410–413.

Zimmer A. 2002. The influence of metallurgy on the formation of welding aerosols. J Environ Monit 17:1–6.

Zwemer FJ, Pratt D, May J. 1992. Silo filler's disease in New York state. Am Rev Respir Dis 146:650–653.

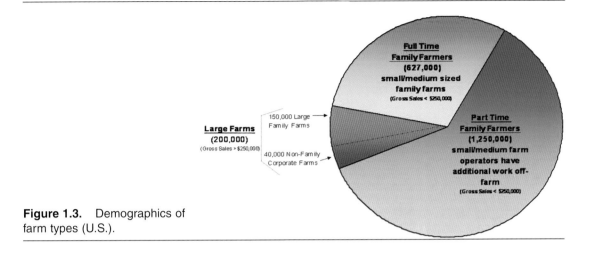

Figure 1.3. Demographics of farm types (U.S.).

Figure 2.1. A woman working with her swine.

Figure 2.2. The men of farm families are increasingly taking off-farm jobs, increasing the pressure on a woman to assume multiple roles on the farm and increasing her risk of injury.

Figure 2.3. A propane-powered radiant heater, which increases the level of CO in swine barns and the risk of prenatal and birthing complications in women.

Figure 2.9. Farmers often keep working well past usual retirement age. The man on the far right in this photo is 91 years old and still works daily.

Figure 2.12. Migrant worker (Source: Photograph by Tierra Nueva, www.peoplesseminary.org/images/013102.a.jpg).

Figure 2.14. Amish men working in their modified traditional ways (Source: iStockphoto.com, photograph by Diane Diederich, www.istockphoto.com).

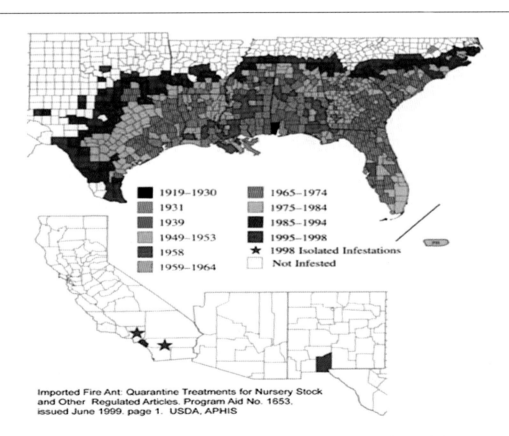

■	1919–1930	▨	1965–1974
▨	1931	▨	1975–1984
■	1939	■	1985–1994
▨	1949–1953	■	1995–1998
■	1958	★	1998 Isolated Infestations
▨	1959–1964	□	Not Infested

Imported Fire Ant: Quarantine Treatments for Nursery Stock and Other Regulated Articles. Program Aid No. 1653. issued June 1999. page 1. USDA, APHIS

Figure 4.1. Range expansion of imported fire ant from 1918–1998 (Source: USDA website, http://www.usda.gov/wps/portal/usdahome).

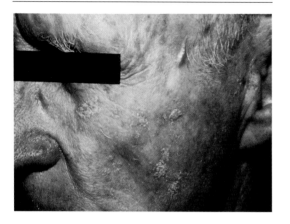

Figure 4.2. Actinic keratoses.

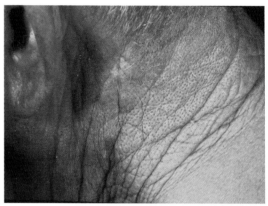

Figure 4.3. Compare the thickened and wrinkled skin from chronic sun exposure to skin protected by the shirt collar on this 65-year old farmer.

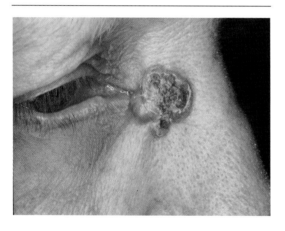

Figure 4.4. Basal cell carcinoma.

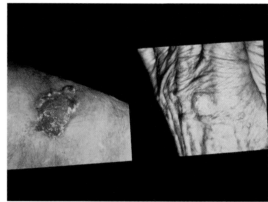

Figure 4.5. Squamous cell carcinoma; two different clinical appearances.

Fgure 6.1. Spray application of insecticides to row crops—applicators are much better protected inside a tractor with a cab, especially if the windows are shut and the cabin filtration system is properly functioning.

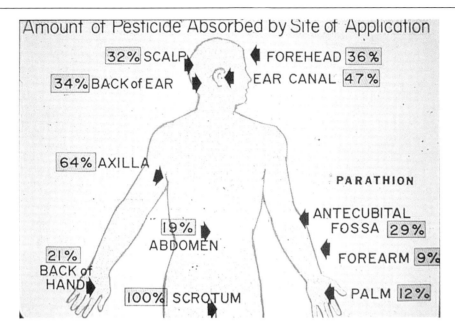

Amount of Pesticide Absorbed by Site of Application

32% SCALP   FOREHEAD 36%

34% BACK of EAR   EAR CANAL 47%

64% AXILLA

PARATHION

ANTECUBITAL FOSSA 29%

19% ABDOMEN

FOREARM 9%

21% BACK of HAND

100% SCROTUM

PALM 12%

Figure 6.2. The most common systemic exposure to pesticide is through skin contact. Skin absorption of organophosphate pesticides is especially efficient in areas of thin skin with high blood supply.

Figure 6.3. Mixing and loading insecticides prior to application can be an operation with a high potential for worker exposure as the chemicals input into the tank are concentrated. Minimum protection during this application includes unlined rubber gloves.

Figure 6.4. In crop dusting applications, there is risk for insecticide exposure to the pilot, the flagger on the ground, who marks the way for the pilot, and the persons who mix and load the plane and clean up the plane.

Figure 6.5. Picking citrus or other fruit or orchard crops presents possible exposure to workers from pesticides that may remain on the foliage. (*Source:* Photograph by Terry Noble, www.gaysmills.org/orchards.html).

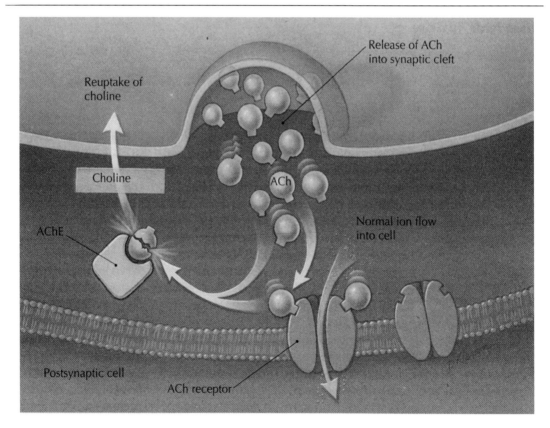

Figure 6.6. Normal transmission of a nerve impulse across a synapse—acetylcholine is released from the end organ at the terminus of the nerve fiber and travels across the synapse to transmit the impulse to the next nerve fiber. The acetylcholine is rapidly broken down at the synapse by acetyl-choline esterase to stop sending the signal.

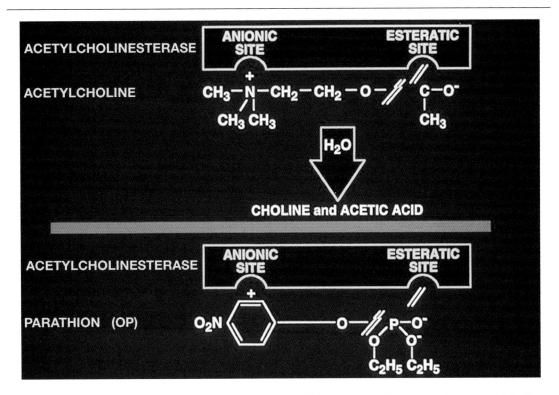

Figure 6.7. Organophosphate (OP) and carbamates (CB) combine with and inactivate acetylcholine esterase (AChE). Excess acetylcholine (ACh) at the synapse overstimulates the receiving nerve fiber. A positive test for poisoning is a decreased amount of detectable AChE. OP–AChE bonds tend to be longer-lasting than CB-AChE bonds.

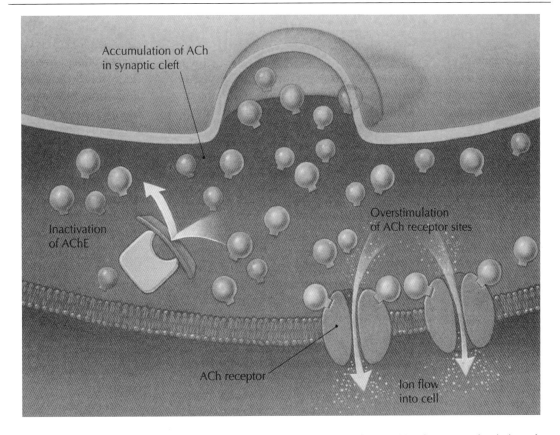

Figure 6.8. As OPs or CBs tie up AChE, there is a build-up of ACh resulting in a overstimulation of the sympathetic system.

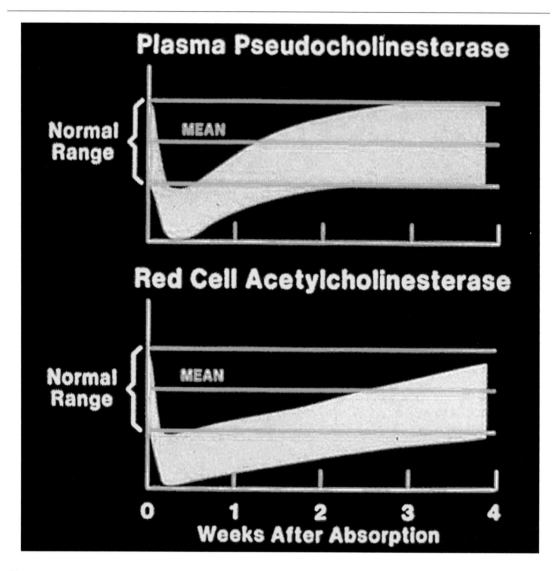

Figure 6.9. There is a much longer time for recovery of RBC AChE (4 or more weeks) as compared to plasma AChE (2–3 weeks). This means that measuring only plasma acetylcholine esterase may detect a rather recent poisoning, but not one that occurred a week or two prior to the test. The RBC test will detect a poisoning that may have occurred up to 3 or more weeks previously. It is best to conduct both plasma and RBC tests simultaneously.

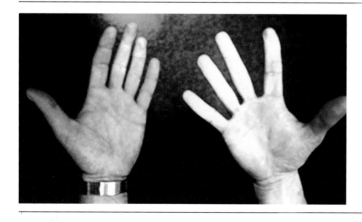

Figure 9.7. Vibration white fingers.

4
Agricultural Skin Diseases

Kelley J. Donham and Anders Thelin

INTRODUCTION

Agricultural workers as well as other rural residents frequently come in contact with agents capable of causing skin diseases, including plants, insects, pesticides, sunlight, heat, and infectious agents (Burke 1997). A study of agricultural producers in the U.S. (383 males and 265 females) revealed an overall prevalence of contact dermatitis of 10% in men and 14% in women (Park and others 2001). In a California study, the highest rates of occupational skin diseases were found in farm workers relative to the comparison population (Mathias 1988). However, 34% of a cohort of North Carolina farm workers reported symptoms of dermatitis at the end of the growing season (Arcury and others 2003). From Finland it has been reported that cow dander, disinfectants and detergents, wet and dirty work, and rubber chemicals were the main causes of farmers' occupational hand eczema (Susitaival and Hannuksela 1995). The hand labor that is common among farmers generally and migrant and season workers in particular, results in close direct contact with plants, which may contain irritants or allergens, chemicals used in production, and livestock that may create increased risk for contact dermatitis.

Farm and rural-associated skin diseases can be classified into five major categories:

1. Contact dermatitis
2. Infectious dermatitis
3. Arthropod-induced dermatitis
4. Sun-induced skin conditions
5. Skin disorders related to heat, cold, humidity

CONTACT DERMATITIS

The most common type of dermatitis (especially in agricultural and rural settings) is **irritant contact dermatitis. Allergic contact dermatitis** is less common, and **photocontact (irritant or allergic) dermatitis** reactions are rare.

Irritant contact dermatitis can be caused either by a single exposure to a strong irritant or by multiple exposures to weak irritants. Allergic contact dermatitis is caused by exposure to an antigen to which the patient has developed cell-mediated immunity. This reaction mostly is delayed as a typical allergic contact dermatitis. The allergic reaction may also be immediate, mediated by IgE antibodies as urticaria, pruritic wheals, and erythema and possibly edematous swelling if the hands or feet are involved. Allergic contact dermatitis is most often a chronic condition resulting in eczematous lesions. These can be distinguished only through allergy testing (patch and skin prick testing). Photocontact dermatitis develops when the offending substance, having contacted the skin, is exposed to sunlight, chemically changing the substance to an irritant (photoirritant contact dermatitis), or allergen/sensitizer (photoallergic contact dermatitis) (Adams 1969; Fisher 1973).

Although the overall risk rate of contact dermatitis among farmers is not known, Finnish studies suggest that it may be as high as 12% (Susitaival and others 1994). However, it can be two to four times higher in atopic individuals. Forty-five percent of atopic women in the previous study reported hand dermatitis (Susitaival and Hannuksela 1995). Respiratory allergies are predictors for allergic skin conditions; e.g., farmers with symptoms of allergic rhinitis have higher risk for contact dermatitis as compared to those without respiratory allergy (Spiewak and Dutkeiwicz 2002; Spiewak and others 2001).

Pesticide-Induced Dermatitis

Most cases of pesticide-induced contact dermatitis are irritants in nature. The dermatitis develops from skin contact when these chemicals are not properly diluted, when protective clothing is not worn, or when work clothes are re-worn without washing. Other exposure circumstances occur when workers such as fruit pickers have close and sustained contact with foliage that has been sprayed with a pesticide. Additional risk factors are apparent when no pesticide training is provided, when the chemicals are mixed, loaded, or applied, or when insufficient care is used in applying chemicals (Arcury and others 2003). Examples of strong irritant pesticides include trichloroacetic acid (a herbicide), the fumigants ethylene oxide and methyl bromide, and propargite (Omite-CR). The latter is a noncholinesterase inhibiting acaricide, used on orange trees. Other irritant chemicals used in fruit and vegetable growing include Dyrene (contains anilazine, an acaricide) and elemental sulfur (commonly used in many areas of fruit and vegetable growing) (Centers for Disease Control 1986).

Herbicide-induced allergic contact dermatitis most commonly is caused by one of the chlorinated acetanilide, such as Propachlor (trade name Ramrod, Bexton) or alachlor. The herbicides maleic hydrazide and thiram (also fungicide and insect repellent) also are sensitizers. Thiram sensitization is notable as this chemical is present in many products in addition to herbicides, including some insecticides, fungicides, rubber additives, soaps, shampoos, paints, and putty.

Insecticide-induced allergic contact dermatitis can be caused by pyrethrum, commonly used in many household insect sprays and powders. A pyrethrum-sensitive individual may also develop sensitivities to chrysanthemums, shasta daisies, ragweed, and other members of the daisy family because they all contain oleoresins similar to pyrethrum. Phenothiazine (in addition to its use as a pesticide) is a feed additive used to control roundworms in pigs, sheep, and cattle. Contact with animal feeds or spraying with substances containing phenothiazine may result in allergic or photocontact dermatitis (Belsito 2003). Sun exposure to people who have been in contact with these chemicals directly or in feed may lead to phototoxic dermatitis on sun exposed areas of the skin.

In a number of epidemiological studies, fungicides have been found to be a cause of allergic contact dermatitis. Products containing formaldehyde, beta-naphthol, thiram, guanidines, captan, captafol, and imidazole derivatives may act as sensitizers (Table 4.1).

Although contact dermatitis is the most common type of dermatitis associated with pesticides, there are several other conditions that have been reported as a result of chronic handling of pesticides, including urticaria, erythema multiforme, ashy dermatosis, parakeratosis variegata, prophyria cutanea tarda, chloracne, hypopigmentation, and nail and hair disorders. Arsenical pesticides (although not licensed today in most

Table 4.1. Common Agricultural Chemicals That Are Skin Sensitizers for Agricultural Workers

Insecticides	Herbicides	Fungicides	Biocide/ Fumigant	Antibiotics	Dusts and Animal Dander
Thiram Pyrethrum Phenothiazine	Thiram Chlorinat Acetanilides • Propachlor (tradename Ramrod, Bexton) • Alachlor Maleic hydrazide	Thiram Captan Captafol Imidazoles Beta-naphthol Guanidines	Formaldehyde	Furazolodin Spectinomycin	Cattle Grain dust Hay dust

developed economies), are a known cause of skin cancers including carcinoma in situ, and basal and squamous cell carcinomas (Spiewak 2001).

Other Chemicals Related to Dermatitis

There are many substances in addition to pesticides that can cause contact dermatitis, including antibiotics, vaccine products, petroleum products, and chemicals/substances in different materials (see Table 4.1). A rural health care provider must attempt to keep up to date about those products currently in use and their potential health hazards, and be familiar with work practices so that an appropriate occupational history may be obtained leading to a diagnosis, as illustrated by the following case reports.

Antibiotics can cause both irritant and allergic contact dermatitis. Tetracyclines are common feed-additive antibiotics for swine, poultry, and cattle. Inhalation of feed dust may create sufficient systemic levels that a photoirritant contact dermatitis may develop on sun-exposed skin surfaces (Belsito 2003). Tars—e.g, creosote (a wood preservative used for fence posts, fence boards, etc.)—can cause photocontact dermatitis as well as irritant-contact dermatitis.

Fertilizers contain irritant nitrogen and phosphate compounds and sometimes cobalt and nickel. Anhydrous ammonia is a fertilizer, which may cause severe chemical burns (Latenser and Lucktong 2000).

Dairy farmers are frequently in contact with chemicals associated with rubber or synthetic rubber products (e.g., hoses of milking machinery and rubber gloves for example). Delayed allergy to rubber chemicals has been found among dairy farmers with eczema (Nurse 1978; Telintum and Nater 1974).

Oils and fuels are common causes of skin problems. Oil acne or oil folliculitis is resulting from exposure to oil especially under oil-soaked clothing. Most common sources today are insoluble cutting oils in machinists and greases and lubricating oils in mechanics. The condition probably is often not reported since most workers know that with better hygiene the condition improves.

Animal-Related Skin Dermatitis

Several reports from Finland indicate that sensitivity to cow dander creates a significant risk for immediate or delayed allergic contact dermatitis

among dairy farmers (Susitaival and others 1994). Eighteen percent of women in dairy farming reported hand dermatitis within the past year (Susitaival and others 1994). The 1-year prevalence of hand dermatitis was 7% among men in dairy farming. Risk factors included previous history of atopic dermatitis (fourfold), and respiratory atopy (twofold). Skin tests among Finnish farmers revealed that 30% were positive for cow dander by either prick or patch testing. Other positive skin tests included animal feeds, rubber and rubber gloves, and udder ointments. A 12-year follow-up revealed 50% of the patients first diagnosed with hand dermatitis were still having problems, suggesting the chronic nature of this disease and the difficulty in controlling it (Susitaival and Hannuksela 1995).

Plant-Induced Dermatitis

There are several plants growing in rural areas and on farmland that cause allergic contact dermatitis. The most frequent reactions are caused by poison ivy, poison oak, and poison sumac. These three plant species cause a reaction commonly called *rhus dermatitis,* which is a classic delayed allergic contact dermatitis, taking from 48 to over 100 hours for lesions to develop. The sap of these plants is capable of inducing cell-mediated hypersensitivity in two out of three people. The plants are of the Anacardiaciae family, which shares the characteristics of bearing groups of whitish berries, and contain highly sensitizing pentadecylcatachols (PDCs). Urushiol is the specific PDC associated with the poison ivy, oak, and sumac.

The rhus plants have a varying geographic distribution. Poison ivy is by far the most common and is found in Eastern Asia, the Kurile Islands, and the Sakhalin Islands of Russia. The plant in the American continent exists in the Eastern two-thirds of Canada and the United States below the 44th parallel. It grows throughout Mexico and Guatemala and in some of the Caribbean countries. Poison oak has a more limited distribution to the West Coast of the U.S. Poison sumac is primarily in the eastern one-third of the U.S. and Canada from Quebec to Florida, and extending west into Texas (Agriculture and Agri-Food Canada 1990).

Urushiol is located in the sap of nearly all parts of the plant. Therefore, dermatitis from these plants can be contracted when any part of the plant is broken, allowing sap to escape directly onto the skin of a sensitized person. The sap from all three plants maintains its antigenicity for months (American Academy of Dermatology 2005).

A much less prevalent plant-induced allergic reaction is caused by ragweed (Ambrosia species). Ragweed is much more commonly maligned as a cause of allergic rhinitis (or hay fever). However, the pollen from this plant carries two allergenic materials, a protein that causes allergic rhinitis, and an oleoresin (which contains the allergen sesquiterpene lactone) that causes an allergic skin condition, from an airborne exposure. The exposure can produce a diffuse pattern presentation of eruption similar to that caused by poison ivy smoke. It can be differentiated from photocontact dermatitis by presence of the eruptions on shaded areas of skin, such as under the chin and behind the ear lobule. The rash may be accentuated in flexural creases and at the edges of clothing. Fortunately, only a small percentage of individuals exposed to ragweed oleoresin will develop allergic contact dermatitis, compared to the high percentage of risk from poison ivy exposure (Belsito 2003).

Initially, ragweed-induced allergic contact dermatitis is present only in late summer and fall. However, annual recurrences tend to become longer lasting.

There are a number of plants in addition to those presented above that have been known to cause allergic contact dermatitis. Numerous reports in the literature discuss these vegetable-related skin conditions. Table 4.2 lists those that have been incriminated as causative agents. One example is an allergic contact dermatitis that may occur in workers who pick and handle asparagus. The allergen appears to be a sulfur-containing growth inhibitor (1,2,3-Trithiane-S-carboxyli-cacid), which is more prominent in the early growing season. The lesions typically occur on the fingertips, but may spread to the whole hand (Hausen and Wolf 1996; Rademaker 1998; Society NZD 2004).

Other plants can cause contact or photocontact dermatitis. Vegetables in the family Umbelliferae (carrots being the most notorious) are capable of producing contact dermatitis. Vegetable-induced dermatitis is most likely to affect persons who

Table 4.2. Plants and Their Offending Substances That Cause Allergic Contact Dermatitis

Plants Causing Allergic Contact Dermatitis	Offending Plant Chemicals
Bulbs of tulips, hyacinths, and other bulb flowering plants	NK
Carrots	Furocoumarins
Parsnips	Furocoumarins
Asparagus	1,2,3-Trithiane-5-carboxylic acid
Cucumber	dibromodicyanobutane
Tomato	dibromodicyanobutane
Zucchini	dibromodicyanobutane
Potatoes	NK
Garlic	NK
Leek	NK
Mustard	NK
Kiwifruit	NK
Mangoes	Pentadectylcatechols
Cashew nuts	Pentadectylcatechols
Pecan	NK
Olive oil	NK
Rice	NK

Sources: http://tigmor.com/food/library/articles/contact.htm; Hausen and Wolf 1996; O'Dennell and Foulds 1993.
NK = not known.

harvest, pack, and unpack, for sale, large quantities of cut and wet vegetables. Some of the species in the Umbelliferae family, for example, Heracleum mantegazzianum, common in northern Europe and Asia in pastureland and wetland, can in combination with sunshine cause a phototoxic reaction (phytophoto dermatitis). Sunshine modifies flurocoumarins, present in juices of certain plants, resulting in an irritant-based skin response.

Occasionally, contact dermatitis can result from an airborne exposure to dust, animal feed, mites, animal dander, and plants, and it can be irritant, allergic, or photoirritant/allergic in nature. An example of such a case was reported in a 57-year-old female hops plantation worker (Spiewak and Dutkeiwicz 2002). After 30 years' experience working with hops, she developed an eczematous allergic contact dermatitis on her hands. Additionally, she developed erythema of the neck, edema of the lids, and conjunctiva, typically beginning after 30 minutes aerosol exposure to fresh dried hops. This second condition persisted 48–72 hours after removal from the environment. This woman had to quit working with hops, but remained allergic, with exacerbations on use of beauty creams and herbal sedatives that contained hops (*Humulus lupulus*). Her husband continued to work with hops. The woman would have exacerbations when sleeping with her husband, as he would have small amount of hops attached to his skin or bedclothes (cannubial contact).

Clinical Picture of Contact Dermatitis

Irritant contact dermatitis may be divided into two types. Immediate irritant contact dermatitis results from a single contact with a strong substance causing a toxic reaction often similar to a burn. Erythema, blistering, and ulceration occur soon after contact. Most irritant contact dermatitis, however, is delayed and results after prolonged or repeated contacts. Erythema, increasing dryness, and thickening, as well as patchy hyperkeratosis, are frequent characteristics. Itching and painful fissuring are also common symptoms. There is large individual variation in the clinical picture. Atopics are more likely to develop all kinds of dermatitis after short periods of exposure.

Contact with oil, greases, etc., may result in a pustular irritant dermatitis with acneiform characteristics. Repeated rubbing may result in a thickened psoriasis resembling *lichen simplex*. Polyhalogenated biphenyls and related chemicals may induce chloracne.

Allergic contact dermatitis often begins 24–48 hours after contact among sensitized persons. An erythematous rash develops rapidly. Papule-formation and blistering are common. Itching is always a prominent symptom. After some days a chronic stage evolves. It is often impossible to separate a chronic allergic contact dermatitis from an irritant dermatitis by the clinical picture. The diagnosis of allergic contact dermatitis must be done using patch or prick testing.

Exposure to the rhus plants (poison ivy, poison oak, and poison sumac) mostly results in an erythematous, edematous, very pruritic rash, which appears from 1–7 days following exposure. The extent and severity of reaction varies among persons, and lesions often continue to develop for several days. Vesicles may form and rupture, producing a serous drainage and crusting. Serous drainage (and scratching) does not spread the lesions, as there is no urushiol within the vesicles. However, scratching may lead to a secondary infection. Severity of reaction depends both on the quantity of antigenic sap transferred to the skin and on the individual's sensitivity. Furthermore, the delay of onset and decreased severity is related to slower absorption of the agent in areas of skin with a thicker stratum corneum (e.g., palms, dorsal surface of hands and arms).

Most commonly, the pattern of the skin lesions have a linear or streaked configuration. This is because the urushiol exudes from the broken stems or leaves of the plant, and as the person moves, the plant drags or brushes across the skin and the urushiol is applied in a linear pattern. A more diffuse pattern may appear when the urushiol contaminates the skin secondarily from contaminated clothing or a dog, for example, or when exposed to smoke of burning plant material.

Bulb finger is a chronic inflammatory condition of fingertips that develops in some harvesters, sorters, and packers of tulip, hyacinth, onion, and garlic bulbs. The irritant is thought to be a sesquiterpene lactone present in the bulbs (White and Cox 2000). However, bulb allergens and fungicides used to treat bulbs may also be a portion of the agents that cause this problem. Initially, fingertips become red and inflamed. Further handling of these bulbs will cause a dry, fissured, and scaly dermatitis.

Usually phytophoto dermatitis initially presents with blisters. Lesions heal often leaving a hyperpigmentation, which commonly has a bizarre streaked pattern and persists for months. This streaked pattern is due to contact with the leafy parts of the vegetable (i.e., carrot tops) contacting the skin, leaving an imprint of the agent.

Although the previous conditions are quite common, there are several uncommon skin conditions that may result from contact with substances in agriculture. These conditions are listed in Table 4.3.

Treatment of Contact Dermatitis

Topical corticosteroids are used to treat most localized forms of contact dermatitis. In addition, standard principles of dermatologic therapy—

Table 4.3. Uncommon Skin Conditions of Agricultural Workers Caused by Agricultural Chemicals

Condition	Description	Exposure
Urticaria	Elevated reddened areas, also called wheal and flare reaction	Fungicides: Captan, Chlorothalonil Insect Repellant: Diethyltoluamide
Erythema mutiforme	Concentric rings, "bulls eye" eruptions on the skin	Organophosphate insecticides: Methyl parathion, Dimethoate
Ashy dermatitis	Ashen-colored eruptions (macules) on the skin of varying size. Also called erythema dyschromicum perstans	Fungicide: Chlorothalonil
Parakeratosis variegata	Early ashy dermatoses appearance, progresses to entire skin and poikiloderma (skin atrophy with speckled-like discoloration)	Undetermined pesticide and fertilizer contact
Porphyria cutanea tarda	A skin photosensitivity with large blisters, scarring, excess pigmentation, skin thickening, hair loss	Herbicides: 2,4-D and 2,4,5-T
Chloracne	Very severe and chronic acne	Herbicides: Those chemicals with chlorinated polycyclic aromatics. Probably it is the **dioxin contaminant of** the manufacture of these compounds: pentachlorophenol, propanil, dichloroaniline, methazole
Skin hypopigmentation	Light-colored areas on the skin	Carbamate insecticides: (unspecified)
Nail disorders	Yellowed, brittle, altered growth of the nails	Herbicides: paraquat, diquat, dichloronitrocresol
Hair loss		Chlorinated hydrocarbon: DDT
Skin cancer	Mainly squamous cell carcinoma	Herbicides: paraquat, arsenicals

Spiewak 2001

such as physiologic NaCl; soaks and compresses for weeping lesions; and lubrication for dry, scaling lesions—are important. Cool baths with, e.g., colloidal oatmeal will help alleviate pruritus. Prolonged use of the more potent topical corticosteroids may result in atrophy, striae, and erythema that may be permanent. Systemic absorption may also occur if administered over a broad area or if ultrapotent.

For widespread or severe cases, systemic corticosteroids may be necessary, although risk of side effects is higher. A 7–10-day taper of either topical or systemic corticosteroids may help prevent a "rebound flare" of dermatitis.

There are many ways to help prevent or minimize the skin reaction in people who have had contact with the sap of rhus plants. Washing exposed skin with soap and water within 20 minutes of contact decreases the extent and severity of the subsequent dermatitis. Mild rhus dermatitis can be treated symptomatically with compresses, calamine lotion, potent topical corticosteroid preparations, and orally administered antihistamines. Severe rhus dermatitis is best managed with systemic corticosteroids. For adults, 30–80 mg of prednisone daily, in divided dosages, is appropriate initially. The dose can be reduced by 5 milligrams per day when vesicles are no longer forming. Antihistamines can help alleviate the itch. Compresses, colloidal baths, and calamine lotion are helpful adjuncts.

For patients in whom topical corticosteroids

are contraindicated, two newer agents that are nonsteroidal should be considered. Tactolimus (Protopic® ointment) and pemecrolimus (Elidel® cream) are analogs of cyclosporine, and their mechanism of action is T-cell suppression. They are FDA approved, and generally well tolerated. In addition, they are much more expensive than corticosteroids and long-term safety and efficacy remain to be determined.

Prevention of Contact Dermatitis

All types of contact dermatitis can be prevented by elimination or reduction of exposure to causative agents. Specific measures include use of protective clothing, changing to clean clothes and gloves when they become contaminated, and washing exposed areas of the skin before lunch and at the end of the work day. Agricultural workers should take care to read and follow label directions on the package when applying pesticides, fertilizers, and other agricultural chemicals. Furthermore, there are barrier creams that farmers can use on hands and exposed skin that may prevent or decrease the amount of absorption of irritant or allergenic substances. However, if not properly used, barrier creams may be dangerous. There is a specific barrier cream for rhus exposure that contains octylphenoxy-polythoxy ethanol in propylene glycol, or 5% quanternium-18 bentonite (respective trade names *Oak Ivy* and *Ivy Block*).

Poison ivy and related sensitizing plants can be eliminated by pulling the plants out by their roots (grubbing). Grubbing is most effective if the soil is soft and moist. Grubbing should be done using proper precautions, such as wearing rubber gloves and protective clothing that is washed after use. These plants also can be eliminated by the use of herbicides. The broad spectrum herbicide Round Up (glyphosate) works well on these plants. This chemical is applied on the leaves, absorbs into the plant, and kills the whole plant (roots and all) in a matter of days to weeks. Mowing and cutting of plants should not be attempted, since these actions encourage plant stem proliferation from remaining roots.

For patients who are extremely sensitive to urushiol or because of their job cannot avoid contact with rhus plants, hyposensitization is an option. There are two forms of the antigen used for hyposensitization, oral and subdermal injection.

However, hyposensitization provides partial protection at best. It may reduce the symptoms or protect in mild exposures. Oral hyposensitization may produce side effects such as pruritus ani, because the rhus antigen is not inactivated in the gut, and when it reaches the skin on exit of the gastrointestinal system, it can cause perianal rhus dermatitis.

INFECTIOUS DERMATITIS

The three most important infectious dermatoses of agricultural workers include zoonotic dermatophytic fungi and two animal viral diseases, contagious ecthyma of sheep and goats (orf) and pseudo cowpox of cattle (milker's nodule). These dermatoses, as well as tularemia, anthrax, and other infections with skin manifestations, are discussed more in detail in the chapter on zoonotic diseases.

Dermatophytic fungal infections found on the farm are caused primarily by *Trichophyton* and *Microsporum* species. By far the most common animal dermatophyte infection in humans is *Trychophyton verrucosum*. Cattle are the primary host for this species. Other farm animals have their own dermatophytes that may infect humans. These fungi can be contracted from contact with infected animals or from contaminated objects in the animal environment (e.g., feed bunks, fences, etc.).

Classic dermatophytic infections in people present as red, scaling lesions with a tendency for central clearing, producing a "bulls eye" type of lesion. Generations ago, people thought this lesion must be caused by a worm under the skin, thus the term ringworm. However, animal ringworm infections of humans have many different presentations in humans. Generally, ringworm infections in humans are more severe than the infection in the host animal. This is probably due to better parasite-host adaptation in the primary host. Some of the consequences of this more aggressive infection can be seen if there is an infection of the hair and skin of the scalp. Infections may extend into the follicles, producing hair loss and deeper inflammation. In dairy farms, lesions may be seen on the chin and neck, as this is where the milker (in stanchion or tie stall barns) has contact with the cow as he or she puts the milking machines on and off the cow.

Treatment of dermatophytic fungi ranges from

application of topical preparations to systemic therapy, depending on the extent and depth of infection. Topical preparations may include iodine solutions, or tolnaftate (1%) among other products. The topical imidazole and allylamine classes of antifungals may be most efficacious. Examples include miconazole (Monistat®), Clotrimazole (Lotrimin®) and terbinafine (Lamisil®). Topical use for 1 week after clinical symptoms resolve may help prevent relapse. Systemic therapy (griseofulvin per os) for 4 weeks minimum is required in severe cases. Nonspecific measures to reduce inflammation, such as a burst and taper of prednisone may be helpful for highly inflamed infections, especially those of the scalp, and may help reduce severity of damage to the hair follicles and prevent hair loss (Rippon 1974).

Orf is caused by infection with a poxvirus. This common disease of sheep and goats (contagious ecthyma) may be transmitted to humans by direct contact of abraded skin with infected sheep or goats usually by feeding lambs or baby goats, or from the contaminated environment (barns, stalls, fences, etc.) where sheep or goats have been kept.

Following an incubation period of about 1 week, lesions typically develop on the backs of hands, fingers, or arms. They begin as one or few red papules that enlarge to slightly umbilicated nodules, which become hemorrhagic and pustular. The center of the lesion breaks down to produce a red oozing surface that develops a crust. Secondary bacterial infection is common. Systemic responses, such as regional lymphadenopathy or a mild febrile response may occur. Lesions of orf resolve spontaneously in about 6 weeks, with scarring (Leavell 1974).

Milker's nodules also are caused by infection with a poxvirus. However, this virus is caused by the virus pseudo cowpox. Human infections are contracted through direct skin contact with teats and udders of infected cows.

The lesions in humans are indistinguishable from those of orf. History helps in determining the diagnosis. Like orf lesions, milker's nodules resolve spontaneously in about 6 weeks but usually do not scar (Nomland and McKee 1952).

Treatment of either orf or milker's nodules is aimed at controlling secondary bacterial infections. Topical antibiotic ointments and covering with a dressing may be indicated.

ARTHROPOD-INDUCED DERMATITIS

Because agricultural workers spend a great deal of time outdoors, they are exposed to a large variety of arthropods that bite and sting. Arthropods include wasps, bees, and ants (hymenoptera species), as well as a large variety of spiders, mites, and ticks (arachnids), mosquitoes and biting flies (Diptera), and caterpillars (Lepidoptera).

Cutaneous responses to most of these arthropods' bites vary considerably with the type of material that the insect may inject into the skin and the person's degree of sensitization. Arthropods may inject saliva or other materials just as a means to facilitate a blood meal (hematophagus insects such as mosquitoes, ticks, mites, etc.), or as a defensive toxin that will give an immediate painful sensation (e.g., wasps, bees, ants). Some arthropods possess systemic toxins (e.g., black widow spiders and scorpions), and some inject a cytotoxic substance, causing local necrosis (e.g., brown recluse spiders). Initially, some bites and stings from hematophagus insects are likely to produce little or no reaction. As a person develops sensitivity, general anaphylactic reactions may be a risk and local reactions like erythematous macules, papules, pruritic lesions, or blisters may develop. Finally, after repeated exposure, sensitivity may wane. A good example of this is the often-observed welts that children receive from mosquito or chigger bites. This severe reaction is rare in adults. The systemic or cytotoxic reaction seen with spider bites does not necessarily change with experience and age of the patient (Harves and Millikan 1975).

Hymenoptera Stings

The most common response of a sting from a hymenoptera species (wasp, bee, ant) is an immediate pain and burning sensation at the site of the sting. This is caused by toxic proteins that are largely enzymes, causing severe local inflammation of the surrounding nerve tissue and soft tissues. One member of the *Hymenoptera* family, *Solenopsis invicta*, or commonly known as the *imported red fire ant (IRFA)* has been a particular problem in southern states of the U.S. (Ant Colony Developers Association 2002; Gilbert and Patrock 2002). Accidentally imported from the cargo of ships coming to the port of New Orleans from Argentina in the 1930s, this prolific ant

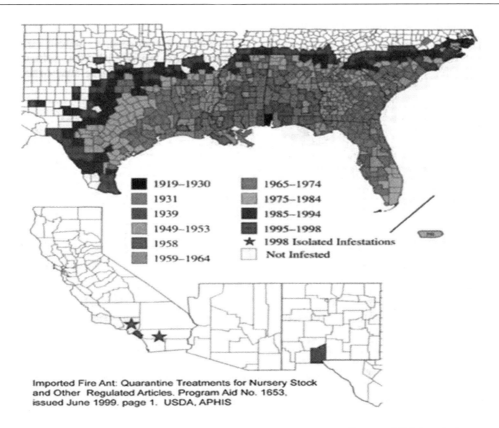

Figure 4.1.   Range expansion of imported fire ant from 1918–1998 (Source: USDA website, http://www.usda.gov/wps/portal/usdahome). (This figure also is in the color section.)

species has spread over eight states from as far east as North and South Carolina and Florida, and westward covering a band of states ending in Eastern Texas (see attached map, Figure 4.1). It is the most common hymenoptera species causing human suffering in the region. Bites from IRFA cause almost diagnostic lesions consisting of what looks like pustules but contain a sterile milky-colored fluid. These lesions may last for up to 10 days. The fire ant toxin contains a protein (which is probably responsible for the allergic response), as well as a cytotoxic alkaloid, which is responsible for the tissue damage. Most other hymenoptera stings result in mild discomfort and temporary skin lesions.

By far, the most life-threatening response is a systemic allergic reaction. Although upwards of 20% of individuals may have IgE antibodies to hymenoptera toxins, only 1–3% of the population will suffer a serious allergic reaction. However, it

is estimated that between 40–100 people die every year in the U.S. from hymenoptera stings. There are certain risk factors to which one should pay attention that may warn of an impending anaphylactic response. First, if the person has experienced a previous systemic response, the risk for that person of developing a severe reaction with a future sting is very high. An affected person should be taken to the nearest medical facility should he/she develop a very large local response or systemic symptoms: hives, general pruritus or burning sensation, swelling of lips and tongue, wheezing, etc. (McLean 1968).

Arachnids

There are several species of mites that may cause skin problems in rural and agricultural people. These mites are both parasitic mites and free-living mites. Most animal species, including humans, have their own specific species of parasitic mites that produce the skin condition called *scabies*. The scabies mite burrows into the epidermis, where the female lays eggs, perpetuates the life cycle, and produces a very pruritic infestation.

Scabies mites of animals such as pigs, horses, cattle, and dogs may temporarily infest people. However, these mites do not burrow into the epidermis and reproduce in humans as do the human-specific scabies mite. They do, however, inflict bites as they feed on their temporary host. They attach by their mouth parts and inject saliva which enzymatically digests the skin. The larva feeds on digested protein for a few hours to 4 days before falling off. The resulting inflammation of the skin may be accompanied with a serous exudate and an annoying pruritus. Resulting lesions most commonly are skin-colored to red edematous papules on exposed skin. However, distribution and severity of the reaction depends on mode of exposure and the victim's degree of sensitivity. Lesions vary from normal-appearing areas of pruritus to macules, papules, petechial lesions, vesicles, and even bullae. The location of animal scabies lesions on humans are typically on the hands and forearms where they have direct contact with the infected animals.

There are a number of free-living mites that may cause skin lesions in humans. Chigger bites (caused by mites of the family *Trombiculidae*) are probably most common and although not serious, are frequent among farm people. The larval form of this mite feeds mainly on vegetation, but requires animal protein for complete development. Grain mites often cause lesions on the hands, and typically on the webbing between the fingers. Similar to chiggers, these mites cause only a temporary infestation, but leave the host with a very pruritic but self-limiting hand and arm dermatoses.

SPIDERS

There are three types of spiders that can cause skin conditions, and some may cause systemic responses. The brown recluse spider (*Latrodectus geometricus*) is present in most of the western hemisphere. The venom of this spider is primarily an enzymatic cytotoxin. A bite from this spider will not cause immediate pain, but will onset in 2–6 hours. The area at the site and surrounding the bite will begin to turn purple to black in a few days, and progress to areas of skin necrosis. It may take up to 3 months or more to heal, depending on the location and size of the area of necrosis, and may result in scarring.

Widow spiders are present in most areas of the Western Hemisphere, Europe, and some areas of Asia and Africa. Species in this group of spiders are mostly black, but there are brown, red, and others. Common to this group of spiders is an hourglass figure on the ventral surfaces of their abdomen. The black widow spider (*Latrocedtus onactans*) is perhaps the most common and well-known spider of this group. Only the female will bite, injecting its combined enzymatic and neurotoxic venom, called *alpha-latrotoxin*. This toxin opens cation channels, with a net result in excess neural stimulation at motor endplates. There may be an immediate but often minor local pain from the bite, but within an hour, a generalized reaction may begin with spreading pain and cholinergic symptoms of sweating, diarrhea, abdominal cramps, and dyspnea. There may also be motor function affect with cramping of large muscle groups and general weakness. Death is very rare, but recovery usually takes several days (Clark and others 1992).

There are over 500 species of scorpions around the world, mainly in warm regions from desert to subtropical climates. All of them might sting, but most are no more severe than the bite of a honeybee. There are species and geographic variation in the type and potency of their toxin (Corona and others 2001). The toxin is most commonly a neurotoxin similar to that of the widow spiders. However, there are some species that also have a cytotoxin, similar to the brown recluse spider. Only one species (*Centroides sculpturatus*) in North America (located in the southwestern portion) contains a potent neurotoxin. It should be noted that some sensitized people may have an anaphylactic reaction in response to a scorpion sting similar to that caused by the hymentoptera species.

Lepidoptera Species

There are several species of caterpillars that may cause dermatitis and/or a systemic reaction. The common one is the puss caterpillar (*Megalopyge urens*). When a person touches this caterpillar, the hairs may stick into the person's skin and release a toxin that contains high levels of histamine and the enzyme hyaluronidase, which produces immediate pain and proteolytic and hemolytic toxic and inflammatory reactions.

Treatment of Arthropod Dermatoses

ANAPHYLACTIC REACTIONS

Treatment of arthropod-induced dermatoses is highly variable. Persons with a history of severe reactions to hymenoptera stings should wear bracelets or tags stating their sensitivity. If stung, these people should have access to powerful drugs (diphenhydramine, ephedrine, and/or corticosteroides) and eventually proceed to the nearest hospital. Regarding prevention, persons who have had a history of systemic allergic reactions following hymenoptera stings (such as generalized hives, wheezing, fainting, or shock), hyposensitization may be considered.

TREATMENT OF SKIN REACTIONS FROM INSECT BITES AND STINGS

Mites from the general environment (such as chiggers) grain mites, dried fruit mites, and animal scabies need not be specifically treated. These mites cannot survive on human skin for more than a few days, and thus die or drop off spontaneously. However, victims are very appreciative of symptomatic, antiinflammatory, and antipruritic treatment. Over-the-counter products such as topical steroids and or antihistamines may be used, but they have little effect in relief of the symptoms. The other more effective over-the-counter products contain camphor, phenol, and menthol. Also, steps need to be taken to eliminate or prevent exposure. Protective clothing and gloves are useful when handling grain or animals that may have scabies. Chigger bites can generally be avoided by putting a barrier, such as a blanket, between yourself and the grass while working or picnicking. Fumigation of grain helps rid the product of vermin and mites, enhances grain quality and decreases the health risk of grain handling as it comes out of storage.

Regarding hymentoptera species stings, only the honeybee will leave its stinger in its victim (Graber 2002). As the stinger will continue to pump venom for up to 30 minutes while in the skin, it is a good idea to locate it and either scrape it away with a fingernail, or pull it out with a forceps. Additional treatment may be appropriate when bites are numerous, when patients react severely, or when secondary infections develop. A

prescription-grade topical steroid often provides good symptomatic relief. For wasp or bee stings, sprinkling papain powder (usually contained in most meat tenderizers and available in most grocery stores) applied on a wetted lesion is effective if used early. The tenderizer enzymatically destroys protein and polypeptide secretions that act as irritants or allergens. Systemic antihistamines may reduce pruritus associated with multiple bites. Antibiotics, usually systemic, may be indicated for treating secondary infections.

Treatment of arachnid bites may require more invasive treatment. There are commercial antivenins for the widow spiders, the brown recluse, and scorpion bites or stings (Allen and Norris 1995). There is wide geographical variation in scorpion venoms, and antivenins are species and geographic specific. For best results, these should be administered within the first 4 hours of acquiring the bite. The brown recluse spider bite may require debridement or excision of the wound site, followed with a skin graft (to prevent scarring). Dapsone has been reported to decrease the tissue necrosis if applied early and may prevent scarring.

Arthropod-related problems can be prevented by taking the following precautions: wear light-colored, nonflowery clothing; do not wear scented preparations; avoid activities where "bugs" are likely to be encountered; and keep an insecticide aerosol handy. Arthropods are best deterred by insect repellants containing diethyltoluamide.

SUNLIGHT-INDUCED DERMATOSES

Both acute and chronic changes result from the sun's ultraviolet rays. With few exceptions, these problems are more common among light-skinned individuals.

The most frequent reaction to sunlight is sunburn. Much less common is photocontact dermatitis presented above under contact dermatitis. Skin cancer is related to farming and a significant risk among old farmers.

Chronic Sun-Induced Skin Disorders

The most common chronic, sun-induced changes to the skin include the triad of thickening, loss of elasticity, and wrinkling, and actinic keratoses (Howell 1960; Willis 1971).

Actinic keratoses, basal cell carcinomas, and squamous cell carcinomas typically do not appear until the patients are in their late 50s or 60s.

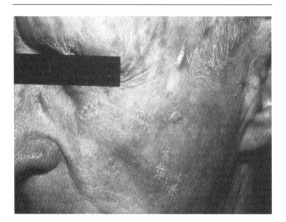

Figure 4.2. Actinic keratoses. (This figure also is in the color section.)

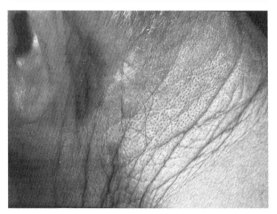

Figure 4.3. Compare the thickened and wrinkled skin from chronic sun exposure to skin protected by the shirt collar on this 65-year old farmer. (This figure also is in the color section.)

Actinic keratoses are precancerous lesions that appear as pink scaly patches on the nose, cheeks, ear lobes, neck, and dorsal surface of the forearms, back of the hands, and other areas that receive extensive sun exposure. They present as red to brown macules, or slightly raised papules with a rough scale (Figure 4.2). Although they are sometimes painful or pruritic, actinic keratoses are primarily of cosmetic concern. However, 5–10% will develop into squamous cell carcinomas.

Although mild sunburns do not require therapy, more serious sunlight-induced dermatoses require a variety of treatments. Patients with moderate sunburn can be helped with cool baths, aspirin or nonsteroidal antiinflammatory drugs, and topical steroid creams. Systemic steroids are beneficial for severe sunburn. Treatment of photocontact dermatitis is similar to that of contact dermatitis, which was discussed in the first section of this chapter.

Most actinic keratoses do not need any special treatment. However, both physical methods—such as curettage, electrodessication, or liquid nitrogen freezing—and chemical approaches (5-fluorouracil, imiquimod) are available.

All sun-induced skin problems can be dramatically reduced by use of protective clothing, such as a wide-brimmed hat, and by use of a broad-spectrum sunscreen cream of at least SPF 25 if both UVA and UVB protection is needed. Note that chemical sunscreens are not recommended for children. There is a risk of photoallergy (e.g., PABA), and consequently the physical sunscreens

are safer. Agricultural workers would be wise to apply a sunscreen each morning during the spring and summer months. A second midday application is wise when sweating is moderate to profuse. Staying or working in a shaded area from 12 noon until 2 p.m. daylight saving time, when possible, is highly recommended.

Skin Cancer

The back of the farmer's neck in Figure 4.3 shows excessive wrinkling and thickening. The collar of his shirt shaded the skin half way down his neck producing the line of sparing. Although this condition is mainly of cosmetic concern, it does represent extensive chronic sun exposure that is a risk for skin cancer. Skin cancer generally has been on the rise since the 1940s, and only recently (around the year 2000) has there been a leveling out of reported skin cancer cases. Currently there are over 1 million cases reported annually in the U.S. alone.

The neoplasms are caused by light energy in the ultraviolet range of 200–400 nm, with the greatest risk with UVB radiation (290–320 nm) (Kripke and Ananthaswamy 2003). Ozone in the stratosphere filters out radiation, naturally blocking a portion of the UVB portion of the light spectrum. However, the release of chlorofluorocarbons from human activities has decreased the ozone layer and therefore increased the amount of UV radiation potential in sunlight. Thus we can expect a continued high prevalence in skin cancers in the future.

Farmers, like persons in other professions with hours of work outdoors under the sun, have a significantly high risk of acquiring a skin cancer. In a study of cancer risks among farmers it was shown that they had a higher risk of all types of skin cancers on head or neck but a lower risk of skin cancer in other parts of the body (Wiklund 1986) demonstrating that life style and protection have great impact on the risk.

It should be noted that in addition to sun exposure, chronic exposure to the formerly commonly used arsenical herbicides can also cause skin cancers, including carcinoma in situ (Spiewak 2001).

The most common skin cancer is basal cell carcinoma (Figure 4.4). If untreated, basal cell carcinomas can be very destructive locally, but they rarely metastasize. Squamous cell carcinomas are neoplastic lesions of the superficial layers of the skin (Figure 4.5). These carcinomas occur at the same areas of predilection as actinic keratoses, and as mentioned above, may arise from them.

Although one specific melanoma (lentigo melanoma) has a similar relationship to chronic sun exposure as nonmelanotic tumors, the sun risk factors for most melanomas are not clear (Desmond 2003). Melanomas do not typically occur on chronically sun-exposed areas of the skin, such as the ears, nose, and cheek ridge.

Most skin cancers (with the exception of melanoma) are not highly malignant. Approximately 2 in 100 cutaneous squamous cell carcinomas metastasize. Metastasis is much more common when the lesions develop on the lips or ears. Usually the spread is to regional lymph nodes, although spread to the lung is next most common. For more information on skin cancers, see Chapter 5.

SKIN DISORDERS RELATED TO HEAT, COLD, AND HUMIDITY

Heat-Induced Dermatoses

Hot, moist environments can cause **miliaria rubra** commonly called *prickly heat*. Miliaria results from inflammation and possible infection and consequent obstruction of sweat ducts. The obstructed and inflamed sweat ducts present as a rash, with uniform, small, red papules or vesicular papules, which are regularly spaced. The regular spacing pattern may result because the le-

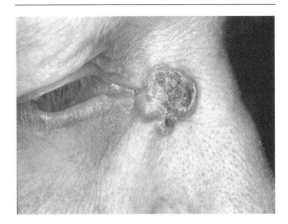

Figure 4.4. Basal cell carcinoma. (This figure also is in the color section.)

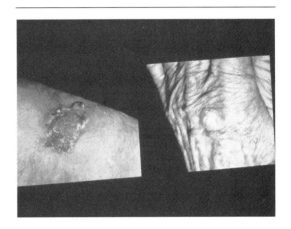

Figure 4.5. Squamous cell carcinoma; two different clinical appearances. (This figure also is in the color section.)

sions are located near hair follicles. However, close observation will reveal that they lie between hair follicles. Miliaria is commonly found in areas where moisture may collect, such as body fold areas, and under the arms. Also, it occurs at sites of pressure and areas of friction such as the belt line (Holtz and Kingman 1978; Lobitz 1965).

Treatment depends on removal of the patient from the hot, humid environment. Application of topical steroid cream may be helpful. Prevention of miliaria involves avoiding sustained heat and humidity by common sense measures such as wearing cool, well-ventilated clothing, and using fans and air conditioners to keep one's home cool.

Regular bathing and application of talc or similar body powders at the beginning of the workday will help absorb moisture, reduce friction, and reduce the risk or severity of miliaria rubra.

Excessive sweating may cause **intertrigo**, which is a erythematous eruption in body folds with maceration and risk for secondary bacterial and candidal infection especially among obese persons.

Cold-Related Skin Disorders

Chilblains (Perniosis) is a mild form of cold injury. Outstanding areas of the body like fingers, toes, nose, and ears are especially affected. The lesions are swollen with bullae and ulcerations and reddish blue discolorations. Treatment is symptomatic.

In **frostbite** there is an impairment of circulation. Superficial frostbite affects only the superficial layers of the skin. In more severe cases deeper layers are affected with risk of gangrene. The consequences of a deep frostbite cannot be evaluated immediately. Several weeks have to pass to estimate the tissue loss accurately. Even minor frostbites may result in a long-term hypersensitivity to cold, including Raynaud-like reactions, paresthesias, and hyperhiderosis. For more information of heat- and cold-related conditions, see Chapter 9.

REFERENCES

Adams R. 1969. Occupational Contact Dermatitis. Philadelphia: J.B. Lippencott.

Agriculture and Agri-Food Canada. 1990. Poison Ivy, Western Poison Oak, Poison Sumac. Agriculture and Agri-Food Publication 1699/E (c)Minister of Supply and Services Canada 1990 Cat.No. A53-1699/1989E ISBN 0-662-17265-5. Printed 1980. Revised 1990 10M-2:90.

Allen R, Norris R. 1995. Delayed use of widow spider antivenin. Annals of Emergency Medicine 26(3): 393–394.

American Academy of Dermatology. 2005. Poison Ivy, Oak and Sumac. http://www.aad.org/public/Publications/pamphlets/Poison_IvyOakSumac.htm.

Ant Colony Developers Association. 2002. Fire Ants. http://www.aphis.usda.gov/ppq/ispm/fireants/.

Arcury T, Quandt S, Mellen B. 2003. An exploratory analysis of occupational skin diseases among Latino migrant and seasonal farm workers in North Carolina. Journal of Agricultural Safety and Health 9(3):221–232.

Belsito D. 2003. Allergic Contact Dermatitis. In: Freedberg I, Eissen A, Wolff K, Austen K, Goldsmith L, Katz S, editors. Fitzpatrick's Dermatology in General Medicine. 6th ed. New York: McGraw-Hill. p 1164–1169.

Burke W. 1997. Skin Diseases in Farmers. In: Langley R, McLymore R, Meggs W, Roberson G, editors. Safety and Health in Agriculture, Forestry, and Fisheries. Rockville, MD: Government Institutes, Inc. p 322–352.

Centers for Disease Control. 1986. Epidemiologic Notes and Reports Outbreak of Severe Dermatitis among Orange Pickers—California. Morbidity and Mortality Weekly Report 35(28):465–467.

Clark R, Wethern-Kestner S, MV V, Gerkin R. 1992. Clinical presentation and treatment of black widow spider envenomation: A review of 163 cases. Annals of Emergency Medicine 21(7):782–787.

Corona M, Valdez-Cruz N, Merino E, Zurita M, Possani L. 2001. Genes and peptides from the scorpion Centruroides sculpturatus. Toxicon 39(12):1893–1898.

Desmond R. 2003. Epidemiology of malignant melanoma. The Surgical Clinics of North America 83(1): 1–29.

Fisher A. 1973. Contact Dermatitis. Philadelphia: Lea and Febiger.

Gilbert L, Patrock R. 2002. Phorid flies for the biological suppression of imported fire ants in Texas: Region specific challenges, recent advances and future prospects. Southwestern Entomologist 25:7–17.

Graber M. 2002. Stingers can continue for inject venom for 20 minutes. EMS Update, An Emergency Medical Services Learning Resources Center Publication 23(2):1–2.

Harves R, Millikan L. 1975. Concepts of therapy and pathophysiology in arthropod bites and stings. International Journal of Dermatology 14:543–562, pp 621–634.

Hausen B, Wolf C. 1996. 1,2,3-Trithiane-5-carboxylic acid, a first contact allergen from Asparagus Officinalis (Liliaceae). American Journal of Contact Dermatitis 7(1):41–46.

Holtz E, Kingman A. 1978. Pathogenesis of miliaria rubra: Role of resident microflora. British Journal of Dermatology 99:117–137.

Howell J. 1960. The sunlight factor in aging and skin cancer. Archives of Dermatology 82:865–869.

Kripke M, Ananthaswamy H. 2003. Carcinogenesis: Ultraviolet radiation. In: Freedberg I, Eissen A, Wolff K, Austen K, Goldsmith L, Katz S, editors. Fitzpatrick's Dermatology in General Medicine. 6th ed. New York: McGraw-Hill. p 371.

Latenser BA, Lucktong TA. 2000.Anhydrous ammonia burns: Case presentation and literature review. Journal of Burn Care Rehabilitation 21:40–42.

Leavell V. 1974. Orf-Reported 19 Human Cases. Journal of the American Medical Association 204:657–666.

Lobitz R. 1965. Miliaria. Archives of Environmental Health 11:460.

Mathias C. 1988. Occupational dermatoses. Journal of American Academy of Dermatology 19(6): 1107–1114.

McLean J. 1968. Management of Insect Sting Reactions. Modern Treatment 5:814–824.

Nomland R, McKee A. 1952. Milker's nodules: Report of ten cases. Archives of Dermatology 65:663–671.

Nurse D. 1978. Dermatitis danger in dairy farmers. Med J Aust 1:223.

O'Dennell BF, Foulds IS. 1993. Contact dermatitis due to dibromodicyanobutane in cucumber eye gel. Contact Dermatitis 29(2):99–100.

Park H, Sprince N, Whitten P, Burmeister L, Zwerling C. 2001. Farm-related dermatoses in Iowa male farmers and wives of farmers: A cross-sectional analysis of the Iowa Farm Family Health and Hazard Surveillance Project. Journal of Occupational and Environmental Medicine 43(4):364–369.

Rademaker M. 1998. Occupational contact dermatitis among New Zealand farmers. Australia Journal of Dermatology 39:164–167.

Rippon JW. 1974. Medical Mycology. Philadelphia: W.B. Saunders.

Society NZD. 2004. Dermnet.

Spiewak R. 2001. Pesticides as a cause of occupational skin diseases in farmers. Annals of Agricultural and Environmental Medicine 8(1):1–5.

Spiewak R, Dutkeiwicz J. 2002. Atopy, allergic diseases, and work-related exposures among students of agricultural schools: First results of the Lublin study. Annals of Agricultural and Environmental Medicine 9:249–252.

Spiewak R, Gora A, Dutkiewicz J. 2001. Work related skin symptoms and type I allergy among eastern-Polish farmers growing hops and other crops. Annals of Agricultural and Environmental Medicine 8(1):51–56.

Susitaival P, Hannuksela M. 1995. The 12-year prognosis of hand dermatosis in 896 Finnish farmers. Contact Dermatitis 32:233–237.A

Susitaival P, Husman L, Hersmanheimo M, Notkola V, Husman K. 1994. Prevalence of hand dermatoses among Finnish farmers. Scandanavian Journal of Work, Environment & Health 20:206–212.

Telintum J, Nater J. 1974. Contact allergy caused by nitrofurazone (furacine) and nifurprazine (carofur). Hautarzt 8:403–406.

White G, Cox, N. 2000. Contact Dermatitis, Chapter 4 in Diseases of the skin. St. Louis: Moseby.

Wiklund K, Holm LE. 1986. Trends in cancer risks among Swedish agricultural workers. Journal of the National Cancer Instituite 77(3):657–664.

Willis I. 1971. Sunlight and skin. Journal of the American Medical Association 217:1088–1093.

5
Cancer in Agricultural Populations

Kelley J. Donham and Anders Thelin

INTRODUCTION

A 68-year-old dairy farmer in Eastern Iowa (U.S.) had been complaining of back pain and difficulty urinating for at least 2 years. At the continued insistence of his wife, he made an appointment with a physician (the first time he had been to a doctor in 15 years). A digital rectal exam revealed an enlarged and somewhat irregular prostate. A follow-up prostate-specific antigen blood test was elevated. He was referred to a urologist; the nearest one was over 200 km away, which required overnight travel. It took several days for him to organize someone to come in and milk and feed the cows in his absence. A biopsy revealed prostate cancer, which had spread beyond the capsule. Additional tests showed the cancer had spread to several pelvic lymph nodes and also to the vertebrae of the backbone. He was given the choice of chemotherapy and radiation, castration, or estrogen therapy. The first two options would mean having to leave the farm, not being able to take care of his dairy operation. The last option would allow him to maintain his farm, but with perhaps less potential for the best possible long-term outcome. He was concerned that he could not afford to pay for the more expensive treatment. He discussed the options with his family physician. Given all the information and his perceived limitations, he chose the estrogen therapy and stayed at home to tend his cows as long as he could. The farmer was able to work for about 8 months longer, until he got too sick. He died at his home under the care of a rural visiting nurse, 1 year after the diagnosis.

His physician expressed regret that he had not come in sooner, because the cancer may have been arrested before it disseminated. The farmer's wife expressed concern that there may be a cancer epidemic, as she knew at least four other persons that live within a 2 mile radius that have cancer. She wants to know whether the pesticides they have used all these years may have been the cause. The physician could not answer her questions.

This case (based on a real case and friend of this author [KJD]) is illustrative of the common problems facing farm patients with cancer. This chapter will cover these issues, though concentrating on the epidemiology and risk factor for selected cancers in the farming population. Causation, prevention, and social medical considerations will also be discussed.

GENERAL DESCRIPTION OF CANCER RATES IN THE FARM POPULATION

Epidemiology

There have been well over 100 articles published on the epidemiology of cancers in farm populations in the past 20 years. There are many similarities in the epidemiology of cancer among farm populations in the industrialized countries. Most studies have concentrated on rate differences of cancers in farm populations relative to comparison populations. There have been two notable studies in the U.S. in the last decade that have revealed consistent trends in the overall farm cancer burden. Cerhan and co-workers (1998) published on the overall cancer deaths by cancer type in Iowa (U.S.). The following are the top 10 leading cancers that cause death in farmers and the percentage of the total caused by these cancers: 1) lung, 37%; 2) colon, 20%; 3) non-Hodgkin's lymphoma, 7%; 4) brain, 7%; 5) prostate, 7%; 6) pancreas, 6%; 7) kidney, 6%; 8) melanoma, 5%; 9) leukemia, 3%; and 10) stomach, 2% (Table 5.1). Blair and co-workers (2005, in press) published mortality data from the Agricultural Health Study in North Carolina and Iowa (N = 84,738). This study revealed an overall Standardized Mortality Ratio (SMR) of 0.6 for all cancers, corroborating other studies that overall cancer is lower among farmers. Lung cancer explains the majority of the overall cancer deficits, accounting for 60% of the overall cancer sparing. Several cancers had an SMR greater than or equal to 1.0, including gall bladder, eye, Hodgkin's disease, thyroid, ovarian, uterine, non-Hodgkin's lymphoma, leukemia, soft tissue sarcoma, stomach, liver, brain, and colon. However, none of these differences reached statistical significance.

In the farm population compared with the general population, there are significant differences in the overall rates and distribution of several types of cancers. One important consistent difference is that farmers have reduced mortality from all cancer sites (Blair and Zahm 1995; Cerhan and others 1998; Kristenson and others 1996a; Pukkala and Notkola 1997). The decrease in overall cancer rates is from 10–25%, depending on the specific study. The overall deaths are lowered, primarily because lung cancer deaths (the most com-

Table 5.1. The 10 Leading Causes of Cancer Mortality in the Agricultural Health Study, Iowa and North Carolina (U.S.)

Cancer Type	Percentage of Total Cancers in This Group
Lung	25%–37%
Colon	11%–20%
Non-Hodgkin's Lymphoma	6%–7%
Brain	7%
Prostate	4%–7%
Pancreas	6%
Kidney	6%
Melanoma	3%–5%
Leukemia	3%–5%
Stomach	2%

(Blair and others 2005, in press; Cerhan and others 1998).

Table 5.2. Cancers with Decreased Mortality Rates in the Farming Population Relative to Comparison Populations

Cancer Type
Lung
Bladder
Liver
Tongue
Esophagus
Colon
Rectum
Kidney

(Blair and Zahm 1995; Faroy-Menciere and Deschamps 2002; Kirkhorn and Schenker 2002; Wiklund and Dich 1995).

mon cause of cancer-related death overall) is decreased by about 30% relative to comparison populations (Cerhan and others 1998). Additional to lung cancer, there are some less common cancers that are also decreased in the farm population, including bladder, liver, tongue, esophagus, colon, rectum, and kidney (Blair and Zahm 1995; Wiklund and Dich 1995) (Table 5.2).

Although farmers may experience decreased overall cancer mortality, there are several cancers for which they suffer excess mortality. The specific cancers vary to some degree between studies. The combined results of these studies corroborate the increased risk for the following cancers: 1) leukemia, 2) non-Hodgkin's lymphoma, 3) Hodgkin's lymphoma, 4) multiple myeloma, 5) skin cancers (including squamous cell and basal cell carcinomas, lip cancer, and melanoma), 6) soft tissue sarcoma, 7) prostate cancer, 8) testicular cancer, 9) brain cancer, and 10) stomach cancer (Blair and Zahm 1995; Faroy-Menciere and Deschamps 2002; Pukkala and Notkola 1997; Wiklund and Dich 1995). Although these findings are quite consistent, the relative risks of these cancers are generally not great, ranging from 20–200% above comparison populations, with the lower limit of this range being more common. Table 5.3 summarizes the cancers found at increased risk in the farming population.

There are several limitations of these epidemiologic studies that should be recognized. First of all, most studies are mortality studies. Thus, there is little data on cancers of low malignancy, such

Table 5.3. Cancers of Increased Risk to the Farming Population

1. Cancers with Significant Evidence for a Farming Occupational Association
   Leukemia
   Skin Cancers (Basal Cell and Squamous Cell Carcinoma)
   Lip Cancer
   Multiple Myeloma
2. Cancers with Varying Evidence for a Farming Occupational Association
   Non-Hodgkin's Lymphoma
   Hodgkin's Lymphoma
   Prostate
   Soft Tissue Sarcoma
   Brain
3. Cancers with Limited Evidence for a Farming Occupational Association
   Testicular
   Stomach

(Blair and Zahm 1995; Faroy-Menciere and Deschamps 2002; Kirkhorn and Schenker 2002; Milham 1999; Pukkala and Notkola 1997; Wiklund and Dich 1995).

as nonmelanotic skin cancers and testicular cancer. Furthermore, there is little information on chronic cancers or treatable cancers where people may die of co-morbidities rather than the cancer, such as prostate cancer, and various leukemias. Therefore, more morbidity data is needed to determine the overall cancer burden on the farm population. Furthermore, the standard incidence data (SIR) that are calculated will provide only ecologic associations and possible hypotheses. To identify risk factors, accurate exposure data is required that may be applied to more detailed analyses, such as case control and logistic regression studies. Few of these studies are available at this time. However, a recent report on morbidity in the 85,000-person Agricultural Health Study has revealed similar patterns to previous mortality and morbidity studies (Alavanja and others 2005, in press). This study revealed decreased SIRs for all cancers (SIR 0.8), with significant decrements in the incidence of lung, buccal, urinary, and bowel cancers. However, there was no decrease in cancers that have shown decrements in other studies, including Hodgkins, non-Hodgkins, lymphoma, multiple myeloma, leukemia, and brain. There

was increased incidence in prostate and lip cancer, and in women an increase was seen in ovarian cancer and melanoma. There was an interesting comparison to the commercial pesticide applicator group of this study, who had almost identical cancer rates and pattern to the general population. However, their age (younger) and tobacco and alcohol use were similar to the general population.

It is interesting to note that although lung cancer is decreased in farming populations, it still is the number one cause of cancer deaths in the farm population, accounting for nearly one-third of the total cancer deaths (Blair and others 2005, in press). With the exception of pancreatic and kidney cancer, the remaining cancers with increased rates in the farming population account for 50–60% of the total cancer deaths. The implications of this are relevant to prevention because a focus on lung and colon cancer will have a much larger potential impact on overall cancer burden than those remaining cancers that make up a minority (40–50%) of the total cancer deaths.

Trends

As there has been tracking of cancers in some populations for over 30 years, some trends have been noted. Since the 1970s, data from the Iowa (U.S.) cancer registry has revealed a decreasing trend in stomach cancer, with relative rates dropping from 1.14 to 1 in the 1990s. However, there has been an increase noted over this time in cancer of the prostate, large intestine and pancreas, and in Hodgkin's disease. There has been no change observed in other hematopoietic cancers (Cerhan and others 1998). While there have been changes noted in cancer patterns over time in the farm population, there have also been certain changes in cancer rates in the general population, including increases in prostate, all skin cancers, non-Hodgkin's, multiple myeloma, and brain cancer. It is interesting to note that all those increasing in the general population are still at increased risk within the farm population. Therefore, it would appear that whatever risk factors there are for these cancers in the general population, there is even excess risk within the farm population. Another trend of note is that in some agricultural populations there had been a risk for lung cancer. Milham (1999) noted an increase in lung cancer in Washington (U.S.) orchard workers until recent times. This excess was related to the use of inorganic arsenical pesticides. However, these have not been used since the 1940s, and that risk factor and the excessive lung cancers in that population have now diminished.

PROTECTIVE AND RISK FACTORS AND MECHANISMS OF CAUSATION

Protective Factors

Understanding the causation of cancer involves analysis of exposures to risk factors and to protective factors. As mentioned previously, most studies show the overall cancer rates are lower for the farm population. Most studies cite the protective factors of lower smoking rates. Farmers in most developed agricultural communities smoke significantly less than the general population. Differences range from 50% less to nearly 66% less. Several authors use this difference to explain the relative decrease in the "smoking-related cancers," including lung, esophagus, head and neck, and bladder (Blair and others 2005, in press; Blair and Zahm 1995; Kirkhorn and Schenker 2002; Kristenson and others 2000; Pukkala and Notkola 1997). An additional life style protective factor is the decreased use of alcohol relative to the general population (Blair and others 2005, in press). It has been shown that the combination of smoking and alcohol are risk factors for head and neck and esophageal cancer. Several other authors have also identified a decreased use of alcohol among farmers, also associated with a decrease in overall cancer rates (Cerhan and others 1998; Kristenson and others 1996b; Pukkala and Notkola 1997). Others propose that increased physical activity of the farming population is also protective for cancer, especially for colon cancer. Other cancers have also shown a decrease relative to exercise including lung, mouth and throat, liver, pancreas, bladder, and kidney (Blair and others 2005, in press).

One interesting hypothesis—or an alternative or additional protective factor for cancer in the farm population—is exposure to endotoxin. As discussed in Chapter 3, endotoxin is a component of the cell membrane of gram-negative bacteria, and it is found in high concentrations in most agricultural dusts. Endotoxin is an exquisite inflammatory substance responsible for respiratory symptoms and generalized symptoms. Mastrangelo and others (1996) conducted a study of cancer in dairy farmers, comparing them to cancer outcomes in crop and orchardists in Italy. They found that overall lung cancer in dairy farmers had been reduced relative to the comparison groups. Furthermore, there was a greater decrease in lung cancer as length of time in dairy farming increased and as the amount of land increased associated with the dairy farms. The authors interpreted these exposure measures as surrogates for increasing endotoxin exposure, and thus protective for lung cancer. They also hypothesized that the increased endotoxin exposure increased tumor necrosis factor from the stimulated alveolar macrophages, which was protective for developing lung cancer cells. Lange (2000) supports this hypothesis, noting that endotoxin is an immunomodulator, which not only stimulates production of tumor necrosis factor but activates lymphocytes and prevents cancer by that mechanism, especially lung cancer.

Risk Factors and Mechanisms of Causation Life Style

Mentioned above, avoidance of smoking, limiting alcohol consumption, and exercise are significant lifestyle factors that are protective for cancer. The farm population in most industrialized countries are beneficiaries of these lifestyle factors, relative to the general population. Regarding dietary factors, it is well known that excess high-fat, low-fiber diets and diets with low amounts of fruits and vegetables create risks for cancers in the general population. Dietary habits in the farm population are not well studied, and therefore little is known about dietary effects on cancer in this population. However, one study suggests that there may be some differences in diets that may have a small effect on cancer incidence (Cerhan and others 1998). Farmers generally have a higher caloric intake, but their body mass index is quite similar when compared to the general population (which suggests that they expend more calories due to greater amounts of exercise/work), and they receive a higher percentage of calories from protein, and proportionally a higher percentage of that protein comes from red meats. They have a lower intake of vegetables, but a higher intake of fruits. Their fiber intake is slightly higher. Regarding intervention strategies, dietary improvement (as with the general population) is an area that could help reduce risks. The exercise protective factor advantage among farmers is probably waning, as increased mechanization reduces the necessity of exercise from manual labor. Furthermore, as manual labor requirements reduce, the culture, available time, and facilities for recreational exercise are not apparent in farm life. Lack of exercise may emerge as a risk factor in future years. These authors agree that an important emerging area for preventive intervention is to address cultural, time, and facilities barriers to recreational and preventive exercise, which are currently emerging and likely increasing in the future.

Toxic Exposures

Specific agents and mechanisms of cancer causation in the farming environment are still not well understood today. Most of the evidence is still in the hypothesis stage. However, there are some agents where there is sufficient scientific evidence to reveal a cause-effect. These include inorganic arsenical pesticides and herbicides that cause lung cancer and skin cancer. However, these chemicals have been banned since the 1940s, and that effect should now be removed from the farming population (Milham 1999). It is clear that chronic sun exposure is related to non-melanotic (basal cell, squamous cell carcinomas, and lip) skin cancers. There is evidence that chronic sun exposure also may be a risk factor for melanomas in some regions (such as in Sweden) or with a particular type of melanoma (lentigo). However, most melanomas generally appear to be associated with frequent sunburns at an early age. Use of phenoxyacetic acid herbicides (2,4-D and 2,4,5-T) is related to soft tissue sarcoma, and non-Hodgkin's lymphoma. However, the data on this subject is not consistent. (It should be noted here that agent orange, the chemical defoliant used during the Vietnam War, was an equal mixture of 2,4-D and 2,4,5-T. It was the contaminant dioxin, a by-product of the manufacturing process for these chemicals, that was the suspected toxicant. Recent manufacturing processes now have reduced this contaminant).

Many other associations between various chemicals and cancer in farmers have been observed in epidemiologic studies, but the weight of evidence in most of these associations is not strong enough to prove a cause-effect relationship at this time. Organophosphate insecticides have shown up in many studies as associated with leukemia, non-Hodgkin's lymphoma, soft tissue sarcoma, and pancreatic cancer. Immunotoxicity is one speculated mechanism for this association. Organophosphates inhibit serine esterase, which in turn inhibits the function of the cellular immune system, thus decreasing "immunity" to cancerous cells. Genotoxicity (causing damage to genetic material) is another potential mechanism of organophosphate cancer causation (Blair and others 2005, in press).

Chlorinated hydrocarbon insecticides (although used very little since the early 1970s) are thought to be endocrine disrupters and thus related to genitourinary tumors. Diesel exhausts contain polycyclic aromatic compounds, which affect the immune system by decreasing IgG and IgA antibody production. Mycotoxins (produced by various mold species) may alter the immune system and play some role in carcinogenesis (Kristenson and others 1996b). As mycotoxins are a common component of agricultural dusts, particularly dust from moldy grain, they may play a role in farmers' cancers. Drinking water from surface sources or shallow wells in rural areas often contains high levels of nitrates and atrazine (a corn herbicide). These may combine in the acid medium of the stomach to produce nitrosamines, which are known carcinogens (Kirkhorn and Schenker 2002).

Animal contact has been associated with a variety of hematopoietic cancers in farm populations. These associations have been seen with cattle, chickens, cats, and pigs (Pukkala and Notkola 1997). The first three of these species have well-described tumor viruses, and several studies have led to hypotheses that these animal tumor viruses may infect humans, causing cancer. Associations with cattle and tumor viruses have been studied most extensively. The most common cancer in cattle is bovine lymphosarcoma (BLS), which is caused by a virus, very similar in genetic sequence to the human type 2 leukemia and the HIV viruses. Ecologic associations of human leukemia (particularly acute lymphocytic leukemia) have been seen with dairy farming (Kristenson and others 1996b). The incidence of leukemia has been shown to be higher in areas of high dairy cattle density, and with increased prevalence of diagnosed BLS in the cattle population (Donham and others 1980). The incidence of BLS-virus infection in dairy herds is quite high, with about 30% of the herds infected in some regions (compared to about 1% of beef herds). However, only a small percentage of infected cattle develop BLS in their lifetime. (It should be noted that beef and dairy cattle are usually killed well before their natural life span, so it is not known how many of these infected animals might have developed BLS if allowed to live longer.) It is also clear that the BLS virus is shed in milk of infected cattle, even in infected animals that are clinically normal. Furthermore, it is common practice that dairy farm families drink milk unpasteurized from their herds (Rubino and Donham 1984). Experimental infections of nonhuman primates with BLS virus have provided evidence that they became infected (antibody titer rise over time) and have shown an altered neutrophil-to-lymphocyte ratio. Although no cancer was documented in these animals over the course of the study, one of the chimpanzees developed leprosy. This was the first case of leprosy diagnosed in any nonhuman animals other than armadillos (Donham and Leininger 1977). The authors of this study hypothesized that this virus damaged the cellular immune system of this animal, allowing a latent form of leprosy to manifest clinically. Therefore, a new hypothesis was developed that these animal tumor viruses (at least BLS virus) may indirectly be a risk factor by damaging the cellular immune system. This hypothesis has not yet been tested.

DETAILS OF EPIDEMIOLOGY AND AGRICULTURAL RISK FACTORS FOR SPECIFIC CANCER TYPES

Cancers of the Hematopoietic and Reticuloendothelial Systems

The excess of hematopoietic cancers in the farm population is highly consistent among epidemiologic studies and across countries with developed agricultural systems. Leukemia, Hodgkin's and non-Hodgkin's lymphoma, and multiple myeloma are typically elevated in most studies. Furthermore, these elevated rates for these specific cancers have been relatively consistent over many years of investigations, with no evidence of a trend upward or downward (Blair and Zahm 1995). Although results are mixed, several studies have shown that hematopoietic cancers are more common among livestock producers relative to grain or other types of farming (Amador and others 1995; Pahwa and others 2003). This has created the hypothesis that the real risk is from chemicals used in livestock production or from animal tumor viruses. However, there appears to be variance in risk factors between specific cancers of the hematopoietic system. For example, BLS virus seems to be more related to acute lymphocytic leukemia, while other studies have shown associations of benzene exposure to myeloid leukemia (Donham and others 1980; Sperati and others 1999).

Skin and Lip Cancer

Basal cell, followed by squamous cell, skin carcinomas are by far the most common of all types of cancer in the farm population. These skin cancers rarely metastasize, rarely cause death, and therefore are not apparent in most cancer epidemiology studies, which are mortality studies. As most cancer epidemiology studies of farmers are mortality studies, these cancers often are not addressed. However, if either of these cancers occurs on the lip, they have a slightly higher probability to metastasize. Many studies show lip cancer mortality higher in farm populations. It is very clear that long-term, chronic sun exposure is the most important risk factor for these skin cancers. They are usually not seen until the sixth decade of life, occur on highly sun-exposed skin, and in people with fair complexion, and often may be preceded by actinic keratoses. Other risk factors proposed for lip cancer include viral infections and immune deficiencies (Khuder 1999).

General studies have shown that melanoma incidence and mortality rates are increasing in the farm population (Khuder and others 1998; Kirkhorn and Schenker 2002). As previously mentioned, the relationship between sun exposure and melanoma are not as clear as with nonmelanotic skin cancers. However, sun does appear to be a risk factor.

Prostate Cancer

Prostate cancer in farmers has been extensively studied in recent years. At least two meta-analyses have been conducted (Keller-Byrne and others 1997; Van Der Gulden and Volgenzang 1996) plus two large morbidity studies of farmers and pesticide applicators (Alavanja and others 2003; Dich and Wiklund 1998). Risk factors have focused on pesticide exposures, with the fumigant methyl bromide showing an association with some evidence of a dose-response relationship (Alavanja and others 2005, in press; 2003). Methyl bromide is used to fumigate grain or other produce, and it is also used as a soil sterilant. In actuality, only a small percentage of farmers would be exposed to this material, as it has very specialized usage. Exposure assessments have been conducted during applications, and air levels can exceed regulatory maximum exposure limits. Methyl bromide is an alkylating agent and is on the U.S. OSHA list of potential carcinogens.

Other prostate cancer risks revealed in studies include several pesticides, including organophosphate, pyrethrins, and pyrethroids (specifically permethrin). One isomer of the latter chemical (in addition to several other chemicals) is known to activate protein tyrosine kinase, which is thought to enhance prostate cancer (Settimi and others 2001).

Another theory of prostate cancer causation is hormonal mediation. Estrogen is known to be protective for prostate cancer, so anything inhibiting estrogen may increase risk. Dioxin is an estrogen inhibitor, and dioxins are found in older formulations of 2,4-D and 2,4,5-T herbicides. Furthermore, dioxins bioaccumulate in fat stores, and as farmers are known to have higher amounts of animal fats in their diets, they may have higher body burdens of dioxin. Grain dust exposure may also slightly increase risk, and this may be due to dioxins within grain dust from contamination with previous herbicide applications (Keller-Byrne and others 1997). Other chemicals also mimic steroidal hormones including mirex, toxaphene, and parathion. Metabolites of DDT interfere with steroid metabolism and alter metabolism of steroids via the mixed oxidative enzyme system. Other potential dietary risks include increased fat and decreased fruit and vegetable consumption (consistent with farmer's diets).

Family history of prostate cancer is a major risk factor. In fact, Alavanja and co-workers (2005, in press; 2003) have shown an interaction between family history and exposure to several pesticides as a risk factor for prostate cancer. Although not consistent, several studies indicate that the risk relative to comparison populations is further increased in elderly farmers (Parker and others 1999).

In summary, prostate cancer appears to be related to multiple risk factors in the farming environment. The fumigant methyl bromide and a variety of other pesticides seem to play a role, with methyl bromide showing the strongest correlation at present. High-fat diets seem to play a role as well, and all risk factors seem to be exacerbated if there is a positive family history of prostate cancer.

Testicular Cancer

There has been a general increase in testicular cancer in the past two decades in the general population (Kristenson and others 1996a). Generally studies have shown a relative increase in testicular cancer in agriculture populations compared with other occupations (Faroy-Menciere and Deschamps 2002). A general hypothesis relates risks to early life exposures. To study this in the farm population, a cohort study of parental exposures was carried out relative to testicular cancer in their sons. Fertilizer usage was found as a risk factor RR = 2.44, which rose to RR = 4.21 in nonseminoma tumors. This relative risk is quite large compared to most farm-related cancers. This is only one study, and there is no obvious biological explanation for this risk factor. Therefore, more studies are needed to confirm this reported association.

Brain Cancer

Khuder and co-workers (1998) conducted a meta-analysis of 33 epidemiologic studies of brain cancer in farmers. They found an overall relative risk of 1.3, with a confidence interval of 1.09–1.56. The risk was greater for studies within the central regions of the U.S. Relatively little research evidence is available for risk factors and mechanisms of brain cancer in farmers. However, Kristensen and colleagues (1996a), in a review of the literature, indicated that there was some evidence of risk related to infection with *Toxoplasma gondii* and general pesticide and insecticide exposures. The mechanism for *T. gondii* infection causing brain cancer is speculative; this intracellular protozoan parasite has a predilection for nervous tissue and that infection of brain cells results in some genotoxic effect, resulting in formation of the tumor. The farm population is known to be at

Table 5.4. Agricultural Risk Factors for Cancer Types

Cancer Type	Reported Risk Factors
1. Hematopoietic Cancers:	1.1. Livestock Exposures (Animal Tumor Viruses) Cattle (Especially Dairy Cattle), Also Chickens, Pigs, Cats, Pesticide Use
• Leukemia	See 1.1 Above
• Non-Hodgkin's Lymphoma	See 1.1 Above, Plus Phenoxyacetic Acid Herbicides (2,4-D, and 2,4,5-T)
• Hodgkin's Lymphoma	See 1.1 Above
• Multiple Myeloma	See 1.1 Above
2. Skin:	
• Basal Cell Carcinoma	Chronic Sun Exposure
• Squamous Cell Carcinoma	Chronic Sun Exposure
• Melanoma	Frequent Sunburns Early in Life, Chronic Sun Exposure
3. Soft Tissue Sarcoma	Phenoxyacetic Acid Herbicides (2,4-D, and 2,4,5-T)
4. Prostate	Methyl Bromide (Fumigant, Soil Sterilant) Organophosphate Insecticides Pyrethrins, Permethrin Endocrine Disruptors: • Dioxin • Mirex • Toxaphene • Parathion • DDT Diet: High Fats and Low Fruits and Vegetables Family History of Prostate Cancer Family History Plus Pesticide Exposure
5. Testicular	Fertilizer Application by Parents
6. Brain	*Toxoplasma gondii* Infection Insecticides
7. Stomach	Nitrates and Atrazine in Drinking Water

an increased risk for *T. gondii* infection. The hypothesized mechanism of the relationship of pesticide exposure and brain cancer revolves around the immunotoxic mechanisms of insecticides (Blair and Zahm 1995). A summary of exposures and reported risk factors are seen in Table 5.4.

CANCER IN SPECIAL AGRICULTURAL POPULATIONS

Although most cancer research among the farming population has been conducted among principal operators (which are predominantly white males), more recent research has been conducted to provide some insight into risks for women, children, and migrant seasonal workers in agriculture.

Cancer in Farm Women

Generally the research is fairly positive regarding cancer experience in farm women. Overall, (similar to farm men) farm women experience less cancer, as reported risk rates have been 8–40% lower than comparison population (Folson and others 1996; Kristenson and others 2000; Pukkala and Notkola 1997). The main cancer risk deficits include lung, breast, cervical, and ovarian (Folson and others 1996; Kristenson and others 2000, 1996a; Settimi and others 1999). The overall decrease is likely related to lowered cigarette smoking. Although the overall risks for breast and ovarian cancer are not elevated, there is some evidence that oganochlorine exposure (breast) and triazine herbicides (ovarian) are associated risks (Blair and

Zahm 1995). A more recent report indicated that female pesticide applicators had an increased risk for ovarian cancer, but this was not seen in the farming or spousal population of farmers (Blair and others 2005, in press). There is little further evidence for other occupational or environmental exposure risks for cancer in farm women, with the exception of one study which suggests about a 50% increase in risk for non-Hodgkin's lymphoma (Folson and others 1996). This suggests that women may experience similar exposure risks as men, and further research is warranted.

Cancer in Farm Children

There is some evidence that farm children may be at increased risk for certain cancers. Studies indicate increased rates of Ewing's bone sarcoma, brain, testicular and Wilm's sarcoma (Kristenson and others 1996a; Valery and others 2002), and leukemia (Gunier and others 2001; Kristenson and others 1996a).

Several studies have investigated risk factors for childhood cancers. Horticulture and pesticides have shown associations with early-age onset cancers, and poultry farming with later-age onset of tumors (Kristenson and others 1996a). Some risk factors have also been identified for tumors that have not necessarily been shown to be at increased risk generally among farm children. Young children of parents working in horticulture and with a history of pesticide exposure may generate increased risk for all cancers, Wilm's tumor, non-Hodgkin's lymphoma, eye, and neuroblastoma (Kristenson and others 1996a). Brain cancer in children has been associated with pig farming and pesticides (Fear and others 1998; Kristenson and others 1996a). Leukemia has been associated with pesticide exposure (Fear and others 1998), and in one study, with exposure to a specific pesticide used in orchards, proparagite (Gunier and others 2001). Kidney cancer has also been associated with pesticides (Fear and others 1998). An extensive Australian study of parental exposures and risk of Ewing's sarcoma in children has revealed farming residency of parents as a possible significant risk factor. However, this study was not specific enough to determine detailed exposure to risk factors. A review of childhood cancers for the Agricultural Health Study revealed slight increases in children for all lymphomas (SIR 2.18, 95% CI 1.36–4.10), and non-Hodgkin's

lymphoma (1.98, 95% CI 1.05–3.76) (Flower and others 2003). There was a small risk associated with these cancers for children of fathers that did not use gloves when handling pesticides.

In summary, there are indications for small risk factors for certain cancers in farm children. Pesticides are the most common exposure source cited, but animal, and presumptive animal tumor viruses, may also be a possibility. The mechanism may be from genotoxic damage to germ cells or congenital damage to the fetus. This hypothesis is due to the studies of associate parental exposures and childhood cancers. However, there could also be associations of environmental or occupational exposures in later-age onset of tumors.

Cancer in Hispanic Farm Workers

There is little data regarding cancer patterns among Hispanic farm workers. Cancer patterns do not mirror farm owner/operators, as lung and prostate cancers seem to be much more important. The available literature is reviewed and summarized in Chapter 2 on special risks populations.

Cancer in Selected Farming Subpopulations

Cancer has been studied in a few agricultural subpopulations revealing cancer risks and patterns usual to the general agricultural community. Sugarcane production workers in Puerto Rico have a high risk of oral cancer. The relative risk is 4.4 compared with other populations (Coble and others 2003). There have been no particular risk factors identified for the high risk of oral cancer in this subpopulation. However, it is likely related to personal behaviors, such as tobacco or other substance use, perhaps in combination with some occupational exposure.

Although bladder cancer incidence is generally low in the agricultural population, researchers in France have noted an increasing bladder cancer gradient from the north to the south of France. As vineyard production is greater in the south of France compared to the north, a hypothesis was investigated that vineyards and pesticide exposure were an explanation for the observed gradient. An epidemiological study did show a slightly elevated association of bladder cancer to vineyard production and pesticide use with a standardized mortality ratio of 1.14 (confidence interval 1.07–1.22). Further research is warranted to further define this relationship.

SOCIAL MEDICINE ASPECTS OF CANCERS IN FARMERS

Rural and agricultural populations have a series of social and psychological concerns that health care providers should take into consideration when caring for them. Cancer is a frightening word to most people. In fact, 25% of the Anglo farmers and 50% of Latinos consider cancer to be the same as a death sentence (Burman and Weinert 1997a, b). There has been much publicity about the adverse health affects of cancer, which weighs on peoples' minds. A Wisconsin (U.S.) study (Perry and Bloom 1998) revealed three major themes of concern regarding pesticide exposure, including

1. Do pesticides cause cancer?
2. What other health risks are related to pesticides?
3. Can I sustain the farm with adverse health consequences, or can I farm without pesticides?

Once cancer is diagnosed in a farm family, there are several relationship issues that occur. There are relationship problems that occur between spouses, friends, and other family members. A major problem is that women living with cancer do not feel spousal support and understanding of their feelings. Furthermore, there are feelings of aloneness and avoidance of friends and family. These latter problems are worse in women than men. On the other hand men are concerned about their job and how they will keep the farm going and support the family (Burman and Weinert 1997a, b).

Although the overall cancer rates are lower in the farming population generally, cultural, physical, and economic barriers to ideal health care services create complications in managing cancer in rural and agricultural populations. These complications are summarized as follows:

1. General difficulties with health care accessibility in rural areas
2. Cultural issues of the farming population who generally tend to not use medical services unless they are very sick
3. Tendency of avoidance or procrastination for seeking medical care because of fear of a possible diagnosis of cancer

4. Cultural and practical considerations that make farm work and taking care of animals come before personal health care
5. Farm population is an older population, so there is more cancer of certain types, creating a concern in the rural population that there are "epidemics" of cancer
6. Concern that pesticides may cause cancer, but still needing to use pesticides to maintain livelihood
7. Little information available for prevention
8. Limited options for home-based or outpatient treatment

Health professionals should consider all the dynamics mentioned above in the face of a cancer diagnosis in a farm family member. These issues should be discussed with the families and solutions considered. As mental and social health care professionals are in short supply in many rural and farm areas, the health care provider can seek nonprofit organizations or support groups to help these families cope. The Extension Service or private nonprofit organizations like churches and Sharing Help Awareness United Network www.shaunnetwork.org or AgriWellness (www.agriwellness.org) aim to give social and psychological support for victims of farm injuries or serious illnesses.

PREVENTION OF CANCER IN FARM POPULATIONS

The basic methods of prevention of cancer are similar to other preventive strategies in occupational health. These methods include deployment and enforcement of regulations, identification and removal of the hazard, substitution of a safer product or process, proper use of personal protective equipment, early detection, and education. As the various cancers and risk factors were reviewed, the specific hazardous exposures are not fully known. The exposures that are likely carcinogenic for the agricultural population include sun and 2,4-D and 2,4,5-T herbicides. Beyond these risk factors the information is not very specific; however, one should choose to limit exposure to all pesticides, assure drinking water is free of nitrates and atrazine, and assure exposure to animal tumor viruses is minimized. As regulations for worker protection are relatively minimal in agriculture, most of the emphasis has to rely on education and cancer screening. One thing to recognize is that lung, colon, and prostate cancers are the most important causes of cancer mortality. Even though lung cancer is not an agricultural risk, there should still be an emphasis on smoking cessation in farm patients, as large gains in cancer reduction are still possible here, and the main risk factor is well established. Dietary factors recommended by the American Cancer Society should be emphasized for prevention of breast, colon, and prostate cancer. Regarding cancer screenings, standard recommendations such as those from the American Cancer Society should be followed. For example, women and men over 50 years of age should receive regular breast and prostate cancer screenings. However, cancer screening among the farm population is not very common. Furthermore, such services are not highly accessible in many rural areas (Muldoon and others 1996), and this population tends to use health services mainly for treatment of acute illnesses. Barriers identified for cancer screening for the farm population include 1) cost of cancer screening, 2) time from work, 3) distance to providers, and 4) self-reliant behaviors (Reading and others 1997). Thus, the rural health care professional must make a special effort to help increase the degree of cancer screening in agricultural patients. Skin cancer prevention is based on prevention of sun exposure. One survey of over 1000 farmers and spouses indicated that although nearly 90% of the population knew the long-term consequences of sun exposure, only 40% of the men and 65% of women adequately protected their skin from the sun. The use of sun protection was directly related to income and education. Few of those surveyed ever had a specific skin exam. Clearly, medical practitioners need to encourage routine skin exams and the routine use of sun protection in their farm patients.

REFERENCES

Alavanja M, Blair A, Sandler D, Hoppin J, Thomas K. 2005 (in press). Prostate Cancer and Agricultural Pesticides.

Alavanja M, Samanic C, Dosmeci M, Lubin J, Tarone R, Lynch C, Knott C, Thomas K, Hoppin J, Barker J, and others. 2003. Use of agricultural pesticides and prostate cancer risk in the Agricultural Health Study. American Journal of Epidemiology 157(9):800–814.

Amador D, Nanni O, Falcini F, Saragoni A, Tison V, Calloa A, Scarpie E, Ricci M, Riva N, Buiatti E. 1995. Chronic lymphocytic leukemias and non-Hodgkins-lymphomas by histological type in farming–animal breeding workers—A population case-control study based on job titles. Occupational Environmental Medicine 52(6):374–379.

Blair A, Sandler D, Tarone R, Lubin J, Thomas K, Hoppin J, Samanic C, Coble J, Kamel F, Knott C, and others. 2005 (in press). Mortality Among Participants in the Agricultural Health Study.

Blair A, Zahm S. 1995. Agricultural exposures and cancer. Environmental Health Perspectives 10(3 (Suppl. 8)):205–208.

Burman M, Weinert C. 1997a. Concerns of rural men and women experiencing cancer. Oncology Nursing Forum 24(9):1593–1600.

——. 1997b. Rural dweller's cancer fears and perceptions of cancer treatments. Public Health Nursing 14(5):272–279.

Cerhan S, Cantor K, Williamson K, Lynch C, Torner J, Burmeister L. 1998. Cancer mortality among Iowa farmers: Recent results, time trends, and lifestyle factors (United States). Cancer Causes Control 9:311–319.

Coble J, Brown L, Hayes R, Huang W, Winn D, Gridby G, Bravo-Otero E, Frarimeni J. 2003. Sugarcane farming, occupational solvent exposures and the risk of oral cancer in Puerto Rico. Journal of Occupational Environmental Medicine 45(8):869–874.

Dich J, Wiklund K. 1998. Prostate cancer in pesticide applicators in Swedish agriculture. Prostate 34(2):100–112.

Donham K, Berg, J, Sawin, B. 1980. Epidemiology relationships of bovine lymphosarcoma and human leukemia. American Journal of Epidemiology 112:80–92.

Donham K, Leininger, JR. 1977. Spontaneous leprosy-like disease in a chimpanzee. Journal of Infectious Disease:132–136.

Faroy-Menciere B, Deschamps F. 2002. Relationships between occupational exposure and cancer of the testis. Annals of De Medecine Interne 153(2):89–96.

Fear N, Roman E, Reeves G, Pannett B. 1998. Childhood cancer and paternal employment in agriculture: The role of pesticides. British Journal of Cancer 77(5):825–829.

Flower K, Hoppin J, Lynch C, Blair A, Knott C, Shore D, Sandler D. 2003. Cancer risk and parental pesticide application in children of agricultural health study participants. Environmental Health Perspectives 112(5):631–635.

Folson A, Zhang S, Sellers T, Zheng W, Kushi L, Cerhan J. 1996. Cancer incidence among women living on farms: Findings from the Iowa Woman's

Health Study. Journal of Occupational Environmental Medicine 38(11):1171–1176.

Gunier R, Harnly M, Reynolds P, Hertz A, Von Behren J. 2001. Agricultural pesticide use in California: Pesticide prioritization, use densities, and population distributions for a childhood cancer study. Environmental Health Perspectives 109(10):1071–1078.

Keller-Byrne J, Khudex S, Schaub E. 1997. Meta-analysis of prostate cancer and farming. American Journal of Independent Medicine 31:580–586.

Khuder S, Mutgi A, Schaub E. 1998. Meta-analysis of brain cancer and farming. American Journal of Independent Medicine 34(3):252–260.

Khuder SA. 1999. Etiologic clues to lip cancer from epidemiological studies on farmers. Scandinavian Journal of Work Environmental Health 25(2): 125–130.

Kirkhorn S, Schenker M. 2002. Current health effects of agricultural work: Respiratory disease, cancer, reproductive effects, musculoskeletal injuries, and pesticide-related illnesses. Journal of Agricultural Safety and Health 8(2):199–214.

Kristenson P, Anderson A, Irgens L. 2000. Hormone-dependent cancer and adverse reproductive outcomes in farmer's families—Effect of climactic conditions favoring fungal growth in grain. Scandinavian Journal of Work Environmental Health 20(4):331–337.

Kristenson P, Anderson A, Irgens L, Bye A, Sundheim L. 1996a. Cancer in offspring of parents engaged in agricultural activities in Norway: Incidence and risk factors in the farm environment. International Journal of Cancer 65(1):39–50.

Kristenson P, Anderson A, Irgens L, Laate P, Bye A. 1996b. Incidence and risk factors of cancer among men and women in Norwegian agriculture. Scandinavian Journal of Work Environmental Health 22:14–26.

Lange J. 2000. Reduced cancer rates in agricultural workers: A benefit of environmental and occupational endotoxin exposure. Medical Hypothesis 55(3):383–385.

Mastrangelo G, Marzia V, Marcer G. 1996. Reduced lung cancer mortality in dairy farmers: Is endotoxin exposure the key factor? American Journal of Independent Medicine 30(5):601–609.

Milham S. 1999. Cancer among farmers: A meta-analysis. Annals of Epidemiology 9:71.

Muldoon J, Schoolman M, Morton R. 1996. Utilization of cancer early detection services among farm and rural non-farm adults in Iowa. Journal of Rural Health 12((Suppl. 4)):321–331.

Pahwa B, McDaffie H, Dosman J, Robson D, McLaughlin J, Spinelli J, Fincham S. 2003. Exposure to animals and selected risk factors among Canadian farm residents with Hodgkin's disease, multiple myeloma, or soft tissue sarcoma. Journal of Occupational and Environmental Health 45(8): 857–868.

Parker A, Cerhan J, Putnam S, Cantor K, Lynch C. 1999. A cohort study of farmers and risk of prostate cancer in Iowa. Epidemiology 10(4):452–455.

Perry M, Bloom F. 1998. Perceptions of pesticide—Associated risks among farmers: A qualitative assessment. Human Organization 57(3):342–349.

Pukkala E, Notkola V. 1997. Cancer incidence among Finnish farmers, 1979–1993. Cancer Causes Control 8(1):25–33.

Reading D, Lappe K, Kreuger M, Kolehouse B, Steneil D, Leer R. 1997. Screening and prevention in rural Wisconsin: The Greater Marshfield Experience. Wisconsin Medical Journal August, 1997:32–37.

Rubino M, Donham K. 1984. Inactivation of bovine leukemia virus infected lymphocytes in milk. American Journal of Veterinary Research 45(8):1553–1556.

Settimi L, Comba P, Bosia S, Ciapini C, Desideri E, Fedi A, Perazzo P, Axelson O. 2001. Cancer risk among male farmers: A multi-site case control studies. International Journal of Occupational Medicine and Environmental Health 14(4):339–347.

Settimi L, Comba P, Carrieri P, Boffelta P, Magnani C, Terracini B, Andrion A, Bosia S, Ciapini C, DeSantis M, and others. 1999. Cancer risk among female agricultural workers: Two multi-center case-control study. American Journal of Independent Medicine 36(1):135–141.

Sperati A, Rapiti E, Quercia A, Terezoni B, Forastiere F. 1999. Mortality among male licensed pesticide users and their wives. American Journal of Independent Medicine 36(1):142–196.

Valery P, McWhirter W, Sleigh A, Williams G, Bain C. 2002. Farm exposures, parental occupation, and risk of Ewing's sarcoma in Australia: A national case-control study. Cancer Causes Control 13:263–270.

Van Der Gulden J, Volgenzang P. 1996. Farmers at risk for prostate cancer. British Journal of Urology 77(1):6–14.

Wiklund K, Dich J. 1995. Cancer risks among male farmers in Sweden. European Journal of Cancer Prevention 4(1):81–90.

6
Health Effects of Agricultural Pesticides

Kelley J. Donham and Anders Thelin

INTRODUCTION

It was a sunny and windy day in May when farmer (C.B.) was spraying insecticides on his corn fields on his farm in central Iowa (U.S.) (see Figure 6.1). He had been working since 7 a.m. Near noon, he began to feel tired so he went to the house for lunch. During lunch, he had a sudden urge for a bowel movement. He went to the bathroom with a severe case of diarrhea. While in the bathroom, he became extremely weak and lost his vision. His wife found him in the bathroom in a near comatose state. She was able to get him into the car, and with great foresight, she took along a container of the chemical he had been applying to his fields. The emergency room physician examined C.B. and then called the state poison control center and read the information regarding the specific ingredients from the product label. Determining that C.B. had been applying an organophosphate (OP) insecticide, the physician treated him with an IV drip of atropine over a course of 8 hours. C.B. began to improve and over the next 3 days made an uneventful recovery.

C.B. was very lucky. An alert and resourceful wife combined with accessibility of knowledgeable emergency medicine personnel and a quality poison control center, allowed C.B. to survive a life-threatening poisoning to farm again.

Although acute pesticide poisonings are not common events in agriculture, they can be life threatening when they happen. C.B. splashed some of the concentrate on his hands while mixing and loading the product and was caught in the pesticide drift due to the wind as he was applying the chemicals.

More than any other occupational health hazard, farmers are highly concerned about possible personal health of the pesticides they apply (Thu and others 1990). Much of this concern is because of the amount of publicity pesticides receive and because there is much we do not know about the chronic health problems that pesticides may cause. This chapter provides an overview of the types of pesticides, how they are used, their acute toxicity, and methods to treat and prevent poisonings. Furthermore, it attempts to remove some of the misunderstandings about this group of chemicals so that health professionals may be better able to manage acute exposures when they are encountered and communicate more fully informed to their agricultural patients/clients.

DEFINITION OF PESTICIDES

Pesticide is a generic term often perceived by the general public as a substance that is very toxic to life and persistent in the environment. This creates illogical thinking and a perception and fear that all pesticides are dangerous. This public perception is enhanced by the extensive media coverage that pesticides receive. This situation promotes an emotional rather than a logical approach to prevention, deterring governmental and nongovernmental organizations from applying amounts of resources for prevention of occupational and environmental problems in agriculture relative to the magnitude of the specific problem. For example, in the U.S., only a few deaths are reported annually from occupational exposure to pesticides, while nearly 800 die from acute injuries; yet the attention and resources given to the

Figure 6.1. Spray application of insecticides to row crops—applicators are much better protected inside a tractor with a cab, especially if the windows are shut and the cabin filtration system is properly functioning. (This figure also is in the color section.)

pesticide problems are many times greater than resources available for acute injuries (National Safety Council 2001).

Pesticides include an incredibly diverse group of chemicals that have a wide range of toxicity, species specificity, and persistence in the environment. In the U.S. alone, there are 675 active pesticide ingredients used and over 16,000 formulations that are marketed for a wide variety of applications (Calvert and others 2004; Farahat and others 2003). One objective of this chapter is to provide information that will help the reader determine the relative degree of hazard of these various products so that informed and appropriate actions may be brought to bear on the problems.

Pesticides are products that kill living things that are economically, socially, or healthfully detrimental to us (pests). The *icide* part of the word means to kill (Morgan 1978). The prefixes (e.g., insect, herb, rodent, and fungi) tell what they kill. Insecticides are pesticides that kill insects, and generally (as a class of chemicals) they are some of the more toxic pesticides and cause most of the human health problems. Insecticides are what the general public thinks about when they hear or say pesticides. For the purpose of this chapter, we will cover only insecticides, herbicides, and fumigants, which are those pesticides that have a primary risk for agricultural workers. Fumigants are chemicals that are very broad in their lethality to life forms and will be discussed as a special class of pesticides.

HISTORY

DDT was developed in the mid 1940s during WW II. This chemical was found extremely effective in controlling insects because of its lethality to them, and because it was persistent (one applica-

tion lasted for several weeks). The use of these chemicals has resulted in many advances against human diseases, such as malaria, and plant diseases, such as corn root worm. The publicity and interest on the part of many public research organizations and pharmaceutical and chemical companies created a vast array of new pesticide products over the ensuing three decades (Mutel and Donham 1983).

As mentioned above, there are over 16,000 formulations of pesticides registered in the U.S. alone. This includes insecticides, herbicides, rodenticides, fungicides, and others. Only insecticides, herbicides, and fumigants will be covered in this chapter as they are by far the most common pesticides used in agriculture and are responsible for most of the pesticide poisonings. Other toxicological problems, such as silo gas and nitrates are covered in other sections of the book, under the appropriate exposure source or anatomical system involved.

FREQUENCY OF PESTICIDE POISONINGS

The available information on the incidence and prevalence of pesticide poisonings are only estimates (similar to most occupational or public health illness reporting). The reason is that mild poisonings may never get medical attention, be recognized, or be reported. In most states or countries, there is no requirement for reporting pesticide poisonings. Furthermore, most agricultural people are not inclined to seek medical treatment unless they are seriously ill. Also, many farm workers, particularly those who might not be documented or may have language barriers, are reluctant to seek medical attention for fear of being fired; discovered by authorities and required to leave the country; or embarrassed financially, socially, or culturally by the medical care system. Finally, recognition of pesticide poisonings by medical care personnel may be difficult, as symptoms may be vague and mimic other conditions. For all these reasons, the actual number of documented cases is underreported. On the other hand, estimates by some advocate groups may overestimate the actual number of poisonings. The available estimates reported here arise from multiple sources, including emergency room and hospital records, poison control centers, workman's compensation records, and special surveys.

Estimates for the U.S. poison control centers indicate that pesticides make up about 5–6% of all poisonings. There are an estimated 2,700 hospitalizations for pesticide poisoning per year, and of these, only about 28% are occupationally related (the majority of which are related to agriculture). However, 55% are accidental exposures, mainly children getting into pesticides that are not kept secure. The remaining 17% are suicides (Mutel and Donham 1983). There are also about 64 fatal poisonings per year. About 15% of these are occupational and are distributed among chemical manufacturers, formulators, packaging and warehouse workers, applicators, and farmers and field workers (Mutel and Donham 1983).

A more recent survey from the Sentinel Event Notification System for Occupational Risks (NIOSH) provides more detail of pesticide poisonings from six states in the U.S. (Arizona, California, Florida, New York, Oregon, and Texas). Only occupationally related cases are presented in this survey (Calvert and others 2004). These states include those where a large number of farm workers are employed. The survey covered the years 1998 and 1999. There was an average of 505 reported incidents each year. The overall rate of poisoning was 1.17 per 100,000 potentially exposed persons (all occupations). However, when considering only agriculture, the rate was 18.2 per 100,000. Of these cases, 69.7% were of low severity, 29.6 were of moderate severity, and 0.4% were of high severity. There was an average of 1.5 fatalities per year over the period. About half of these pesticide poisonings were from insecticides.

One recent report indicated about 500 California agricultural workers are poisoned yearly. Although a few of these poisonings are severe,

most of these poisonings are relatively mild (Reeves and Schafer 2003).

Even though the available data sources are not exact, it is obvious that acute pesticide poisonings in agriculture in the United States are not nearly as serious a problem as are acute physical injuries. The latter accounts for approximately 23 fatalities per 100,000 and 625 nonfatal injuries per 100,000 (data from 1997 census and the National Safety Council (2001)). These injury rates are nearly 35 times higher compared to acute [mostly nonfatal] pesticide poisoning rates as reported by Reeves and Schafer (2003). Furthermore, it is obvious that accidental, nonagricultural poisonings and suicides are more frequent than occupational exposures.

Chronic effects of pesticides are even more difficult than acute poisonings to assess. Measures of relative risk for associations of pesticide exposures to cancers in farm populations are in the range of 1.5–1.6 (see Chapter 5 on agricultural cancers).

The data presented above are only for the U.S., but one could assume relative rates would be similar in other industrialized countries. Developing countries probably have many more acute poisonings because of higher exposures (often with knapsack sprayers); less awareness; fewer control regulations; and lack of diagnostic, clinical, and reporting services.

HOW WORKERS ARE EXPOSED TO PESTICIDES

The predominant way workers are exposed to pesticides is by skin contact. Insecticides by their very nature are readily absorbed through the skin (see Figure 6.2). Clothes, gloves, shoes, and other apparel that become contaminated with insecticides may remain a continual source of exposure until that clothing is removed and carefully washed. Covered skin is even more susceptible to absorption. Ingestion is a less-important exposure mechanism in the agricultural setting. The greatest risks for ingestion are from accidental ingestion by children, contamination of food products, putting pesticides in unmarked or emptied food containers (e.g., soft drink bottles), and using used pesticide containers for drinking water containers. Another mechanism of ingestion exposure is eating in unsanitary conditions around places where pesticides are handled and where

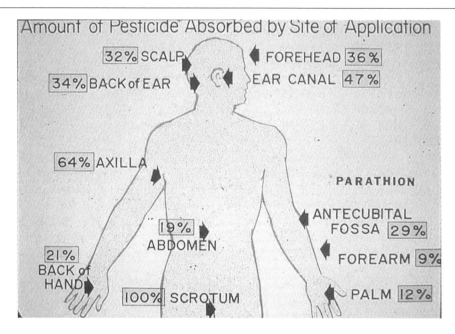

Figure 6.2. The most common systemic exposure to pesticide is through skin contact. Skin absorption of organophosphate pesticides is especially efficient in areas of thin skin with high blood supply. (This figure also is in the color section.)

hand washing is not practiced prior to eating or smoking. Inhalation is probably the least common mechanism of exposure except for fumigants (which will be discussed later in this chapter) and from combustion of pesticides or pesticide containers. A good portion of inhaled (except fumigants) pesticide may be trapped in the trachea, where it is elevated back into the pharynx by the mucociliary apparatus, and swallowed. True inhalation exposure is rare. Particles in most formulations (including powders, microcapsules, and sprays) are larger than respirable size (Morgan and others 1978). Those rare formulations that include mists or fumes are the most likely source of respiratory exposure.

Specific occupations that have pesticide exposure risk follow the various tasks in the fate of the chemicals from manufacturing to the consumer of products to which pesticides have been applied. Some specific jobs that have risk for pesticide exposures include workers in chemical manufacturing plants, warehouses, transportation of pesticides, formulation plants, applicators (includes mixing and loading operations) (Hines and others 2001), farm owner-operators, and farm workers. Livestock producers also have significant expo-

sures (primarily dermal) when applying parasiticides to animals (Stewart and others 1999).

Workers in formulation plants have particularly risky jobs, as these plants are often small operations with poorly trained employees, little occupational health staff, and poor industrial hygiene control, and are located in warm climates, which makes wearing of personal protective equipment uncomfortable and possibly creates an additional risk of heat stress. These workers also handle concentrated active ingredients, which increases their exposure risk. Mixing and loading operations involve diluting and mixing pesticides for transfer into the application equipment (Figure 6.3). These workers also handle concentrated products, and the work processes of chemical transfer have a high risk of spillage. Applicators typically do not handle concentrates but are still at risk of exposure from seepage, spray drift, maintenance, and cleaning of equipment. Reeves and Schafer (2003) estimated that 51% of occupational poisoning cases in California are caused by pesticide drift and 25% from exposure to pesticide residues on plant foliage. Workers who load and clean crop dusting planes have a high risk of exposure, as the contaminated wash water often splashes on the worker.

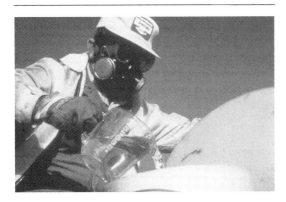

Figure 6.3. Mixing and loading insecticides prior to application can be an operation with a high potential for worker exposure as the chemicals input into the tank are concentrated. Minimum protection during this application includes unlined rubber gloves. (This figure also is in the color section.)

Figure 6.4. In crop dusting applications, there is risk for insecticide exposure to the pilot, the flagger on the ground, who marks the way for the pilot, and the persons who mix and load the plane and clean up the plane. (This figure also is in the color section.)

Furthermore, the flagger on the ground, who marks where the next pass of the airplane should be, is also at risk of exposure (Figure 6.4). Farm workers who pick fruits and vegetables that have been sprayed are at risk because of contact with the dried insecticide on the foliage (Figure 6.5). In order to reduce this exposure, there are regulations for specific applied chemicals regarding the time delay necessary following application to safely reenter fields for harvest (reentry times). Fields must be posted to advise people to keep out of these fields until it is safe to reenter. It must be noted (especially in smaller farms) that farm owner-operators or workers may perform (and thereby be exposed) through several of the tasks described above, including mixing and loading, applying, and harvesting. However, there has been a trend in many areas for commercial firms to be hired to conduct much of the agricultural pesticide applications. This is good in one way, as more experienced and knowledgeable people are conducting the application. Licensed applicators do have a lower risk of high-exposure events than other workers (Alavanja and others 1999).

Dosemeci and others (2002) have developed an algorithm that is predictive of the extent of chronic exposure farmers receive. They reported on six factors that are related to lowered exposures, including 1) use of closed mixing systems,

Figure 6.5. Picking citrus or other fruit or orchard crops presents possible exposure to workers from pesticides that may remain on the foliage. (*Source:* Photograph by Terry Noble, www.gaysmills.org/orchards.html). (This figure also is in the color section.)

2) use of tractors with enclosed cabs and charcoal air filters, 3) decreased frequency of washing application equipment, 4) increased frequency of changing gloves, 5) increased frequency of bathing and hand washing, and 6) increased frequency of changing clothes after spills.

One analysis of acute toxic exposures found three consistent risk factors: 1) failure to follow labeled directions, 2) inexperience of the worker, and 3) unpredictable random events (i.e., broken hose or accidental spill) (Mage and others 2000).

Other occupationally exposed persons include grain elevator workers who are involved in grain fumigation and fire fighters and bystanders where pesticides are burning. The latter are at risk from exposure to volatilized pesticides or products of pyrolysis.

There is a low risk of exposure to handlers, retailers, or consumers of food products treated with pesticides. Most industrialized countries have special agencies (e.g., USDA or FDA in the U.S.) that test agricultural products for pesticide content. However, there have been a few cases where consumers have been poisoned from pesticides that are taken up within the plant tissue (e.g., aldicarb or temik) (Baron 1994).

INSECTICIDES: CLASSES AND TOXIC MECHANISMS

Four primary classes of insecticides are discussed here, including 1) organochlorines (OCLs), 2) organophosphates (OPs), 3) carbamates (CBs), and 4) pyrethroids (PYs). (Additionally a few commonly used compounds not fitting with any particular group will also be discussed). Compounds within the four primary classes are linked by similar chemical structures and similar mechanisms of action. These classes of insecticides vary regarding their toxicity, persistence in the environment, and other toxicological features. However, they also vary within class, depending on the specific chemical structure and the specific formulation. For example, the lethal dose where 50% of exposed test animals die (LD_{50}) varies thousandfold among OPs products used in agriculture (Reigart and Roberts 1999). Additionally different formulations of products affect their toxicity; for example, products of high volatility, those formulated as aerosols, gases, or fine dusts, present a higher poisoning risk compared to granular or slow-release microcapsules.

DDT (mentioned previously) was developed in the mid-1940s, and for the following two decades became the most highly used product of the OCL class of chemicals. Several other OCLs developed during the interim have also been used extensively. These chemicals were very effective insecticides, as they were readily absorbed by the chitin exoskeleton of insects, they readily killed insects, and they persisted in the environment so that they continued to kill insects for extended periods with one application (Morgan and others 1978). They were used in most areas of row crop production, fruit and vegetable production, control of ectoparasites on livestock, and vector control in public health applications. Some of the positive results of OCLs have been the control of many insect pests and dramatic increase in crop production; for example, control of corn root worm and the cotton bowl weevil have dramatically increased production of those two important commodities. Examples of increased livestock and poultry production include control of lice and mange mites in many species. Control of mosquitoes has resulted in a decline in malaria in many regions of the world. (However, malaria has now returned as an important disease in many tropical countries as the use of OCLs has been diminished).

The toxic mechanism of OCLs is disruption of nerve impulse transmission. A nerve impulse is created following a stimulus that results in an exchange of sodium for potassium across the nerve cell membrane, facilitated by a biochemical "sodium pump." The exchange of ions results in a wavelike shift of polarity traveling along the nerve axon, innervating the end organ or tissues (muscle), or connecting other parts of the brain. OCLs inhibit the sodium pump, ceasing nervous transmission. The target and nontarget species (including animals and human beings) are affected in the same manner (Mutel and Donham 1983). Fortunately, insects are much more susceptible to these chemicals than mammals. In fact, OCLs have relatively low mammalian acute toxicity compared to other insecticides (e.g., OPs) we discuss later in this chapter. This is shown by the fact that cattle, sheep, or swine producers who used OCLs on their stock in past years often got soaked with the chemicals along with the animals they were treating. Rarely would acute toxic symptoms develop with this exposure. The work practices to which producers had become accustomed when working with OCLs was later a factor in many acute poisonings with newer and much more acutely toxic OPs that began to emerge in the late 1960s and 1970s. The reason was because when producers handled these new OPs as they did the OCLs (not very carefully) they were poisoned because they are much more acutely toxic.

A major shift in the public's thinking about insecticide usage was influenced by the book *Silent Spring*, written by Rachel Carson (1962). She eloquently communicated to the world a vision of what it might be like to wake up to a spring season with no songbirds. She painted this scenario as a result of the observed effect on the lowered hatchability of bird eggs resulting from the eggshell-thinning effects of persistent OCLs in the environment. Her book rapidly created a large public awareness and concern about the use of OCLs. Not only are these chemicals persistent in the environment they are stored in fat tissues of animals, and thus bioaccumulate in food webs, resulting in species on the higher end of the food chains having the highest exposures and the highest levels of OCLs in their fat stores. A tremendous amount of research has backed up the initial findings regarding bioaccumulation of OCLs in many animal species (including human beings) and has documented the negative effects on reproduction of especially large carnivorous birds (raptors such as the bald eagle). This resulted in legislation that severely limited the use of OCLs in many applications around the world. (Note that human exposures to OCLs may still occur, because they are still used in certain applications, and that old stores of OCLs may still be found on farms.)

In recent years, there has been greater concern about more chronic health effects of OCLs in people. Cancer, endocrine disruption, and health concerns have been investigated (Garcia and others 1999). Cancer issues from pesticides are covered in Chapter 5 of this text. Increased risk associated with OCLs for a variety of congenital and reproductive problems have been speculated, including hypospadia, oral clefts, nervous system anomalies, and abnormal sperm. However, the sum of these studies does not give a clear picture regarding the extent of risk and what specific OCLs may be implicated(Garcia and others

1998). Georgellis (1999) suggests there are significant methodological limitations of epidemiology to determine these apparent low-level risks, and new research methods need to be applied to this area to enable a clearer picture of risk.

Because of the broader environmental concerns and new regulations, the second class of insecticides mentioned above (OPs) began to replace the OCLs in the late 1960s and early 1970s. One primary difference between these two classes of insecticides is that OPs degrade relatively rapidly in the environment, and they do not bioaccumulate. However, they are much more acutely toxic to human beings and other mammals than are OCLs. OP and carbamate (CBs, a third class of insecticides) insecticides like OCLs, are neurotoxins. OPs and CBs interfere with the transmission of nerve impulses across a synapse (the space where one nerve cell ends and another begins) (Morgan 1978; Mutel and Donham 1983; Reigart and Roberts 1999). Normally, when a nerve impulse reaches the end of a nerve, there are vesicles in the end organ that release acetylcholine (ACh), which carries a chemical message across the synapse to initiate the nerve impulse in the connecting nerve. Shortly after the impulse is transmitted, a second enzyme is released from the end organ, acetylcholine esterase (AChE). This enzyme breaks down ACh so that the nerve impulse will not continue to fire (Figure 6.6). OPs and CBs tie up AChE by competing for reaction sites, effectively disabling the enzyme (Figure 6.7). Therefore, the toxic principle is an overstimulation of the nerves, while OCl poisoning is a lack of nerve firing (Figure 6.8).

CBs differ somewhat from OPs in chemical structure and the chemical linkage CB-AChE is short lived, whereas the OP-AChE linkage is much more persistent. In fact, there is some evidence that after 5 or more hours (aging), the OP-AChE linkage is permanent and new AChE must be produced by the end organ to counteract the poisoning. This later toxic principle has implications for differences in treatment of OP and CB, poisonings which will be discussed below.

There has been increased interest recently regarding the chronic health effects of OPs. (Cancer risks are discussed in Chapter 5.) Other studies suggest there are chronic neurological effects from AChE-inhibiting insecticides. The most

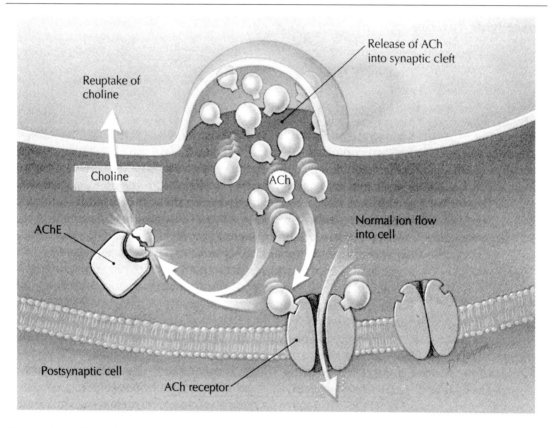

Figure 6.6. Normal transmission of a nerve impulse across a synapse—acetylcholine is released from the end organ at the terminus of the nerve fiber and travels across the synapse to transmit the impulse to the next nerve fiber. The acetylcholine is rapidly broken down at the synapse by acetylcholine esterase to stop sending the signal. (This figure also is in the color section.)

common neurological deficits found include visual motor speed, verbal abstraction, attention, and memory loss (Farahat and others 2003). Associations of OCLs with deficits in male reproductive hormones and thyroid hormones have also been found (Garry and others 2003).

The fourth class of insecticides discussed here are pyrethrum products. Pyrethrins are chemical esters that make up pyrethrum, the insecticidal component of the extract of flowers of the chrysanthemum family. These products are highly absorbed through the chitin of insects, paralyzing their nervous system and causing a rapid knockdown effect (Reigart and Roberts 1999). Pyrethrins are often combined with other chemicals to enhance their lethal effect on insects. Pyrethrin combined with piperonylbutoxide and octylbicycloheptenedicarboximide are the most common constituents of household insect sprays or bombs.

They are used in dairy barns for fly control, but they are not used on crops as they degrade rapidly in the sun and heat.

Pyrethroids are synthetically derived chemicals similar to natural pyrethrum. Pyrethroids are much more potent chemicals than pyrethrum and are used in agricultural and commercial pest control applications. They do not readily degrade with sunlight and heat. Generally, pyrethroids are much less toxic to mammals, including humans, relative to OCLs, OPs, and CBs. These chemicals are dermal and respiratory irritants and allergens. They cause contact dermatitis, both irritant and allergic types. They also cause allergic rhinitis and asthma (Reigart and Roberts 1999). These chemicals have low systemic toxicity. They are rapidly degraded by the liver and excreted via the kidneys. High doses (usually seen only with ingestion of large doses and with the cyano-

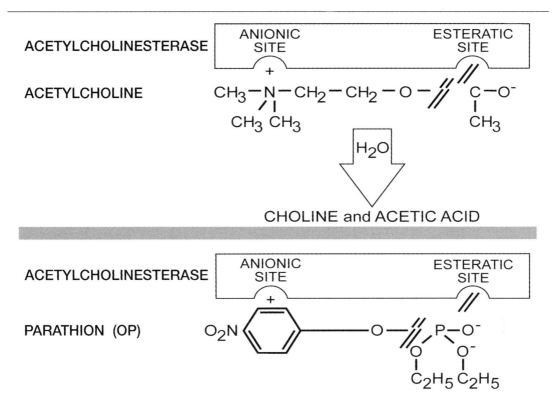

Fgure 6.7. Organophosphate (OP) and carbamates (CB) combine with and inactivate acetylcholine esterase (AChE). Excess acetylcholine (ACh) at the synapse overstimulates the receiving nerve fiber. A positive test for poisoning is a decreased amount of detectable AChE. OP–AChE bonds tend to be longer-lasting than CB-AChE bonds. (This figure also is in the color section.)

pyrethroids) may induce seizures. Other related symptoms may include dizziness, salivation, headache, fatigue, vomiting, diarrhea, and irritability to sound and touch. (Cyano-pyrethroids include flucythrinate, cypermethrin, delta permethrin, and fluvalinate.) Unusual sensations in the skin (paresthesia) noted by stinging, burning, itching, and numbness, may be seen, primarily on the face but also on hands and arms. This effect is increased by sun, heat, and water.

Washing the skin with soap and water and flushing the eyes is the first response. Treatment of allergic reactions is with antihistamines and topical and/or systemic steroids. Asthma may be treated with beta agonists. Vitamin E oil preparations are effective against the paresthesia. Corn oil and petroleum jelly are less effective, and zinc oxide may increase the symptoms. Removal from the digestive tract is warranted only if a large amount has been ingested and it can be done

within 1 hour following ingestion. As seen above, some of the generalized symptoms may mimic OP poisoning. However, these chemicals are not ChEase inhibitors. Atropine administered due to a wrong diagnosis may be harmful, even fatal to the patient (Reigart and Roberts 1999).

Elemental sulfur is a very common chemical used on orchard, vine, and vegetable crops to control mites (acaraside) and fungi. These products are moderate dermal irritants, resulting in irritant contact dermatitis, irritation of the eyes, and respiratory tract. Some of the sulfur on the foliage may be oxidized to sulfur oxides. These gases can increase eye and respiratory irritation for those working directly with the plant foliage. Large amounts of sulfur (e.g., 100 gm), may be ingested with relatively little systemic toxicity. It is absorbed readily and excreted by the kidneys (Reigart and Roberts 1999). There is a minor concern that excess hydrogen sulfide may be formed

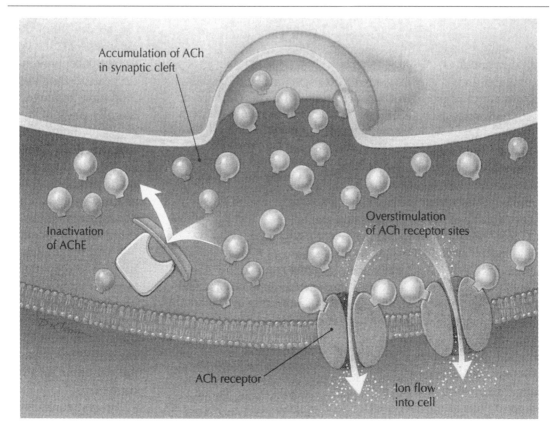

Figure 6.8. As OPs or CBs tie up AChE, there is a build-up of ACh resulting in an overstimulation of the sympathetic system. (This figure also is in the color section.)

in the gut, creating a secondary toxic exposure. Sulfur is an excellent laxative (cathartic), and extended diarrhea may induce dehydration and electrolyte imbalance. Treatment of dermal contamination is removal of the sulfur from the skin with soap and water and flushing the eyes with balanced saline or clean water.

Proparagite is a common acaricide used primarily on citrus crops. There have been no systemic poisonings with this product, but it is an irritant and sensitizer causing eye irritation and irritant and allergic contact dermatitis (Reigart and Roberts 1999). Control and treatment is as with sulfur described above.

CLASSES AND TOXIC MECHANISMS OF HERBICIDES

As mentioned previously, the chemical classes and toxicology of herbicides are much more diverse relative to insecticides. Nine primary classes of herbicides discussed here include: 1) phenolic compounds, 2) chorophenoxy compounds, 3) arsenical compounds, 4) dipyridil compounds, 5) organonitrogen compounds, 6) phalate compounds, 7) propenal, 8) triazine compounds, and 9) phosphonate compounds (Morgan 1979; Reigart and Roberts 1999).

Generally speaking, herbicides are much lower in acute toxicity than are insecticides. Mechanisms of toxicity range widely. Chronic health problems have been noted with some concerns, particularly associated with triazine and arsenical herbicides. The latter have been linked to certain cancers (non-Hodgkin's lymphoma and soft tissue sarcoma for triazine, and liver, skin, and lung cancers for arsenicals). Furthermore, poor semen quality has been linked to atrazine, 2,4-D, and metolachlor (Swan and others 2003).

Phenolic Compounds

Phenolic compounds include chloro and nitrophenols, products that function as preemergent herbicides and some that function as contact herbicides. These products have been used for many years for applications, such as along railroad right-of-ways and field borders. Pentachlorophenol has been widely used as a wood preservative and as a germicide for hide preservation. Although these compounds have caused poisonings and occasional deaths, they are not strong systemic toxins (relative to insecticides). However, they are stable compounds and may be commonly found in the blood and urine of most animal species (including humans), likely because of eating contaminated foods, inhalation of the volatilized chemical, or contact with wood (such as home interiors using wood logs or wood paneling) that has been treated with the chemicals.

Their toxic principle is the uncoupling of oxidative phosphorylation in cellular respiration, preventing the energy-storing reaction of ADP to ATP. This results in excess energy being released as heat, rather than being stored. Therefore, hyperthermia (fever) and sweating is a principal clinical sign, and symptoms include feeling warm. Poisoned persons working in hot environments are at special risk for heatstroke. (Nitrophenols are more likely to cause hyperthermia than chlorophenols.) Workers in hot environments are more susceptible to the effects of these compounds, as heat stress is the main factor that has resulted in deaths from these compounds. Hyperthermia (in addition to a direct cellular effect) may result in cellular necrosis of muscles, liver, kidney, and brain tissues. These chemicals are irritants, and inhalation of vapors or direct contact with pentachlorophenol may cause respiratory and skin irritation.

Chlorophenoxy Compounds

These systemic herbicides have been widely used as brush and weed killers since the early 1950s. The primary chemicals in this class include 2,4-D, and 2,4,5-T. These chemicals are effective mainly on broad-leafed plants. These are used in many agricultural operations for weed and brush control. These products were used extensively in an equal mixture combined with an orange dye as a defoliant (Agent Orange) in the Vietnam War. As these chemicals are applied as an emulsion, occupational exposure may occur by inhalation of droplets or direct skin contact during mixing-loading operations, application or application equipment repair, or clean-up.

These chemicals have low acute toxicity. They are moderate irritants to the skin and mucous membranes. Heavily exposed persons may experience headache, dizziness, and airways irritation. Dioxins (highly toxic compounds) were a contaminant of the early manufacturing process, as the products were synthesized under too high a temperature. (Regulations now limit no more than 0.1 ppm of dioxins in chlorphenoxy compounds.) Several cases of dioxin toxicity in manufacturing plant workers (but not agricultural workers) have been noted, including chloracne and neurological injury.

Several epidemiologic studies in the past two decades are conclusive that 2,4-D and 2,4,5-T are related to the increase in non-Hodgkin's lymphoma and soft tissue sarcoma in the farming community (see Chapter 5). It is unclear whether these cancers are related to the dioxin in the preparations or the active ingredient.

Arsenicals

There are three groups of arsenicals that have been used as pesticides or animal antiprotozoal or antibacterial agents. The various forms include inorganic trivalent (arsenites), inorganic pentavalent (arsenates), and various forms of organic (methylated) pentavalent forms (Peterson and Talcott 2001). Inorganic trivalent forms are by far the most toxic of this group (4–10 times more toxic than pentavalent forms). The organic pentavalents are by far the least toxic. Prior to 1950, the inorganic trivalent arsenic trioxide was a commonly used herbicide and insecticide. A common trade name was Paris green, and was responsible for many human deaths as it is a general cellular toxin, with a predilection for the central nervous system. Chronic effects include lung and skin cancers. One may still find the chemical stored on farms or orchards, left over from years of past use resulting in occasional poisonings. Ingested arsenic is a general cellular toxin, which disrupts general cellular metabolism by substitution in phosphorylation reactions. Arsenic also disrupts various enzymes and coenzymes. Arsenic is highly toxic to blood vessels, causing dilatation and increased permeability of the capillaries. Ingested arsenic first attacks the intestinal lining and blood vessels supplying the intestines (resulting in stasis and insufficient blood supply [ischemia] causing abdominal pain and gastrointestinal bleeding). It also causes necrosis of the liver, kidneys, and central nervous system. Chronic poisoning is marked by hair loss, hyperpigmentation of the skin, and hyperkeratosis.

Because of its high toxicity with many poisonings in nontarget species, trivalent arsenicals have been replaced by pentavalent forms, either inorganic, or the organic form, called cacodylic acid. They largely are used as defoliants in cotton, and in home use as a crabgrass killer. Other forms are used as intestinal antibacterial and antiprotozoal agents in poultry and swine. It is a fairly safe chemical in nontarget species, as it is poorly absorbed from the gut or through the skin.

Dipyridils

Dipyridils are broad-spectrum herbicides. Two of the primary chemicals in this group are paraquat and diquat. That is, they kill all vegetation to which the chemical has been applied. One important feature of these chemicals is that they adhere strongly to about anything they come into contact with. For example, when they are applied to the soil, they adhere strongly to the soil particles, and they are of little environmental or personal risk from that point on.

The acute toxic features of these chemicals are due to the extreme irritant nature of these chemicals. If a worker is exposed to a spray of one of these chemicals, one or more of the following may occur: nosebleed, corneal opacity, yellowing and disfiguring of finger nails, or dry cracked skin. None of these features are life threatening.

If this chemical is ingested, there may be mucosal burns, GI irritation, nausea, vomiting, and diarrhea. One week following ingestion there may be liver and kidney necrosis. There is an additional unique toxic feature of these chemicals (seen only if blood levels of paraquat rise to a certain level) which usually results from exposure by ingestion of the concentrate. Four to ten days after ingestion, there is lung involvement that ends in a malignant proliferation of fibrous connective tissue in the alveoli, which almost invariably ends in death of the patient.

Organonitrogens

This group of chemicals includes most preemergent herbicides, some soil sterilants, and some contact herbicides. Generally, these are low systemic toxic chemicals. These chemicals are mainly irritants or sensitizing agents. Ingestion of these chemicals results in GI irritation, but no liver or kidney involvement. The chlorinated acetanilides (especially propachlor) are extreme sensitizers. Some individuals become so sensitized that complete avoidance is necessary.

Phtalic Acid

This group of chemicals is used primarily on cotton, strawberries, beans, and other vegetable crops, and weeds in turf. An example of this group of chemicals includes dacthal, or DCPA. These chemicals are of low systemic toxicity. Ingestion may result in blood in the urine, but no other symptoms or signs appear unless a large amount of the substance is ingested.

Propenal

This product is an irritant gas. One specific use for this chemical is as tear gas. Other product names include acrolein, aqualin, and acrylaldehyde. They are typically applied by bubbling the gas into water to control water weeds, for example, in irrigation ditches. The chemicals are extremely strong irritants. Asthmatics are at increased risk from exposure to these substances.

Triazines

This group of chemicals has been one of the most highly used herbicides in the past 40 years. It is used as a preemergent in corn (e.g., atrazine) and many small grain crops. Although it is found as a contaminant in surface and groundwaters, and is a suspect carcinogen, there are relatively few acute health concerns. These chemicals (especially pramitol) are strong irritants to the skin, eyes, and respiratory tract (Reigart and Roberts 1999). However, they have very low systemic toxicity.

Phosphonates

Over the past decade, phosphonates have become the dominant broad-spectrum herbicide. They have replaced paraquat and diquat in many applications. The most common chemical of this group glyphosate (Round up) is available for both home and agricultural use. Genetic engineering of plants has developed many crops that can tolerate these chemicals, increasing the usage of this product tremendously in agriculture (e.g., Round up Ready soy beans). Similar to the triazines, these chemicals have low systemic toxicity. However, they are moderate irritants to the skin, eyes, and upper respiratory tract.

DIAGNOSIS AND TREATMENT OF INSECTICIDE POISONING

The basic principles for treatment of acute poisoning of pesticides (Reigart and Roberts 1999) can be applied to most of the insecticides and herbicides discussed in this chapter. However, there are some differences and special situations that may require careful judgment for the particular situation. One of the first things to do, if dermal exposures to insecticides is evident or strongly suspected, is to remove the clothes and wash the body and hair with soap and water to prevent fur-

ther pesticide absorption. Secondly, if there is evidence of an ingestion exposure, one should strongly consider inducing emptying of stomach contents either by stimulating the patient to vomit (emesis), or conducting gastric lavage. The next step is instillation of gastrointestinal absorbents (e.g., activated charcoal) followed by inducing emptying of the bowels by use of laxatives (catharsis). Good judgment must be used to determine which, if any, of these procedures should be used. One cannot treat all suspected pesticide exposures the same way. One must determine if the particular chemical and amount ingested is potentially toxic enough to warrant the risk of aspiration pneumonia and discomfort for the patient. Furthermore, emesis and/or lavage are likely only of benefit within the first 60 minutes following ingestion. Risk of aspiration of stomach contents must be evaluated, and managed. Table 6.1 summarizes the general principles discussed above.

Table 6.1. General Principles for Treatment of Pesticide Exposures (Reigart and Roberts 1999)

Objective of Treatment	Actions
Skin decontamination	Remove clothing, completely wash and shampoo the patient. (Attendants should use protective gloves and aprons if highly toxic chemicals like OPs or CBs are suspected).
Eye decontamination	Continually flush eyes for 15 minutes, preferably with sterile balanced saline, or water.
Airway protection	Clear airway, intubate if ventilatory support needed, administer O_2 as necessary (not indicated with paraquat poisoning).
Gastrointestinal tract clearing	Lavage if less than 60 minutes from ingestion, and only if relatively large amounts ingested. Contraindicated with ingestion of hydrocarbons or caustics. Use oralgastric tube, saline infusion, and aspiration. Intubate if unconscious or airway not protected.
Administering adsorbents	Give within 60 min. of ingestion. Keep giving even if patient vomits. May have to give antiemetic or administer with an oralgastric or nasogastric tube. Intubate if unprotected airway. May repeat in 2–4 hours.
Inducing vomiting	May use ipecac. Effective in 30 min. Use only in alert, conscious people, and within 60 min. of ingestion. Do not use if there is aspiration risk or ingestion of hydrocarbons or corrosives.
Inducing catharsis	Sorbitol is the drug of choice often contained with activated charcoal. Give 1–2 g/kg, 1–2 ml/kg of 70% sorbitol; children, 1.5–2.3 ml/kg up to 50 g. Sorbitol is not likely needed in poisoning with OPs, CBs, arsenicals, or sulfur, as these are cathartics. Do not give if any signs of bowel stasis, e.g., obstruction, recent surgery, or ingestion of diquat or paraquat (the latter may cause bowel stasis).
Controlling seizures	Give Lorazepam. Adults 5–10 mg/kg–IV/ 5–10 min., 30mg max; children, 0.2–0.5 mg/kg IV/5 min, 10 mg max.

The following paragraphs describe signs and symptoms of specific poisonings to assist in specific diagnosis. The symptoms of acute OCL toxicity include a depressed affect and convulsions. As OCLs are hepatotoxic, and induce liver enzymes, one would expect to see elevated transaminases in blood serum tests.

Acute symptoms associated with OP poisoning vary with the degree of toxicity (Reigart and Roberts 1999). Mild poisonings may mimic alcohol intoxication or heat exhaustion. Symptoms may include fatigue, nausea, and vomiting. More severe symptoms can be divided into the portion of the nervous system that is affected, i.e., muscarinic, nicotinic, or central nervous system. Muscarinic symptoms appear as classical salivation, lacrimation, urination, and defecation (SLUD, or all faucets on cholinergic signs). The two rather unique clinical signs related to OP poisoning that should help to differentiate from other poisonings are myosis or pinpoint pupil (some patients report periods of blindness or difficulty seeing), and bradycardia (slow heart rate). There is very little else that would cause this combination of symptoms. An additional clinical sign would be moist rales (breath sounds) on auscultation of the chest (because of excess mucous secretions into the airways caused by the muscarinic response). Nicotinic symptoms appear as muscle fasciculations (twitches), often subtle, but seen in the face or extremities. It should be noted that some individuals have an increased genetic risk for certain OP insecticides. Those individuals with reduced paraoxonase enzyme activity are more susceptible to parathion and related OPs (Lee and others 2003).

An appropriate occupational history of exposure should include details of work practices and processes as described above. The principal diagnostic test for acute OP toxicity is the acetylcholinesterase (AChE) test. The following principles of the use and interpretation of this test are essential in its utility in diagnosing a poisoning (Reigart and Roberts 1999; Wilson 2001). OP and CB insecticides tie up AChE (a low AChE value is a positive test for poisoning). As it is impossible to obtain and test for AChE in the nerve synapse, other sources of AChE in the body such as blood plasma and red blood cells (RBCs) serve as surrogates. Plasma AChE is produced by the liver continually, and therefore in an otherwise healthy person, the AChE remains depressed only for 1–2 weeks, as the liver will replace that which is tied up rather rapidly. RBC AChE is depressed much longer (4–8 weeks), as it is only replaced as new RBCs are produced by the bone marrow (Figure 6.9). This point is very important, as it is best to test for both RBC and plasma AChE if possible, because otherwise a positive diagnosis may be missed if the test is at the wrong time in the disease course. Another important point is that normal blood values of AChE are widely variable between individuals. Therefore, it is best to have individual baseline test results, to compare a second (exposed sample) to determine whether there is real depression of AChE. A 20% depression from baseline is suspect, and a 40% depression is considered compatible with a diagnosis of an acute poisoning. Without baseline values for a patient, one has to rely on published normal ranges provided by the manufacturer of the particular test, which may not be relevant to every person.

There are some situations when highly exposed workers may take atropine for prophylaxis. They may do so on their own volition, on advice from a health care provider, or under direction of an employer. The source of atropine may be illicit. Crop dusters are one group that has been known to practice the use of prophylactic atropine (Figure 6.4). There are several medical reasons why this is not a good idea. First of all, if the person is exposed, and the atropine in the system is insufficient, clinical OP toxicity may develop at a hazardous time, such as when flying the crop dusting airplane, resulting in a risk of a serious crash. Furthermore, atropine depresses the ability of the worker to sweat, increasing the risk of hyperthermia. Eye pupil dilation can decrease visual acuity making operation of equipment more hazardous and creating a risk for retinal damage from bright light.

DIAGNOSIS AND TREATMENT OF HERBICIDE POISONINGS

The following paragraphs summarize the use, toxic principles, and treatment for the major classes of herbicides (Morgan 1979; Reigart and Roberts 1999).

Phenolic Compounds

Symptoms of phenolic compound toxicity include high body temperature, weakness, sweat-

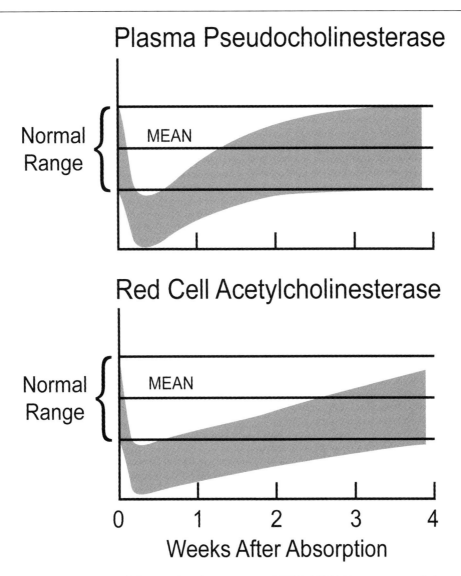

Figure 6.9. There is a much longer time for recovery of RBC AChE (4 or more weeks) as compared to plasma AChE (2–3 weeks). This means that measuring only plasma acetylcholine esterase may detect a rather recent poisoning, but not one that occurred a week or two prior to the test. The RBC test will detect a poisoning that may have occurred up to 3 or more weeks previously. It is best to conduct both plasma and RBC tests simultaneously. (This figure also is in the color section.)

ing, flushed appearance, and depressed mental function. Headache, dizziness, and peripheral neuropathy may also be present. Symptoms of severe poisonings may include toxic psychosis, manic behavior, convulsions, and coma. Chronic poisoning by both nitrophenols and pentachlorophenol results in fatty liver, toxic nephrosis, and weight loss. Nitrophenols have a characteristic bright yellow color, and diagnosis may be aided by a yellow staining of the skin (direct contact), or sclera of the eye (absorption).

Laboratory confirmation is by testing the urine for metabolites of this chemical group, using spectrophotometric or chromatographic methods. Levels in excess of 1000 ppb and compatible history, symptoms, and clinical signs are consistent with an acute poisoning.

There is no specific antidote for these chemicals. The most effective treatment has been the use of physical methods of cooling the hyperthermic patient, such as cool baths or cooling blankets. The antipyretic activity of aspirin and Tylenol has not been effective and may be contraindicated in acute poisoning.

Chlorophenoxy Compounds

Laboratory confirmation of absorption is by urine analysis with gas-liquid chromatography. As with the case of many herbicides, chlorophenoxy compounds are excreted rapidly (within 24 hours). Therefore, analysis for confirmation must be accomplished soon after exposure. Secondary dioxin exposures are difficult to determine as levels of dioxin in today's chlorphenoxy herbicides are usually below levels that can be easily determined as a poisoning.

Skin and eyes contacted by chlorophenoxy herbicides should be flushed with copious amounts of water. Ingestion should include evacuation of the stomach, followed by administration of a slurry of activated charcoal. The herbicide itself often induces catharsis, and therefore additional cathartics are often not needed.

Arsenical Compounds

Poisoning by inorganic or trivalent arsenicals is very rare today as they have almost been entirely replaced by the pentavalent or organic arsenicals. However, there are still occasional poisonings by the former, and ingestion of the inorganic trivalent arsenicals is manifested by acute gastrointestinal symptoms, including abdominal pain, nausea, and diarrhea, caused by apparent reduced blood flow of the splanchnic vessels (ischemia) to the intestines, rather than direct irritation to the mucosa. Subsequent symptoms are related to the direct cellular damage to the liver, kidneys, and nervous system. Chronic clinical signs include hair loss, hyperpigmentation of the skin, chronic liver and kidney disease, and anemia. Chronic symptoms include weakness, incoordination, pain in extremities, persistent excess blood and protein in the urine, and loss of appetite and weight. Lung, liver, and skin cancer have been demonstrated as a risk for workers having applied inorganic or pentavalent arsenicals to grapes and other fruit and orchard crops.

Recent concerns have been expressed by scientists because of the natural presence of low-level arsenic in some drinking water and from low levels found in homes and in play areas of children living near agricultural areas where arsenic was applied years in the past (Wolz and others 2003). However, health effects of these low arsenic levels are undetermined.

Confirmation of poisoning is by measuring arsenic in the urine or blood, at more than several hundred (typically over 1,000) micrograms. A meal of seafood may interfere with test results, causing a temporary increase of arsenic in the urine. Therefore a second measure may be indicated to confirm poisoning, recognizing however that arsenic is cleared very rapidly by the kidney.

Treatment must include washing the skin of exposed persons, followed by administration of activated charcoal and catharsis. Dimercaprol (British anti-lewisite, B.A.L.) is a chelating agent that increases the rate of excretion of arsenic, especially important in chronic poisoning.

Although there is a wide range of toxicity of the various arsenicals, it is recommended that acute exposures of any arsenical should be treated in the same manner. However, it is important to know to which specific product the patient was exposed, because prognosis is highly related to the specific chemical.

Dipyridil Compounds

Paraquat and diquat are the principal chemicals in this group. Dermal damage from exposure is a result of the irritant properties of these chemicals and includes nosebleed, corneal opacity, discoloration, and atrophy of fingernails, drying, cracking, and discoloration of the skin. Dermal exposures are relatively minor compared to the consequences of ingestion. The immediate effects of ingestion include superficial burns of the oral and pharyngeal mucosa, and gastrointestinal irritation, which often causes nausea, vomiting, diarrhea, and sometimes dark-colored stools from an upper GI bleed (melena). These symptoms may be mild and distract the patient and health care provider from the serious consequences that follow. During the first week following ingestion, there are indications of liver and kidney injury, but these are not usually fatal injuries. Also at this time there are subclinical lung parenchymal changes, manifest by intraalveolar hemorrhage and edema with necrosis of alveolar pneumocytes. Some patients may exhibit only pulmonary edema, and these usually survive with appropriate critical care management. However, 4–10 days following ingestion, most patients proceed to a second malignant stage of the disease. Extensive proliferation of fibrous connective tissue begins within the alveoli. This reaction continues, accompanied by progressive edema and difficult breathing (dyspnea). Most all cases result in death by asphyxiation.

There is individual variability of susceptibility to the malignant pulmonary toxic reaction described above. Poisoning has resulted from a wide range of reported amounts of ingested concentrated product (the product used to dilute for field application), which is about 20% of active ingredient. Therefore, any ingestion of the concentrated product should be treated as a potentially lethal exposure. (Note that only ingestion exposures are considered a risk for the malignant pulmonary reaction, and applicators have generally not been a risk group).

Any history of ingestion of paraquat (especially the concentrate) should be considered an emergency. Emergency treatment should begin prior to laboratory tests to prove absorption. As paraquat is absorbed relatively slowly from the gut, prompt recognition of the hazard and removal of the chemical from the gut has saved many lives. Immediate evacuation and lavage of the stomach should be initiated, followed by instillation of adsorbents, preferably fuller's earth (bentonite) and montmorillonite. Activated charcoal is a good substitute for fuller's earth. Absorbents should be given to the limits of tolerance of the patient. Saline catharsis (enema) should follow.

Following emergency treatment, laboratory diagnosis of ingestion and absorption of paraquat can be accomplished by analysis of urine. (Product manufacturers and several EPA-supervised laboratories can conduct this analysis.) This analysis will help to confirm absorption, and/or the success of emergency treatment, and will aid in prognosis. Once the chemical is absorbed into the bloodstream, excretion can be accelerated by diuresis with diuretics such as manitol and furosemide. Hemodialysis and hemoperfusion over charcoal have been used successfully to lower the blood levels of paraquat.

A conundrum in medical treatment of paraquat poisoning is that oxygen stimulates the proliferation of fibrous connective tissue in the alveoli. Administration of supplemental oxygen is not indicated unless the arterial oxygen falls to levels that may cause cerebral anoxia. Some physicians have even practiced putting patients (early in the course of disease) in a hypoxic environment (10–12 % oxygen) to retard pulmonary fibrosis. Corticosteroids and immunosuppressants have been used in some cases to retard the malignant proliferation of connective tissues. The effectiveness of these latter two practices is uncertain.

Organonitrogen Compounds

Similar to most herbicides, these chemicals have relatively low acute systemic toxicity. They are primarily dermal irritants and sensitizers. Propachlor is one chemical of this class that is a potent allergen, causing a number of disabling cases of allergic contact dermatitis. Ingestion of these chemicals has resulted in irritation of the gastrointestinal tract, manifested as nausea, vomiting, and diarrhea. No other systemic or organ systems toxicity has been observed.

One subgroup of organonitrogen herbicides is carbamates. Although these herbicides are chemically similar to the carbamates insecticides, they do not tie up AChE. Therefore, individuals exposed to these chemicals should not be treated with atropine.

Ingestion of small amounts of these chemicals may warrant administration of a few ounces of activated charcoal followed by saline catharsis. Ingestion of large amounts of these products may warrant gastric lavage, followed by administration of activated charcoal and saline catharsis. However, each case should be evaluated carefully, as these chemicals are generally of low systemic toxicity, and the risk of aspiration pneumonia during lavage should be weighed in relation to the risk of toxicity.

Confirmation of absorption may be indicated in individuals with a history of ingestion, especially if symptoms occur following occupational exposure. Analytical methods to determine urinary metabolites of many of the chemicals in this class are available.

Skin should be washed thoroughly if dermal exposure occurs. Also, contaminated clothing should be removed and replaced with clean clothing. Eyes should be flushed thoroughly with water or a balanced saline solution. Immediate or delayed hypersensitivity reactions may occur. Immediate hypersensitivity may be treated with antihistamines and steroids as necessary. Delayed or chronic allergic contact dermatitis cases may be treated with topical steroids, combined with systemic steroids in severe cases. As most people who develop sensitivity to these chemicals, complete avoidance by substitution of products or removal of the worker to different jobs is often necessary as personal protective equipment is usually not adequate to protect these people.

Phthalate Compounds

Dacthal is an example of a derivative of phthalic acid used as an herbicide to control weeds in cotton, strawberries, beans, and a variety of vegetables and turf. Similar to many other herbicides, these chemicals are low in systemic toxicity. Similar to organonitrogens, they are primarily irritants to the skin, eyes, and mucous membranes. Treatment for either dermal or ingestion exposure to these chemicals is the same as for the organonitrogens. Analytical methods for identification of Dacthal in blood or in urinary metabolites are available if indicated.

Propenal

This herbicide (similar to tear gas) is administered as a gas. It is extremely irritating primarily to the eyes and mucous membranes. It is contained in gas cylinders, and exposures have primarily been associated with accidental mishandling of equipment or equipment failure (leaky or failed valve or hose). The gas causes tearing, runny nose, irritation of the throat and bronchi, and occasionally pulmonary edema. A positive history of exposure is usually sufficient for diagnosis. Ordinarily the only treatment necessary is to remove the victim from the exposure source to fresh air for a period of time until symptoms dissipate. Asthmatics may have more serious problems, as this irritant may initiate an asthmatic attack. Treatment with bronchodilators may be necessary, in conjunction with administration of oxygen with positive pressure assistance.

CLASSES AND TOXIC MECHANISMS OF FUMIGANTS

The following paragraphs summarize the toxic principles, applications, and treatment of fumigants (Morgan 1978; Reigart and Roberts 1999).

Fumigants are gases or liquids used to control insects, rodents, and other vermin. In agriculture, they are largely used on stored grains or other agricultural produce (grapes, citrus fruits, peaches, etc.) to prevent loss from insects or rodents. They may also be used inside livestock buildings (when empty of livestock) to control harmful bacteria, flies, rodents, and birds. They may also be used as soil sterilants to control nematodes that may be harmful to vegetable crops.

Fumigants have one important characteristic in common; they are low molecular weight, and they have enormous powers of diffusion and penetration. This accounts for their effectiveness in preserving stores of grain and other produce even when the agent is released at only a few sites in the storage structure.

Most poisonings have occurred when workers reentered fumigated structures before the buildings were adequately ventilated. Occasionally, fumigant liquid has been spilled on skin. This represents a life-threatening exposure. In most instances, workers have not realized how severely toxic most fumigants are. This is especially true among occasional users, such as farmers who attempt to treat small amounts of stored grain on their farms. Various factors, including structural and work processes, must be analyzed to determine risk of exposure to fumigants (Reed 2001).

Although injuries vary with particular chemicals, six general types of injury are caused by fumigants: 1) eye, mucous membrane, and skin effects; 2) respiratory effects; 3) central nervous system effects; 4) cardiac effects; 5) liver damage; and 6) peripheral nerve damage (Morgan 1978.) Generally, all these chemicals are strong irritants and can cause minor to severe damage to the eyes, mucous membranes, and skin. Eye damage may range from minor irritation to corneal desquamation and ulceration. Dermal effects may manifest as minor irritation or more severe bullous dermatitis with extensive necrosis of skin. Upper respiratory effects may include laryngeal edema and bronchospasm. As with inhalation of any strong irritant gas, acute or delayed (4–12 hours) pulmonary edema may be seen. Central nervous system effects may manifest as stupor or unconsciousness and respiratory depression or various forms of involuntary motor activity. Respiratory irritation is most common. Acute or delayed pulmonary edema may be the most common life threatening effect. Cardiac arrhythmia may result in sudden death. Liver damage is usually reversible, but sometimes results in lethal necrosis or chronic cirrhosis. Peripheral neuropathy may be protracted or even permanent.

The high capacity for fumigants to diffuse and penetrate complicates protecting exposed workers. Systemic poisoning from fumigants can easily occur as they cross the skin barrier readily. Except for the self-contained breathing apparatus, respiratory protective devices are ineffective. Masks and respirators are of little or no value. Fumigants penetrate natural rubber and neoprene. It is generally recommended that workers not wear rubber suits or gloves because severe skin injury can result if small amount of gas or liquid are occluded by the covering. The only effective protection against most fumigant compounds is complete avoidance.

The effects of two commonly used fumigants, methyl bromide and carbon tetrachloride, are discussed here in detail. Exposure to methyl bromide has caused a significant number of acute deaths from pulmonary edema. Prolonged or recurrent low-level exposures produces a bizarre neurological syndrome in exposed workers, apparently due to selective effects on the basal ganglia in the brain. The victims show marked incoordination, disturbed speech, cerebellar ataxia, and awkward involuntary movements of the extremities. These symptoms sometimes develop hours or even days following the last contact with methyl bromide. Manifestation persists for weeks or even months. So far, all victims of this type of poisoning have ultimately recovered normal neurological function.

Pulmonary edema must be treated by limiting physical activity, by IV manitol or 50% dextrose, by diuretics, and by positive pressure breathing apparatus. There is no antidote for the neurological syndrome.

Carbon tetrachloride is less irritating than methyl bromide to the respiratory tract. However, it can cause cardiac arrhythmia if inhaled in high concentrations, and it is severely hepatotoxic. There also may be direct and indirect toxic effects on the kidneys. Many of the chlorinated gases and liquids used as fumigants cause poisonings similar to that induced by carbon tetrachloride.

Damage to the liver commonly becomes apparent 1–3 days after exposure. It is often manifest first as jaundice. Massive necrosis usually results in postnecrotic cirrhosis, if not in death.

There is no specific antidote for carbon tetrachloride poisoning. Intravenous glucose and vitamin infusion help protect the liver. Hemodialysis may be necessary to sustain life if renal failure occurs.

Although there are some generalities associated with fumigant exposures, there are also specific toxic principles with the various chemicals. Table 6.2 lists specific toxic principles of commonly used fumigants.

TREATMENT OF FUMIGANT POISONINGS

The first step is to remove victims from the exposure to fresh air and have them kept quiet and in a reclined or sitting position with a backrest to help prevent pulmonary edema. Assure respiration and cardiac function and administer oxygen and/or artificial respiration as necessary. Caution must be taken in rescuing people from confined spaces where fumigants have been used. Confined space entry procedures must be used to assure that rescuers do not become victims. This means that unless it can be absolutely assured that the environment is free of toxic exposures, rescuers should use a self-contained breathing apparatus (SCBA), with a rescue harness and adequate power to remove a would-be rescuer remotely.

The initial basic treatment of fumigant exposure is skin and eye decontamination, clearing of the airway, and gastrointestinal tract clearing (if ingestion has occurred). The same procedures should be followed as described in Table 6.1, treatment of insecticide poisonings. One difference from insecticide exposures is that skin exposure may lead to severe dermal or eye damage that will need medical attention following acute treatment.

If pulmonary edema is evident, intermittent positive pressure oxygen may be given, but only as needed, as excess oxygen may increase pulmonary damage from the fumigant. Oxygen saturation or pO_2 should be monitored. A diuretic (e.g., furosemide 40 mg IV, or .5–1 mg/kg for children, up to 20 mg) may be given to help reduce the edema fluid.

If shock occurs, the patient should be placed on an incline with the head lower than the feet (the Trendelenburg position) to help prevent blood pooling in the lower extremities.

Table 6.2. Toxic Principles of Fumigants

Fumigant Product	Toxic Principles
Acrolein	Irritating gas, used on aquatic weeds. Eye, skin and respiratory tract irritant.
Acrylonitrile	Eye, skin, and respiratory tract irritant. Transformed to hydrogen cyanide in the gut, resulting in cyanide poisoning.
Carbon Disulfide[1]	Moderate eye and respiratory tract irritant, strong skin irritant. "Rotten egg" or "sweet" odor. CNS toxicant, headache, nausea, delirium, paralysis, respiratory failure, death.
Carbon tetrachloride[1]	Minor eye, skin, respiratory irritant. Potent liver and kidney toxin. Possible cardiac arrhythmias.
Chloroform	Mild irritant to eyes, skin, and respiratory tract. CNS depressant (formerly used as an anesthetic). Large doses may lead to cardiac arrhythmia, liver and kidney damage.
Chloropicrin	Strong irritant to eyes, skin, and respiratory tract. Ingestion may cause gastric ulceration.
Dibromochloropropane	Strong irritant to eyes, skin, and respiratory tract. Symptoms (sx) include nausea, vomiting, CNS signs. Liver and kidney toxin. Chronic exposure may lead to male infertility due to necrosis of seminiferous tubules.
Dichloropropane/Dichloropropene	Strong irritant to eyes, skin, and respiratory tract. Possible risk of liver, kidney, and cardiac toxicity from ingestion of large quantities.
Ethylene oxide/Propylene oxide	Strong irritant to eyes, skin, and respiratory tract. May cause arrhythmias.
Formaldehyde[1]	Strong irritant to eyes, skin, and respiratory tract. Potent sensitizer, leading to allergic contact dermatitis. Asthma-like sx. Inhalation may cause acidosis.
Hydrogen cyanide	See acrylonitrile above. General cellular toxin, disrupting cytochrome oxidase system. Predilection for the CNS.
Methyl bromide[1]	Severe irritant to the eyes, skin, and especially the lower respiratory tract, causing acute or delayed (4–12 hours) pulmonary edema. CNS toxin resulting in varying sx including incoordination, tremors, slurred speech, and seizures. Epidemiologic associations with prostate cancer. Ozone-destroying chemical. Scheduled for phase-out in 2005 under the Montreal Protocol.
Methyl chloride	Similar to methyl bromide, but less irritating, and mild systemic toxicity.
Naphthalene	Used as moth balls. Mild irritant to eyes, skin, and respiratory tract. Genetic susceptibility in some due to glucose-6 –phosphate-dehydrogenase deficiency, mainly in Mediterranean and African ethnicity, resulting in hemolysis and renal tubular damage.
Phosphine[1]	Used as aluminum phosphide (phostoxin) tablets when exposed to air, slow release of phosphine gas (increases safety factor). Extreme irritant to eyes, skin, and respiratory tract. Odor of dead fish. Arrhythmias, cardiac failure, and pulmonary edema are most serious hazards.
Sulfur dioxide	Severe irritant to eyes, skin, and respiratory tract. May cause pulmonary edema, or asthma-like condition.

[1]Most common products used in agriculture.

192

If convulsions occur, treat as indicated in Table 6.1. It should be noted that convulsions are most likely to occur with poisonings from methyl bromide, cyanide, acrylonitrile, phosphine, or carbon disulfide. Convulsions from methyl bromide may be refractory to benzodiazepam and phenylhydantoin treatment, and barbiturates may have to be used.

Kidney and liver function should be monitored. Urine analysis with evidence of urinary casts suggests kidney tubular damage. Hemodialysis may be needed if there is significant kidney damage. Regarding liver function, the following enzymes should be monitored: alkaline phosphatase, LDH, ALT, and AST.

There are certain additional recommendations for some specific chemicals. There may be some initial severe CNS signs with carbon disulfide, and in most instances, the patient will recover from these signs spontaneously.

Actions can be taken to help limit the amount of liver damage from carbon tetrachloride exposure. Initial treatment could include a hyperbaric chamber with oxygen, which is thought to limit the liver damage with acute exposure. N-acetyl cysteine (Mucormyst) may be administered to decrease free radical damage. It may be administered orally at 20% (1:4) in a carbonated beverage, for a total of 140 mg/kg, followed by 70 mg/kg every 4 hours for 17 doses. Mucormyst may also be given via stomach tube or IV.

Phosphine gas exposure has been treated with magnesium sulfate to retard the cardiac effects. The dosage given is 3 grams in the first 3 hours by IV, followed by 6 grams in the following 24 hours over the following 3–5 days.

Hydrogen cyanide and acrylonitrile (same toxic principle) is treated with nitrite, in one of the following forms: amyl nitrate, sodium nitrite, or sodium thiosulfate (available as Lilly cyanide kits, Eli Lilly, Indianapolis, Ind.). It is administered at the rate of 0.55–1 mg/kg, depending on the hemoglobin concentration.

GENERAL STRATEGIES FOR PREVENTION OF PESTICIDE POISONING

Prevention strategies can be broken down into seven basic components: 1) hygienic work practices; 2) avoidance, proper storage, handling, and separation from the chemicals; 3) use of personal protective equipment; 4) medical monitoring; 5) regulations and enforcement; 6) training; and 7) proper storage and handling of pesticides.

Hygienic Work Practices

As dermal absorption is by far the highest risk for occupational exposure to pesticides and hands are the parts of the body that come into contact with pesticides most often, frequent hand washing is a very important component of prevention. It is especially important to wash hands before eating or smoking, so as not to transfer pesticides from the hands to food or cigarettes and create an oral route of exposure. This implies that there must be accessible hand washing facilities at the work site. For field workers, this can be accomplished by having an accessible water supply for hand washing. For individual farmers, the most convenient way to accomplish hand washing is to carry a supply of disposable prepackaged disposable towels that contain soap and are premoistened. These are readily available commercially. Skin barrier creams may be helpful to retard absorption, especially as washing may reduce natural protective skin oils.

There are often emergency situations in the field where an accidental spill, mechanical malfunction, or misdirected spray may result in a heavy exposure to the skin and clothing. There must be accommodations for rapid bathing and changing of clothes. For field workers, the management must have protocols set in place to handle these situations. This may mean having a shelter with portable shower facilities and access to clean clothing on site, or readily available transportation to such facilities.

Chronic exposure may result from buildup of pesticides on the skin over several days. Therefore, daily showering or bathing and shampooing is very important. For hired field workers, housing is often substandard, and access to showering or bathing facilities at their living quarters may not be readily available. Managers must understand the importance of daily personal cleaning, and they should provide showering/bathing/shampooing facilities as part of their worker protection programs.

During application and working in fields where pesticide residues may still be present, wearing clean clothing daily is important (Gladen and others 1998). Nonwashable gloves and shoes should be avoided as they tend to soak up pesticides and become a constant source of exposure. Clothing that may have been contaminated should be kept out of the house and separate from clothing of other household members. Such clothing should be washed three times and separately from other clothing.

Eating and break places for those working with pesticides should be kept separate from the workplace and kept clean and sanitary. Also, there should be access to hand washing facilities at these sites.

Avoidance, Proper Storage, and Separation

Methods to separate people from the pesticides are extremely important. Engineering methods to accomplish this have advanced over the past decade. For example, a system has been developed by a North American machinery manufacturer called "lock and load." This system incorporates pesticides that are sealed in containers that are attached to the applicator by the operator who has no need to open the container directly or to handle the pesticide; this leaves postapplication as the only possible exposure. Further development and use of these integrated systems are important elements of applicator protection.

There are also management techniques of avoidance—for example, storing pesticides outside the house and in a secure (locked) storage place inaccessible to children and unauthorized persons. Pesticides should not be mixed and should always be stored in their original containers. Empty containers should be rinsed three times and the rinse applied with the regular application. Empty pesticide containers should never be reused and either disposed of according to directions or returned to the seller. Mixing/loading operations should be at least 50 m from the house (Gladen and others 1998) and in well-ventilated areas. Following field application, each chemical has a certain degradation time after which contact with plant foliage is not hazardous. Therefore implementation and posting of fields for safe reentry times is important. In some states this reentry posting is a regulated standard.

Personal Protective Equipment

For those working in areas of low risk (field work, etc.), regular cotton clothing is probably sufficient. Treatment of cotton pants or shirts with commercial sprays with a stain-resistant or water-repellant material (e.g., the trade named product ScotchGuard), provide much better protection than untreated clothing against pesticide exposure. Daily washing of clothes during application season is a very important preventive procedure. As mentioned before, best-practice recommendations include washing clothes three times and separate from other clothing (Phillips and others 2003).

For those workers involved in mixing, loading, or application operations, additional personal protective equipment is necessary. The use of rubber or synthetic rubber gloves is extremely important, as dermal adsorption is such a high risk, and the hands are the most common site of exposure. These gloves should be unlined, as contamination of these linings will be a source of continuous exposure and may create a greater risk than no gloves at all. Reusable gloves should be routinely washed inside and out to prevent contact of contaminated glove surfaces with the hands (on the inside of the gloves) and other parts of the body that may be contacted by the gloves. Plastic aprons and rubber footwear also should be used in mixing and loading operations. Respirators afford little protection in most of these types of operations unless a fine powder or highly volatile chemical is used. There are certain situations such as where orchard "blast" applicators are used when respirators and a complete rubber suit along with rubber gloves are necessary. In these operations, pesticide sprays are "blasted" into the foliage with a powerful fan attached to the pesticide reservoir that is pulled through the field by a tractor. Operators without a protective cab, or who otherwise have exposure to this blast, are highly exposed via both the dermal and respiratory routes, and therefore need extensive protection.

It is a common observation by this author that workers are either underprotected or overprotected. The hazard of underprotection is obvious. The hazard of overprotection, especially in hot climates and with elderly or workers with co-morbidities, may result in heat stress or cardiorespiratory embarrassment. Achieving the right level of protection requires an understanding and analysis of the exposure risks, including the toxicity and formulation of the specific product, mixing/loading and application processes, degree of training and compliance of workers, and climate. There is not a single specific formula for proper worker protection. Understanding and integration of all of the information areas mentioned above, along with the labeled directions or material safety data sheet (MSDS) of the particular products will help to achieve the appropriate protection.

Worker Monitoring

There are several ways that workers exposed to pesticides may be monitored. Perhaps the most common way (and one that has been included in some regulatory practices) is monitoring of AChE (medical monitoring) for OP- and CB-exposed workers (Bolognesi 2003).

MEDICAL MONITORING

As discussed above, OPs and CBs cause a lowering of the blood AChE. Monitoring of this blood parameter can help prevent acute toxicities. There are various methods of measuring AChE in the blood. Because AChE levels are highly variable among individuals, it is important to obtain a preexposure baseline level, and then follow-up measures are compared to the person's baseline. If the AChE levels fall to 20% of baseline of the serum level, and 30% of the RBC level from baseline, the person should be removed from exposure until the levels return to their baseline. A 40% drop from baseline warrants medical attention (Wilson 2001). The following principles should be employed in AChE monitoring of workers:

1. Take a sample at least 2 weeks before start of application or other exposure.
2. Record the baseline measure in a secure manner and one that can be matched to the person and subsequent measures.
3. Conduct both serum and RBC cholinesterase levels.
4. Conduct follow-up tests at midseason and at the end of the application season.
5. Make sure the same lab procedure and, preferably, the same lab is used for baseline and follow-up tests.

EXPOSURE MONITORING

Monitoring of exposure is usually conducted as a research technique, rather than as a preventive procedure. A common practice has been to have the workers wear cotton gloves and/or have cotton patches taped in standardized locations on the body and clothing. Following work exposure, the gloves are removed and taken to a laboratory where they are extracted with solvents and analyzed for pesticide content and concentration. This procedure provides a relative measure of dermal exposure. Another method is the use of fluorescent tracers in the pesticide. Following contact with the pesticide, a person is exposed to an ultraviolet light, and the degree of fluorescence is related to the total amount of exposure. This is a qualitative test; however, recent research has involved the measure of fluorescence based on a computer integration of fluorescence from several angles and body locations (the so-called VITAE method). These fluorescent methods have proven to be a great teaching device for workers to see how much of the body becomes exposed during work with pesticides, but they have not proven sufficiently quantitative for health monitoring (Stewart and others 1999).

Measure of urinary metabolites has shown to have field adaptations. Many pesticide breakdown products are excreted in the urine. Spot tests using these procedures have shown promise as field monitoring methods (Castorina and others 2003). They are more specific and somewhat more quantitative than other tests, when standardizing the concentration of urine by measuring the creatinine concentrations of the urine sample.

Training

A proper understanding on the part of owner-operators, managers, and workers of the toxicity of the chemicals and method of protection is extremely important. This involves first of all a basic understanding. Most industrial countries have programs to train applicators. For example, in the U.S., all applicators of so-called "restricted use" chemicals (chemicals that are moderately to highly toxic to human beings or the environment) are required by law to take pesticide applicator training. However, other factors in addition to training are needed to create effective prevention (Perry and others 2000) as little relationship has been shown between training and personal protective equipment (PPE) use (Mandel and others 1996).

Regulations and Enforcement

Most industrial countries also have regulations that require the safe use of pesticides both from

Table 6.3.   Summary of U.S. Federal Regulations That Effect Pesticide Usage (Rowe 2003)

Endangered Species Act	Prohibits use of pesticides in areas that may harm endangered species. Product label required to list where the product is not to be used.
Federal Insecticide, Fungicide, and Rodenticide Act (FIFRA)	Product label required to assure proper use of product is followed. Applying product not in accordance with label is in violation. Pesticide uses and pesticide merchants must be registered with EPA. Established class of "restricted use" pesticides (more toxic chemicals), that require special training for use.
Field Sanitation Standard	Employers of more than 10 employees must provide toilet, hand washing facilities, clean drinking water, and training about safe hygienic practices.
Food Quality Protection Act	Sets standards for pesticide residues in food. EPA must assess each pesticide for aggregate public exposure (food, water, etc.), determine cumulative effects, health in infants and children, and hormonal effects.
Food, Agriculture, Conservation, and Trade Act (FACT)	Requires applicators to keep detailed records of specific products, use, amounts and areas applied of all pesticides.
Occupational Safety and Health Act	Requires employers to take reasonable steps to protect workers' safety and health. Includes education, access to material safety data sheets of all products, and reporting of any pesticide-related illnesses.
Resource Conservation and Recovery Act (RCRA)	Regulates manufacture, transportation, treatment, storage, and disposal of hazardous substances (including many pesticides).

environmental aspects as well as employee's health. In the U.S., pesticide regulations are split between the Environmental Protection Agency (EPA) and the Occupational Health Administration (OSHA). Each state must follow these or more strict regulations if they so choose. Table 6.3 lists the various regulatory acts at the federal level that have jurisdiction in pesticide usage. An excellent review of each of these regulations is seen at http://entweb.clemson.edu/pesticid/Document/lawsregs.htm.

It is interesting to note that even with all these regulations, Reeves and Schafer (2003) found that in California, 41% of pesticide poisonings are due to violations of worker safety and health laws.

SUMMARY

Only selected pesticides, from the broad array of chemicals available, of particular importance to the agricultural workers have been covered in this chapter. Further details can be found in books and other references listed below. Many toxic syndromes are complex, and they are not easy to rec-

ognize from symptoms and clinical signs alone. The most important element in diagnosis is an awareness of the potential hazard presented by these chemicals and the appropriate history for exposure potentials. The alert health professional will not fail to inquire into the possibility of toxic exposure when confronted with complex patterns of clinical disease in persons exposed to agricultural pesticides.

NOTE

Treatment of acute OP and CB poisonings is to counteract excess ACh. Atropine is the primary antidote. It blocks the effect of excess AChE on muscarinic receptors. Therapeutic dosages are to effect atropinization (e.g., pupil dilation, decreased bronchial secretions, dry mouth, tachycardia, etc.). Based on the extent of poisoning, dosages may range from 2–4 mg every 15 minutes and up to 300 mg in a day. Children under 12 should be dosed at 0.05–0.1 mg/kg every 15 minutes to attain the effect. Dosage is by IV preferably, but intratracheal infusion may be used.

Pralidoxime may be used early (first 48 hours) in OP poisoning cases, because it can reverse the Op-ChE bond. The dosage for adults and children over 12 is 1.0–2.0 g in 100 ml saline by intravenous infusion over a 30 minute period. For children under 12, the dosage is 20–50 mg/k (in 100 ml saline) over 30 minutes. Pralidoxime is not indicated in CB poisonings because the CB-ChE bond is temporary (Riegart and Roberts 1999).

REFERENCES

Alavanja M, Sandler D, McDonnell C, Mage D, Kross B, Rowland A, Blair A. 1999. Characteristics of persons who self-reported a high pesticide exposure event in the Agricultural Health Study. Environ Res 80(2 Pt 1):180–186.

Baron R. 1994. A carbamate insecticide: A case study of aldicarb. Environ Health Perspect 102 Suppl 11: 23–7.

Bolognesi C. 2003. Genotoxicity of pesticides: A review of human biomonitoring studies. Mutat Res 543(3):251–272.

Calvert G, Plate D, Das R. 2004. Acute occupational pesticide-related illnesses in the US, 1998–1999: Surveillance findings from SENSOR-Pesticides Program. Am J Ind Med 45:14–23.

Carson R. 1962. Silent Spring. New York: Fawcett Crest.

Castorina R, Bradman A, McKone T, Barr D, Harnly M, Eskenazi B. 2003. Cumulative organophosphate pesticide exposure and risk assessment among pregnant women living in an agricultural community: A case study from the CHAMACOS cohort. Environ Health Perspect 111(13):1640–1648.

Dosemeci M, Alavanja M, Rowland A, Mage D, Zahm S, Rothman N, Lubin J, Hoppin J, Sandler D, Blair A. 2002. A quantitative approach for estimating exposure to pesticides in the Agricultural Health Study. Ann Occup Hyg 46(2):245–260.

Farahat T, Abdelrasoul G, Amr M, Shebl M, Farahat F, Anger W. 2003. Neurobehavioral effects among workers occupationally exposed to organophosphorous pesticides. Occup Environ Med 60:279–286.

Garcia A, Benavides F, Fletcher T, Orts E. 1998. Paternal exposure to pesticides and congenital malformations. Scand J Work Environ Health 24(6):473–480.

Garcia A, Fletcher T, Benavides F, Orts E. 1999. Parental agricultural work and selected congenital malformations. Am J Epidemiol 149(1):64–74.

Garry V, Holland S, Erickson L, Burroughs B. 2003. Male reproductive hormones and thyroid function in pesticide applicators in the Red River Valley of Minnesota. J Toxicol Environ Health A 66(11): 965–986.

Georgellis A, Kolmodin-Hedman B, Kouretas D. 1999. Can traditional epidemiology detect cancer risks caused by occupational exposure to pesticides? J Exp Clin Cancer Res 18(2):159–166.

Gladen B, Sandler D, Zahm S, Kamel F, Rowland A, Alavanja M. 1998. Exposure opportunities of families of farmer pesticide applicators. Am J Ind Med 34(6):581–587.

Hines C, Deddens J, Tucker S, Hornung R. 2001. Distributions and determinants of pre-emergent herbicide exposures among custom applicators. Ann Occup Hyg 45(3):227–239.

Lee B, London L, Paulauskis J, Myers J, Christiani D. 2003. Association between human paraoxonase gene polymorphism and chronic symptoms in pesticide-exposed workers. J Occup Environ Med 45(2): 118–122.

Mage D, Alavanja M, Sandler D, McDonnell C, Kross B, Rowland A, Blair A. 2000. A model for predicting the frequency of high pesticide exposure events in the Agricultural Health Study. Environ Res 83(1):67–71.

Mandel J, Carr W, Hillmer T, Leonard P, Halberg J, Sanderson W, Mandel J. 1996. Factors associated with safe use of agricultural pesticides in Minnesota. J Rural Health 12(4 Suppl):301–310.

Morgan D. 1978. Rural Health Series: Toxicology of Fungicides, Rodenticides, and Fumigants. Washington, D.C.: National Library of Medicine.

——. 1979. Rural Health Series: Toxicology of Commonly Used Herbicides. Institute for Agricultural Medicine and Rural Health, University of Iowa.

Morgan D, Donham K, Mutel C, Cain C. 1978. Pesticide poisonings and injuries: Where, when, and how. Institute of Agricultural Medicine and Rural Health, Department of Occupational and Environmental Health, University of Iowa.

Mutel C, Donham K. 1983. Medical Practices in Rural Communities. New York: Springer-Verlag.

National Safety Council. 2001. Injury Facts, 2001 Edition. Itasca, IL: NSC Press. p 46–48.

Perry M, Marbella A, Layde P. 2000. Association of pesticide safety knowledge with beliefs and intentions among farm pesticide applicators. J Occup Environ Med 42(2):187–93.

Peterson M, Talcott P. 2001. Small Animal Toxicology. Philadelphia: W.B. Saunders Co. p 420–428.

Phillips T, Belden J, Stroud M, Coats J. 2003. Evaluation of a cold-water hand-washing regimen in removing carbaryl residues from contaminated fabrics. Bull Environ Contam Toxicol 71:6–10.

Reed C. 2001. Influence of environmental, structural, and behavioral factors on the presence of phosphine in worker areas during fumigations in grain elevators. J Agric Saf Health 7(1):21–34.

Reeves M, Schafer K. 2003. Greater risks, fewer rights: U.S. farmworkers and pesticides. Int J Occup Environ Health 9(1):30–39.

Reigart J, Roberts J. 1999. Recognition and Management of Pesticide Poisonings. Office of Pesticide Progams, USEPA, Washington, D.C. p 68–69, 87–89.

Stewart P, Fears T, Nicholson H, Kross B, Ogilvie L, Zahm S, Ward M, Blair A. 1999. Exposure received from application of animal insecticides. Am Ind Hyg Assoc J 60(2):208–212.

Swan S, Kruse R, Liu F. 2003. Semen quality in relation to biomarkers of pesticide exposure. Environ Health Perspect 111(12):1478–1484.

Thu K, Donham K, Yoder D, Ogilvie L. 1990. The farm family perception of occupational health: A multistate survey of knowledge, attitudes, behaviors, and ideas. Am J Ind Med 18(4):427–431.

Wilson B. 2001. Cholinesterases. Handbook of Pesticide Toxicology. Germany: Academic Press. p 967–985.

Wolz S, Fenske R, Simcox N, Palcisko G, Kissel J. 2003. Residential arsenic and lead levels in an agricultural community with a history of lead and arsenate use. Environmental Research 93(3):293–300.

7
General Environmental Hazards in Agriculture

Kelley J. Donham and Anders Thelin

INTRODUCTION

"The natural environment is often cited as one of the major advantages of rural over urban life, and many a migrant to the city has yearned for the sight of green fields, the smell of fresh-cut hay, the smell of a clear running brook, or simply a deep breath of clear country air. . . . By and large, rural residents have been little more beneficent in their relation with nature than the denizens of our concrete and asphalt jungles. Indeed, profligate waste, and destructiveness have marked the exploitation of our land and other natural resources since the earliest white settlements in America" (Ford 1978).

Mr. Ford's somewhat pessimistic quote from nearly three decades ago might still be considered fresh and relevant by environmentalists and controversial by many. The debate goes on, and progress (both politically and in actual environmental change) is measured incrementally. Since the ending of our hunter-gatherer societies and the beginning of our agricultural societies, there has been a trade-off between the process of producing food for the society and stress on the natural environment that provides the food. Furthermore, the advent of the industrial revolution in the early 1800s charted a path of increasing intensification of agricultural and nonagricultural industry that has further stressed the natural environment that we live in. The latter not only challenges the urban environment, but also our rural environments as urban discharges and emissions reach the streams and air that connect urban and rural landscapes. As our industrialized nations' economies grew strong and basic necessities of life were generally cared for, timing was right to support a consciousness raising about our environment during the late 1960s and 1970s. During this period, the industrialized nations developed similar national regulations and standards for air and water quality.

Research has indicated that water and air pollution are now global issues, projecting environmental issues into the international arena. In the 1980s, research began to emerge indicating that rural environmental problems may be as great as or even greater than some urban environmental problems. Furthermore during this time, more people left the farm and city people moved out to the country for a lifestyle change, and farming began to become more concentrated into larger and more intensive operations. These factors contributed to an increased awareness and concern for water, air, and soil contamination contributions from our agricultural operations. Furthermore, research provided evidence for contributions of agricultural pollution to degraded water quality for drinking and recreational use, as well as regional and global changes in natural ecosystems. A large cross section of citizens have become concerned about these issues, some of these individuals having had or perceived to have had direct health consequences from these exposures. Emotions are high among people who are concerned, complicating specific diagnoses for persons who claim to suffer health problems from these exposures. On the other hand, many in the agricultural community feel threatened that their industry has been negatively portrayed, and they fear excessive regulation will unnecessarily burden their operations economically, making it impossible to farm. Because of all these reasons, rural health professionals must be as aware and concerned about environmental issues and their potential health effects as are their urban counterparts.

This chapter provides an overview of the major rural environmental pollutants and their sources. We will focus on those pollutants that have a known direct or indirect effect on human health, and those that have attracted a high profile of public concern. Our objective is to create awareness in rural health practitioners so that they may be better prepared to anticipate and diagnose environmental illness, participate in informed public debate, and provide information regarding the nature and prevention of resulting health effects in their communities.

Figure 7.1. Row crop cultivation that has exposed bare soil to wind and rain, combined with practices to speed drainage, has created difficulties to manage soil erosion. However, land management practices in the last three decades have slowed the problem.
(*Source:* Medical Practice in Rural Communities, Chapter 3, Figure 3.5, page 48, book authors: Cornelia F. Mutel, Kelley J. Donham, Rural Health and Agricultural Medicine Training Program, Department of Preventive Medicine and Environmental Health, College of Medicine, The University of Iowa, Iowa City, Iowa 52242, U.S.A., © Springer-Verlag New York Inc., 1983. Image reprinted with kind permission of Springer Science and Business Media.)

WATER QUALITY

Introduction

Water pollution is by far the most important concern in rural areas. The U.S. Environmental Protection Agency (USEPA) (2000) reports that 36% of all river miles in the U.S. are impaired for certain specific uses (e.g., recreation or drinking water sources), and that at least 60% of that pollution comes from agricultural sources. Howarth (2002) reports that much of the nitrogen fertilizer applied to soils ends up in our surface and groundwater systems (mean of 26% with a range of 3–80% depending on the model applied, regional soil types, and land use). This directly adversely effects rural well water, as indicated by surveys in the midwestern and eastern sectors of the U.S., revealing that from 12–46% of water wells are contaminated with nitrates in excess of the EPA limits of 10 mg/l (Hamilton and Helsel 1995; U.S. Environmental Protection Agency 2000). Although nitrogen may be the most common contaminant and perhaps the most serious direct threat to human health, there are additional water contaminants from fertilizers, including phosphorus, chloridion, calcium, and magnesium. Pesticides used in crop production find their way to water sources. Animal wastes may also contribute to water contamination in the form of antibiotics, antibiotic-resistant genetic material, infectious agents, veterinary pharma-

ceutical residues, and trace elements (Boxall and others 2004).

Natural systems have the capacity to mitigate certain loads of pollutants put on it (if the system is not overloaded, and if there is time). However, time to degrade contaminants has been short circuited in many areas as millions of acres have been tiled, streams have been straightened, wetlands have been drained, and earth put under concrete for parking lots and roads. The former has been ongoing since the mid-1800s in North America and even earlier in Europe to enhance the conditions for crop cultivation. Figure 7.1 demonstrates the effects of increased drainage speed. Water very rapidly runs off the land, not allowing time for recharging (the breakdown of contaminants by soil microbes) to occur, causing soil erosion, particulate contamination, and ultimately increasing the probability of contaminating our surface and groundwater sources (Wu and Babcock 1999). (It should be noted that to help

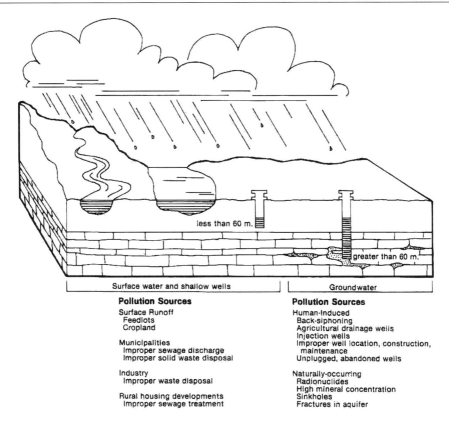

Figure 7.2.   Surface waters and shallow wells experience the same sources of pollution. Ground water may contain harmful substances from natural sources as well as from human actions. (*Source:* Medical Practice in Rural Communities, Chapter 3, Figure 3.1, page 44, book authors: Cornelia F. Mutel, Kelley J. Donham, Rural Health and Agricultural Medicine Training Program, Department of Preventive Medicine and Environmental Health, College of Medicine, The University of Iowa, Iowa City, Iowa 52242, U.S.A., © Springer-Verlag New York Inc., 1983. Image reprinted with kind permission of Springer Science and Business Media.)

counter this problem, there have been recent government programs to slow water runoff, such as building ponds, retaining wetlands, retiring highly erodable cropland, etc.). Figure 7.2 demonstrates that runoff not only creates a risk for surface water quality, but also makes groundwater accessible by shallow (less than 60 m) water wells.

National water quality regulations passed in most industrial countries in the 1970s have helped to decrease major urban or point sources of water pollution. However, only regulations promulgated in the past two decades (Centner 2000; Matisziw and Hipple 2001) have been relevant to non-point source and agricultural pollutants. These newer regulations are attempting to address issues of

agricultural concentration (which identifies some large facilities as point sources) and considering non-point (or contamination of regional watersheds from the sum pollution of multiple sources). In addition to regional concerns, there is a global agricultural pollution concern because of air emissions from industrialized and developing countries (Follett and Delgado 2002; Ham 2002).

The direct contributions of water pollution to human health are difficult to quantify, and vary with the water source and population. Although acute health concerns are occasionally reported, chronic health problems are often multifactorial and difficult to attach directly to poor water quality. The health effect will be covered in more detail later in this chapter.

SOURCES AND SUBSTANCES OF WATER POLLUTION FROM AGRICULTURAL OPERATIONS

Animal Wastes and Inorganic Fertilizers

A major source of water contaminants from agricultural activities is from livestock production and include products found in animal wastes (nitrogen compounds, phosphorus, particles, chloride, calcium, magnesium, antibiotics, microbes, antibiotic-resistant genetic material, and veterinary pharmaceuticals). Inorganic fertilizers (e.g., anhydrous ammonia and potash) are important sources of nitrates, phosphorus, chloride, calcium, and magnesium contamination. These combined substances can be called the "footprint" of agricultural water pollution (Hamilton and Helsel 1995). Nitrates and phosphorus are the most important elements of this footprint, but the latter three have relatively little environmental health consequence and will only be covered in summary in this text. Manure is typically directly applied to land as a plant nutrient, and it contains a relatively high level of nitrogen (depending on animal species and prior treatments). Other pollutants associated with manure include potassium, microbes (some of which might be antibiotic-resistant), and veterinary pharmaceuticals. A more detailed discussion of each of these follows.

NITRATES

As nitrogen is a fundamental element of life, recycling of nitrogen (the nitrogen cycle) is a fundamental ecologic principle. The nitrogen in the air (and soil) is "fixed" by special bacteria into nitrate or ammonia, which can be taken up by plants as a food source. Plants fed to animals are digested and the nitrogen is incorporated into animal proteins. The nitrogen is then recycled to the air and soil as animal waste as biomass is applied to soil. The process of nitrification and denitrification by bacteria in our soils and waters convert this nitrogen to ammonia, nitrate, and other oxidized forms (primarily nitrous and nitric oxide), and finally to nitrogen gas (N_2) and back to the atmosphere. However, when nitrogen sources are added to fields faster than plants can utilize it, one of the major breakdown products (nitrate, NO_3) accumulates. Nitrate is very susceptible to run off into rivers and lakes, or finally the oceans,

or into our groundwaters. Nitrate, when consumed, will be reduced to nitrite in the GI tract of animals, which combines with hemoglobin to form methemoglobin, which cannot carry oxygen. Furthermore, there are concerns of nitrites combining with amino acids or the herbicide atrazine in the gut to form carcinogens (these health effects are discussed in more detail below and in Chapter 5).

The nitrogen cycle is out of balance in a large part of agriculture in industrialized and some developing countries. Although nonagricultural industry and auto exhaust are significant nitrogen sources, agriculture makes up 86% of the human-generated nitrogen. The quantity of applied fertilizer is increasing—50% of all fertilizer applied to crops has been applied since 1984. Only 50% of the applied nitrogen ever reaches plant tissues (Follett and Delgado 2002). The remainder escapes by runoff or volatilization to overburden the nitrogen cycle. The problem has advanced strongly as our agricultural systems have evolved from small diversified low input-operations up through the 1950s to larger more intensive monoculture operations dependent on larger amounts of fertilizers and pesticides, exacerbated by faster water runoff, less surface foliage, and more soil erosion (Novotny 1999).

PHOSPHORUS

Second to nitrate, phosphorus (P) is the most ubiquitous agricultural environmental pollutant. It travels with nitrate because its main sources are from inorganic crop fertilizers and animal manures. Phosphorus is a component of manure and exists in a ratio of around 1:1 with nitrogen. However, plants utilize about 7 parts of N to about 1 part of P resulting in a buildup of P in soils (Donham 2000). An excess of P is left, which is not highly soluble, and it tends to be stored in soil, to the point where it reduces water filtration capacity and degrades the fertility of the soil for many years. Furthermore, P may leave the farm as runoff, primarily with eroded soil, to contaminate surface waters. An example of how serious degraded soil fertility can become, 1 million acres of the Netherlands (a country with high pig production and small land mass) is degraded because of excess P from excessive animal manure applications (Vos and Zonneveld 1993).

PARTICLES
Soil erosion is a major source of particulate contamination of surface water sources worldwide (Ford 1978). Each year in the U.S., an estimated 4 billion tons of soil contaminates our rivers, streams, and lakes as a result of soil erosion (Miller 1975). This loss is equal to 4 million tons of topsoil, 15 cm deep, and an area equal to one-sixth of the land farmed in Iowa. Particulates alone do not necessarily pose a direct health hazard, but they do produce aesthetic problems, such as poor taste, turbidity, and odors. Particulates also are a signal of potential contaminants (such as nitrates and microbial organisms) from surface sources and may carry potential toxic and carcinogenic substances (such as dieldren and the herbicide 2,4,5-T) adsorbed to the particulate surface.

TRACE ELEMENTS
Sodium (Na), potassium (K), copper (Cu), and zinc (Zn) are found in animal manure because they are additives to animal feeds, often at levels higher than the animal is capable of metabolizing. Although there is probably no toxic health hazards to humans from this kind of exposure, there are problems with soil fertility degradation or eutrophication and from toxicity to grazing animals (mainly Cu).

MICROBES
Microbial contamination of ground and surface waters can occur from livestock operations (Bitton and Harvey 1992). Organisms that have been associated with animal waste and that have human health implications include *Helicobacter pylori* (Lee 1993), *Campylobacter, Salmonella, Cryptosporidium,* and *Listeria* (Fayer 1998); (Fone and Barker 1994). Although there may be hundreds of species of organisms found in swine waste, it is important to note that most pathogens do not survive in animal wastes very long because they are not well suited to survive desiccation, sunlight, low pH, high osmosity, and high ammonia concentrations in stored swine waste slurry (Donham and Dauge 1985; Donham and others 1988). For example, *Salmonella* and *Leptospira* species were found to survive only 3 days in swine waste (Will and Diesch 1972) or 19 days (*Salmonella*) in poultry manure (Berkowitz and others 1974; Bitton and Harvey 1992). Survival

of organisms after land application is only a few days and is retarded by low temperatures, low soil moisture, low pH, sunlight, and competition with other organisms.

ANTIBIOTICS
Field studies in the vicinity of poultry and swine confinement facilities have revealed the presence of antibiotics in a variety of water sources, including lagoons, monitoring wells, field tiles, streams, and rivers. The following antibiotics were found at a concentration of around 100 micrograms/liter of water: tetracyclines, sulfa, beta lactams (e.g., penicillin), macrolids (e.g., erythromycin), and fluroquinalones (e.g., enterofloxacin) (Campagnolo and others 2002). It is fairly evident that there is a risk for consumption of antibiotics and antibiotic-resistant organisms in well or surface waters, from runoff or seepage into the groundwaters from (Confined Animal Feeding Operation) CAFOs. The health significance of consumption of these levels of antibiotics in water is uncertain. However, as mentioned in Chapter 12, the risk is comprised of cumulative exposures, included with other sources of antibiotics and antibiotic-resistant organisms.

VETERINARY PHARMACEUTICALS
Small quantities of antibiotics, paraciticides, and growth enhancers (or their by-products) pass in the urine or feces of animals and find their way to soils and water sources through manure application. However, relatively little is known about the fate and environmental impact of these substances. Perhaps the primary concern for environmental health is from excessive antibiotic use and possible influence on emerging antibiotic-resistant organisms. Generally, pharmaceuticals are a potential concern, but are secondary in importance to nitrogen, volatile organic compounds, and pathogens.

Pesticides

Rain may wash pesticides (including insecticides, herbicides, and fungicides) from plants or soil into bodies of surface water and into shallow wells. Organochlorine insecticides and some herbicides persist in the environment, and these are most likely to contaminate rural water. However, many organochlorine insecticides are being replaced by the less-persistent organophosphate and carbamate insecticides, which are less likely to contaminate water supplies. Certain herbicides (e.g., atrazine) and the chlorinated phenoxyacetic acid products (2,4-D, and 2,4,5-T) and fungicides (e.g., hexachlorobenzene and pentachlorophenol) are also of concern (Craun and McCabe 1973; Miller 1975).

The quality and quantity of data are only adequate to support very gross generalizations about health consequences of pesticide-contaminated water. Certain pesticides such as the insecticides aldrin, deilrin, toxaphene, and doxins, (contaminants of the herbicide 2,4,5-T) are suspected risk factors for hematopoetic and other cancers as demonstrated in animal studies and in epidemiologic studies (National Academy of Sciences 2003). Several epidemiologic studies have shown associations between indicators of pesticide exposure and hematopoetic cancers (Burmeister 1981; Donham and others 1980; McKee 1974). Other possible chronic health problems attributed to insecticides (none conclusively shown) include aplastic anemia (lindane), peripheral neuropathy (organophosphates), hypertension, and elevated cholesterol (organochlorines) (D'Ercole and others 1976). Breast-fed infants may also be exposed to pesticides if their mothers have ingested contaminated water (U.S. Environmental Protection Agency 1980a).

Several cases of acute pesticide poisoning occur each year from drinking water, when a spilled pesticide has entered an unprotected water well (unsealed, uncased, or broken-cased) or from a mixing-loading operation. During mixing and loading operations the pesticide is placed in a tank and diluted to the proper concentration by adding water from a hose connected to a hydrant. Should the hydrant be turned off, with the hose under the liquid surface, the contents of the tank will siphon back down the hose into the well (unless there is a check valve in the hydrant) (Figure 7.3). Organophosphate or carbamate insecticides are most likely to cause acute poisonings from back-siphoning cases.

Water Pollutants from Urban Sources

Improperly treated municipal sewage may pollute surface water supplies with infectious agents and nitrates. Furthermore, urban municipalities and industries usually locate their waste disposal sites in rural areas. If these sites (sanitary landfill or toxic waste contaminate facilities) are not properly en-

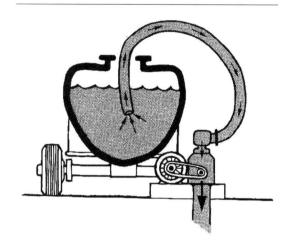

Figure 7.3. Back siphoning of agricultural chemicals into wells occurs when a hose filling a mixing/application tank is connected to a hydrant that does not have a check valve. (*Source:* Medical Practice in Rural Communities, Chapter 3, Figure 3.3, page 45, book authors: Cornelia F. Mutel, Kelley J. Donham, Rural Health and Agricultural Medicine Training Program, Department of Preventive Medicine and Environmental Health, College of Medicine, The University of Iowa, Iowa City, Iowa 52242, U.S.A., © Springer-Verlag New York Inc., 1983. Image reprinted with kind permission of Springer Science and Business Media.)

gineered and managed, nitrates, infectious agents, and other potential toxic substances may leach into shallow aquifers or run off into surrounding streams, thereby contaminating local water supplies. Despite governmental regulations, these incidents happen sporadically throughout rural areas of industrialized countries. The overall health effects on the rural population are not known, since reporting is rare and each situation may have different health manifestations (Delfino 1977).

Recent research suggests that chlorination of public water supplies presents a dilemma in environmental health practice. Most municipalities chlorinate their drinking water to kill infectious agents. However, chlorine may combine with organic molecules in the water to produce several trihalomethane compounds which are suspected carcinogens. There is growing concern that rural (or urban) populations may have a higher risk of cancer if their surface water source is downstream from or in the vicinity of a community that discharges chlorinated water. (Details of this health concern appear later in this chapter.)

Water Pollutants from Rural Industrial Sources

Many urban and rural industries intermittently discharge (unintentionally and possibly intentionally) improperly treated waste products into streams or lakes or improperly dispose of waste chemicals in the ground, resulting in groundwater contamination (U.S. Environmental Protection Agency 1980a; Metzler 1982). Packing plants and rendering plants that do not have adequate waste treatment facilities, for example, can add nitrates, infectious agents, and complex organic materials to water supplies.

Methylmercury, a second example of a rural industrial water pollutant, is found in many industrialized countries, including areas of the north central and northeastern U.S. and in eastern Canada. Mercury is used in paperpulp mills as a fungicide. It is also a byproduct of chloralkali plants and a contaminant of coal-fired power plants. Mercury is transformed by microflora in the environment into the neurotoxin methylmercury, which accumulates in fish and other aquatic life, endangering people who eat fish. A classic case of chronic methylmercury poisoning occurred among Japanese people who consumed seafood from the contaminated waters of Minimata Bay. Northern Canadian Indians of Quebec and Ontario are at risk because they acquire much of their food (protein) from fish in methylmercury-contaminated lakes. Studies are in progress to investigate the full extent of this problem (National Academy of Sciences 2003).

Agricultural drainage wells are similar to injection wells, but they are designed to remove excess surface water to allow crop production. Field tiles in wet areas collect the surface water and transport it to wells which drain the water into the groundwater sources. Though new drainage wells are illegal, those remaining (to be phased out) carry surface contaminants directly to groundwaters that may also serve as the same source of drinking water for area residents (see Figure 7.4).

A variety of other industrial chemicals may be accidentally spilled or discharged into rural water supplies. Examples include a spill of carbon tetrachloride that contaminated water for miles downstream in the Ohio River, and a train wreck that

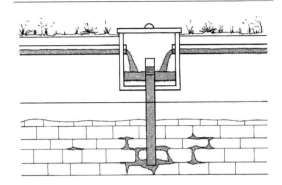

Figure 7.4. Drainage wells are designed to remove excess surface waters to ground waters. This expedient method of water removal risks polluting ground water sources with agricultural chemicals, nitrates, and infectious agents, among other surface substances; drainage wells are slowly being phased out in most countries. (*Source:* Medical Practice in Rural Communities, Chapter 3, Figure 3.2, page 45, book authors: Cornelia F. Mutel, Kelley J. Donham, Rural Health and Agricultural Medicine Training Program, Department of Preventive Medicine and Environmental Health, College of Medicine, The University of Iowa, Iowa City, Iowa 52242, U.S.A., © Springer-Verlag New York Inc., 1983. Image reprinted with kind permission of Springer Science and Business Media.)

spilled 13,000 liters of phenol, contaminating rural wells for miles around the accident scene (Delfino 1977; Reynolds and others 1997). A veterinary pharmaceutical company in a small midwest town had a landfill containing orthonitroanaline that leached into the adjacent river, jeopardizing water quality for several miles downstream. Underground leaking petrochemical tanks and pipelines also have contaminated groundwater in rural areas (Harris and others 1982; Metzler 1982).

Petroleum companies and many other chemical companies have used injection wells to dispose of a variety of wastes. Waste material is forced under pressure deep into the ground, below the aquifers from which drinking water normally is drawn. However, if there is a mechanical failure or a geophysical connection between the two levels, the injected material may contaminate a drinking water aquifer. Petroleum companies frequently inject saltwater into oil wells to increase their yield or to dispose of excess saltwater (often a by-product of oil wells) from active wells to nearby inactive wells. Most injection wells are in Texas and Kansas oil fields and in the upper Midwest industrial belt. The health consequences of injection wells have not been quantified, and specific health consequences depend on the particular contaminant (Musterman and others 1980).

Water Pollutants from Inappropriate Land Management

Water contamination commonly results from inappropriate land use and from lack of pollution controls. Examples include rural subdivisions with too high a density of septic tanks or poorly functioning leach fields. Strip mines often result in excessive erosion and acid leakage into water sources. Water seeping through abandoned deep-shaft mines may pick up hazardous minerals or particulates, and working mines may deposit wastes directly into lakes or onto land where runoff can carry contaminants into surface waters. In Minnesota, taconite mining companies have dumped tailings in Lake Superior for many years, contaminating the lake with asbestos. Many rural as well as urban residents obtain their water from that lake (U.S. Environmental Protection Agency 1980a; Lamka and others 1980).

Impending problems in areas west of the Mississippi River (particularly the Platte and Colorado River basins) include increased levels of mineral salt in soil and groundwater, cross-aquifer contamination, and a decline in local water tables. All these problems have resulted from intensive irrigation of arid lands that have been brought into agricultural production (Chanlett 1979; McKee 1974). Excessive salinity of water has a laxative effect on people drinking this water because of its sodium and magnesium salts. Furthermore, some people who drink this water may have difficulty controlling hypertension and hypocalcemia secondary to pregnancy because of its sodium salts (National Academy of Sciences 1977). Furthermore, as salinity (and therefore osmolarity) of soils increases, more irrigation water is required to allow availability of water to the plants, creating an environmental conundrum.

Water Pollutants from Natural Sources

Several natural substances may contaminate rural water supplies. For example, some aquifers natu-rally contain high concentrations of fluoride (over 8 million people in 44 states of the U.S. have water high in fluoride) (National Academy of Sciences 1977). The states most commonly having wells with high fluoride levels included Arizona, Colorado, Illinois, Iowa, New Mexico, Ohio, Oklahoma, South Dakota, and Texas. At levels of 1.2 mg/l or greater, fluorine can cause a brown-colored mottling of teeth in children, and at 3 mg/l, fluorine may result in abnormal bone formation (skeletal fluorosis). The safe upper limit is considered to be 0.7 mg/l (National Academy of Sciences 1977; World Health Organization 1972).

On the other hand, fluoride in water at the proper level (0.5–0.7mg/l) has beneficial health effects in preventing dental caries in children. Many rural water supplies have less than 0.5 mg/l of fluorine, creating a risk of excess dental caries in children.

Arsenic (Ar) is a heavy metal that occurs naturally in certain groundwater aquifers around the world as a natural result of filtration of water through rock formations that contain high concentrations of this element. Furthermore, there are manmade sources of Ar concentration due to mining and manufacturing (World Health Organization 1972). Chronic exposure to Ar in water can increase the risk for skin, bladder, and lung cancers; diabetes; and anemia. Furthermore, there is evidence of risk for reproductive, developmental, immunologic, and neurological effects (U.S. Environmental Protection Agency 2000a). There are many areas around the world where Ar may be of concern in groundwater. In North America, there are areas of Ar concern in the far west, parts of Texas, the upper Midwest, and the New England States (U.S. Geological Survey 2004). Based on new toxicological information, the USEPA has recently proposed a lowering of the regulated limits of Ar in drinking water from 50 ppm to 5 ppm. The EPA states this change will add additional protection to over 22 million Americans.

As most water regulations are related to public water supplies, most rural and farm residents are not regulated. Health professionals in rural areas should be aware of the areas (U.S. Geological Survey 2004) where high Ar occurs in the groundwaters, and advise their patient community to have their water tested and include Ar in the test battery.

Many rural well waters have excessive quantities of calcium and magnesium ions, a condition commonly referred to as "hard water." This "problem" actually may have a positive health effect. Epidemiologic studies have suggested that there is an inverse correlation between water hardness and cardiovascular disease. People who drink soft water have slightly higher rates of atherosclerotic disease, hypertension, and several other cardiovascular problems. Many rural residents with deep wells with hard water buy water softeners. Although scientists have not fully described the relationship between hard water and cardiovascular disease, it may be prudent to install a water softening system for hot water only (not drinking water). Also, most water softeners use an ion-exchange system, replacing magnesium and calcium with sodium. People who must restrict sodium intake for control of cardiovascular disease should not drink water softened by this type of method (National Academy of Sciences 1977; World Health Organization 1972). A reverse osmosis water purifier would be preferred for these people.

Water contamination from radionuclides is an increasing concern worldwide. The normal background radiation exposure for humans in the United States is about 100 mrem/year. A small portion of this radiation (estimated at 0.24 mrem/year on the average) comes from water that contains naturally occurring radionuclides (National Academy of Sciences 1977). The largest contributor of such radiation is K40, which is not thought to be of any health consequence. However, certain radionuclides, (e.g., Sr90, Ra226, and Ra 228) are deposited in bones, and these may increase the risk of bone cancer. Of these radionuclides, Ra226 is thought to be of greatest importance because as an alpha emitter, it has a greater chance of producing cellular change than substances that emit beta radiation. Certain areas of the country have high concentration of Ra in the water. Some epidemiologic studies suggest that populations in these areas have slightly higher rates of bone cancer (National Academy of Sciences 1977). In addition, recent epidemiologic studies in Iowa have shown an association between high Ra levels in water and lung cancer (Bean and others 1982). Over a million people have water sources with greater than the recommended maximum concentration of Ra, which is

5 pCi/l. Over 120,000 people have wells with higher than 9 pCi/l (National Academy of Sciences 1977).

Many rural wells have naturally occurring high sulfate levels. Sulfates are not removed by normal water treatment or softening processes. The recommended maximum level for sulfates is 250 mg/l. Above 350 mg/l sulfates impart a bad taste and odor to the water. The only documented direct human health effect of high sulfate is diarrhea, which does not usually occur until the sulfate level is around 1000 mg/l (Glanville 1981).

ENVIRONMENTAL HEALTH EFFECTS OF WATER POLLUTION

Nitrogen

Ingested nitrite is absorbed into the blood and complexes with hemoglobin to produce methemoglobinemia, which cannot carry oxygen (Mensinga and others 2003). Infants are much more susceptible to methemoglobinemia than are adults, because of the dose relative to their size and their minimal ability to compensate for compromised oxygen-carrying capacity. The typical history of a poisoning is an infant who has been fed formula remixed with water containing high levels of nitrates. Babies may present with a blue-gray color of the skin (thus the common name of "blue baby" for methemoglobinemia in infants). These infants may present with an irritable or lethargic disposition, depending on the degree of toxicity (Knobeloch and others 2000). The infant's condition may progress rapidly to coma and death if not recognized and treated readily. The treatment for methemoglobinemia is methylene blue, IV, at 1–2 mg/kg, in a 1% solution. A clinical response should be seen in 20 minutes. Repeated doses if necessary should not go beyond 7 mg/kg and be given over at least a 4-hour period.

Other causes of methemoglobinemia are a congenital metabolic problem (usually benign and managed with daily doses of ascorbic acid) and local anesthetics of the lidocain family (Martin 1995). Furthermore, GI infections and inflammation may increase production of nitric oxide, which can be reduced to nitrite in the gut, causing a low-level methemoglobinemia (Avery 1999). History, including ingestion of high nitrate water, is crucial to anticipation of methemoglobinemia.

A simple bedside test for methemoglobinemia

is to take a drop of venous blood and expose it to room air (or better, pure oxygen). Methemoglobin (and sulfhemoglobin, another condition that can occur on a farm, associated with hydrogen sulfide poisoning) will not return from blue to red (Schoffstall and others 2002). A second simple test that may be used is to combine some of the patient's blood with potassium cyanide, which will cause the methemoglobin blood to turn from blue to red. This test will also distinguish methemoglobin from sulfhemoglobin, as sulfhemoglobin will not change color with potassium cyanide. The preferred laboratory test is Co-oximetry, which is a spectrophotometer that measures at four wavelengths, and can differentiate from sulfhemoglobinemia.

The EPA guideline for nitrate in drinking water is 10 mg/l. Two cases of blue baby investigated in the U.S. revealed a water concentration of 22.9 and 27.4 mg/l (Knobeloch and others 2000).

Additional suspected perinatal risks from nitrate consumption include suppressed intrauterine growth, prematurity, and spontaneous abortion. Laboratory animal studies and case studies of eight spontaneous abortions in four women in Indiana (U.S.) suggest there is a risk of abortion with nitrate-contaminated well water (Anonymous 1996). Confirming the health risks of excess nitrate for pregnant women requires further study (Levallois and Phaneuf 1994).

Another suspected risk for nitrate ingestion is an increased risk for stomach cancers and perhaps other cancers (Levallois and Phaneuf 1994). The concern is associated with the formation of N-nitroso carcinogens by the reaction of nitrates with amino acids and/or N-alkyl-amides (such as the herbicide atrazine). Van Leeuwen and others (1999) have shown in Ontario (Canada) that atrazine and nitrate in the water are associated with excess stomach cancer. Several studies have shown that the farming population is at increased risk for stomach cancer (see Chapter 5). There is greater total intake of nitrates in the European Union (EU) (50–140 mg/day) than in the U.S. (40–100 mg/day) (Mensinga and others 2003). The World Health Organization (WHO) and the EU have set acceptable daily limits of ingested nitrate at 3.7 mg/kg/day (259 mg/day for a 70 kg person) and the USEPA recommended limit is 1.6 mg/kg/day (112 mg/day for a 70 kg person) (Mensinga and others 2003). Therefore, people

who drink two liters of water per day, containing 10 mg/l of nitrate-nitrogen (45 mg/l nitrate), consume 90 mg of nitrate per day (close to the maximum recommended consumption by the USEPA). Surveys in the U.S. have shown that from 12–46% of wells tested have nitrates greater than 10 mg/l (depending on the land use and geology of the region) (Hamilton and Helsel 1995). Highest levels are found in irrigated areas with sandy or porous soils. Wells of less than 200 feet in depth are more susceptible to contamination. Water is not the only source of ingested nitrates, as vegetables fertilized with high amounts of nitrogen fertilizer may also contribute to the total dietary burden (Mensinga and others 2003).

Although specific mechanisms and dose-response are not fully known for nitrate toxicity, the weight of evidence clearly indicates both acute and chronic health effects are real. Limits on ingestion of nitrates are clearly warranted.

Phosphorus

Phosphorus (P) at levels in drinking water is not directly toxic to humans or animals, but it contributes along with nitrogen to eutrophication of our surface waters and saltwater estuaries (Jawson and others 1998). Results of this contamination include brown and red (red tide) algal blooms, which are toxic to shellfish, sea mammals, and humans who eat these toxic shellfish (Carpenter and others 1998; Howarth and others 2002). The increased oxygen demand caused by the algal blooms can result in fish kills, decreased aquatic diversity, decreased quality of waters for recreation, and undesirable odors and tastes in drinking water.

Eutrophication of our ocean estuaries on the East Coast of the U.S. has been speculated as the source of an overgrowth of the dinoflagellate *Pfiesteria piscicida.* The changes in the aquatic ecology because of eutrophication favors the overgrowth of this organism.

Trace Elements

There is a low-level concern (as previously mentioned) with chorine in the water, and formation of trihalomethane compounds (Carpenter and others 1998). Additional to chlorine from municipal treatment systems (Lantagne 2001), chlorine comes from inorganic fertilizers, rural homes, and animal feeding operations (AFOs) that have chlorination systems, and rural water supply systems. Although the benefits of chlorination likely exceed the risks, trihalomethanes (formed in the environment from chlorine) are carcinogens in laboratory animals, and there are epidemiologic studies that reveal an association between the presence of these substances in drinking water and increased bladder, colon, and rectal cancers, and reproductive problems in women. The USEPA has set a limit of 80 ppb trihalomethanes in drinking water. As there are few studies that have looked for these substances in rural water supplies, the extent of potential hazard is not known.

Copper and zinc toxicity for some plants has been studied, and soils with long-term applications of manure may surpass these levels. Although nitrates and Na will leach into ground or surface waters to cause excess eutrophication, K, Cu, and Zn are not highly water soluble and tend to build up in soils. K is a crop nutrient, as are N and P, and may be removed by uptake into plants. However, the others (Cu, Zn, Na) tend to accumulate, which degrades the productivity of the soils and leads to possible CU toxicity to grazing sheep (Chanlett 1979; National Academy of Sciences 1977, 2003).

All elements in water are not necessarily bad. For instance, magnesium (Mg) may have a positive heart health effect. Magnesium effects heart muscle contraction and reduces arterial constriction (National Institutes of Health 2004).

Particles

Particulates are an indicator for other potential contaminants, such as phosphorus, nitrates, pesticides, and microbial organisms. Particulates produce aesthetic problems but not direct health hazards. They are the reason why the streams and rivers in many agricultural areas are dark and murky, rather than clear and bright. Regarding personal use, they may produce unpleasant tastes, and odors. Clothes washed in this water may not come clean.

Microbes and Antibiotics

Additional to gastrointestinal infections, another concern of microbes in water is that they carry and transmit genetic material coding for antibiotic resistance. Animals are commonly fed low levels of antibiotics for growth promotion, which enhance

the selection for antibiotic-resistant organisms. Humans may acquire resistant zoonotic pathogens directly from an infected animal or pick up a nonpathogenic-resistant organism that may transfer that resistant gene to a pathogen in the gut of an individual (Velasquez 1995). (Details of this hazard are seen in this chapter below, and in Chapter 12 on veterinary therapeutics.)

Although there have been documented cases of gastrointestinal (GI) illnesses, and other waterborne illnesses such as leptospirosis and salmonellosis (from animal-to-water-to-person) (Fone and Barker 1994; Fuortes and Nettleman 1994), it has been difficult to characterize or quantify the overall health risk of these exposures (Moore 1993). However, the following information suggests potential risks are present.

Fecal bacterial species growing in lagoons of confined animal feeding operations (CAFOs) have been found in groundwater in the vicinity of CAFOs (Krapac and others 1998).

An estimated 400,000 persons in Milwaukee (U.S.) in 1993 contracted *Cryptosporidium Parvum* infection from drinking city water (National Resources Defense Council 1998). One hundred deaths were attributed to this outbreak. The infectious agent was thought to originate from runoff from dairy farms that contaminated city water supply reservoirs. Young ruminants are especially susceptible to this infection and shed the organism in their feces. The filtering component of the water treatment plant of most cities is not capable of removing this organism, and it is resistant to usual municipal water treatments (Robens 1998). The usual human infection persists for 4–7 days with typical gastrointestinal symptoms. Ninety-four animal species are susceptible, with eight different cryptosporidia species, *C. parvum* being the most common in livestock and humans. Feral birds may serve as environmental disseminators of this organism. There are no data at present on CAFOs as sources of human infection with *C. parvum* (Fayer 1998).

Pfiesteria piscicida appears to have thrived in (U.S.) water sheds, associated with a high density of CAFOs in the region. *Pfiesteria* has been implicated as an agent causing skin irritation and neurological affects, such as short-term memory loss and other cognitive impairments. It has also been incriminated as a cause of numerous fish kills (National Resources Defense Council 1998).

Although there is a great deal of public concern about *Pfiesteria* on the east-central coastal plains of the U.S., a great deal about the health concerns of this organism remain unknown.

Veterinary Pharmaceuticals

The fate, transport, and impact of veterinary pharmaceuticals in the environment were recently discussed in a review article (Bennish 1999). It is apparent that veterinary pharmaceuticals are a potential health concern, but secondary in importance to nitrogen, volatile organic compounds, and pathogens.

Control of Water Pollution

There is adequate scientific and technical information available to control water pollution from agriculture. However, there is a political-economic struggle in most industrialized countries to pass and enforce effective legislation that will curb pollution without creating an economic liability to production agriculture. Voluntary changes in production practices have had profound environmental benefits. Minimal or no tillage practices in row crop production leaves plant material on top of the ground, dramatically decreasing soil erosion, and thus water pollution. Special diets and new corn varieties have been developed for animals, leading to less P in the manure. The USDA is sponsoring programs to build or preserve wetlands (Knight and others 2000) and riparian buffers or vegetation filter strips that will decrease runoff into streams. One researcher advocates the use of polyacrilamide addition to soils, which acts to stabilize soils so that it will hold on to contaminants, preventing runoff (Entry and others 2002). Special taxes on pesticides and fertilizers have shown some benefit in pollution control (Archer and Shogren 2001). "Precision agriculture" uses global positioning systems to control the amount of fertilizer and pesticides applied to cropland according to specific localized needs. These systems have the potential to reduce excess farm chemical applications that may end up as water contamination.

Agricultural Water Pollution Summary

Table 7.1 summarizes the various water contaminants important in rural and agricultural environments, their sources, health hazards, and control measures. Water pollution is the most important environmental problem in agriculture. It is important in all industrialized countries and in many developing countries. Nitrates are the most important and most common substance in water that may produce health hazards. The problem is overuse of organic and inorganic fertilizers and animal manure added to farm land that overburdens the natural nitrogen cycle, resulting in excess N running into surface and groundwaters. Direct health hazards include methemoglobinemia which causes blue baby syndrome in infants. Potential chronic health problems include possible cancer, as nitrates can combine in the GI tract to form nitrosamines, which may be related to excess stomach cancer seen in agricultural populations. Control of water contamination is by the control of erosion and runoff and a decrease in the overuse of nitrogen and phosphorus that outstrips the natural nitrogen cycle.

AIR QUALITY

Introduction

Human health effects from rural and agricultural air pollution occur sporadically, but the overall magnitude of health concern is generally less than water pollution. There are numerous sources of air pollutants that may originate in the rural environment, or they may be blown in from urban areas. There are acute exposures that can be traced directly to air pollution (such as eye burns from direct exposure to anhydrous ammonia, or poisonings from pesticide drift) but are rarely reported. Most health problems are more chronic and less severe; they may complicate other health conditions or act in an additive or synergistic manner with other air pollutants. For example, in rural areas, as in urban, any one of the low-level pollutants (such as ozone, nitrogen oxides, or particulates) may have inconsequential health effects. However, each may precipitate severe health problems in people who have concurrent respiratory or cardiovascular conditions, who smoke, or who are exposed to other industrial or agricultural air pollutants. Table 7.2 summarizes the major air pollutants in rural and agricultural settings.

Air Pollution from Livestock and Crop Operations

The most common air pollution problems include odors from livestock operations, dust from tilled

Table 7.1. Rural Water Pollutants

Substance	Pollution Source	Water Source — Surface	Water Source — Ground	Health Effect	Prevention
Nitrates	Runoff and seepage into groundwater, from heavily fertilized cropland livestock feedlots, or earthen manure storage structures	X	X	Infant methemoglobinemia (blue baby)	Periodically have drinking water tested yearly for nitrates. Use bottled water for infant's formula or breastfeed during first year of infant's life
	Wells improperly constructed, maintained, or located		X	Biologic conversion to nitrosamines (suspected carcinogens)	Educate and encourage farmers to establish runoff control measures
	Septic tanks, cesspools	X			Install a deep well
	Improperly treated sewage discharged from municipalities or industries	X			Zoning laws to control rural development
Pesticides	Runoff from agricultural fields	X	X	Acute poisonings	Monitor and enforce point source pollutants
	Spillage	X	X	Suspected effects: Carcinogenic	Proper location and construction of wells
	Back-siphoning into well		X	Aplastic anemia, Peripheral neuropathy, Hypertension, Elevated cholesterol	Have wells tested for specific pesticides applied on adjacent fields
Infectious agents	Human Source Septic tanks	X	X	Enteritis	Zoning laws and public health laws to prevent high-density rural development
Salmonella E. coli	Cesspools	X	X	Gastroenteritis	
Campylobacter	Improperly treated sewage	X		Hepatitis	Ensure proper construction of septic systems (permit system)
F. tularensis	Sludge from sewage dumped in landfills		X	Tularemia	
Giardia				Giardiasis	
Entamoeba	Animal Source	X	X	Echinococcosis	Ensure proper location and construction of wells
Viruses	Domestic animals (runoff from feedlots or from fields where manure has been applied)				Ensure proper location and maintenance of feedlots to prevent runoff
Helminths	Wildlife (runoff or defecation directly into streams)	X			
Cryptosporidia					
Helicobacter pylori					

(continued)

Table 7.1. Rural Water Pollutants (*continued*)

Substance	Pollution Source	Water Source Surface	Water Source Ground	Health Effect	Prevention
					Boil or chemically treat water from streams or lakes for drinking when on camping or hiking trips
Methylmercury	Fossil fuel power plants Pulp mills Chloralkali plants	X X		Mercury contamination of fish Nervous system disorders	Prevent discharge into streams and lakes
Mineral salts	Irrigation		X	Mainly aesthetic problems of poor taste Possible diarrhea; complicates control of hypertension, hypocalcemia secondary to pregnancy	Zoning laws to limit irrigation in marginal areas
Fluorine	Added to municipal water supplies Natural geologic formations	NA X	NA X	Low levels; prevents dental caries High levels: mottling of teeth, skeletal fluorosis	Test water if newborn in family, use alternative water source if level is too high
Calcium and magnesium (water hardness)	Natural geologic formations	X	X	Aesthetic (taste) and property damage, scale in pipes, and the like May reduce the risk of cardio-vascular disease People on reduced sodium diets should not drink water from sodium ion-exchange water softeners	Home water softener on hot water source only
Radiation	Natural geologic formations		X	Uncertain, possibly increased risk for bone cancer and lung cancer	Home water softener Drill well to different aquifer (usually shallow wells are radiation free) Use treated surface water

212

Table 7.1. Rural Water Pollutants (*continued*)

Substance	Pollution Source	Water Source — Surface	Water Source — Ground	Health Effect	Prevention
Sulfates	Natural geologic formations		X	Bad taste, diarrhea	Drill well to different aquifer Install special water treatment process (i.e., aeration or chill) Use treated surface water
Particulates	Soil erosion from agricultural cropland or surface mining	X		Direct health effects unknown, but may carry pesticides, microbes, nitrates Signals runoff problem Aesthetic effects: objectionable odor, turbidity, odors	Promote soil conservation practices, such as conservation tillage and terracing
Hazardous chemical wastes	Industrial waste disposal sites that are improperly located, constructed, or maintained Industrial injection wells	X	X X	Variable depending on specific substance	Monitor old dump sites Provide properly located, engineered, and maintained hazardous waste disposal sites Contact local environmental authority if hazardous waste site suspected
Trihalomethane compounds	Chlorination of municipal water supplies	X		Reduced risk for waterborne infections Suspected carcinogens	None indicated at present

Table 7.2. Summary of Rural Air Pollution

Pollutant	Source	Health Effect	Control/Prevention
Odors	Agriculture: livestock production	Rural neighbors may experience somatic symptoms from odors, including respiratory symptoms and nausea; complaints of having to stay indoors	Good management practices and use of many available source control measures Zoning laws to separate agricultural operations from residences or place limits on concentration of animals in a region Promulgate and enforce odor source regulations (where they exist) Nuisance lawsuits
Particles Fugitive dust	Agriculture: tilled fields Surface mining Unpaved roads	Difficult to measure a direct health hazard An additive irritant to other air pollutants, especially in individuals with respiratory or cardiovascular disease	Promote soil conservation practices, such as conservation (minimum) tillage Install and preserve windbreaks Oil or calcium chloride on gravel roads
Other suspended particulates	Agriculture: burning of crop residue Forestry: managed forest burning Rural manufacturing Rural energy production Mining Burning in open dumps	Similar to fugitive dust, but mainly localized in southeastern United States Occurs sporadically during certain times of the year Similar to fugitive dust Similar to fugitive dust	None recommended at present time Enforce guidelines from state and federal environmental offices Eliminate open dumps Public education
Pesticides	Aerial spraying to control insects that damage agricultural crops or forests	No health hazard recognized for area residents	Education of applicators Regulations to assure that proper pesticide is applied in the proper amounts Education of agricultural businesses

Table 7.2. Summary of Rural Air Pollution (*continued*)

Pollutant	Source	Health Effect	Control/Prevention
Anhydrous ammonia	Accidental leakage from storage tanks in towns Leakage from soil fertilizer applications Leakage from refrigeration units Animal and poultry urine and manure degradation	Chemical burns to eyes and skin Chemical pneumonitis and pulmonary edema At low levels, respiratory tract inflammation, synergistic with particulate exposure	Periodic checks for proper functioning of storage tanks and transfer equipment Periodic checks for proper function and maintenance of application equipment Manage source control from livestock and poultry operations
Nitrogen oxides	Rural manufacturing Rural fossil fuel energy plants Rural motor vehicles Drift from urban centers Ensilage processing of forage crops Oxidation of ammonia from livestock and poultry manure	Irritant to eyes and respiratory tract	Support current regulations and their environment
Sulfur oxides	Rural coal-fired energy plants Drift from urban centers Oxidation of sulfur compounds from animal manure degradation	Irritant to eyes and respiratory tract	Support current regulations and their environment
Carbon dioxide	Rural industry Rural fossil fuel energy plants Drift from urban centers	Unknown	Support current regulations and their environment
Hydrocarbons	Rural industry and manufacturing	Unknown	Support current regulations and their environment
Radioactivity	Nuclear energy plants	Unknown	Support current regulations concerning nuclear plants
Carbon monoxide	Automobiles and trucks Drift from urban areas	Unknown at levels found in the rural environment Possibly enhances arteriosclerosis Possibly decreased birth weight and retarded development in newborn	See below

(*continued*)

215

Table 7.2. Summary of Rural Air Pollution (*continued*)

Pollutant	Source	Health Effect	Control/Prevention
Ozone	Photochemical oxidation of combustion products	Irritation to eyes and respiratory tract Prolonged exposure may lead to chronic bronchitis or obstructive lung disease People with concurrent heart or lung disease are at greatest risk	Reduce reliance on automobiles for transportation Enforce regulations to decrease emissions from internal combustion engines
Acid rain	Conversion of nitrous and sulfur oxides to sulfuric and nitric acids, which are washed to the ground by precipitation	Direct human health effects not fully evaluated Quality of surface water decreased	Enforce regulations for emissions from industries, cars and trucks

fields, particulates from burning crop residues, pesticides from aerial application operations, and anhydrous ammonia from accidents at storage sites (Ford 1978). Polluting gases that are emitted from livestock operations and fertilizers applied to cropland, include nitrogen containing gases, carbon dioxide, hydrogen sulfide, and methane. Local or community health effects of these operations are subtle and difficult to quantify, but are of great concern to residents living in areas of high livestock density. These local concerns, as well as regional and global consequences, will be discussed later in this chapter in the section title "Air Quality Concerns from CAFOs."

Fate of Emissions and Atmospheric Effects

There are local (neighbor) health concerns relative to hydrogen sulfide and ammonia (Thu and others 1997). There are also regional and global concerns relative to ammonia and methane (Donham 2000). Western Europe, particularly the Netherlands, has taken the lead in research in the global impact from ammonia, methane, and carbon dioxide emitted from livestock operations. Ammonia volatilizes into the atmosphere from manure stored in lagoons and manure applied to land. Once in the atmosphere, it can be transported downwind to be deposited into surface waters and on soils as ammonia or as other oxidized nitrogen compounds (Hutchinson and Viets 1969). About 50% of the total emitted ammonia comes from land application, and about 40% from storage of liquid manure (e.g., lagoons). Luebs (1973) used acid traps to quantify atmospheric ammonia in the vicinity of a concentrated dairy area in California. He found that 8.5 kg/hectare per week of ammonia was deposited within 1 km, and .25 kg/hectare/week 8 km downwind of the facility.

In addition to the nitrogen from water pollution, deposition of excess ammonia nitrogen from the air causes increased risk for eutrophication of ponds, lakes, and streams in the region. It can also bring excess nutrients to natural areas, causing overgrowth by undesirable plant species and nitrate leaching through soil (Voorburg 1991). Ammonia in the atmosphere may also react with acids already in the air, such as hydrochloric, sulfuric, and nitrous acids. This results in ammonium aerosols, which are then transported and can return to earth with precipitation. Apsimon and Kruse-Plass (1991) have reported that these compounds may be more strongly acidifying to soils and water than strong acids. Denitrification and nitrification of ammonia produces nitrous oxide (N_2O), which contributes to greenhouse gases and ozone depletion. From 1–2% of total N loss from land-applied manure is lost as N_2O (Whalen and others 2000). Denitrification also produces nitrous oxide (NO), which can contribute to acid rain (Follett and Delgado 2002; Petersen 1999).

Methane is a greenhouse gas, contributing to global warming (Sharpe and others 2002). Methane is 21 times stronger on a molecular basis, as a greenhouse gas, than carbon dioxide. It is 58 times stronger on a mass basis (Communities 2001) because of its relative greater ability to absorb long wavelengths of light energy. Total world yearly methane emission is 354 million metric tons. The U.S. emits 27 million metric tons, and livestock (wastes and ruminant eructation) account for 7.4% of this total (Sharpe and others 2002). Intensive confinement systems with anaerobic storage of manure along with large surface area lagoons increase the amount of methane emissions.

Specific Sources of Air Contaminants in Rural Areas

PARTICULATES
Fugitive dust from tilled fields (Miller 1975) and emitted particles from burning of crop residues or forest land are two of the major sources of particulate hazards in the air of agricultural communities. For example, fugitive dust in the San Joaquin Valley of California often results in violations of air particulate standards of 50 micrograms per cubic meter and the EPA ambient air quality standards of 150 micrograms particulates per cubic meter (Clausnitzer and Singer 1996). Specific health hazards of these two pollutants are not certain, but when added to other respiratory insults, they may cause overt symptoms in predisposed individuals (e.g., asthmatics). Mold board plowing tillage methods expose soils to drying winds and pulls hundreds of tons of topsoil into the atmosphere. These particulates may have regional as well as global affects. Minimum tillage practices and windbreaks decrease wind erosion.

Residues of cropland and forest land are burned as a regular agricultural practice in some areas of the world, creating air contamination with suspended particles. In Georgia (U.S.) for example, nearly 600,000 acres of agricultural land and over 500,000 acres of forest land are burned off yearly (Ward and Elliot 1976), producing more that 26,000 metric tons of suspended particulates. Brazil and other countries in South America particularly are bringing thousands of acres annually into production by burning indigenous forests. This practice produces very significant particulate pollution in a wide area of South America.

Other fine air particulates (PM 2.5) may form from ammonia reacting with oxides of nitrogen and sulfur to form fine particles (PM 2.5), which can be deposited into the deep portions of the lungs to increase morbidity and mortality especially when combined with co-morbidities (Anderson and others 2003).

ODORS

Odors may be a concern in many different agricultural enterprises. The majority of the concern in recent years has been in community members who live in the vicinity of concentrated animal feeding operations (CAFOs). Odors and odorants are not only a nuisance concern, but they also cause unpleasant somatic physiologic responses (Schiffman and others 1995). Odors and their health consequences are discussed in detail later in this chapter.

MICROBES

High levels of microbes may be generated into the local air environment from a variety of crop and livestock operations. Air downwind from open feed lots or confined animal facilities contains many times higher microbial content than air in comparison agricultural operations. Fugitive dust from open cattle or dairy feed lots, or exhausted air from confined poultry or swine operations contains much higher levels of microbes than air in comparison communities. Furthermore, areas downwind from fields that are irrigated (especially with spray booms) with effluent from animal manure storage or anaerobic treatment lagoons contain microbes at a factor of 10^3–10^5 times microbe concentrations in comparison areas. Air sampling indicates that the vast majority of these organisms in the air are not viable. It is difficult to find cases where an infection has

been acquired from aerosolized organisms in the vicinity of agricultural operations. Therefore, infections may not be as important a risk as exposure to inflammatory substances such as endotoxins and glucans that are a component of the microbes in the dust.

ENDOTOXIN AND GLUCANS

Many microbes contain toxins (e.g., endotoxin, glucans) which are potent inflammatory substances. These substances may cause asthma-like symptoms, bronchitis, mucus membrane irritation, and organic dust toxic syndrome (a systemic influenza-like illness) (Rylander 1994). Thu and others (1997) have shown that residents in the vicinity of CAFOs experience respiratory symptoms similar to workers inside these facilities (see Chapter 3). However, there is some evidence that exposure to endotoxins, particularly early in life, may protect from atopic asthma at a later time in life (see Chapters 2 and 3) (Roy and Thorne 2003).

ANTIBIOTICS

Antibiotics build up in the dust inside livestock buildings (Hamscher and others 2003). Additionally, antibiotics are found in the exhaust air of swine CAFOs (Zahn and others 2001). Streptomycin and other antibiotics are sprayed on fruit trees to prevent or treat certain microbial infections (Khachatourians 1998). Obviously, there is ample opportunity for antibiotics to contaminate the air in the vicinity of certain livestock and orchard operations. However, this situation has had little study to determine the extent of human exposure and health risks (see Chapter 12).

AIR POLLUTION FROM AGRICULTURAL CHEMICALS

During adverse weather conditions or when spraying residential neighborhoods, herbicides and insecticides may contaminate the air. Pesticides attach to soil particulates or volatilize into the atmosphere, where they are transported regionally and globally. Pesticides may be found in many places around the world in air and rain, even in remote areas of the world (Van Duke 1999). From 1–50% of applied pesticides may enter the atmosphere (VandenBerg and others 1999). Air pollution from agricultural pesticides presents nominal acute health hazards. However, certain serious occupational hazards may exist for the applicators and agricultural workers who mix and load chemicals and maintain application equipment (see Chapter 6) (International Labour Organization 1979).

Agricultural service and supply businesses commonly store large quantities of anhydrous ammonia fertilizer. Occasionally large clouds of this very irritating chemical escape when a tank or valve leaks (Figure 7.5). Area residents and workers may receive severe burns when the chemical contacts their eyes and skin. If they inhale ammonia, they may die of acute chemical pneumonitis and rapidly developing pulmonary edema (see Chapter 3). Rural physicians and emergency medical personnel should be aware of the first aid treatment for anhydrous ammonia exposure: flushing affected areas with copious amounts of water and administering oxygen if necessary (Helmers and others 1971; U.S. Department of Health, Education, and Welfare 1979). In the case of large ammonia leaks, the cloud may be washed from the air by spraying it with water from a fire control hose.

Rural Industry

Expansion of rural manufacturing, mining, and/or processing has increased industrial contributions to rural air pollution. Moreover, the increasing demand for energy has resulted in construction of new coal-fired and nuclear plants in rural areas, where they are less costly to build and where air pollution control standards are more easily met. As a result, air pollution from particles, carbon dioxide, nitrogen oxides, sulfur oxides, hydrocarbons, mercury, and radioactivity is potentially increasing (Ford 1978). Like other individual rural pollutants, these have not been demonstrated to be the direct cause of any particular health problems but are assumed to have an additive effect with other pollutants.

Since the 1980s, concern has been expressed by environmental scientists about the potential effects of high-voltage electrical fields surrounding high-power transmission lines. Currently, between a quarter-million and a half-million miles of high-voltage lines (765 kV or greater) cross the rural countryside between new generating plants and urban centers (Electrical Power Research Institute 1981). Environmental scientists have found ecologic relationships between high-voltage electrical fields and increased rates of in-

Figure 7.5. Anhydrous ammonia is an important fertilizer in many areas of the world. Excess nitrogen can run off with soil erosion, escape into the air, and be a source of chemical damage to eyes and respiratory tract of applicators. (Note that this farmer is operating a tractor without a rollover protective structure. Injury from a possible overturn of this tractor is a much greater immediate risk to life than the anhydrous ammonia he is applying.)
(*Source:* Medical Practice in Rural Communities, Chapter 3, Figure 3.4, page 46, book authors: Cornelia F. Mutel, Kelley J. Donham, Rural Health and Agricultural Medicine Training Program, Department of Preventive Medicine and Environmental Health, College of Medicine, The University of Iowa, Iowa City, Iowa 52242, U.S.A., © Springer-Verlag New York Inc., 1983. Image reprinted with kind permission of Springer Science and Business Media.)

fant mortality, fetal malformation, and abnormal behavior in test animals (Electrical Power Research Institute 1981). This information is not strong or consistent; therefore, implications for human health are uncertain.

Dust from surface mining processes and from roads heavily traveled by trucks transporting ore create concern of fugitive dusts in some rural areas (Ford 1978). However, direct acute health problems of dust from mining are difficult to document.

Control

Gaseous nitrogen and methane emissions are perhaps the two most important substances to control, relative to regional and global effects.

Control of air pollution from nitrogenous products in agricultural operations involves regulations and incentives for producers not to apply more nitrogen to the soil than the crops can utilize (do not overburden the natural nitrogen cycle). This requires a combination of best management practices (soil testing and calculation of plant needs) and application of the appropriate amounts to the appropriate soils (precision farming). Regulations, monitoring, and enforcement are also needed to assure best management practices. Furthermore, there must be assurance that soil preparation and nitrogen application methods are conducted in a manner that prevents escape of nitrogen products into the air. Methane capture from anaerobic digestion of manure and utilizing this as an energy source has an important potential for control of this greenhouse gas. However, cost of this practice has prevented wide-scale application.

Summary of Agricultural Air Pollution

Relative to water pollution, air pollution is secondary in terms of direct human health consequences of agricultural pollution. Urban and rural industries contribute to the air quality degradation in rural areas in addition to agriculture. There are environmental concerns of health consequences of air quality from a local or community concern, as well as regional and global concern. Nitrogenous gases are of greatest concern, both regionally and globally, followed by methane. Particulate matter and agricultural chemicals are of less concern. Direct acute health hazards due to these exposures are difficult to define; however, vulnerable populations such as children, the elderly, and those with co-morbidities are of greatest concern.

Solid Waste

The primary problem of solid wastes in rural areas is that it contributes to contamination of groundwater. Secondary problems include contamination of surface water; attraction of filth-laden insects and rodents; and pollution of air by odors and vapors, and by particulates from open burning (U.S. Environmental Protection Agency 1980; Harris and others 1982). Socioeconomic and demographic trends have increased the problems of rural solid wastes during the past two decades. The more humans intensely use the land, the greater the quantity of solid waste produced.

Figure 7.6. Empty agricultural chemical containers should be rinsed three times and returned to the retailer or otherwise disposed of in an approved manner. These empty containers could be a source of poisoning if residual contents leaked out or if these containers were used for another purpose, exposing humans or animals. (*Source:* Medical Practice in Rural Communities, Chapter 3, Figure 3.6, page 56, book authors: Cornelia F. Mutel, Kelley J. Donham, Rural Health and Agricultural Medicine Training Program, Department of Preventive Medicine and Environmental Health, College of Medicine, The University of Iowa, Iowa City, Iowa 52242, U.S.A., © Springer-Verlag New York Inc., 1983. Image reprinted with kind permission of Springer Science and Business Media.)

Increased waste results from farming, mining, and manufacturing in rural areas; from residential areas encroaching on the rural lands; and from consumer products used by those who live there. Even hazardous wastes are transported from urban areas to rural landfills (Ford 1978; U.S. Environmental Protection Agency 1976).

Safe disposal of chemical containers poses a solid waste problem. They may be used unsafely for unintended purposes (such as water containers) or thrown into an open ditch in an outlying area of the farm (see Figures 7.6 and 7.7). Direct contact with or consumption of residual chemicals may cause human or animal illness directly, or these chemicals may contaminate surface or groundwater. Used pesticide containers should be rinsed three times. The empty containers should then be returned to the seller or disposed of in a sanitary landfill. The rinse water should be ap-

Figure 7.7.   Empty agricultural chemical containers disposed of improperly could lead to a source of water or soil pollution.
(*Source:* Medical Practice in Rural Communities, Chapter 3, Figure 3.8, page 58, book authors: Cornelia F. Mutel, Kelley J. Donham, Rural Health and Agricultural Medicine Training Program, Department of Preventive Medicine and Environmental Health, College of Medicine, The University of Iowa, Iowa City, Iowa 52242, U.S.A., © Springer-Verlag New York Inc., 1983. Image reprinted with kind permission of Springer Science and Business Media.)

plied to the soil or crop as specified for the intended use of the specific chemical.

Rural manufacturing, mining, and energy production facilities increase the need for special facilities to store potentially hazardous wastes. Urban manufacturing and energy production also create solid waste problems in rural areas. Less populated rural areas are chosen as disposal sites for hazardous chemical and radioactive wastes. Often, local officials and area residents may not be aware of potential hazards, allowing construction of improperly located or constructed hazardous waste disposal facilitates (Harris and others 1982). One survey found that 34 of 50 industrial waste sites caused local groundwater contamination (Agency UEP 1980). As more people have moved to some rural areas, and as farm dwellers have become less self-sufficient, use of consumer items and resulting packaging wastes have increased.

Generalizations are difficult to make about the types of human health problems caused by solid waste pollution because these problems depend on specific circumstances and the type and quantity of substance to which humans are exposed.

Most rural areas need more and better waste storage, collection, and disposal systems. Some rural solid waste systems have developed, such as regional collection boxes or weekly collection at mailboxes. These systems help deter rural residents from maintaining open dumps on their own acreage. New mechanisms are needed to ensure proper disposal of pesticide containers. The use of returnable containers and large bulk containers are two promising practices (U.S. Environmental Protection Agency 1976). Zoning laws could assure an adequate sewage treatment capacity in rural areas that are currently in place. New techniques could provide safe methods for recycling animal wastes, and regulations could govern proper disposal of industrial radioactive wastes.

ANIMAL FEEDING OPERATIONS
Introduction

The proliferation of animal feeding operations (AFOs) has been a feature of the increased intensity, concentration, specialization, and consolidation that has been occurring in agriculture in industrialized countries over the past three decades. The USEPA has a specific definition of AFOs that includes the confinement of animals for at least 45 days in a 12-month period, and in a structure or lot where vegetation or crops will not grow. The term *Confined Animal Feeding Operation (CAFO)* is used to describe AFOs over a specified size that are a risk as a point source for water pollution, and therefore require regulation (U.S. Environmental Protection Agency 2002). In regard to water, air, and solid waste pollution, CAFOs have many of the same concerns as described above for other agricultural operations. However, there are special concerns regarding CAFOs because of the sheer size and the concentration of animals; feed; manure (usually handled in liquid form); dead animals; flies; and associated gases, particulates, odors and odorants, and infectious diseases all concentrated on a small land area. There are concerns that the manure cannot be recycled without pollution in such a small area, and that local and regional air and water quality suffers.

Public concerns relative to adverse consequences of livestock production have been increasingly voiced since the late 1960s (Thu and Durrenberger 1998). Numerous regional, national and international conferences have been held on the subject since 1994 (Merchant 2002; Thu 1995). An in-depth review of the literature has been recently published (Donham 2000).

This section of this chapter concentrates on occupational and community health consequences of large-scale livestock operations. Physical health as well as social and economic concerns of individuals and communities are considered. A broad definition of health is used here as defined by the World Health Organization (WHO). WHO states that health is ". . . a state of complete physical, mental, and social well-being and not merely the absence of disease or infirmity" (WHO 1948).

One reason large-scale livestock production has raised concern is that it has separated from family farming and has developed like other industries in management, structure, and concentration. The magnitude of the problem is highlighted by the following facts relative to the U.S.:

• Nationwide, 130 times more animal waste is produced yearly than human waste.
• Animal waste is not treated as is human waste before it is returned to the environment (Esteban 1998).

In the following pages the scientific literature is reviewed relative to environmental impacts of air and water emissions on community.

ENVIRONMENTAL HEALTH ISSUES

Air Quality Concerns from CAFOs

Merkel and others (1969) published the first assessment of the content of gases from swine manure. Nearly 200 compounds have been measured as emitted from animal manure (O'Neill and Phillips 1992). The health risk of most of these substances is not known. Some of these substances have a very low odor threshold (e.g., 1 part per billion). Ammonia and hydrogen sulfide are the most important emittants from livestock waste regarding direct human health risks. Methane and carbon dioxide are important greenhouse gases.

The primary sources of gaseous compounds originate from the degradation of feces and urine applied to land and from animal wastes stored in liquid or solid phase. Ammonia is a by-product of almost any treatment method of animal waste. Other fixed compounds and trace compounds come primarily from the anaerobic decomposition of manure in liquid storage systems. The emitted compounds can be grouped into the following classes of chemicals: mercaptans, sulfides, disulfides, amines, organic acids, phenols, alcohols, ketones, indole, skatole, carbonyls, esters, and nitrogen heterocycles.

In addition to gases, often overlooked are the particulate substances that are emitted from livestock feeding operations. There is a large quantity of organic dust generated from feed sources and the pigs (hair, dander, and dried feces). This dust contains many bioactive substances, including endotoxin and glucans (two very important inflammatory substances) (Donham and others 1986). Also, there is a bioaerosol component of this dust. Many gram-negative and gram-positive bacteria, fungi, and molds have been identified (Thorne and others 1992). Some of these organisms also grow within confinement buildings, contributing to concentrations of organisms that are also found in the air outside the building (Kiekhaefer 1995). The vast majority of organisms identified in the air are saprophytic and very few pathogens are identified. They are combined with dust that becomes a part of the total aerosolized particulates. The size of the particulates emitted is relatively small. About 50% are less than 10 microns, which means they are inhalable.

Odors and Odorants

Most of the public concern on CAFOs has been about odor. An odor is an unpleasant sensation in the presence of an odorous substance (odorant). The odorant may or may not have additional harmful toxic effects. Ritter and Chirnside (1984) identified the classes of compounds from animal manure which are odorants (previously mentioned). Ammonia and hydrogen sulfide among the fixed gases are also odorants.

Riskowski (1991) described an odor phenomenon associated with livestock (mixed gas) environments where ammonia and hydrogen sulfide odors are detectable at much lower odor threshold concentrations levels than previously published.

It is likely that in this mixed environment, other less concentrated chemicals and particles interact to enhance the detection of these odorants.

Researchers have looked at the fixed gases (e.g., ammonia and hydrogen sulfide) as potential surrogates for emissions and odors. However, the results of several researchers have shown that there is a poor correlation between ammonia and hydrogen sulfide and odor strength (Payne 1994; Smith and Kelly 1996).

The particulate emissions interact synergistically with gases to cause important occupational health hazards. Although a good deal is known about health effects of dust and gases on workers inside buildings, little is known about health effects of living in the neighborhood of CAFOs. Goodrich and others (1975) have shown (relative to background) a very high level of viable organisms downwind of manure sprinklers, as well as inside beef and turkey facilities. Pickrell (1991) has shown swine barn environments to have significantly higher particle and microbe concentrations compared to other livestock environments, and 103–106 times higher inside swine buildings compared to outside (Kiekhaefer 1995). Very little is known about hazardous concentrations of odorants in outdoor air around CAFOs. We do know there are serious worker health problems caused by H_2S and ammonia in the interior. However, it is difficult to infer health risks outside these buildings based on interior studies. Available data (Reynolds 1996) suggests ammonia and hydrogen sulfide are on the order of 10^3 times higher inside buildings compared to outside.

Concerns about odors from livestock facilities can be considered a nuisance, which is often how courts treat them. However, there is growing evidence that odors may cause physical illness. Overcash and others (1983) indicated odors may cause nausea; vomiting; headache; shallow breathing; coughing; sleep disorders; upset stomach; appetite depression; irritated eyes, nose, and throat; and mood disturbances, including agitation, annoyance, and depression. Ackerman (1990) reported odors can result in strong emotional and physical responses, particularly after repeated exposures. Odors can result in a mixture of emotional and physical responses. Schiffman and others (1995) studied the profile of mood states (POMS) of 44 persons living near large animal facilities. Compared to controls she found

that people living near the facilities were more angry, confused, tense, depressed, and fatigued. In order to determine acceptable odors relative to distance from the source, Walsh and others (1995) surveyed persons living in a 5 km area surrounding a large cattle feedlot. They measured odors according to an odor panel and found acceptable odor levels within the 5 km radius. Results of more recent research suggest that the physical health problems may also be caused by long-term inflammation, secondary to inhaled dust and gases (Donham and others 1995; Pickrell 1991).

COMMUNITY HEALTH ISSUES

Physical Health

When considering the health hazards of residents living in the vicinity of CAFOs, one has to look beyond direct toxic explanations, especially when considering air emissions. The reason is that in many environmental cases (e.g., Three Mile Island, Love Canal, etc.) where there are community health complaints, they often cannot be explained by measured concentrations of hazardous substances and standard toxicological mechanisms (Shusterman 1992). For example, Jacobson and others (Davidson and others 1987) reported H_2S levels in the vicinity of CAFOs, well under the threshold limit value (TLV) for occupational health set at 10 ppm. In a study by Reynolds and others (1997) levels of ammonia in the vicinity of swine CAFOs were generally less than 1 ppm. Concentration of endotoxin and dust were near the lower limits of detection of the instrumentation used (which was around 10 endotoxin units per cubic meter of air, and dust less than 0.5 mg/cubic meter). However, these residents had a high prevalence of respiratory symptoms compared to controls.

Possible reasons for this observation include the fact that residents live in the area more than 8 hours per day. Also, there may be vulnerable populations who react at much lower levels than occupational limits. For example, several states have limits for hydrogen sulfide at 20–50 ppb, three orders of magnitude below the occupational exposure limit and most federal agencies' limits for environmental exposure.

Kilburn (1997) reported on neurobehavioral effects of hydrogen sulfide gas. This small study reported neurological deficits in 16 exposed persons, including decreases in balance, reaction times, visual field performance, color discrimination, hearing, cognition, motor speed, verbal recall, and mood states. H_2S is a toxin with several effects, including tissue irritation and poisoning of cellular respiration mechanisms with a predilection for brain cells. However, health effect of chronic low level H_2S exposure remains uncertain.

Results from studies of physical and mental health concerns of residents near CAFOS have been reported in only three studies to date (1997). These were controlled studies of self-reported symptoms, and no attempts were made to document objective correlates of health impairment. Thu and others (1997) (Schiffman and others 1995; Wing and Wolf 2000) reported respiratory symptoms (significant relative to comparison populations controls) almost identical in type and pattern to workers in CAFOs. Schiffman and others (1995) (mentionned previously) reported excessive mood alterations in CAFOs neighbors. There are numerous instances of similar studies in other environmental settings, including community concerns around paperpulp mills, hazardous waste sites, refineries, and solid waste disposal sites (Shusterman 1992). Although most of these studies have not documented objective findings of toxic physical insult to humans, a few studies have reported subtle findings such as increased concentrations of urinary catecholamines. Additionally, most of these studies have not shown evidence of known toxic levels of substances in the environment.

Extra-Toxic Mechanisms

In addition to the literature regarding direct toxic effects, there is also literature on extra-toxic effects of a low level of emissions (Shusterman 1992). This literature has focused on trying to explain symptoms of community members who may be exposed to waste sites, sewage treatment plants, and other large population-based community exposures. Medical research and regulatory agencies have difficulty dealing with these situations, as they are not clear-cut. Clear-cut situations would include an objective finding of an adverse health effect, measured toxic substances at known toxic concentration, and an obvious dose/response relationship. These community ex-

posures are much more complex, as they mix physical, mental, emotional, and social environment. "Genetic memory" and other very basic limbic-level self-preservation mechanisms may be involved. The following paragraphs will review some of the literature that helps explain the reality of adverse health symptoms in community environmental concerns where there is the absence of objective toxicological data.

The Somatasization of Adverse Odors

There are two cranial nerves involved in innervating the nasal mucosa: the first cranial nerve (olfactory nerve) and the fifth cranial nerve (trigeminus) (Shusterman 1992). The olfactory nerve is primarily responsible for odor detection. The trigeminus nerve has many branches that penetrate the oral mucosa and provides additional information to the brain on odor sensation, such as irritation or pungency, which trigger protective responses, including decreased respiratory rate, rhinitis, tearing, cough, gag reflex, and bronchoconstriction. These are all warning indicators that something associated with the odor may be harmful and our genetic-based "instinctive protective" mechanisms are telling us to make physiologic changes to meet the impending insult or to get out of the area (Shusterman 1992; Shusterman and others 1988). Therefore, it makes sense that odors can result in symptoms of mucosal irritation, nausea, and feelings of "disease" (Shusterman 1992; Shusterman and others 1988).

Complexed with these physiological responses to low-level irritants and odors, there are behavioral interactions that may explain health symptoms of illness associated with odors. There are five possible mechanisms for extra-toxic odor-related symptoms (Shusterman 1992).

INNATE ODOR AVERSIONS

As a basic protective mechanism, our body wants to avoid certain odors that may signify potential harm. For example, odors in "putrefaction" gas—e.g., H_2S, mercaptans, and other sulfur-containing chemicals—are common substances that stimulate physiologic effects at lower than toxic levels. These gases may be associated with spoiled food, but are also associated with animal manure, as are many of the odors associated with these innate odor responses.

PHEROMONAL PHENOMENA

Pheromones stimulate physiologic responses, especially around sexual reproduction. These are most overt for insects, but many mammals, including humans, also respond to them. Some odors might destroy normal positive pheromone responses resulting in impaired sexual function, as reported by Schiffman and others (1995) (Shusterman 1992; Shusterman and others 1988) for people living in the vicinity of CAFOs.

EXACERBATION OF UNDERLYING CONDITIONS

Previous research has shown that workers with underlying conditions (asthma, atopy, bronchitis, heart conditions) are more susceptible to the CAFO environment than others (Schiffman and others 1995). Furthermore, it is now evident that individuals genetically differ in their susceptibility to endotoxins. Research by Meggs and others (1996) (Rylander 2004) (previously discussed) also lends strength to the theory that underlying conditions may amplify exposures.

AVERSIVE CONDITIONING

Some persons previously exposed to high levels of gases causing toxic effects may respond physiologically to less than toxic levels of this substance in future exposures. This author has observed this in several cases of CAFO workers who experience symptoms and anxiety when smelling CAFO odors following a severe CAFO gas exposure episode. This conditioned stimulus is probably an innate protective mechanism. This can also happen with lower level exposures over a long period of time (acquired odor intolerance) (Meggs 1997; Meggs and others 1996).

STRESS-INDUCED ILLNESS

Odor-related stress-induced illness has been discussed as a component of "environmental stress syndrome." This phenomenon has been seen following disaster sites such as Three Mile Island (Shusterman and others 1988). Studies have shown there are increased urinary chatecholamines in the affected individuals. They also have a feeling of depression, helplessness, and a high degree of environmental worry, which is exacerbated by detection of the offending odor. Community members may be excessively worried that their property values are falling because of the odor source in their neighborhood. Odors can act as a cue for these individuals, stimulating adverse physiologic risks relative to an associated exposure. Long-term stress can be associated with muscle tension and headaches, coronary artery disease, and peptic ulcers.

Summary of Extra-Toxic Mechanisms

In studies of physical health complaints in communities around CAFOs, it would be expected that objective findings of toxicity would be difficult to obtain. However, that does not, and should not, discount the fact that people experience valid symptoms. The reasons have to do with complex interactions of the brain and somatic systems. First of all, odors may initiate somatic symptoms based on enervations of the trigeminus nerve. Furthermore, odors may initiate physiologic activity as response to primordially acquired aversions to toxic substances. These responses may be modulated by people as they generally worry about environmental threats and the frequency with which odors are experienced. Furthermore, these conditions may be exacerbated by previous toxic exposures to the substances in question, creating a learned response of avoidance even when very low exposures are present. Further exacerbation may occur when combined with a feeling that a person has no control over the situation, with fears about declining property values, and with the resulting "environmental stress." If an individual has an underlying health condition, such as asthma, further complications may be present.

REFERENCES

Ackerman D. 1990. Smell. A Natural History of the Senses. London: Chapmans. p 5–59.

Agency UEP. 1980. Solid Waste Facts. Report nr SW-694. p 1–8.

Anderson N, Strader R, Davidson C. 2003. Airborne reduced nitrogen: Ammonia emissions from agriculture and other sources. [Review]. Environmental Int 29(2–3):277–286.

Anonymous. 1996. Spontaneous abortions possibly related to ingestion of nitrate-contaminated well water—LaGrange County, Indiana, 1991–1994. MMWR Morbidity Mortality Weekly Report 45(26):569–72.

ApSimon H, Kruse-Plass M. 1991. The role of ammonia as an atmospheric pollutant. In: Nielsen V, Voorburg JH, L'Hermite P, editor. Odour and Ammonia Emissions from Livestock Farming. Barking, England: Elsevier Science Publishers Ltd. p 17–20.

Archer D, Shogren J. 2001. Risk-indexed herbicide taxes to reduce ground and surface water pollution: An integrated ecological economics evaluation. Ecological Economics 38:227–250.

Avery A. 1999. Infantile methemoglobinemia: Reexamining the role of drinking water nitrates.[see comment]. Environmental Health Perspectives 107(7): 583–6.

Bean J, Isaacson P, Hahne R. 1982. Drinking water and cancer incidence in Iowa. II. Radioactivity in drinking water. American Journal of Epidemiology 116:924–932.

Bennish M. 1999. Animals, Humans, and Antibiotics: Implications of the Veterinary Use of Antibiotics on Human Health. Advances in Pediatrics Infections 14:269–290.

Berkowitz J, Kraft D, Finstein M. 1974. Persistence of salmonellae in poultry excreta. Journal of Environmental Quality 3:158–161.

Bitton G, Harvey R. 1992. Transport of Pathogens through soils and aquifers. In: Mitchell R, editor. Environmental Microbiology. New York: Wiley-Liss Inc. p 103–124.

Boxall A, Fogg L, Blackwell P, Kay P, Pemberton E, Croxford A. 2004. Veterinary medicines in the environment. Reviews of Environmental Contamination and Toxicology 180:1–91.

Burmeister L. 1981. Cancer mortality in Iowa farmers, 1971–1978. Journal of the National Cancer Institute 66:461–464.

Campagnolo E, Johnson K, Karpati A, Rubin C, Kolpin D, Meyer M, Esteban J, Currier R, Smith K, Thu K, and others. 2002. Antimicrobial residues in animal waste and water resources proximal to large-scale swine and poultry feeding operations. Science Total Environment 299(1–3):89–95.

Carpenter S, Caraco N, Correll D, Howarth R, Sharpley A, Smith V. 1998. Nonpoint pollution of surface waters with phosphorus and nitrogen. Ecological Applications 8(3):559–568.

Centner T. 2000. Animal feeding operations: Encouraging sustainable nutrient usage rather than restraining and proscribing activities. Land Use Policy 17(3):233–240.

Chanlett E. 1979. Environmental Protection. New York: McGraw-Hill. 33, 81–85 p.

Clausnitzer H, Singer M. 1996. Respirable-dust production from agricultural operations in the Sacramento Valley, California. Journal of Environmental Quality 25:877–884.

Communities CotE. 2001. Third Communication from the European Community under the UN Framework Convention on Climate Change. Brussels. Report nr SEC (2001) 2053.

Craun G, McCabe L. 1973. Review of the causes of waterborne disease outbreaks. Journal of the American Waterworks Association 65:74–83.

D'Ercole A, Arthur R, Cain J. 1976. Insecticide exposure of mothers and newborns in a rural agricultural area. Pediatrics 57:869–874.

Davidson L, Fleming R, Baum A. 1987. Chronic stress, catecholamines, and sleep disturbance at Three Mile Island. Journal of Human Stress 13:75–83.

Delfino J. 1977. Contamination of potable groundwater supplies in rural areas. In: Pojasek R, editor. Drinking Water Quality Enhancement Through Source Protection. Ann Arbor, Michigan: Ann Arbor Scientific Publications.

Donham K. 2000. The concentration of swine production. Effects on swine health, productivity, human health, and the environment. Veterinary Clinics of North America: Food Animal Practice 16(3): 559–597.

Donham K, Berg J, Will L. 1980. The effects of long-term ingestion of asbestos on the colon of F344 rats. Cancer 45:1073–1084.

Donham K, Reynolds S, Whitten P, Merchant J, Burmeister L, Popendorf W. 1995. Respiratory dysfunction in swine production facility workers: Dose-response relationships of environmental exposures and pulmonary function. American Journal of Industrial Medicine 27(3):405–418.

Donham K, Scallon L, Popendorf W. 1986. Characterization of dusts collected from swine confinement buildings. American Industrial Hygiene Association Journal 47:404–410.

Donham K, Yeggy J, Dauge R. 1988. Production rates of toxic gases from liquid manure: Health implications for workers and animals in swine buildings. Biological Wastes 24:161–173.

Donham KJ, Dauge R. 1985. Chemical and physical parameters of liquid manure from swine confinement facilities: Health implications for workers, swine, and the environment. Agricultural Wastes 14:97–113.

Electrical Power Research Institute. 1981. Electrical field research continues. Electrical Power Research Institute May:34–35.

Entry J, Sojka R, Watwood M, Ross C. 2002. Polyacrylamide preparations for protection of water quality threatened by agricultural runoff contaminants. Environmental Pollution 120(2):191–200.

Esteban E. 1998. The Confinement Animal Feeding Operation Workshop, June 23–24, Washington DC.

Fayer R. 1998. Sources of environmental contamination with *Cryptosporidium:* The role of agriculture, November 4–5, St. Louis, MO.

Follett R, Delgado J. 2002. Nitrogen fate and transport in agricultural systems. Journal of Soil and Water Conservation 57(6):402–408.

Fone D, Barker R. 1994. Associations between human and farm animal infections with *Salmonella typhimurium* DT 104 in Herefordshire.

Ford T. 1978. Contemporary rural America: Persistence and change. In: Ford T, editor. Rural USA: Persistence and Change. Ames, IA: Iowa State University Press. p 4–5.

Fuortes L, Nettleman M. 1994. Leptospirosis: A consequence of the Iowa flood. Iowa Medicine 84: 449–450.

Glanville T. 1981. Water Quality for Home and Farm. Cooperative Extension Service Pm 987. Ames, IA: Iowa State University.

Goodrich P. 1975. Airborne health hazards generated while treating and disposing waste, St Joseph, MI, ASAE Publication PROC-275. p 7–10.

Ham J. 2002. Seepage losses from animal waste lagoons: A summary of a four-year investigation in Kansas. Transactions of the ASAE 45(4):983–992.

Hamilton P, Helsel D. 1995. Effects of agriculture on ground-water quality in five regions of the United States. Ground Water 33(2):217–226.

Hamscher G, Pawelzick H, Sczesny S, Nau H, Hartung J. 2003. Antibiotics in dust originating from a pig-fattening farm: A new source of health hazard for farmers? Environmental Health Perspectives 11(13).

Harris G, Garlock C, LeSeur L. 1982. Groundwater pollution from industrial waste disposal. A case study. Journal of Environmental Health 44:1073–1084.

Helmers S, Top F, Knapp L. 1971. Ammonia injuries in agriculture. Journal of the Iowa Medical Society May:271–280.

Howarth R, Boyer W, Pabich J, Galloway N. 2002. Nitrogen use in the United States from 1961–2000 and potential future trends. Ambio 31(2):88–96.

Hutchinson G, Viets F. 1969. Nitrogen enrichment of surface water by absorption of ammonia volatilized from cattle feedlots. Science 166:514–515.

International Labour Organization. 1979. Guide to Health and Hygiene in Agricultural Work. Geneva: International Labour Organization. p 94–147.

Jacobson L. 1997. No correlation between hydrogen sulfide and odor, Nov-Dec, 1997, Swine Center, University of Minnesota, St. Paul, MN.

Jawson S, Wright R, Smith L. 1998. U.S. Department of Agriculture's National Program on Manure and Byproduct Utilization, November 4–5, 1998, St. Louis MO. p 1–4.

Khachatourians G. 1998. Agricultural use of antibiotics and the evolution and transfer of antibiotic-resistant bacteria. Canadian Medical Association Journal 159(9):1129–1136.

Kiekhaefer M. 1995. Cross seasonal studies of airborne microbial populations and environment in swine buildings: Implications for worker and animal health. Annals of Agricultural and Environmental Medicine 2:37–44.

Kilburn K. 1997. Exposure to reduced sulfur gases impairs neurobehavioral function. Southern Medical Journal 90:997–1006.

Knight R, Payne V, Borer R, Clarke R, Pries J. 2000. Constructed wetlands for livestock wastewater management. Ecological Engineering 15(1–2):41–55.

Knobeloch L, Salna B, Hogan A, Postle J, Anderson H. 2000. Blue babies and nitrate-contaminated well water. Environmental Health Perspectives 108(7): 675–8.

Krapac I, Dey W, Roy W, Smyth C. 1998. Impacts of bacteria, metals, and nutrients on groundwater at two hog confinement facilities, November 4–5, 1998, St Louis, MO. p 29–50.

Lamka K, Le Chavallier M, Seidler R. 1980. Bacterial contamination of drinking water supplies in a modern rural neighborhood. Applied Environmental Microbiology 39:734–738.

Lantagne D. 2001. Trihalomethane Formation in Rural Household Water Filtration Systems in Haiti. Massachusetts Institute of Technology.

Lee A. 1993. Pathogenicity of *Helicobacter pylori*: A perspective. Infection and Immunity 61:1601–1610.

Levallois P, Phaneuf D. 1994. Nitrate contamination of drinking-water—An evaluation of health risks. Canadian Journal of Public Health—Revue Canadienne De Sante Publique 85(3):192–196.

Luebs R. 1973. Enrichment of the atmosphere with nitrogen compounds volatilized from a large dairy area. Journal of Environmental Quality 2:137–141.

Martin DG WC, Gold MB, Woodard CL Jr, Baskin SI. 1995. Topical anesthetic-induced methemoglobinemia and sulfhemoglobinemia in macaques: A comparison of benzocaine and lidocaine. Journal of Applied Toxicology 15(3):153–158.

Matisziw T, Hipple JD. 2001. Spatial clustering and state/county legislation: The case of hog production in Missouri. Regional Studies 35(8):719–730.

McKee W. 1974. Environmental Problems in Medicine. Springfield, Illinois: CC Thomas. p 273, 314–324, 628–629.

Meggs W. 1997. Hypothesis for induction and propagation of chemical sensitivity based on biopsy studies. Environmental Health Perspectives 105: 473–478.

Meggs W, Elshick T, Metzger W. 1996. Nasal pathology and ultrastructure in patients with chronic airway inflammation (RADS and RUDS) following an irritant exposure. Clinical Toxicology 34:383–396.

Mensinga T, Speigers G, Meulenbelt J. 2003. Health implication of environmental nitrogenous compounds. Toxicology Reviews 22(1):41–51.

Merchant J, Ross R, Hodne C. 2002. Iowa Concentrated Animal Feeding Operations Air Quality Study. Iowa City: University of Iowa and Iowa State University.

Merkel J, Hazen T, Miner J. 1969. Identification of gases in a confinement swine building atmosphere. Transactions of the American Society of Agricultural Engineers 12:310–315.

Metzler D. 1982. Health implications of organics in groundwater. American Journal of Public Health 72:1323–1324.

Miller T. 1975. Living in the Environment: Concepts, Problems, Alternatives. Wadsworth Co., Belmont, California. 108–128 p.

Moore A. 1993. Surveillance for waterborne disease outbreaks—United States, 1991–1992. Morbidity and mortality weekly report. Surveillance summaries/CDC. 42(SS-5):1–22.

Musterman J, Fisher R, Drake L. 1980. Annual Progress Report.

National Academy of Sciences. 1977. Drinking Water and Health. Washington, DC. p 2410–2414, 270–275, 369–400, 440–446, 781–804, 857–899.

———. 2003. Drinking Water and Health. Washington DC. p 2, 410–414, 270–275, 369–400,440–446, 781–804, 857–899.

National Institutes of Health. 2004. Heart Disease Nutrition and Diet. Personal Health Zone.

National Resources Defense Council. 1998. NRDC Pro: America's Animal Factories—Chapter 1. p 1–7.

Novotny V. 1999. Diffuse pollution from agriculture. Water Science and Technology 39(3):1–13.

O'Neill D, Phillips VR. 1992. A review of the control of odor nuisance from livestock buildings. Journal of Agricultural Engineering Research 53.

Overcash M, Humenik F, Miner J. 1983. Livestock Wastes Management, vol I & II. Boca Raton, FL: CRC Press Inc. p 1–11 (vol I), 165–173 (vol II) p.

Payne H. 1994. Maximum exposure limits for gases and dust in pig buildings, June 3, 1994, Adelaide, Australia.

Petersen S. 1999. Nitrous oxide emissions from manure and inorganic fertilizers applied to spring barley. Journal of Environmental Quality Sep/Oct 1999 28(5):1610.

Pickrell J. 1991. Hazards in confinement housing—Gases and dusts in confined animal houses for swine, poultry, horses, and humans. Veterinary and Human Toxicology 33:32–39.

Reynolds S. 1996. Longitudinal evaluation of dose-response relationships for environmental exposures and pulmonary function in swine production workers. American Journal of Industrial Medicine 29:33–40.

Reynolds S, Donham KJ, Stookesberry J, Thorne PS, Subramanian P, Thu K, Whitten P. 1997. Air quality assessment in the vicinity of swine production facilities. Journal of Agromedicine 4:37–45.

Riskowski G. 1991. Methods for evaluating odor from swine manure. Applied Engineering in Agriculture 7:248–253.

Ritter W, Chirnside A. 1984. Impact of land use on ground-water quality in southern Delaware. Ground Water 22:38–47.

Robens J. 1998. Research needs to prevent ground water contamination from animal feeding operations, November 4–5, 1998, St. Louis, MO. p 105–107.

Roy C, Thorne, P. 2003. Exposure to particulates, microorganisms, beta (1-3) glucans, and endotoxins during soybean harvesting. *American Industrial Hygiene Association Journal.* 64(4):487–495.

Rylander R. 1994. Symptoms and mechanismsI—Inflammation of the lung. American Journal of Industrial Medicine 25:19–34.

Rylander R. 2004. Organic Dust and Disease: A Continuous Research Challenge. American Journal of Industrial Medicine 46:323–326.

Schiffman S, Miller E, Suggs M, Graham B. 1995. The effect of environmental odors emanating from commercial swine operations on the mood of nearby residents. Brain Research Bulletin 37:369–375.

Schoffstall J, Bouchard M, Schick P. 2002. Methemoglobinemia. In: Aboulafia D, editor. eMedicine.

Sharpe R, Harper L, Byers F. 2002. Methane emissions from swine lagoons in Southeastern US. Agriculture Ecosystems & Environment 90(1):17–24.

Shusterman D. 1992. Critical review: The health significance of environmental odor pollution. Archives of Environmental Health 47:76–87.

Shusterman D, Balmes J, Cone J. 1988. Behavioral sensitization to irritants/odorants after acute overexposures. Journal of Occupational and Environmental Medicine 30:565–567.

Smith R, Kelly J. 1996. A comparison of two methods for estimating odour emissions from area sources, February 7–9, 1996, Kansas City, MO. p 263–269.

Thorne P, Keikhaefer M, Donham K. 1992. Comparison of bioaerosol sampling methods in barns housing swine. Applied Environmental Microbiology 58:2543–2551.

Thu K. 1995. Understanding the Impacts of Large-Scale Swine Production. In: Thu K, editor. June 29–30, 1995, Des Moines, Iowa. University of Iowa.

Thu K, Donham K, Ziegenhorn R. 1997. A control study of the physical and mental health of residents living near a large-scale swine operation. Journal of Agricultural Safety and Health 3:13–26.

Thu K, Durrenberger P. (eds). 1998. Pigs, Profits, and Rural Communities,. Albany: State University of New York.

U.S. Department of Health Education, and Welfare (NIOSH). 1979. Working Safely with Anhydrous Ammonia: National Institute for Occupational Safety and Health. p 4–13.

U.S. Environmental Protection Agency. 1976. Decision Makers Guide in Solid Waste Management. USEPA. p 60–63, 70, 75, 109.

——. 1980a. Ground Water Protection. Davis J, editor. USEPA. p 1–19.

——. 1980b. Solid Waste Facts: SW-694 USEPA. p 1–8.

——. 2000a. Proposed Revision to Arsenic Drinking Water Standard. USEPA.

——. 2000b. The Quality of our Nation's Waters. A summary of the national water quality inventory: 1998 Report to Congress. Washington DC: USEPA. Report nr EPA 841-S00-001.

——. 2002. Animal Feeding Operations. National Pollutant Discharge Elimination System: USEPA.

U.S. Geological Survey. 2004. National Analysis of Trace Elements—Arsenic in Ground Water of the U.S.

Van Duke H. 1999. Atmospheric dispersion of current-use pesticides: A review of evidence from monitoring studies. Water Air and Soil Pollution 115:21–70.

Van Leeuwen J, Waltner-Toews D, Abernathy T, Smit B, Shoukri M. 1999. Associations between stomach cancer incidence and drinking water contamination with atrazine and nitrate in Ontario (Canada) agroecosystems, 1987–1991. International Journal of Epidemiology 28(5):836–840.

VandenBerg F, Kubiak R, Benjay W, Majewski M, Yates S, Reeves G, Smelt J, VanderLinden A. 1999. Emission of pesticides into the air. Water Air Soil Pollution 115(1–4):195–218.

Velasquez J. 1995. Incidence and transmission of antibiotic resistance in *Campylobacter jejuni* and *Campylobacter coli*. Journal of Antimicrobial Chemotherapy 35:173–178.

Voorburg J. 1991. Achievements of the odour group. In: Nielsen V, Voorburg JH, L'Hermite P, editor. Odour and Ammonia Emissions from Livestock Farming. Barking, England: Elsevier Science Publishers Ltd. p 212–213.

Vos C, Zonneveld J. 1993. Patterns and Processes in a landscape under stress: The study area landscapes ecology of a stressed environment. Vos C, Opdam, editors. London: Chapman and Hall.

Walsh P, Lunney C, Casey K. 1995. Odour impact survey. Milestone Report No. 18, Meat Research Corporation Project No. DAQ-079, Department of Primary Industries, Beef Industry Group, Toowomba, Qld, Australia.

Ward D, Elliot E. 1976. Georgia rural air quality: Effect of agricultural and forestry burning. Journal of Air Pollution Control Association 26:216–220.

Whalen S, Phillips R, Fischer E. 2000. Nitrous oxide emission from an agricultural field fertilized with liquid lagoonal swine effluent. Global Biogeochemical Cycles 14(2):545–558.

WHO (World Health Organization). 1948. http://www.who.int/about/definition/en/.

——. 1972. Health Hazards of the Human Environment, Geneva. p 50–65,119,158,201–202,205–209.

Will L, Diesch S. 1972. Leptospires in animal waste treatment—Animal health problem? Proceedings, 76th Annual Meeting of the US Animal Health Association, 1972, p 138–149.

Wing S, Wolf S. 2000. Intensive livestock operations, health and quality of life among East North Carolina residents. Environmental Health Perspectives.

Wu J, Babcock B. 1999. Metamodeling potential nitrate water pollution in Central United States. Journal of Environmental Quality 28(6):1916–1918.

Zahn J, Anhalt J, Boyd E. 2001. Evidence for transfer of tylosin and tylosin-resistant bacteria in air from swine production facilities using sub-therapeutic concentrations of tylan in feed. Journal of Animal Science 79:189.

8
Musculoskeletal Diseases in Agriculture

Anders Thelin and Kelley J. Donham

INTRODUCTION

Low back pain is a very common disorder. More than 80% of the population in western industrialized countries experience back problems at sometime. In most cases the pain will not be fully evaluated and no diagnosis will be made. Musculoskeletal symptoms of the neck, shoulder, and extremities may be nearly as common as back pain.

DEFINITIONS

Musculoskeletal injuries include diseases of the bone, joints, and structures around the joints (tendons, ligaments, and muscles). Secondary nerve involvement is also included in musculoskeletal disorders.

The following are common musculoskeletal conditions:

- Acute injuries and delayed effects of acute injuries
  - Fractures and dislocations
  - Sprain (an injury of a ligament that has been partially or completely torn; fibers or insertions)
  - Strain (a strained muscle, tendon, or ligament that has been pushed or pulled to its maximal limit)
- Tendonitis and related conditions
  - Tendonitis, (inflammation of a tendon)
  - Tenosynovitis (inflammation of a tendon sheath)
  - Enthesitis (inflammation of a tendinous insertion)
  - Bursitis (inflammation of a bursa)
- Myositis (inflammation of a muscle, which may be primary—e.g., polymyositis—or secondary to mechanical injuries)
- Arthritis
  - Post-traumatic arthritis (arthritis after acute trauma)
  - Infectious arthritis (arthritis due to direct infection of a joint)
  - Reactive arthritis (inflammation of a joint due to an immunologic process or reaction)
  - Rheumatic arthritis (arthritis due to a rheumatic disease)
  - Osteoarthritis (sometimes also called *arthrosis*; a degenerative process in joint cartilage of partly unknown causes)

Repetitive Strain Injuries

Cumulative trauma disorders or overuse syndrome are terms applied to the same group of morbid entities (Cassvan and others 1997). Not only conditions related to repetition are included but sometimes also problems related to static work.

In this chapter trauma and acute injuries will not be covered. Rheumatic disorders will be briefly discussed regarding work-related problems.

Musculoskeletal injuries have a number of characteristic symptoms: pain, tenderness, stiffness, edema, and disuse. Pain is an early symptom. Nociceptive and neurological pain are probably the most common types of pain. However, chronic pain conditions related to psychological and psychosocial factors are a rapidly increasing problem.

WORK-RELATED DISORDERS

Determining cause-effect association between work exposures and musculoskeletal injuries is often difficult—both for the patient and the health care provider. However, treatment, prevention and worker's compensation claims are dependent on the associations between work and injury.

It is necessary to take account of the potential confounding dimensions, given the problems of diagnosis, management, prognosis, and insurance or employer liability. The effect of work activity or sports activity may be primary or secondary, positive or negative, relative to the injury, for current or chronic degenerative musculoskeletal disorders.

Musculoskeletal disorders may have a complex etiology with multiple risk factors like age, genus, and obesity. A number of risk factors for musculoskeletal disease may be found in farm work (Table 8.1).

EPIDEMIOLOGY

Specialization has been one of the megatrends in production agriculture over the past two decades (see Chapter 1). Today, many farmers are engaged in routine tasks for long periods, rather than constantly moving from one task to another. Thus the ergonomic design of the work environment is of greater importance compared to the past. The risk for static and/or repetitive work is significant in many modern farming operations. Different typical disorders and possible related activities are listed in Table 8.1.

A number of reports over the years have indicated a significant risk of low back pain among farmers (Rosegger and Rosegger 1960; Magora 1974; Penttinen 1987; Holmberg and others 2002, 2004a). Most of these studies have used methods that do not adequately address causality. More and especially prospective longitudinal studies are needed for a better understanding of work-related low back pain among farmers.

A fairly significant relationship between hip joint osteoarthritis and farm work has been reported in a number of studies over the last 25 years (Vingård and others 1991; Croft and others 1992; Axmacher and Lindberg 1993; Thelin and others 1997, 2004). Although the occupational association is clear, the specific causes of this problem are not known.

Tendonitis of the shoulders, epicondylitis, and pronator syndrome, and especially carpal tunnel syndrome are all common disorders related to physical work among farmers (Monsell and Tillman 1992; Stål and others 1999). Female dairy farmers might be at special risk for these conditions (Stål and others 1998).

Farming may or may not have a causal relationship in any particular individual to many of the disorders described in this chapter. However, many patients with these complaints will consult occupational health physicians and rural doctors, and will bring up questions regarding farm tasks and possible relation to their injury, the consequences of further similar work, and compensation. A knowledge of the work of farmers is important to enable health care providers to answer their patients' questions and be able to recommend a specific treatment, prevention, and rehabilitation of their farmer patients.

CHRONIC PAIN CONDITIONS

A number of factors affect the perception of pain as described by Waddell (1987) in the Glasgow illness model. Waddell's model describes how physical problems transcend into distress, illness, and sick leave. His concept of illness behavior describes how pain and other symptoms may be related to hope of compensation in terms of moving to another job, financial gain, or sympathy. The neurologist Henry Miller (1961) coined the term *accident neurosis* to describe a category of patients with chronic pain related to litigation processes.

Other psychosocial issues might predispose to chronic pain, including depression, childhood deprivation, family difficulties, and personality disorders. The point here is that the health professional may often be challenged to differentiate work-related injuries in the absence of objective signs of physical injury or disease. Psychosocial conditions may be important etiologic factors of chronic pain or effect modifiers, or have no relationship (Linton 2000; Bigos and others 1991; Palmer and others 2001b; Mannion and others 1996; Linton and others 1999; Dworkin 1990; Cherkin and others 1996; Jonsson and Nachemson 2000).

It is possible that psychosocial conditions may act as an important etiologic factor, but mostly the impacts of psychosocial conditions are regarded as effect-modifying factors (see Figure 8.1).

Table 8.1. Musculoskeletal Disorders Associated with Physical Factors

Disorder	Causal relationship	Typical activity
Cervical degenerative disk disease	no	Ceiling work, lift fork driving
Neck pain	no	Ceiling work, lift fork driving, load carrying in hand or on shoulder, assembling work
Whiplash injury	yes	Trauma, especially traffic accident
Spinal degenerative disk disease	no	High work load, bending, twisting, difficult working positions, manual material handling, whole body vibration
Spinal stenosis	no	Standing and walking
Spondylolysis and spondylolisthesis	no	High work load, bending, twisting, difficult working positions, manual material handling, whole body vibration
Ankylosing spondylitis	no	Inactivity
Low back pain	possible	High work load, bending, twisting, difficult working positions, manual material handling, whole body vibration
Coccygodynia	no	Sitting, trauma
Impingement syndrome	yes	Working with elevated arms especially with vibrating tools, repetitive and static work with arms in abduction
Shoulder tendinitis	yes	Working with elevated arms especially with vibrating tools, repetitive and static work with arms in abduction
Shoulder dislocations	no	Trauma
Thoracic outlet syndrome	no	Overhead activities, lift fork driving, material handling
Frozen shoulder	no	All kinds of movements in the shoulder
Osteoarthritis of the acromioclavicular joint	probably not	Manual material handling, lifting material, transporting material on the shoulder
Cervicobrachial syndrome	no	Lift fork driving, load carrying in hand or on shoulder, assembling work, manual material handling, work with chainsaw
Epicondylitis	possible	Turning screws, hammering, assembling parts
Olecranon bursitis	yes	Trauma, repetitive or continuous pressure
Flexor pronator syndrome	yes	Manual material handling, milking
De Quervain's tenosynovitis	yes	Manual material handling, sawing, forceful hand wringing
Ulnar cubital tunnel syndrome	yes	Resting forearm on hard surface or sharp edge
Stenosing tenosynovitis	possible	Manual material handling
Scaphoid fractures and nonunion of the scaphoid	yes	Trauma
Carpal tunnel syndrome	possible	Manual material handling, vibrating tools, working with handheld tools and equipment
Dupuytren's contracture	no	Working with handheld tools and equipment
Ganglion	possible	Repetitive manual material handling
Herbeden-Bochards osteoarthritis	no	Manual material handling

(continued)

233

Table 8.1. Musculoskeletal Disorders Associated with Physical Factors (*continued*)

Disorder	Causal relationship	Typical activity
Osteoarthritis of the first carpo-metacarpal joint	possible	Trauma, screw/unscrew, vibration
Trochanteritis	possible	Trauma, repetitive kneeling
Osteonecrosis of the femoral head	no	Walking, running
Osteoarthritis of the hip	yes	Unspecific farm work
Knee ligament injuries	yes	Trauma
Bursitis of the knee	yes	Kneeling, trauma
Chondromalacia patellae	possible	Running, high physical activity, walking, climbing stairs
Injuries of the meniscus	yes	Trauma
Baker's cyst	no	Walking, kneeling
Osteoarthritis of the knee	probably not	Walking, running, kneeling, climbing stairs and ladders, working on rough ground
Periosteitis	yes	Training, running
Ankle sprains	yes	Trauma
Avulsion fractures of the fifth metatarsal	yes	Walking, running
Plantar fasciitis	no	Standing, walking, running
Morton's neuroma	no	Dorsiflexion of toe
Hallux valgus	no	Walking, running
Pes planus	no	Walking, standing
Rheumatoid arthritis	no	No
Psoriasis	no	No
Crohn's disease	no	No
Salmonella	possible	Infection from animals
Shigella	possible	Infection from animals
Yersinia	possible	Infection from animals
Other immunologic factors	possible	Contacts with material and/or animals

A number of studies have reviewed the association of psychosocial factors to low back problems, neck problems, and shoulder problems (Theorell and others 1991; Linton 2000). Such factors are demand/control imbalance, job content, social support, job dissatisfaction, shift work/overtime, and stressful work environment. A strong evidence for an association between low job satisfaction and back and neck pain has been reported (Bigos and others 1991; Bongers and others 1993; Hemingway and others 1997). Even more and complex psychosocial factors may be relevant for back and neck problems (Croft and others 2001). As development of chronic pain is not well understood (Freund and Schwartz 2002), a number of models have been presented to describe how this kind of chronic pain develops and persists (Fransen and others 2002). One of these models is the concept of myofascial pain. This syndrome is described as a regional muscle pain disorder characterized by localized muscular tenderness and pain. So-called trigger points are pathognomonic. A trigger point is a hyperirritable spot, painful on pressure. Tenderness and referred pain are common as well (Travell and Simons 1983). Fibromyalgia may be described as a more generalized myofascial pain syndrome, accompanied by fatigue and sleep disturbance (Reiffenberger and Amundson 1996; Wolfe and others 1992; Quintner and Cohen 1994; Campbell 1989; Smythe 1994; Hubbard and Berkoff 1993; Hanock 1995). Alternative models have been presented of how chronic pain emerges and how repeated pain signals may be transformed into persisting hyperexcitability (Boureau and others 2000; Hong 2002). Pain research has also demonstrated that nociceptive receptors (peripheral, cutaneous receptors) are inhibited by central mechanisms activating other receptors. Disturbance of these descending inhibitory signals may play an important role in chronic pain syndromes (Yunus 1992).

INJURIES OF THE NECK

Neck pain is second only to low back pain as the most common musculoskeletal disorder and is frequently reported among workers. Although neck pain may be related to cervical degenerative disease or whiplash injury, years of research on chronic neck pain plus advanced radiological methods has resulted in little understanding for most chronic neck patients. There is some hope, however, that a more psychological approach, focusing not only on the pain but also on counseling and general physical conditioning, may be helpful (Ferrari and Russell 2003).

Cervical Degenerative Disk Disease

The cause of cervical degenerative disk disease is unknown. It is common in both men and women with a rising prevalence with advancing age. Middle-age persons may have subclinical degenerative changes. Degenerative changes are most common at C5–6. More than one disk frequently is affected. Disk protrusion can affect the nerve roots and thus account for radioculopathy resulting in pain and other symptoms in the arms. Increasing levels of spinal degeneration are related to increasing chronicity of patient complaints (Marchiori and Henderson 1996).

The first symptom of disk disease is pain associated with the neck flexed or extended for long periods. Sometimes the symptoms are first noticed after trauma. Different work histories may be reported, such as working with forklifts, tractor driving, or computer work (Skov and others 1996). A typical problematic farm work practice is driving tractors with equipment on the rear and looking back over the shoulder (Figure 8.3). As with other chronic musculoskeletal problems, psychological and psychosocial issues may be related to onset, severity, and chronicity of symptoms (Figure 8.1) (Bongers and others 1993; Riihimäki and others 1989; Bigos and others 1992).

Neck pain or high interscapular pain related to movements is the most frequently reported symptom, but pain during the night also is common. Patients often make efforts to find a good pillow. Physical examination is often negative. Some restrictions of neck flexion or rotation may be noticed. Localized tenderness is a nonspecific finding. Only in severe cases will there be findings such as a reflex change or local sensory/tactile change.

If the symptoms from the neck are combined with other observations, such as weight loss, fever, intravenous drug use, or history of urinary tract infection, other serious condition of the spine must be considered. In case of trauma a spinal fracture should be suspected. Arthritis (rheumatoid or ankylosing spondylitis) can cause neck symptoms. C reactive protein (CRP) and erythrocyte sedimentation rate are useful tests to help differentiate traumatic or degenerative disease from a systemic condition.

Standard plain film x-ray is indicated in

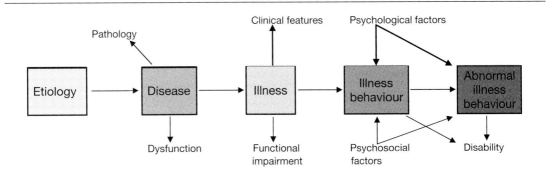

Figure 8.1. Psychosocial factors and illness behavior.

chronic cases and when serious underlying conditions are suspected. Narrowed disk space is a common finding together with degenerative changes such as osteophytes of the surrounding vertebrae. Magnetic resonance imaging (MRI) is not indicated unless there is a suspicion of serious spinal pathology.

In cases with no x-ray findings other diseases must be considered such as neck tension syndrome, whiplash injury, tumors, and entrapment problems of the brachial plexus. Neuropathy related to diabetes or other metabolic abnormalities may also affect the nerve function of the upper extremities.

Patients with neck problems should avoid prolonged sitting with the neck in a fixed position or working with the head in extreme positions. Driving may be difficult and especially tractor driving on rough ground. Early contact with a physical therapist is often recommended. Active therapy, including specified exercises and cervical traction, may be helpful. Heat, massage, and ultrasound may reduce pain and stiffness. A soft cervical collar supports the head and gives some rest for the neck muscles. It may be helpful initially but is of no help or has a negative effect in the longer term. Nonsteroidal antiinflammatory drugs (NSAID) are helpful in reducing pain and stiffness. Oral corticosteroids, muscle relaxants, and sedatives are not usually warranted. Only in very few cases might more effective painkillers be essential.

Modification of the job sometimes must be promoted. The measures may include ergonomic changes of the workstation and changes of work routines and job rotation. Psychosocial difficulties must be considered and measures taken if deemed appropriate.

Some patients with neck problems may develop neurological deficits. They may need more qualified or aggressive therapeutic efforts. Those who have upper extremity radiculopathy and do not respond to conservative therapy should be transferred to specialists for further evaluation.

Neck Tension Syndrome

Neck pain that persists without any objective findings is sometimes labeled as tension myalgia, chronic neck disease, neck-shoulder syndrome, or neck-tension syndrome. Some authors report that tension conditions of the neck result from static sustained muscle contraction (Weiss 1997). The

symptoms are said to be frequent among secretaries, computer operators, and small-part assemblers, as well as among those who carry loads in the hands or on the shoulders. It might also be seen after repeated or sustained overhead work. Other researchers (Palmer and others 2001a) conclude that psychosocial conditions may be as important as occupational physical stress in relation to neck pain (Linton 2000; Holte and others 2003; Croft and others 2001; Marchiori and Henderson 1996).

To help a person with a neck tension problem one must consider the potential psychosocial situation. It may be important to involve physiotherapists as well as psychotherapists in a program to assist the patient to handle his or her situation.

Whiplash Injury

A sudden jerk to the neck may result in a complex injury picture. A whiplash injury is the most characteristic type. Typically a whiplash trauma is an acceleration-deceleration loading of the cervical spine, usually generated by a rear-end car collision. Other types of car crash, injuries related to diving, and fall trauma also may result in a neck distortion giving the same type of injury. Whiplash injuries are probably very common and differ in terms of the severity of the injury. Usually, the symptoms decline and disappear within a short time, but sometimes pain and other problems—such as vertigo, numbness, vision disturbances, difficulty concentrating, and memory impairment—may linger. These symptoms are labelled whiplash-associated disorders (WAD). After many years of research there is still controversy about whether WAD complications are related to physical changes, biopsychological reactions, or insurance claims questions (Benoist and Rouaud 2002; Lovell and Galasko 2002; Ferrari 2003; Kwan and Friel 2003).

The Quebec task force analyzed more than 10,000 publications concerning whiplash and whiplash-related disorders (Quebec task force 1995). Objective signs of injury like fractures, ligament strains, disk fragmentation, and bleeding are reported after studies on animals and postmortem studies (Rauschning and others 1989; Taylor and Twomey 1993). The Quebec classification with four grades is widely accepted (Atlas and others 1996; Pennie and Agambar 1991; Barnsley and others 1994; Ferrari 1999; Pearce 1999) (Table 8.2).

Table 8.2. The Quebec Classification of Whiplash Injuries

Grade	Clinical presentation
0	No functional manifestations or physical signs
1	Neck pain with no physical signs
2	Neck pain, range-of motion limitation, and tender point
3	Neck pain and neurologic impairment
4	Neck pain and fracture or dislocation

The percentage of exposed persons reporting WADs varies among countries (Schrader and others 1996; Giebel and others 1999; Partheni and others 1999). Studies in Australia, New Zealand, and Saskatchewan, Canada, indicate that the insurance system has a considerable effect on the number of WADs (Mills and Horne 1986; Awerbuch 1992; Cassidy and others 2000).

Psychosocial considerations are less questioned in patients with grade 3 and 4 symptoms on the Quebec classification system. These patients should be carefully evaluated at a traumatological clinic. Those with lower grade symptoms might be followed as outpatients. It is important that the patient starts working as soon as possible, at least part time. Some patients may need support to tackle posttraumatic stress reactions.

INJURIES OF THE SPINE

The annual incidence of low back pain episodes is reported to be about 50% of working age adults. More than 10% seek medical care for their back condition. The cost of back conditions is enormous and includes costs of medical treatment, lost productivity, disability, work absence, and disability pensions.

Low back problems are defined as acute or subacute the first few weeks and chronic if they persist more than 3 months (Deyo and others 1988; Hansson 1989). Most patients with low back pain recover spontaneously within a month. However, most people suffer recurrent episodes of low back pain. Risk factors for low back injury include heavy lifting, bending and lifting, bad work positions, whole body vibration, and prolonged sitting in a fixed position. Farming is also a risk factor, and several research studies of farmers have reported high rates of low back pain

within a week or a month. New episodes of low back pain are a reality for many persons (Magora 1974; Bigos and others 1992; Walker-Bone and Palmer 2002). Tractor driving has been identified as a major factor (Rosegger and Rosegger 1960; Schultze and Polster 1979) as well as whole body vibration and prolonged sitting (Wickstrom 1978; Boshuizen and others 1990a, 1990b).

Most of the studies reporting a relationship between farming and low back pain have been based on cross-sectional studies. Furthermore, few studies have included any results on physical diagnosis (e.g., degenerative disk, etc.). More definitive prospective and case control studies are lacking (Lings and Leboeuf-Yde 2000). Part-time farmers have been reported to have more problems than full-time farmers (Park, Sprince and others 2001). Physical factors as well as psychological factors have been demonstrated as risks, but explain less than half of the cases of low back problems (Holmberg and others 2003). Despite frequent symptoms of low back pain farmers do not often seek medical care and are generally not away from their work for very long (Holmberg and others 2003).

Spinal Degenerative Disk Disease

Intervertebral disk degeneration is a normal aging process. However, hereditary factors are important in the development of degenerative changes in the spine. Some differences between men and women have been demonstrated. Females experience a greater number of affected joints compared to men, while the latter have more lumbar spine degeneration (Kellgren and Lawrence 1958). Sciatic symptoms (pain in the back of the legs) are more prevalent among men than in females of the same age (Hirsch and others 1963). Men experience symptoms at a younger age, but between 35 and 54 years there is almost no gender difference. The highest prevalence overall is reported for the middle age groups (Krämer 1973).

People in general as well as many physicians think that disk degeneration and disk hernia (with or without sciatic symptoms) emerge because of heavy work, heavy lifting, or turning and twisting at the waist. However, research of workers in traditionally heavy jobs shows that they have less back problems and less sick leave due to sciatic symptoms than a comparison population. Numerous studies have failed to show a relation between

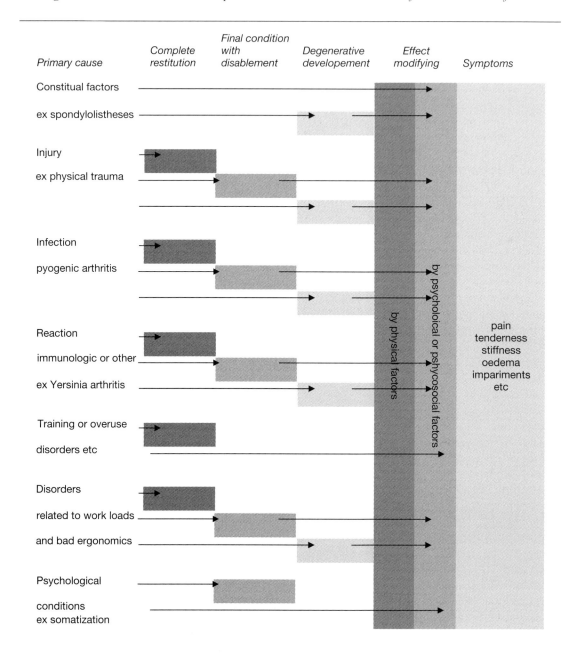

Figure 8.2. Musculoskeletal disorders—causes and effects.

mechanical stress on the back to degenerative disk changes (Farfan 1977).

During the last two decades modern techniques using MRI and statistical regression methods have expanded the knowledge in this area. No specific risk for heavy lifting work was apparent compared to people with jobs involving sitting during long periods (Evans and others 1989;

Videman and others 1990). Other studies have found only a weak relation between disk degeneration and occupation. Almost half of the general population of middle age and beyond have MRI-objective disk changes (Modic and Herfkins 1990). One-third of individuals without symptoms have disk changes and 50% of those with symptoms have normal MRIs (Boos and others

1995; Savage and others 1997). In a study of twins, hereditary factors were related to disk problems but not occupation (Battie and others 1995). Experimental studies did not indicate any relation between whole body vibration and disk compression and degeneration (Hutton and others 2000; Videman and others 2000).

Based on available research, it appears that disk degeneration is mainly dependent upon hereditary factors. However, sitting for long periods appears to be a greater risk factor than heavy work. This may be related to the fact that there is less space around the nerve roots when sitting. A fixed position without motion may also promote swelling if there is an ongoing inflammation, and prolonged sitting may promote a weaker muscle capacity. In any case disk degeneration in relation to work is very difficult to define.

Sciatic pain, weakness, and numbness are mostly caused by pressure on the nerves exiting the spinal cord from local inflammation or a bulging or herniated disk and from an associated inflammation.

The onset of symptoms may be sudden or gradual. Pain is associated with movement of the back. Forward bending may be very difficult. Pain is usually located in the lumbar region but may extend to the buttock and is usually not well defined. Patients with typical nerve root symptoms avoid sitting and prefer standing or walking.

The degree of sciatic pain or radiculopathy is proportional to the level of root pressure. Pain location is as follows:

L4: pain below the knee in the medial part of the lower leg
L5: pain in the lateral calf and dorsal part of the foot and/or in the great toe
S1: pain in the posterior calf and lateral border of the foot.

In case of nerve root pressure, the straight leg raising test (Lasègue's sign) is generally positive. The smaller the angle associated with pain, the more positive is the test. Affection of the S1 root causes absent or decreased ankle reflex and involvement of the L4 root causes absent or decreased knee reflex. Skin sensation may be affected as follows:

L4: decreased sensation in the medial part of the leg

Figure 8.3. Tractor driving while looking back over the shoulder may be problematic in case of neck disorder.

L5: dorsal part of the foot and the great toe
S1: lateral border of the foot and/or the three lateral toes

Muscle force may be decreased so that the patient has difficulty raising onto the toes or back on the heels. The ability to dorsiflex the great toe may also be weak. Sometimes the patient reports hyperesthesia.

Plain film x-ray views are usually obtained with the patient standing. Scans or MRI should be obtained in case of insecurity concerning diagnosis or prior to surgery.

Figure 8.4. Tractor driving is associated with whole-body vibration and prolonged sitting.

The physician examining a patient with back pain must differentiate other causes that may be related to pathologic process involving abdomen, the kidneys, or pelvis. Bone infection, ankylosing spondylitis, and tumors can be differentiated with bone scans, X-ray, and MRI exams, respectively (Figures 8.5 and 8.6).

The great majority of patients with disk degenerative disorder with low back pain with or without radiculopathy will resolve without any treatment. Bed rest should be avoided (Wilkinson 1995). If possible the patient should work and perform daily living activities. The prognosis is better if the patient is physically active (Coomes 1961; Deyo and others 1986). Patients should have instructions to adjust their work according to their ability. In some cases a lumbosacral corset may be helpful but should not be used for long periods. Antiinflammatory drugs such as nonsteroidal antiinflammatories may be prescribed.

Exercise should be implemented as soon as possible. Persons with recurrent problems or when recovery is delayed may be referred to a physical therapist. In some places it is possible to join a "back school." Such schools have specially designed exercise schemes and personnel who can teach the patient about prevention.

Patients with heavy or persistent pain as well as patients with very significant nerve root engagement may be candidates for surgery. Surgery is performed to relieve nerve root pressure. The long-term prognosis is not better for patients who have had surgery. Thus it is necessary to reflect carefully on the patient's situation before a decision of surgery.

Spinal Stenosis

Spinal stenosis is mostly the result of progressive degenerative disease. Deformation of the facet joints as well as deformation of other structures may encroach on the spinal canal as well as the neuroforamina. Spinal stenosis may also be a late complication to disc surgery hernia or congenital conditions.

Spinal stenosis is a common cause of leg pain, especially among old persons. Typically individuals with spinal stenosis have problems with standing and walking, and sitting therefore may be preferable. The pain in the lower extremity is variable between and within patients over time. Patients should be observed walking and reflexes of the lower extremity should be examined at each visit.

Plain film x-rays may show a number of degenerative changes: degenerative low disks with associated osteophytic changes and scoliotic changes. A CT scan or MRI may show the impingement of the nerve roots.

Most patients with spinal stenosis are older and have other degenerative disorders, such as disturbed circulation of the legs from arteriosclerosis (intermittent claudication) which may give symptoms similar to spinal stenosis.

If spinal stenosis is suspected it is a good idea to refer the patient to a specialist of orthopedics for evaluation and possible treatment.

Spondylolysis and Spondylolisthesis

Spondylolysis is a defect in the continuity of the superior and inferior articular processes of the neural arch of a vertebra; almost always the fifth lumbar vertebra. The bone of the articular process is replaced by fibrous tissue, usually the result of a congenital defect, or trauma such as an improperly healed stress fracture. Spondylolysis may cause instability and displacement of a vertebra or spondylolistheseis. Spondylolisthesis may also be the result of osteoarthritis or malformation of the articular processes.

These conditions cause variable symptoms unless there is significant displacement (Levin 1989; Lipson 1989). On examination, there may be a visible or palpable "step" above the sacral crest. Plain x-ray demonstrates the defect on lateral views, and there may be secondary degenerative changes (Figure 8.7). Most patients do not need any treatment, and they should continue to be physically active. Individuals with symptoms may be helped with conservative measures similar to patients with degenerative disk disease. Surgery is justified only when disability is severe.

Available literature suggests it is not necessary to exclude workers with spondylolisthesis from moderate physical work (Postacchini 1989; Matsunaga and others 1990; Osterman and others 1993; Floman 2000). Studies indicate that the prognosis for young persons with spondylolysis is good (Muschik and others 1996).

Ankylosing Spondylitis

Ankylosing spondylitis (Bechterew's disease) is a chronic and often disabling disease. It is often necessary to intervene with persons with ankylosing spondylitis to facilitate their working life.

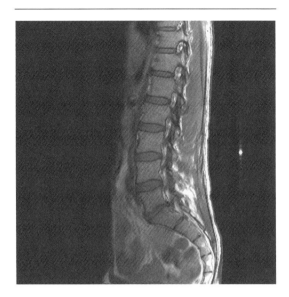

Figure 8.5. Degenerative disk disease—
deformed vertebral bodies of the lower back.
(Source: MRI low back, Department of Radiology,
Central Hospital, Wexiö, Sweden)

Approximately 1% of the white population has
ankylosing spondylitis. The frequency is lower in
Japan and among African black populations. The
variation is strictly related to the variation of the
HLA-B27 gene. Almost 90% of patients with
ankylosing spondylitis have this gene and between 10–20% of those who are HLA-B27-posi-
tive will develop the disease. Ankylosing
spondylitis has primarily been recognized in men;
recently women have also been recognized as susceptible (Russell 1985). Physicians and radiolo-
gists may miss the diagnosis in mild disease; females usually have a milder disease (Marks and
others 1983; Gran and others 1985).

The following five characteristics or manifestations suggest a diagnosis of ankylosing spondylitis (Calin, Porta et al. 1977):

1. Occurrence of symptoms in patients younger
   than 40
2. Insidious onset of back discomfort
3. Persistence of back pain for more than 3 months
4. Morning stiffness
5. Improvement of back pain after exercise

The physical examination may reveal a reduced
mobility in both anterior and lateral planes and a
loss of the normal lumbar lordosis. Lateral x-ray
views of the lumbar spine show that the normal
concavity of the vertebral bodies is lost and they
develop a square form. Severe degenerative bony
changes progress over most portions of the spinal
column (Murray and Persellin 1981).

Other organs may become involved with amy-
loidosis, including the eye, lung, and cardiovascular system. It is not possible to make an accurate
prognosis for an individual patient. The treatment
must focus on a good spinal posture and exercise.
Antiinflammatory medication (NSAID) may be
helpful. Daily physical activity and work activity
should be encouraged but risk of fractures should
be noted in severe cases. The combination of a
more fragile skeleton and difficulties with motion
must be observed as a significant risk for farmers
handling animals and exposed to other kinds of
physical stress.

Coccygodynia

Coccygodynia is persistent pain in the region of
the coccyx (tailbone). Usually there is no evident
pathology. A local injury as a direct fall onto the
coccyx or a blow to the area is typically reported.
The injury in most cases is probably a strain or
sprain in one coccygeal joint.

Pain is worse when sitting. There is localized
tenderness. Plain films are typically normal. Infections or tumors should be ruled out. Treatment
is usually not required. A small pillow to sit on
may be helpful as well as antiinflammatory drugs.
In most cases the condition will resolve spontaneously over time.

Low Back Pain

Low back pain is a frequently used term. It is im-
precise and sometimes it is used to describe many
kinds of low back problems, or when no other
terms apply. Some patients with low back pain
have objective findings such as x-ray findings or
relevant clinical observations (specific low back
pain). Other patients have pain without objective
findings (nonspecific low back pain) (Borenstein
1996). Most back pain early in the disease course
is nonspecific. Pain may be generated by nocicep-
tors in most of the anatomic structures of the
lower back.

Recent studies indicate that a nociceptive stimulus may start a muscle contraction in different
spinal muscles (Indahl and others 1995; Kaigle

and others 1998; Solomonow and others 1998). The contraction may be for the purpose of stabilizing a painful segment and may be facilitated by higher centers in the brain or spinal cord, which may continue the contraction even after the initial nociceptive stimulus has healed (Dahl and others 1992). This biomechanical loading may be acute, repeated, and prolonged, leading to the low back pain. The scientific medical understanding of most low back pain is still limited despite modern technology. For details see biomechanical textbooks and reviews (Farfan 1977; Chaffin and Andersson 1991; Frymoyer and Gordon 1992; Wiesel and others 1996; Mayer and others 2000).

Patients with low back pain and without any objective signs of disease may need the same support and training as patients with a degenerative disk disease (see above).

INJURIES OF THE SHOULDER

The shoulder is the most movable mechanical system in the body. Three joints, the glenohumoral joint, the acromioclavicular joint and the sternoclavicular joint, and more than 20 muscles integrate as a complex unit that creates stability and force in a wide range of motion.

Shoulder pain is a very common symptom often related to occupational activity (Hagberg and Wegman 1987; Tola and others 1988; Frost and Andersen 1999). Shoulder pain affects ability to work in a wide range of jobs from heavy manual jobs like farming or construction to low physical work jobs (e.g., computer terminal job). Repetitive motion, fixed static work, prolonged sitting, poor postures, excessive force, and vibrations are kinds of mechanical stress related to shoulder (Mani and Gerr 2000; O'Neil and others 2001; Buckle and Devereux 2002). Psychosocial conditions (as with other chronic musculoskeletal conditions) are strongly related to shoulder pain and disability (Bergenudd and others 1988; Johansson and Rubenowitz 1994; Dyrehag and others 1998; Linton 2000; Andersen and others 2003).

Occupational groups working with their arms elevated (e.g., welders at shipyards, painters, and construction workers) have significant risks for shoulder pain, especially if they use vibrating tools (Herberts and others 1981; Stenlund and others 1993). As with other musculoskeletal diseases, work exposure in relation to clinical findings are difficult to evaluate.

Regarding farmers, studies in Finland indicate they have a higher risk of osteoarthritis of the glenohumoral joint relative to a comparison population (Katevuo and others 1985; Hagberg and Wegman 1987). Other studies have revealed farmers report a significant association between difficult working positions and shoulder pain. However, farmers overall report a low risk of neck and shoulder problems (Holmberg and others 2003).

Impingement Syndrome or Shoulder Tendonitis

Impingement syndrome has previously been referred to as rotator cuff syndrome, supraspinatus tendonitis, or shoulder tendonitis. The supraspinatus tendon is perhaps the prominent structure for impingement as it passes beneath the acromion process of the scapula (shoulder blade). Individual factors such as the anatomic space, age, and intensity of localized inflammation are related to the risk and the magnitude of symptoms. The bursa between acromion and humerus may also be affected, resulting in a bursitis. In chronic cases especially among older persons the tendon may rupture resulting in a cuff rupture. A supraspinatus tendonitis will produce a "painful arc" which is described as pain when abducting from 45–90 degrees. Repeated or maintained abduction with pressure on the acromion process is also painful. Pain relief with release of pressure is a positive impingement sign (Fongemie and others 1998). Pain is felt in the anterolateral aspect of the upper arm sometimes radiating distally and/or to the base of the neck (Gorski and Schwartz 2003). Lying on the affected side provokes pain.

The clinical evaluation will note a minor weakness as a result of pain inhibition on resisted abduction. The *impingement sign* is usually positive when the impingement is caused by a localized inflammation or injured supraspinatus tendon. Palpation may confirm the site of a rotator cuff lesion, revealing tenderness, or a defect when the supraspinatus tendon is completely torn at its insertion to the greater tuberosity of humerus.

Plain X-rays may show sclerotic changes of the acromion, the greater tuberosity of humerus or osteoarthritis of the acromioclavicular joint. Ectopic calcification is a common finding. Osteoarthritis of the glenohumoral joint is not very common. Rupture of the cuff will be demon-

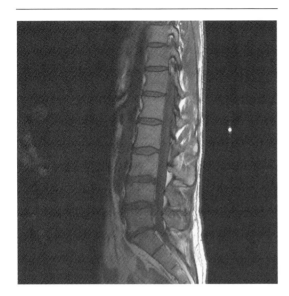

Figure 8.6. Disk herniation with nerve root pressure. (Source: MRI low back, Department of Radiology, Central Hospital, Wexiö, Sweden)

strated by arthrography or MRI (Figure 8.8) (Rossi 1998; Stevenson and Trojian 2002).

Patients with less severe symptoms may be referred to physiotherapy (ultrasound, heat, pendulum exercise) (Ludewig and Borstad 2003). NSAIDs are useful. A local injection of corticosteroids in combination with anesthetics into the subacromial space is a rapid way to resolve pain. Patients who do not respond or who respond only temporarily may be candidates for surgery. This is generally performed by arthroscopy and the aim is to restore the subacromial space. Cuff disruption can be identified and sometimes repaired by arthroscopic surgery.

The long-term outcome of rotator cuff tendonitis is not always positive (Chard and others 1988; Brox and others 1993). Manual work following surgery is not related to a worse prognosis. The level of degeneration and psychosocial factors are good predictors of outcome. Reduction in extreme stress, like some sports, leads to more rapid healing (Hutson 1996; Anderson and others 1999; Brox and others 1999).

Shoulder Dislocations

The shoulder joint is anatomically unstable as the surface area of the glenoid fossa covers only 20%

of the humeral head, and the stability therefore depends on the surrounding tissues. Recurrent anterior subluxation usually occurs in patients who have had an acute full luxation previously (Goss 1988). Luxations are commonly related to contact sports such as football and rugby but are also seen among swimmers. Farming confrontations with cattle, for example, may pose a risk for shoulder luxation.

Patients with recurrent dislocations are well aware of the situation and can reestablish the normal position of the humerus on their own. In other cases the situation is not clear and the symptoms may be more diffuse.

Instability may be associated with some loss of range of rotation. The instability can be demonstrated by applying pressure to the head of the humerus with the patient's arm in abduction which may result in pain and a noticeable laxity of the joint. Plain x-rays are often normal. MRI and examination under anesthesia may be necessary to demonstrate the instability of the joint (Cofield and Irving 1987; Horsfield and Stutley 1988; Kieft and others 1988; Adolfsson and Lysholm 1989). Conservative treatment consists of muscle strengthening and exercises of flexibility. In cases of recurrent subluxation surgery may be needed.

Thoracic Outlet Syndrome

Thoracic outlet syndrome (TOS) is a compression of neurovascular structures as they pass out from the chest and neck under the clavicle to the axilla. Normally the passage is spacious even in case of extreme movements. Most patients with TOS probably have an abnormal anatomy. This may be due to a previous injury to the clavicle, a hypertrophic anterior scalene muscle, or an extra cervical rib (or a fragment of a rib). The incidence of TOS is not well known (Hagberg and Wegman 1987; Cuetter and Bartoszek 1989).

TOS is not a strict work-related disorder but may be aggravated in jobs where heavy loads are carried with extended arms (Guidotti 1992; Kroemer 1992).

Some patients have pain from the shoulder radiating down to the forearm and the hand, and have difficulties with overhead work. Numbness, tingling, and weakness are reported as well as symptoms of vascular TOS as insufficient circulation.

X-rays should be obtained to evaluate anatomic abnormalities. Electrodiagnostic testing might be performed to evaluate involvement at the brachial plexus (Poole and Thomae 1996; Lindgren 1997). The treatment is usually conservative and should focus on postural strength training and education to teach the patient to avoid movements resulting in compression. General physical fitness and reduction of obesity should be encouraged. Rarely patients may require surgery.

Frozen Shoulder

Frozen shoulder (Adhesive Capsulitis) is a condition with restricted active and passive movement in the shoulder. A diffuse inflammatory process is the apparent direct cause of immobility (Kessel and others 1981). Frozen shoulder (may or may not start following trauma) is a process of gradually increasing pain and immobility of the glenohumeral joint. Most patients spontaneously recover within a period of 2 years. Frozen shoulder is frequently used by physicians as a generic term incorrectly used for a variety of different shoulder problems.

Rehabilitation of patients with frozen shoulder is often a problem in occupational medicine. Many frozen shoulder patients with manual tasks will have significant problems to continue their job. However, the long-term prognosis is good as most patients recover within a 2-year period.

Osteoarthritis of the Acromioclavicular Joint

Osteoarthritis or arthrosis of the acromioclavicular (AC) joint is a disease of the cartilage. The cartilage is degenerated and the joint is deformed with osteophytes at the joint margins. Although studies of AC in relation to work exposures are variable, heavy physical work, especially work with vibrating tools or in conjunction with contact sports, appears as a risk factor (Stenlund and others 1992; Stenlund 1993; Worcester and Green 1968).

Pain is localized to the joint area and aggravated by especially overhead use of the limb. Plain X-rays show narrowing of the cartilage space and osteophytes.

Treatment is often not needed. NSAID and/or a local injection of corticosteroids may be helpful. In severe cases surgery is justified.

Cervicobrachial Syndrome

Cervicobrachial pain syndrome is a nonspecific term used for pain in the shoulder and neck. Localized inflammation of the muscles (myositis or myalgia) or myofascial sheaths may be the cause.

Repetitive work strain, bad working positions especially work with elevated arms, fixed static work, and computer terminal work have been associated with cervicobrachial syndrome (Jonsson 1988; Fernstrom and Ericson 1997; Punnett and Bergquist 1997; Backman 1983; Kilbom 1986; Krapac and others 1992; O'Neil, Forsythe and others 2001; Buckle and Devereux 2002).

The treatment of persons with nonspecific symptoms is problematic. Job rotation and other arrangements at the workplace combined with physiotherapy are recommended. Psychosocial factors must be regarded and the way the work is organized is strongly related to the risk of these kinds of shoulder symptoms (Kvarnström 1983).

INJURIES OF THE ELBOW, WRIST, AND HAND

Some studies indicate a risk for forearm problems among milkers (Stål and others 1996, 1997). Vibration is related to a variety of symptoms from the hands and arms in different occupations also among farmers (Thelin 1981; Raffi and others 1996) and lumberjacks (Pyykkö and others 1978; Farkkila and others 1986). All common work-related hand and arm problems might also be seen among farmers.

Epicondylitis

Epicondylitis is the result of inflammation of a tendon at its insertion (enthesis) onto bone at an epicondyle. Lateral and medial humoral epicondylitis are prevalent enthesopathies.

Lateral epicondylitis (also called *tennis elbow*) (Kivi 1982; Allander 1974; Gruchow and Pelletier 1979; Major 1883) is much more common than medial epicondylitis and affects more than 1% of the population.

The prevalence of lateral epicondylitis has been studied in a number of cross-sectional studies (Roto and Kivi 1984; Punnett, Robins and others 1985; Ohlsson and others 1995; Haahr and Andersen 2003) indicating significant risks related to repetitive and manual work. Other studies do not support this finding (McCormack and others 1990; Chiang and others 1993). No studies of the prevalence of epicondylitis among farmers have been found.

Pain radiates to the proximal forearm, and weakness of grip is a common complaint. Most patients have no symptoms at rest.

The typical finding at examination is localized tenderness over the lateral humeral epicondyle. In more severe cases there is a soft tissue swelling overlying the epicondyle.

The lesion usually heals untreated if harmful activity is eliminated. Physiotherapeutic measures are often helpful, especially stretching (Solveborn 1997). NSAID may be helpful and in more established cases a localized steroid injection is effective. Repeated injections may however be problematic and the effectiveness may decline with a rising number of injections. In chronic cases surgery might be recommended. To prevent further problems a brace for the forearm is useful to alter the leverage on the forearm muscles. Different types are available. They should be used only during activities.

The long-term prognosis is good over a course of 2 or 3 years (Haahr and Andersen 2003).

Medial epicondylitis is a similar condition but much less prevalent. It is also called *golfers elbow* but mostly occurs in people who have never played golf. Pain is experienced over the medial elbow and may radiate distally. Wrist movements are painful. Symptoms are mostly not so severe as experienced in lateral epicondylitis. The management is similar to that outlined for lateral epicondylitis.

Olecranon Bursitis

Olecranon bursitis presents as a localized swelling over the olecranon process of the elbow. It is a common condition and can result from trauma. The trauma may be direct or repeated following recurrent friction and pressure. Penetration of the bursa may result in an infection. Some patients do not have any other symptoms than the swelling; others have pain and localized tenderness. Two common generalized conditions, gout and rheumatoid arthritis, may be associated with olecranon bursitis.

Signs of increased warmth suggest an infection. After infection has been ruled out a corticosteroid injection into the bursa usually is beneficial. A protective pad to avoid reinjury may be recommended to persons with activities, which are irritating the bursa.

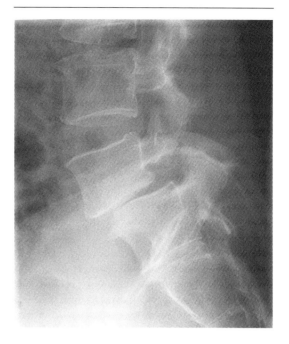

Figure 8.7. Spondylolisthesis. (Source: Plain film X-ray low back, Department of Radiology, Central Hospital, Wexiö, Sweden)

Flexor Pronator Syndrome

The median nerve may be subject to entrapment because of hypertrophy of the pronator teres muscle in the forearm (Figure 8.9) (Kopell and Thompson 1963). This syndrome has been related to playing musical instruments (Bejjani and others 1996). It is also found among female cow milkers who have reported frequent problems of the forearms (Stål and others 1997).

Sensory and motor manifestations may arise, but usually the symptoms are mild with some weakness and/or diffuse pain. The sensory disturbances may be similar to carpal tunnel syndrome. Local infiltration of steroids or anesthetics may be used. In chronic or refractory cases surgical exploration might be necessary.

De Quervain's Tenosynovitis

Two of the tendons to the thumb, the abductor pollicis longus and the extensor pollicis brevis, pass in a common synovial sheath in a bony groove across the styloid process of the radius. At this point a stenosing tendosynovitis sometimes appears related to repeated or heavy friction. The condition was described in the 19th century by De Quervain. The typical patient is a middle-aged female or a young mother with a child (Harvey and others 1990). Women milking cows and working with gardening might be at risk. Instant or repeated use of the thumb may start an inflammation of the tendon sheath resulting in a thickening and constriction of the two tendons.

Pain is associated with a pinch grip and may radiate proximally and distally. Pain is easily demonstrated at resisted contraction and on stretching. Swelling and crepitus over the affected area are sometimes noted.

X-ray will differentiate De Quervain's tenosynovitis from osteoarthritis of the first carpometacarpal joint.

Some patients do not need any special therapy, and rest from movements that aggravate the pain will benefit the patient. Antiinflammatory drugs may be prescribed. Corticosteroid injection into the region is helpful in most cases. Surgery is indicated only in those patients who have recurrent symptoms.

Ulnar Nerve Entrapment or Compression

The most common site of ulnar nerve compression is behind the elbow where the nerve passes behind the medial epicondyle (Stewart 1987). An additional location of entrapment can result from repeated pressure on the palm of the hand, which can cause a compression of the ulnar nerve at the wrist. The latter is reported in bicycle riders after gripping the handlebars for long periods and in workers frequently using a hammer. Symptoms are mainly sensory with pain, numbness and paresthesiae. Diagnosis might be confirmed by nerve conductivity tests. Rest to the elbow or hand is the first step of management. An injection of corticosteroides might be given, and in severe cases surgery is indicated.

Trigger Finger

Stenosing tenosynovitis of the fingers or the thumb occurs as a result of nodule formation within a flexor tendon. Movement is restricted although the nodule usually is small. The finger is often locked and to unlock the finger the patient may extend it passively. A sudden snapping or "triggering" may be felt. As this occurs repeatedly the tendosynovitis progresses so that the movements will be even more difficult.

The formation of the nodule may be a complication of a simple tenosynovitis related to frequent grasping of hard objects. Tenosynovitis and nodule formation may also be related to rheumatoid arthritis and/or diabetes (Chammas and others 1995; Saldana 2001).

Often surgery to incise the thickened tendon sheath is needed. Sometimes an injection of corticosteroids at the site of the lesion provides some relief.

Fractures and Nonunion of the Scaphoid

A young person falling on their outstretched hands may suffer a fracture of the scaphoid. The same type of trauma among elderly probably will generate a Colle's fracture. X-rays must be conducted carefully. If the radiography is negative the wrist should be immobilized for a week and repeat x-rays obtained. In case of a fracture immobilization is needed for 4 months.

A nonunion may generate symptoms long after the original injury. Surgical treatment is often necessary to deal with nonunions.

Carpal Tunnel Syndrome

Carpal tunnel syndrome (CTS) is a very common disorder. The finger flexor tendons, the radial artery, and the median nerve pass through the carpal tunnel, which is a fibroosseous channel, bounded by the carpal bones and the transverse carpal ligament (Figure 8.10). The median nerve innervates the muscles of thumb opposition and is sensory to the first three fingers and half the fourth (Figure 8.11).

CTS is produced by entrapment of the nerve as it passes through the tunnel. CTS was first described in 1947 by Brain and others (1947), and further described by Phalen and others (1959) (Phalen 1966). The associations between CTS and pregnancy, rheumatoid arthritis, myxoedema, gout, and diabetes have been known for some time (Dorwart 1984). The apparent common anatomic feature to the various etiologic factor of CTS is the thickening of the flexor synovialis. Anatomic abnormalities and conditions after traumatic injuries, such as Colles' fracture, have also resulted in median neuropathy (Browne and Snyder 1975; Dekel and others 1988). Recently a significant relation between CTS and obesity has been demonstrated (suggesting the reason for the recent increase in CTS) (Werner and others 1997; Leclerc and others 1998).

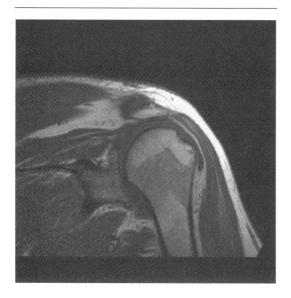

Figure 8.8.   Rotator cuff rupture. (Source: MRI shoulder, Department of Radiology, Central Hospital, Wexiö, Sweden)

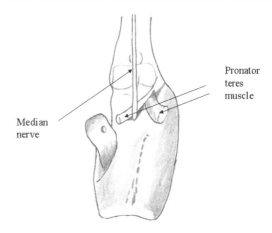

Figure 8.9.   Compression of the median nerve during its passage through the pronator teres muscle.

Numerous studies have demonstrated a strong relation of CTS to work (Hagberg and others 1992; Rossignol and others 1997). Repetitive work especially in combination with vibration has been associated with CTS (Feldman, Travers and others 1987; Lundborg and others 1987; Wieslander and others 1989; Bovenzi 1994; Tanaka and others 1997). Other studies have revealed continuous and powerful gripping as an etiologic factor (Silverstein and others 1987; Stetson and others 1993). As with other musculoskeletal injuries, psychological and psychosocial factors may affect complaints of CTS (Leclerc and others 1998), especially since worker's compensation is a frequent question regarding CTS. Repetitive keyboard use is probably the most reported precipitation factor in litigation cases but the evidence for that kind of exposure as a causal factor is weak (Nordström and others 1997).

In farming relevant risk factors for CTS include being a middle-aged female. Cow milking appears to be a risk for CTS (Stål and others 1999) as has the occupation of professional sheep shearing (Monsell and Tillman 1992).

Generally patients suffer from pain and paresthesia in the region innervated by the median nerve (Figure 8.9). Pain is usually more severe at night. It may radiate up the arm and commonly disturbs sleep in early mornings. Most patients report the need to shake or move their hands or arms to make the pain and numbness abate. The diagnosis should never be delayed until signs of motor dysfunction appear.

Phalen's sign may be positive (hyperflexion of the wrists for a short period provokes paresthesias or dysesthesias in the median distribution). Also *Tinel's sign* may be positive (tapping over the area of entrapment provokes paresthesias). The definitive diagnosis of CTS is based on slowing of the nerve conduction velocity across the carpal tunnel. Prompt relief after injection of a steroid into the carpal tunnel may be used as a diagnostic tool. Signs of disturbed motor function are late symptoms.

If underlying medical conditions such as obesity, diabetes, or rheumatoid arthritis are identified, these should be properly treated. Reduction of provocating work activities is recommended, which may require movement to another job or modifications of the work. In early stages this might be the only measure necessary. Injection of a steroid gives relief for only a limited period. Many patients have to be referred to surgery.

Dupuytren's Contracture

Dupuytren's contracture is a fibrous proliferation in the palmar faschia of the hand and sometimes the plantar faschia of the foot. The proliferation results in a progressive thickening and contracture. The skin becomes hypertrophic (Hill 1985). The affected fingers develop a flexion deformity and cannot be extended. The etiology is unknown but a number of conditions such as epilepsy and chronic alcoholism are associated with Dupuytren's contracture (Noble and others 1992). Some observations indicate that multiple minor traumatic injuries may be etiologically associated with development of this condition (Nemetschek and others 1976; Mikkelsen 1978). Genetic factors are probably the paramount factors in this condition. Surgery is the only therapy.

Ganglion Cysts

A ganglion cyst is a fluid-filled swelling expanding from a joint or a tendon sheath. These cysts are commonly found at the dorsal part of the wrist, but sometimes also found on the volar aspect of the wrist or around the ankle. They are more common among young women.

Ganglions are painless and may be easily observed as they protrude over the dorsum of the hand. Sometimes ganglions are not visible and they may also be painful in these cases. Many ganglions disappear without any special treatment. They may be punctured and the fluid aspirated, or the viscous fluid may be displaced by firm manual pressure. In chronic cases, surgery to remove the sac may be required.

Physical and repetitive work have been related to the genesis of ganglions or to their progress (Barton and others 1992; Kroemer 1992).

Herbeden-Bochards Osteoarthritis

Osteoarthritis of the finger joints is a very common problem among elderly women. The English physician William Heberden first described it during the 18th century. The degenerative process often starts in the distal interphalangeal joints but may progress to most of the small joints of the hand. The condition is usually believed to be a hereditary form of osteoarthritis and it is much more common among females. In most cases no treatment is required and only in very few cases is work capacity reduced.

Osteoarthritis of the First Carpometacarpal Joint

Osteoarthritis of the first carpometacarpal joint (CMC I) is a very common disorder. The cause is unknown in most cases and very often the os-

teoarthritis is asymptomatic. Some patients, however, have pain when using the thumb and differential diagnosis may include De Quervain's disease. A relation to exposure in farming has been discussed. Older farmers often relate injuring their thumbs on the tractor steering wheel before improved steering mechanisms for tractors were introduced in the late 1940s. The direct mechanical link between the tractor wheels and the steering wheel in old tractors sometimes made the steering wheel quickly and unpredictably rotate with great force when the tractor would hit a bump. The driver could catch his thumb on the wheel causing injury. It is possible that this common "historical" trauma may be related to later problems of osteoarthritis of the thumbs.

Plain film x-rays will demonstrate the diagnosis. An orthosis to immobilize the thumb may be helpful. NSAID may also be helpful. A few patients have more recurrent difficulties and more pain. They may be candidates for surgery, mostly an arthroplasty.

INJURIES OF THE HIP

Trochanteritis

The greater trochanter of the femur is the site of insertion of powerful muscles (gluteus medius and gluteus minimus). A bursa is located between these two muscles, and a second bursa is located between the gluteus medius and the tensor faschia lata (Figure 8.12). Inflammatory reactions such as bursitis and tendonitis and/or enthesopathy are common soft tissue lesions in this region and it is often difficult to separate between the different locations (bursa, tendon, tendon insertion). Often there is a combination of structures inflamed.

Pain and tenderness are located to the lateral aspect of the trochanteric area, and in more severe cases the pain may radiate distally. The problems may follow a direct trauma (Haller and others 1989) but more commonly there is a gradual onset. Trochanteritis is related to sporting activities (e.g., football or rugby), but it is also reported in middle-aged and elderly, often overweight females (Little 1979).

Tenderness is characteristically localized over the trochanter and is palpated with the patient lying on the unaffected side. Pain may be reproduced by resisted contraction of the muscles. Arthritis of the hip joint is excluded by radiogra-

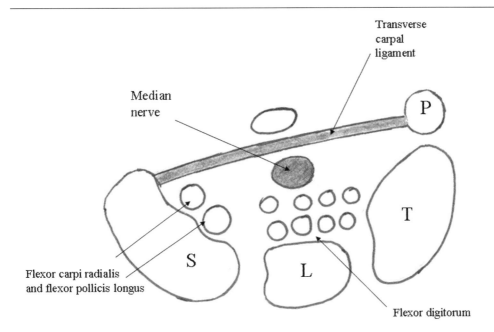

Transverse
carpal
ligament

Median
nerve

P

T

S

L

Flexor carpi radialis
and flexor pollicis longus

Flexor digitorum

Figure 8.10.   The carpal tunnel.

phy. Physiotherapy in combination with NSAID may be helpful, and the patient should rest from activities that produce pain. An injection of a corticosteroid in combination with local anesthetic may also help the patient.

Osteoarthritis of the Hip

Osteoarthritis of the hip can result from several factors, including 1) congenital hip disease, 2) Legg-Perthes disease, 3) slipped upper femoral epiphysis, and 4) trauma with fracture of the femoral head or acetabulum. A large number of studies have demonstrated a relationship between osteorarthritis of the hip and farming (Thelin 1985; Vingård and others 1991; Croft and others 1992; Thelin and others 1997). A comparison of colon ragiographs and urograms on farmers and on an urbanized population demonstrated a tenfold overrepresentation of hip osteoarthrtis among the farmers (Axmacher and Lindberg 1993). The high risk of osteoarthritis of the hip among farmers has been proposed to be associated with tractor driving (Torén and others 2002). However, a recent study demonstrated a significant risk of osteoarthritis associated with animal contacts (dairy farming and swine production) with no special risk to tractor drivers (Thelin and others 2004).

Apparently the risks include both tractor driving and raising livestock in combination. In some rural areas osteoarthritis of the hip is such a common disorder that most farmers know one or more fellow farmers with the disease.

Osteoarthritis of the hip in relation to other occupations has been investigated (Vingård and others 1991; Maetzel and others 1997). However, the results have not shown a consistent risk, as with farming.

Osteoarthritis of the hip as well as the knee are also related to obesity and negatively related to osteoporosis (Felson and others 1988). Smoking is also negatively associated with osteoarthritis (Felson and others 1989; Thelin, Jansson et al. 1997).

Osteoarthritis is a disease of the articular cartilage. The cartilage surface degenerates with fibrillation, fissures, and a general loss of cartilage. A low-grade inflammatory reaction occurs, and as the disease progresses the subchondral bone is affected and osteophytic changes appear at the margin of the cartilage. In severe cases the cartilage has disappeared, bone attrition is established, and the bone structure degenerates. The disorder may be one-sided at first but often the other hip joint is affected later (Figure 8.11).

Pain is first noticed in the groin and anteriome-

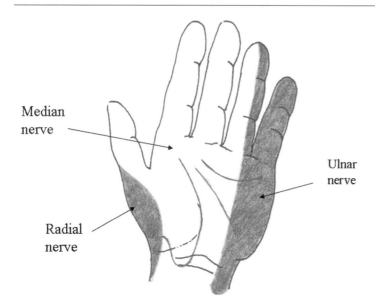

Median
nerve

Ulnar
nerve

Radial
nerve

Figure 8.11. The sensory distribution of the median nerve.

dial part of the thigh. The onset is insidious. The correlation between the radiological changes and the degree of pain is not consistent. Most patients first notice problems after they have been walking on rough ground, in snow or during some unaccustomed activity. Later the patients may have pain not only during activity but also at rest and during the night. Initially the patient observes only slight restriction of movement; but this progresses and bending at the waist becomes difficult, especially in the morning. Patients may have problems putting on socks or tying the shoelaces. Walking distance is reduced and the patient develops a limp. The rate of impairment is variable. Some patients pass from first symptoms to surgery in a 2-year period while others just have small problems for decades.

In severe cases range of motion is restricted in all directions and often there is a reduction of the length of the leg. A plain x-ray gives the diagnosis (Figure 8.13).

In early cases the patient may be helped by NSAID therapy. Physiotherapy and systematic exercise reduce the functional problems and prepares the patient for a better and more rapid recovery when surgery is to be performed (Klassbo and others 2003). If the patient is obese, weight loss is desirable. In severe cases a total hip replacement operation is the standard procedure.

After hip replacement surgery, most patients rapidly recover. Their pain is gone and the range

of motion is in most cases dramatically restored. Farmers are anxious to return to full-time work following surgery. This may be problematic and most orthopedists recommend the patient take a low-risk, physically light job to avoid problems with the prosthesis.

INJURIES OF THE KNEE, ANKLE, AND FOOT

Knee Ligament Injuries

Ligament sprains and strains are common soft tissue injuries prevalent in athletes and in workers. Farmers active in handling cattle or pigs are at risk for these injuries. Knee injuries are also frequently reported in dairy farmers who work daily directly with 600–700-kg animals. Accurate diagnosis of knee ligament sprains is difficult. MRI and arthroscopy may be essential tools.

Longtime difficulties may be related to untreated ligament injuries (Simonet and Sim 1984). Especially rupture of the cruciate ligaments may not be properly diagnosed. Defects may result in a feeling of instability, pain, and a secondary development of osteoarthritis. Patients with knee instability usually report a previous trauma, but not always.

Patients with chronic knee problems related to ligament injuries should be referred to orthopedic surgeons and specialists in sports medicine.

Bursitis of the Knee

Working on the knees (e.g., milking cows, repairing machinery, cleaning, etc.) or a direct blow can produce a prepatellar bursitis. The prepatellar bursa as well as the infrapatellar bursa is exposed to pressure in many types of jobs. Localized swelling with no warmth is noted in front of the patella. The treatment is usually symptomatic. Aspiration of the bursa may be done but often the bursa refills quickly.

Chondromalacia Patellae

A softening, fibrillation, and roughening of the under-surface of the patellar articular cartilage is called *chondromalacia*. The pathological changes have some differences compared to those in osteoarthritis (Bentley and Dowd 1984). Chondromalacia is generally not a progressive disorder.

Chondromalacia patients report pain as a deep-seated ache in the retropatellar area. Sitting for a prolonged time in cars, planes, and tractors with the knee flexed is problematic. The patient may have to get up and walk around for a while. Stiffness and a sensation of instability of the knee are frequently reported. Exercise to strengthen the quadriceps muscles is essential therapy. Also stretching may be helpful and activities that can provoke pain should be avoided. NSAIDs may be given. The long-term prognosis is good.

There are other forms of retropatellar dysfunction that may need to be differentiated by use of MRI or other diagnostic procedures (Merchant 1988; Reid 1992).

Injuries of the Meniscus in the Knee Joint

Tears of the meniscus are a common injury among males 20–30 years old. It may occur in any type of sports; contact sports are most common. Farmers handling cattle (especially dairy farmers) are at risk as well as people in occupations with prolonged squatting or kneeling and elderly with degenerative changes. Frequently patients with defective meniscis do not remember any specific trauma related to their injury. Symptoms of meniscal injuries may arise long after the initial trauma. A history of the knee "locking up" is characteristic. This indicates that a fragment of the meniscus is trapped between the condyles of the femur and tibia, preventing further motion. Some patients report painful "clicks" when they

walk or a "snapping" sensation when they stand up. Swelling in the knee is often present and the patient commonly has a feeling of instability of the knee. Quadriceps atrophy is a common finding. MRI or arthroscopy confirms the diagnosis. Most patients are effectively helped by arthroscopic surgery. Injuries of the meniscus and/or excision of the meniscus predispose to later development of osteoarthritis (Graf and others 1987; Holmberg and others 2005). Therefore early surgical repair may reduce osteoarthritis in the affected joints in later years.

Baker's Cyst

Baker's cyst (popliteal cyst) in the popliteal fossa communicates with the synovial cavity of the knee. Therefore inflammation of the knee, which generates excess synovial fluid, may result in swelling of the popliteal fossa. The patient reports a feeling of tension in the back of the knee and sometimes pain. Examination and treatment of this condition should focus on the knee joint.

Osteoarthritis of the Knee

Farmers appear to be at elevated risk of degenerative osteoarthritis (OA) of the knee (Vingård and others 1991; Sandmark and others 2000), although results have not been reproduced in all studies (Sahlström and Montgomery 1997; Holmberg and others 2004).

OA of the knee is a very prevalent disorder among elderly. As farmers are often elderly, they have an exacerbated risk. The condition is more common among women and there is a strong relation to obesity (Felson and Chaisson 1997; Holmberg and others 2005). Previous injuries such as tears of the meniscus, ligament sprains and knee instability may all contribute to a future OA of the knee. The process may start in any of the cartilage areas and eventually expand to all of these compartments. Medial tibiofemoral OA is most common (Figure 8.14).

Besides farming, other occupations have an increased risk for OA of the knee, including mining, dock work, and shipyard work (Kellgren and Lawrence 1952; Partridge and Duthie 1968; Lindberg and Montgomery 1987). Work that requires squatting, sitting, and kneeling have in other studies been related to OA of the knee (Felson and others 1991; Cooper and others 1994; Coggon and others 2000).

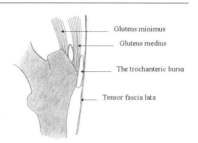

Figure 8.12. The trochanteric bursa: (A) the trochanteric bursa; (B) tensor fascia lata; (c) gluteus medius tendon; (d) gluteus minimus.

OA of the knee has pathology similar to osteoarthritis in other major joints. In severe cases of knee OA, a varus (or valgus) deformity may develop. A disparity in leg length may develop and the patient may have acquired a limp. The disorder may be in one leg first, but often the other knee joint is affected later on. As the degenerative process progresses, the knee starts to deform. Excess of synovial fluid, resulting in swelling and crepitous, is a common finding. Movement of the joint is restricted. Plain X-ray gives the diagnosis but it should be carried out while standing so as not to miss the narrowing of the cartilage space. In early cases tears of the meniscus, chondromalacia, or loose pieces of cartilage in the joint space may be included in a differential diagnosis.

Physical therapy and activity are important to maintain the quadricep's muscle capacity and the stability of the joint (Bischoff and Roos 2003). Weight loss is necessary if the patient is overweight. Antiinflammatory drugs may be used as well as localized steroid injections. A brace for the knee may be used temporarily.

Several surgical interventions are often very successful. In cases with significant and dominating medial tibiofemoral osteoarthritis, a wedge osteotomy may help the patient for many years. In

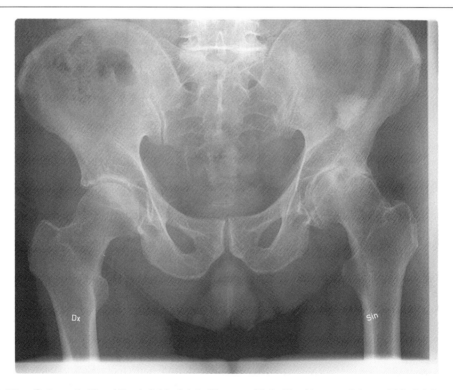

Figure 8.13. Osteoarthritis of the left hip joint. (Source: Plain film X-ray pelvis and hip joints, Department of Radiology, Central Hospital, Wexiö, Sweden)

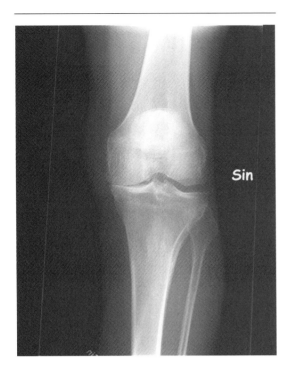

Figure 8.14. Osteoarthritis of the knee. (Source: Plain film X-ray left knee, Department of Radiology, Central Hospital, Wexiö, Sweden)

other cases a unilateral or total knee replacement athroplasty is the best option.

Ankle Sprains

Ankle sprains and malleolar fractures are common in farming and are often the result of slipping on icy or muddy grounds or after struggle with cattle or horses.

The ankle is supported by a large number of ligaments connecting the tibia and fibula with talus and calcaneus. The ligament most often injured is the anterior talofibular ligament, but often several of the ligaments are sprained. A sprain is often combined with a malleolar fracture. Plain x-rays can rule out a fracture (but this may take several views and an experienced reader to identify an ankle fracture). Pain may vary from minimal to severely limiting walking. Local swelling and tenderness are found at the site of ligament damage. If not properly treated ankle fractures and complicated ligament injuries may result in later development of OA and other disabling conditions.

GENERAL DISORDERS, INFECTIONS, AND REACTIVE ARTHRITIS

Rheumatoid diseases are alterations of the immune system for which the complete etiologies are not fully known. Exposure to certain microbes, hereditary factors, certain occupational activities, and certain organic dust exposures are possible risk factors related to rheumatoid diseases (Fingeroth and others 1984; Klockars and others 1987). No studies have been found indicating any association between farming and rheumatoid diseases.

Rheumatoid arthritis—and in some instances maybe psoriasis—are both chronic disabling diseases.

Inflammatory Bowel Diseases

Ulcerative colitis and Crohn's disease are both chronic inflammatory bowel diseases, which sometimes are accompanied by arthritic symptoms (Holden and others 2003). A special relationship between Crohn's disease and ankylosing spondylitis has been described (Breban and others 2003). It has also been shown that Crohn's disease is more common among people with sedentary work (Sonnenberg 1990; Boggild and others 1996) and less prevalent among farmers (Cucino and Sonnenberg 2001). The predicted mortality risk for farmers was 30% below comparison populations.

Reactive Arthritis

Reactive arthritis is classically seen following infection with enteric pathogens such as Salmonella, Shigella, Campylobacter, and Yersinia (Hill Gaston and Lillicrap 2003; Soderlin and others 2003). The etiologic mechanisms may be related to those previously discussed for ankylosing spondylitis (Yu and Kuipers 2003). It is worth noting that many of these infectious agents may be found in a farming environment. Most concerns of controlling these infections are a veterinary responsibility but all these infections may also be a zoonotic risk (Cover and Aber 1989; Seuri and Granfors 1992).

In children younger than 10 years, postinfective arthritis may be more severe than the enteric infection (Bottone 1977). Symptoms of arthritis may arise a few days to a month after the onset of gastrointestinal symptoms. Often the symptoms persist for months. Joint abnormalities are reported more often in Scandinavian countries than in the U.S. (Winblad 1975; Vantrappen and others 1977). Arthritic complications are more common and more severe among individuals who have the HLA-B27 histocompatibility type. Patients with autoimmune abnormalities and arthritis symptoms should be screened for serologic or cultural evidence for a Yersinia infection (Laitinen and others 1977).

Direct septic infections caused by Yersinia and Salmonella are occasional causes of spondylitis and focal osteomyelitis (Mahlfeld and others 2003). Salmonella osteomyelitis may involve any part of the skeleton but most commonly affects the long bones, the chondrosternal junctions, and the spine. Persons with sickle cell disease or prosthesis have a special risk (Rubin and others 1991).

REFERENCES

Adolfsson, L. and J. Lysholm (1989). "Arthroscopy and stability testing for anterior shoulder instability." Arthroscopy 5(4):315–320.

Allander, E. (1974). "Prevalence, incidence, and remission rates of some common rheumatic diseases or syndromes." Scand J Rheumatol 3(3):145–153.

Andersen, J. H., A. Kaergaard, et al. (2003). "Risk factors in the onset of neck/shoulder pain in a prospective study of workers in industrial and service companies." Occup Environ Med 60(9):649–654.

Anderson, N. H., J. O. Sojbjerg, et al. (1999). "Self-training versus physiotherapist-supervised rehabilitation of the shoulder in patients treated with arthroscopic subacromial decompression: A clinical randomized study." J Shoulder Elbow Surg 8(2):99–101.

Atlas, S. J., R. A. Deyo, et al. (1996). "The Quebec Task Force classification for Spinal Disorders and the severity, treatment, and outcomes of sciatica and lumbar spinal stenosis." Spine 21(24):2885–2892.

Awerbuch, M. S. (1992). "Whiplash in Australia: Illness or injury?" Med J Aust 157(3):193–196.

Axmacher, B. and H. Lindberg (1993). "Coxarthrosis in farmers." Clin Orthop(287):82–86.

Backman, A. L. (1983). "Health survey of professional drivers." Scand J Work Environ Health 9(1):30–35.

Barnsley, L., S. Lord, et al. (1994). "Whiplash injury." Pain 58(3):283–307.

Barton, N. J., G. Hooper, et al. (1992). "Occupational causes of disorders in the upper limb." Bmj 304(6822):309–311.

Battie, M. C., D. R. Haynor, et al. (1995). "Similarities in degenerative findings on magnetic resonance images of the lumbar spines of identical twins." J Bone Joint Surg Am 77(11):1662–1670.

Bejjani, F. J., G. M. Kaye, et al. (1996). "Musculoskeletal and neuromuscular conditions of instrumental musicians." Arch Phys Med Rehabil 77(4):406–413.

Benoist, M. and J. P. Rouaud (2002). "Whiplash: Myth or reality?" Joint Bone Spine 69(4):358–362.

Bentley, G. and G. Dowd (1984). "Current concepts of etiology and treatment of chondromalacia patellae." Clin Orthop(189):209–228.

Bergenudd, H., F. Lindgarde, et al. (1988). "Shoulder pain in middle age. A study of prevalence and relation to occupational work load and psychosocial factors." Clin Orthop(231):234–238.

Bigos, S. J., M. C. Battie, et al. (1991). "A prospective study of work perceptions and psychosocial factors affecting the report of back injury." Spine 16(1):1–6.

—— (1992). "A prospective evaluation of preemployment screening methods for acute industrial back pain." Spine 17(8):922–926.

Bischoff, H. A. and E. M. Roos (2003). "Effectiveness and safety of strengthening, aerobic, and coordination exercises for patients with osteoarthritis." Curr Opin Rheumatol 15(2):141–144.

Boggild, H., F. Tuchsen, et al. (1996). "Occupation, employment status and chronic inflammatory bowel disease in Denmark." Int J Epidemiol 25(3):630–637.

Bongers, P. M., C. R. de Winter, et al. (1993). "Psychosocial factors at work and musculoskeletal disease." Scand J Work Environ Health 19(5):297–312.

Boos, N., R. Rieder, et al. (1995). "1995 Volvo Award in clinical sciences. The diagnostic accuracy of magnetic resonance imaging, work perception, and psychosocial factors in identifying symptomatic disc herniations." Spine 20(24):2613–2625.

Borenstein, D. G. (1996). "Chronic low back pain." Rheum Dis Clin North Am 22(3):439–456.

Boshuizen, H. C., P. M. Bongers, et al. (1990a). "Self-reported back pain in tractor drivers exposed to whole-body vibration." Int Arch Occup Environ Health 62(2):109–115.

Boshuizen, H. C., C. T. Hulshof, et al. (1990b). "Long-term sick leave and disability pensioning due to back disorders of tractor drivers exposed to whole-body vibration." Int Arch Occup Environ Health 62(2):117–122.

Bottone, E. J. (1977). "Yersinia enterocolitica: A panoramic view of a charismatic microorganism." CRC Crit Rev Microbiol 5(2):211–241.

Boureau, F., T. Delorme, et al. (2000). "[Mechanisms of myofascial pain]." Rev Neurol (Paris) 156 Suppl 4:4S10–14.

Bovenzi, M. (1994). "Hand-arm vibration syndrome and dose-response relation for vibration induced white finger among quarry drillers and stonecarvers.

Italian Study Group on Physical Hazards in the Stone Industry." Occup Environ Med 51(9): 603–611.

Brain, W. R., A. D. Wright, et al. (1947). "Spontaneous compression of both median nerves in the carpal tunnel: Six cases treated surgically." Lancet 1:277–282.

Breban, M., R. Said-Nahal, et al. (2003). "Familial and genetic aspects of spondyloarthropathy." Rheum Dis Clin North Am 29(3):575–594.

Browne, E. Z., Jr. and C. C. Snyder (1975). "Carpal tunnel syndrome caused by hand injuries." Plast Reconstr Surg 56(1):41–43.

Brox, J. I., E. Gjengedal, et al. (1999). "Arthroscopic surgery versus supervised exercises in patients with rotator cuff disease (stage II impingement syndrome): A prospective, randomized, controlled study in 125 patients with a 2 1/2-year follow-up." J Shoulder Elbow Surg 8(2):102–111.

Brox, J. I., P. H. Staff, et al. (1993). "Arthroscopic surgery compared with supervised exercises in patients with rotator cuff disease (stage II impingement syndrome)." Bmj 307(6909):899–903.

Buckle, P. W. and J. J. Devereux (2002). "The nature of work-related neck and upper limb musculoskeletal disorders." Appl Ergon 33(3):207–217.

Calin, A., J. Porta, et al. (1977). "Clinical history as a screening test for ankylosing spondylitis." Jama 237(24):2613–2614.

Campbell, S. M. (1989). "Regional myofascial pain syndromes." Rheum Dis Clin North Am 15(1): 31–44.

Cassidy, J. D., L. J. Carroll, et al. (2000). "Effect of eliminating compensation for pain and suffering on the outcome of insurance claims for whiplash injury." N Engl J Med 342(16):1179–1186.

Cassvan, A., L. D. Weiss,et al. (1997). Cumulative Trauma Disorders. Boston, Butterworth-Heinemann.

Chaffin, D. and G. Andersson (1991). Occupational Biomechanics. New York.

Chammas, M., P. Bousquet, et al. (1995). "Dupuytren's disease, carpal tunnel syndrome, trigger finger, and diabetes mellitus." J Hand Surg [Am] 20(1): 109–114.

Chard, M. D., L. M. Sattelle, et al. (1988). "The long-term outcome of rotator cuff tendonitis—a review study." Br J Rheumatol 27:385–389.

Cherkin, D. C., R. A. Deyo, et al. (1996). "Predicting poor outcomes for back pain seen in primary care using patients' own criteria." Spine 21(24): 2900–2907.

Chiang, H. C., Y. C. Ko, et al. (1993). "Prevalence of shoulder and upper-limb disorders among workers in the fish-processing industry." Scand J Work Environ Health 19(2):126–131.

Cofield, R. H. and J. F. Irving (1987). "Evaluation and classification of shoulder instability. With special reference to examination under anesthesia." Clin Orthop(223):32–43.

Coggon, D., P. Croft, et al. (2000). "Occupational physical activities and osteoarthritis of the knee." Arthritis Rheum 43(7):1443–1449.

Coomes, E. N. (1961). "A comparison between epidural anaesthesia and bed rest in sciatica." Br Med J 5218:20–24.

Cooper, C., T. McAlindon, et al. (1994). "Occupational activity and osteoarthritis of the knee." Ann Rheum Dis 53(2):90–93.

Cover, T. L. and R. C. Aber (1989). "Yersinia enterocolitica." N Engl J Med 321(1):16–24.

Croft, P., C. Cooper, et al. (1992). "Osteoarthritis of the hip and occupational activity." Scand J Work Environ Health 18(1):59–63.

Croft, P. R., M. Lewis, et al. (2001). "Risk factors for neck pain: A longitudinal study in the general population." Pain 93(3):317–325.

Cucino, C. and A. Sonnenberg (2001). "Occupational mortality from inflammatory bowel disease in the United States 1991–1996." Am J Gastroenterol 96(4):1101–1105.

Cuetter, A. C. and D. M. Bartoszek (1989). "The thoracic outlet syndrome: Controversies, overdiagnosis, overtreatment, and recommendations for management." Muscle Nerve 12(5):410–419.

Dahl, J. B., C. J. Erichsen, et al. (1992). "Pain sensation and nociceptive reflex excitability in surgical patients and human volunteers." Br J Anaesth 69(2):117–121.

Dekel, S., T. Papaioannou, et al. (1980). "Idiopathic carpal tunnel syndrome caused by carpal stenosis." Br Med J 280(6227):1297–1299.

Deyo, R. A. (1988). "Measuring the functional status of patients with low back pain." Arch Phys Med Rehabil 69(12):1044–1053.

Deyo, R. A., A. K. Diehl, et al. (1986). "How many days of bed rest for acute low back pain? A randomized clinical trial." N Engl J Med 315(17):1064–1070.

Dorwart, B. B. (1984). "Carpal tunnel syndrome: A review." Semin Arthritis Rheum 14(2):134–140.

Dworkin, R. H. (1990). "Compensation in chronic pain patients: Cause or consequence?" Pain 43(3): 387–388.

Dyrehag, L. E., E. G. Widerstrom-Noga, et al. (1998). "Relations between self-rated musculoskeletal symptoms and signs and psychological distress in chronic neck and shoulder pain." Scand J Rehabil Med 30(4):235–242.

Evans, W., W. Jobe, et al. (1989). "A cross-sectional prevalence study of lumbar disc degeneration in a working population." Spine 14(1):60–64.

Farfan, H. F. (1977). "A reorientation in the surgical approach to degenerative lumbar intervertebral joint disease." Orthop Clin North Am 8(1):9–21.

Farkkila, M., S. Aatola, et al. (1986). "Hand-grip force in lumberjacks: Two-year follow-up." Int Arch Occup Environ Health 58(3):203–208.

Feldman, R. G., P. H. Travers, et al. (1987). "Risk assessment in electronic assembly workers: Carpal tunnel syndrome." J Hand Surg [Am] 12(5 Pt 2): 849–855.

Felson, D. T., J. J. Anderson, et al. (1988). "Obesity and knee osteoarthritis. The Framingham Study." Ann Intern Med 109(1):18–24.

—— (1989). "Does smoking protect against osteoarthritis?" Arthritis Rheum 32(2):166–172.

Felson, D. T. and C. E. Chaisson (1997). "Understanding the relationship between body weight and osteoarthritis." Baillieres Clin Rheumatol 11(4):671–681.

Felson, D. T., M. T. Hannan, et al. (1991). "Occupational physical demands, knee bending, and knee osteoarthritis: Results from the Framingham Study." J Rheumatol 18(10):1587–1592.

Fernstrom, E. and M. O. Ericson (1997). "Computer mouse or Trackpoint—Effects on muscular load and operator experience." Appl Ergon 28(5–6):347–354.

Ferrari, R. (1999). The Whiplash Encyclopedia: The Facts and Myths of Whiplash. Gaitersburg, Aspen Publishers.

—— (2003). "Re: Whiplash disorders—A review." Injury 34(10):803–805; author reply 805–806.

Ferrari, R. and A. S. Russell (2003). "Regional musculoskeletal conditions: Neck pain." Best Pract Res Clin Rheumatol 17(1):57–70.

Fingeroth, J. D., J. J. Weis, et al. (1984). "Epstein-Barr virus receptor of human B lymphocytes is the C3d receptor CR2." Proc Natl Acad Sci U S A 81(14):4510–4514.

Floman, Y. (2000). "Progression of lumbosacral isthmic spondylolisthesis in adults." Spine 25(3):342–347.

Fongemie, A. E., D. D. Buss, et al. (1998). "Management of shoulder impingement syndrome and rotator cuff tears." Am Fam Physician. 57(4):667–74,680–82.

Fransen, M., M. Woodward, et al. (2002). "Risk factors associated with the transition from acute to chronic occupational back pain." Spine 27(1):92–98.

Freund, B. and M. Schwartz (2002). "Post-traumatic myofascial pain of the head and neck." Curr Pain Headache Rep 6(5):361–369.

Frost, P. and J. H. Andersen (1999). "Shoulder impingement syndrome in relation to shoulder intensive work." Occup Environ Med 56(7):494–498.

Frymoyer, J. and S. Gordon (1992). New Perspectives on Low Back Pain. American Academy of Orthopaedic Surgeons.

Giebel, G. D., A. D. Bonk, et al. (1999). "Whiplash injury." J Rheumatol 26(5):1207–1208; author reply 1208–1210.

Gorski, J. M. and L. H. Schwartz (2003). "Shoulder impingement presenting as neck pain." J Bone Joint Surg Am 85-A(4):635–638.

Goss, T. P. (1988). "Anterior glenohumeral instability." Orthopedics 11(1):87–95.

Graf, B., T. Docter, et al. (1987). "Arthroscopic meniscal repair." Clin Sports Med 6(3):525–536.

Gran, J. T., G. Husby, et al. (1985). "Spinal ankylosing spondylitis: A variant form of ankylosing spondylitis or a distinct disease entity?" Ann Rheum Dis 44(6):368–371.

Gruchow, H. W. and D. Pelletier (1979). "An epidemiologic study of tennis elbow. Incidence, recurrence, and effectiveness of prevention strategies." Am J Sports Med 7(4):234–238.

Guidotti, T. L. (1992). "Occupational repetitive strain injury." Am Fam Physician 45(2):585–592.

Haahr, J. P. and J. H. Andersen (2003). "Prognostic factors in lateral epicondylitis: A randomized trial with one-year follow-up in 266 new cases treated with minimal occupational intervention or the usual approach in general practice." Rheumatology (Oxford) 42(10):1216–1225.

Hagberg, M., H. Morgenstern, et al. (1992). "Impact of occupations and job tasks on the prevalence of carpal tunnel syndrome." Scand J Work Environ Health 18(6):337–345.

Hagberg, M. and D. H. Wegman (1987). "Prevalence rates and odds ratios of shoulder-neck diseases in different occupational groups." Br J Ind Med 44(9):602–610.

Haller, C. C., P. A. Coleman, et al. (1989). "Traumatic trochanteric bursitis." Kans Med 90(1):17–18,22.

Hanock, J. (1995). "Comments on Barnsley et al." Pain 61:487–495.

Hansson, T. H. (1989). Ländryggsbesvär och arbete. Stockholm, Arbetsmiljöfonden.

Harvey, F. J., P. M. Harvey, et al. (1990). "De Quervain's disease: Surgical or nonsurgical treatment." J Hand Surg [Am] 15(1):83–87.

Hemingway, H., M. J. Shipley, et al. (1997). "Sickness absence from back pain, psychosocial work characteristics and employment grade among office workers." Scand J Work Environ Health 23(2):121–129.

Herberts, P., R. Kadefors, et al. (1981). "Shoulder pain in industry: An epidemiological study on welders." Acta Orthop Scand 52(3):299–306.

Hill Gaston, J. S. and M. S. Lillicrap (2003). "Arthritis associated with enteric infection." Best Pract Res Clin Rheumatol 17(2):219–239.

Hill, N. A. (1985). "Dupuytren's contracture." J Bone and Joint Surgery 67:1439–1443.

Hirsch, C., B. E. Ingelmark, et al. (1963). "The anatomical basis for low back pain. Studies on the presence of sensory nerve endings in ligamentous, capsular and intervertebral disc structures in the human lumbar spine." Acta Orthop Scand 33:1–17.

Holden, W., T. Orchard, et al. (2003). "Enteropathic arthritis." Rheum Dis Clin North Am 29(3): 513–530, viii.

Holmberg, S., E.-L. Stiernström, et al. (2002). "Musculoskeletal symptoms among farmers and non-farmers: A population-based study." Int J Occup Environ Health 8(4):339–345.

Holmberg, S., A. Thelin, et al. (2003). "The impact of physical work exposure on musculoskeletal symptoms among farmers and rural non-farmers." Ann Agric Environ Med 10(2):179–184.

—— (2004a). "Psychosocial factors and low back pain, consultations, and sick leave among farmers and rural referents: A population-based study." J Occup Environ Med 46(9):993–938.

—— (2004b). "Is there an increased risk of knee osteoarthritis among farmers? A population-based case-control study." Int Arch Occup Environ Health 77(5):345–350.

—— (2005). "Knee osteoarthritis and body mass index. A population-based case control study." Scand J Rheumatol, in press.

Holte, K. A., O. Vasseljen, et al. (2003). "Exploring perceived tension as a response to psychosocial work stress." Scand J Work Environ Health 29(2): 124–133.

Hong, C. Z. (2002). "New trends in myofascial pain syndrome." Zhonghua Yi Xue Za Zhi (Taipei) 65(11):501–512.

Horsfield, D. and J. Stutley (1988). "The unstable shoulder A problem solved." Radiography 54(614): 74–76.

Hubbard, D. R. and G. M. Berkoff (1993). "Myofascial trigger points show spontaneous needle EMG activity." Spine 18(13):1803–1807.

Hutson, M. A. (1996). Sports Injuries: Recognition and Management. Oxford, Oxford University Press.

Hutton, W. C., T. M. Ganey, et al. (2000). "Does long-term compressive loading on the intervertebral disc cause degeneration?" Spine 25(23): 2993–3004.

Indahl, A., A. Kaigle, et al. (1995). "Electromyographic response of the porcine multifidus musculature after nerve stimulation." Spine 20(24):2652–2658.

Johansson, J. A. and S. Rubenowitz (1994). "Risk indicators in the psychosocial and physical work environment for work-related neck, shoulder and low back symptoms: A study among blue- and white-collar workers in eight companies." Scand J Rehabil Med 26(3):131–142.

Jonsson, B. (1988). "The static load component in muscle work." Eur J Appl Physiol 57:305–310.

Jonsson, E. and A. Nachemson (2000). Ont i ryggen. Ont i nacken. Stockholm, SBU.

Kaigle, A. M., P. Wessberg, et al. (1998). "Muscular and kinematic behavior of the lumbar spine during flexion-extension." J Spinal Disord 11(2):163–174.

Katevuo, K., K. Aitasalo, et al. (1985). "Skeletal changes in dentists and farmers in Finland." Community Dent Oral Epidemiol 13(1):23–25.

Kellgren, J. H. and J. S. Lawrence (1952). "Rheumatism in miners. II. X-ray study." Br J Ind Med 9(3):197–207.

—— (1958). "Osteo-arthrosis and disk degeneration in an urban population." Ann Rheum Dis 17(4): 388–397.

Kessel, L., I. Bayley, et al. (1981). "The upper limb: The frozen shoulder." Br J Hosp Med 25(4):334, 336–337, 339.

Kieft, G. J., J. L. Bloem, et al. (1988). "MR imaging of recurrent anterior dislocation of the shoulder: Comparison with CT arthrography." AJR Am J Roentgenol 150(5):1083–1087.

Kilbom, Å. (1986). "Disorders of the cervicobrachial region among femal workers in the electronics industry." Int J Indust Ergonomics 1:37–47.

Kivi, P. (1982). "The aetiology and conservative treatment of humeral epicondylitis." Scand J Rehabil Med 15:37–41.

Klassbo, M., G. Larsson, et al. (2003). "Promising outcome of a hip school for patients with hip dysfunction." Arthritis Rheum 49(3):321–327.

Klockars, M., R. S. Koskela, et al. (1987). "Silica exposure and rheumatoid arthritis: A follow up study of granite workers 1940–81." Br Med J (Clin Res Ed) 294(6578):997–1000.

Kopell, H. P. and W. A. L. Thompson (1963). Peripheral Entrapment Neuropathies. Baltimore, Williams & Williams.

Krämer, J. (1973). Biomechanische Veränderungen im lumbalen Bewegungssegment. Die Wirbelsälue in Forschung und Praxis. Stuttgart, Hippokrates Verlag.

Krapac, L., A. Krmpotic, et al. (1992). "Cervicobrachial syndrome: Work and disability." Arh Hig Rada Toksikol 43(3):255–262.

Kroemer, K. H. (1992). "Avoiding cumulative trauma disorders in shops and offices." Am Ind Hyg Assoc J 53(9):596–604.

Kvarnström, S. (1983). "Occurrence of musculoskeletal, disorders in a manufacturing industry with special attention to occupational shoulder disorders." Scand J Rehabil Med Suppl 8:1–114.

Kwan, O. and J. Friel (2003). "A review and methodologic critique of the literature supporting 'chronic whiplash injury'. Part II. Reviews, editorials, and letters." Med Sci Monit 9(9):RA230–236.

Laitinen, O., M. Leirisalo, et al. (1977). "Relation between HLA-B27 and clinical features in patients with yersinia arthritis." Arthritis Rheum 20(5): 1121–1124.

Leclerc, A., P. Franchi, et al. (1998). "Carpal tunnel syndrome and work organisation in repetitive work: A cross sectional study in France. Study Group on Repetitive Work." Occup Environ Med 55(3): 180–187.

Levin, D. B. (1989). The painful back. Arthritis and Allied Conditions: A Textbook of Rheumatology. D. J. McCarty. Philadelphia, Lea & Febiger.

Lindberg, H. and F. Montgomery (1987). "Heavy labor and the occurrence of gonarthrosis." Clin Orthop(214):235–236.

Lindgren, K. A. (1997). "TOS (thoracic outlet syndrome)A challenge to conservative treatment." Nord Med 112(8):283–287.

Lings, S. and C. Leboeuf-Yde (2000). "Whole-body vibration and low back pain: A systematic, critical review of the epidemiological literature 1992–1999." Int Arch Occup Environ Health 73(5): 290–297.

Linton, S. J. (2000). "A review of psychological risk factors in back and neck pain." Spine 25(9): 1148–1156.

Linton, S. J., N. Buer, et al. (1999). "Are fear-avoidance beliefs related to a new episode of back pain?" Psychol Health 14:1051–1059.

Lipson, S. J. (1989). Low Back Pain. Textbook of Rheumatology. W. N. H. Kelley, E.D. Jr.; Ruddy, S. et al. Philadelphia, WB Saunders.

Little, H. (1979). "Trochanteric buritis: A common cause of pelvic girdel pain." Canadian Med Ass J 120:456–458.

Lovell, M. E. and C. S. Galasko (2002). "Whiplash disorders—A review." Injury 33(2):97–101.

Ludewig, P. M. and J. D. Borstad (2003). "Effects of a home exercise programme on shoulder pain and functional status in construction workers." Occup Environ Med 60(11):841–849.

Lundborg, G., C. Sollerman, et al. (1987). "A new principle for assessing vibrotactile sense in vibration-induced neuropathy." Scand J Work Environ Health 13(4):375–379.

Maetzel, A., M. Makela, et al. (1997). "Osteoarthritis of the hip and knee and mechanical occupational exposure—A systematic overview of the evidence." J Rheumatol 24(8):1599–1607.

Magora, A. (1974). "Investigation of the relation between low back pain and occupation. 6. Medical history and symptoms." Scand J Rehabil Med 6(2): 81–88.

Mahlfeld, K., J. Franke, et al. (2003). "[Spondylitis due to Salmonella typhimurium]." Unfallchirurg 106(4): 334–338.

Major, H. P. (1883). "Lawn-tennis elbow." Br Med J ii:557.

Mani, L. and F. Gerr (2000). "Work-related upper extremity musculoskeletal disorders." Prim Care 27(4):845–864.

Mannion, A. F., P. Dolan, et al. (1996). "Psychological questionnaires: Do "abnormal" scores precede or follow first-time low back pain?" Spine 21(22): 2603–2611.

Marchiori, D. M. and C. N. Henderson (1996). "A cross-sectional study correlating cervical radiographic degenerative findings to pain and disability." Spine 21(23):2747–2751.

Marks, S. H., M. Barnett, et al. (1983). "Ankylosing spondylitis in women and men: A case-control study." J Rheumatol 10:624.

Matsunaga, S., T. Sakou, et al. (1990). "Natural history of degenerative spondylolisthesis. Pathogenesis and natural course of the slippage." Spine 15(11): 1204–1210.

Mayer, T., R. Gatchel, et al. (2000). Occupational musculosceletal disorders. Function, outcomes and evidence. Philadelphia, PRIDE Research Foundation Scientific Publications.

McCormack, R. R., Jr., R. D. Inman, et al. (1990). "Prevalence of tendonitis and related disorders of the upper extremity in a manufacturing workforce." J Rheumatol 17(7):958–964.

Merchant, A. C. (1988). "Classification of patellofemoral disorders." Arthroscopy 4(4):235–240.

Mikkelsen, O. A. (1978). "Dupuytren's disease: The influence of occupation and previous hand injuries." Hand 10(1):1–8.

Miller, H. (1961). "Accident neurosis." Br Med J 5230:919–925.

Mills, H. and G. Horne (1986). "Whiplash—Manmade disease?" N Z Med J 99(802):373–374.

Modic, M. T. and R. J. Herfkens (1990). "Intervertebral disk: Normal age-related changes in MR signal intensity." Radiology 177(2):332–3; discussion 333–334.

Monsell, F. P. and R. M. Tillman (1992). "Shearer's wrist: The carpal tunnel syndrome as an occupational disease in professional sheep shearers." Br J Ind Med 49(8):594–595.

Murray, G. C. and R. H. Persellin (1981). "Cervical fracture complicating ankylosing spondylitis: A report of eight cases and review of the literature." Am J Med 70(5):1033–1041.

Muschik, M., H. Hahnel, et al. (1996). "Competitive sports and the progression of spondylolisthesis." J Pediatr Orthop 16(3):364–369.

Nemetschek, T., A. Meinel, et al. (1976). "[Aetiology of dupuytren's contracture (author's transl)]." Virchows Arch A Pathol Anat Histol 372(1): 57–74.

Noble, J., M. Arafa, et al. (1992). "The association between alcohol, hepatic pathology and Dupuytren's disease." J Hand Surg [Br] 17(1):71–74.

Nordstrom, D. L., R. A. Vierkant, et al. (1997). "Risk factors for carpal tunnel syndrome in a general population." Occup Environ Med 54(10):734–740.

Ohlsson, K., R. G. Attewell, et al. (1995). "Repetitive industrial work and neck and upper limb disorders in females." Am J Ind Med 27(5):731–747.

O'Neil, B. A., M. E. Forsythe, et al. (2001). "Chronic occupational repetitive strain injury." Can Fam Physician 47:311–316.

Osterman, K., D. Schlenzka, et al. (1993). "Isthmic spondylolisthesis in symptomatic and asymptomatic subjects, epidemiology, and natural history with special reference to disk abnormality and mode of treatment." Clin Orthop(297):65–70.

Palmer, K. T., C. Cooper, et al. (2001a). "Use of keyboards and symptoms in the neck and arm: Evidence from a national survey." Occup Med (Lond) 51(6): 392–395.

Palmer, K. T., K. Walker-Bone, et al. (2001b). "Prevalence and occupational associations of neck pain in the British population." Scand J Work Environ Health 27(1):49–56.

Park, H., N. L. Sprince, et al. (2001). "Risk factors for back pain among male farmers: Analysis of Iowa Farm Family Health and Hazard Surveillance Study." Am J Ind Med 40(6):646–654.

Partheni, M., G. Miliaras, et al. (1999). "Whiplash injury." J Rheumatol 26(5):1206–7; author reply 1208–1210.

Partridge, R. E. and J. J. Duthie (1968). "Rheumatism in dockers and civil servants. A comparison of heavy manual and sedentary workers." Ann Rheum Dis 27(6):559–568.

Pearce, J. M. (1999). "A critical appraisal of the chronic whiplash syndrome." J Neurol Neurosurg Psychiatry 66:273–276.

Pennie, B. and L. Agambar (1991). "Patterns of injury and recovery in whiplash." Injury 22(1):57–59.

Penttinen, J. (1987). Back Pain and Sciatica in Finnish Farmers. Social insurance institution. Helsinki, The Research Institute for Social Security:97.

Petersson, C. J. (1983). "Degeneration of the acromioclavicular joint. A morphological study." Acta Orthop Scand 54(3):434–438.

Phalen, G. S. (1966). "The carpal-tunnel syndrome. Seventeen years' experience in diagnosis and treatment of six hundred fifty-four hands." J Bone Joint Surg Am 48(2):211–228.

Phalen, G. S., L. McCormack, et al. (1959). "Giant-cell tumor of tendon sheath (benign synovioma) in the hand. Evaluation of 56 cases." Clin Orthop 15:140–151.

Poole, G. V. and K. R. Thomae (1996). "Thoracic outlet syndrome reconsidered." Am Surg 62(4): 287–291.

Postacchini, F. (1989). "The evolution of spondylolysis into spondylolisthesis during adult age." Ital J Orthop Traumatol 15(2):210–216.

Punnett, L., J. M. Robins, et al. (1985). "Soft tissue disorders in the upper limbs of female garment workers." Scand J Work Environ Health 11(6):417–425.

Punnett, L. and U. Bergquist (1997). Visual display unit work and upper extremity musculoskeletal disorders. Solna, Arbetslivsinstitutet.

Pyykkö, I., E. Sairanen, et al. (1978). "A decrease in the prevalence and severity of vibration-induced white fingers among lumberjacks in Finland." Scand J Work Environ Health 4(3):246–254.

Quebec task force. (1995). "Quebec task force report on whiplash associated disorders. Section 4. Best evidence synthesis." Spine 85:24–33.

Quintner, J. L. and M. L. Cohen (1994). "Referred pain of peripheral nerve origin: An alternative to the "myofascial pain" construct." Clin J Pain 10(3): 243–251.

Raffi, G. B., V. Lodi, et al. (1996). "Cumulative trauma disorders of the upper limbs in workers on an agricultural farm." Arh Hig Rada Toksikol 47(1): 19–23.

Rauschning, W., P. C. McAfee, et al. (1989). "Pathoanatomical and surgical findings in cervical spinal injuries." J Spinal Disord 2(4):213–222.

Reid, D. C. (1992). Sports Injury Assessment and Rehabilitation. Edinburgh, Churchill Livingstone.

Reiffenberger, D. H. and L. H. Amundson (1996). "Fibromyalgia syndrome: A review." Am Fam Physician 53(5):1698–1712.

Riihimäki, H., S. Tola, et al. (1989). "Low-back pain and occupation. A cross-sectional questionnaire study of men in machine operating, dynamic physical work, and sedentary work." Spine 14(2): 204–209.

Rosegger, R. and S. Rosegger (1960). "Health Effects of Tractor Driving." J Agric Eng Res 5:241–275.

Rossi, F. (1998). "Shoulder impingement syndromes." Eur J Radiol 27(suppl 1):S42–48.

Rossignol, M., S. Stock, et al. (1997). "Carpal tunnel syndrome: What is attributable to work? The Montreal study." Occup Environ Med 54(7): 519–523.

Roto, P. and P. Kivi (1984). "Prevalence of epicondylitis and tenosynovitis among meatcutters." Scand J Work Environ Health 10(3):203–205.

Rubin, M. M., R. J. Sanfilippo, et al. (1991). "Vertebral osteomyelitis secondary to an oral infection." J Oral Maxillofac Surg 49(8):897–900.

Russell, M. L. (1985). "Ankylosing spondylitis: The case of the underestimated femal." J Rheumatol 1985:1.

Sahlström, A. and F. Montgomery (1997). "Risk analysis of occupational factors influencing the development of arthrosis of the knee." Eur J Epidemiol 13(6):675–679.

Saldana, M. J. (2001). "Trigger digits: Diagnosis and treatment." J Am Acad Orthop Surg 9(4):246–252.

Sandmark, H., C. Hogstedt, et al. (2000). "Primary osteoarthrosis of the knee in men and women as a result of lifelong physical load from work." Scand J Work Environ Health 26(1):20–25.

Savage, R. A., G. H. Whitehouse, et al. (1997). "The relationship between the magnetic resonance imaging appearance of the lumbar spine and low back pain, age and occupation in males." Eur Spine J 6(2): 106–114.

Schrader, H., D. Obelieniene, et al. (1996). "Natural evolution of late whiplash syndrome outside the medicolegal context." Lancet 347(9010): 1207–1211.

Schultze, K.-J. and J. Polster (1979). "Berufsbedingte Wirbelsäulenschäden bei Traktoristen und Landwirten." Beit Orthoped u Traumatol 26:356–362.

Seuri, M. and K. Granfors (1992). "Antibodies against Yersinia among farmers and slaughterhouse workers." Scand J Work Environ Health 18(2):128–132.

Silverstein, B. A., L. J. Fine, et al. (1987). "Occupational factors and carpal tunnel syndrome." Am J Ind Med 11(3):343–358.

Simonet, W. T. and F. H. Sim (1984). "Symposium on sports medicine: Part I. Current concepts in the treatment of ligamentous instability of the knee." Mayo Clin Proc 59(2):67–76.

Skov, T., V. Borg, et al. (1996). "Psychosocial and physical risk factors for musculoskeletal disorders of the neck, shoulders, and lower back in salespeople." Occup Environ Med 53(5):351–356.

Smythe, H. A. (1994). "The C6-7 syndrome—Clinical features and treatment response." J Rheumatol 21(8):1520–1526.

Söderlin, M. K., H. Kautiainen, et al. (2003). "Infections preceding early arthritis in southern Sweden: A prospective population-based study." J Rheumatol 30(3):459–464.

Solomonow, M., B. H. Zhou, et al. (1998). "The ligamento-muscular stabilizing system of the spine." Spine 23(23):2552–2562.

Solveborn, S. A. (1997). "Radial epicondylalgia ('tennis elbow'): Treatment with stretching or forearm band. A prospective study with long-term follow-up

including range-of-motion measurements." Scand J Med Sci Sports 7(4):229–237.

Sonnenberg, A. (1990). "Occupational distribution of inflammatory bowel disease among German employees." Gut 31(9):1037–1040.

Stål, M., C. G. Hagert, et al. (1998). "Upper extremity nerve involvement in Swedish female machine milkers." Am J Ind Med 33(6):551–559.

Stål, M., G. A. Hansson, et al. (1999). "Wrist positions and movements as possible risk factors during machine milking." Appl Ergon 30(6):527–533.

Stål, M., U. Moritz, et al. (1996). "Milking is a high-risk job for young females." Scand J Rehabil Med 28(2):95–104.

—— (1997). "The Natural Course of Musculoskeletal Symptoms and Clinical Findings in Upper Extremities of Female Milkers." Int J Occup Environ Health 3(3):190–197.

Stenlund, B. (1993). "Shoulder tendonitis and osteoarthrosis of the acromioclavicular joint and their relation to sports." Br J Sports Med 27(2):125–130.

Stenlund, B., I. Goldie, et al. (1993). "Shoulder tendonitis and its relation to heavy manual work and exposure to vibration." Scand J Work Environ Health 19(1):43–49.

—— (1992). "Radiographic osteoarthrosis in the acromioclavicular joint resulting from manual work or exposure to vibration." Br J Ind Med 49(8):588–593.

Stetson, D. S., B. A. Silverstein, et al. (1993). "Median sensory distal amplitude and latency: Comparisons between nonexposed managerial/professional employees and industrial workers." Am J Ind Med 24(2):175–189.

Stewart, J. D. (1987). "The variable clinical manifestations of ulnar neuropathies at the elbow." J Neurol Neurosurg Psychiatry 50(3):252–258.

Stevenson, J. H. and T. Trojian (2002). "Evaluation of shoulder pain." J Family Practice 51(7).

Tanaka, S., D. K. Wild, et al. (1997). "Association of occupational and non-occupational risk factors with the prevalence of self-reported carpal tunnel syndrome in a national survey of the working population." Am J Ind Med 32(5):550–556.

Taylor, J. R. and L. T. Twomey (1993). "Acute injuries to cervical joints. An autopsy study of neck sprain." Spine 18(9):1115–1122.

Thelin, A. (1981). "Work and health among farmers. A study of 191 farmers in Kronoberg County. Sweden." Scand J Soc Med Suppl 22:1–126.

—— (1985). "Osteoarthritis of the hip joint. A common disorder among farmers." Läkartidningen 82: 3994–3999.

Thelin, A., B. Jansson, et al. (1997). "Coxarthrosis and farm work: A case-referent study." Am J Ind Med 32(5):497–501.

Thelin, A., E. Vingård, et al. (2004). "Osteoarthritis of the hip joint and farm work." Am J Ind Med 45(2):202–209.

Theorell, T., K. Harms-Ringdahl, et al. (1991). "Psychosocial job factors and symptoms from the locomotor system—A multicausal analysis." Scand J Rehabil Med 23(3):165–173.

Tola, S., H. Riihimaki, et al. (1988). "Neck and shoulder symptoms among men in machine operating, dynamic physical work and sedentary work." Scand J Work Environ Health 14(5):299–305.

Torén, A., K. Öberg, et al. (2002). "Tractor-driving hours and their relation to self-reported low-back and hip symptoms." Appl Ergon 33(2):139–146.

Travell, J. G. and D. G. Simons (1983). Myofascial pain and dysfunction. The Trigger Point Manual. Baltimore, Williams & Wilkins.

Vantrappen, G., E. Ponette, et al. (1977). "Yersinia enteritis and enterocolitis: Gastroenterological aspects." Gastroenterol 72(2):220–227.

Videman, T., M. Nurminen, et al. (1990). "1990 Volvo Award in clinical sciences. Lumbar spinal pathology in cadaveric material in relation to history of back pain, occupation, and physical loading." Spine 15(8):728–740.

Videman, T., R. Simonen, et al. (2000). "The long-term effects of rally driving on spinal pathology." Clin Biomech (Bristol, Avon) 15(2):83–86.

Vingård, E., L. Alfredsson, et al. (1991). "Occupation and osteoarthrosis of the hip and knee: A register-based cohort study." Int J Epidemiol 20(4):1025–1031.

Waddell, G. (1987). "1987 Volvo award in clinical sciences. A new clinical model for the treatment of low-back pain." Spine 12(7):632–644.

Walker-Bone, K. and K. T. Palmer (2002). "Musculoskeletal disorders in farmers and farm workers." Occup Med (Lond) 52(8):441–450.

Weiss, L. D. (1997). Occupation-Related Cumulative Trauma Disorders. Cumulative Trauma Disorders. A. W. Cassvan, L.D.; Weiss, J.M.; Rook, J.L.; Mullens, S.U. Newton, Butterworth-Heinemann.

Werner, R. A., A. Franzblau, et al. (1997). "Influence of body mass index and work activity on the prevalence of median mononeuropathy at the wrist." Occup Environ Med 54(4):268–271.

Wickstrom, G. (1978). "Effect of work on degenerative back disease. A review." Scand J Work Environ Health 4 Suppl 1:1–12.

Wiesel, S., J. Weinstein, et al. (1996). The Lumbar Spine. Philadelphia, WB Saunders.

Wieslander, G., D. Norbäck, et al. (1989). "Carpal tunnel syndrome (CTS) and exposure to vibration, repetitive wrist movements, and heavy manual work: A case-referent study." Br J Ind Med 46(1):43–47.

Wilkinson, M. J. B. (1995). "Does 48 hours' bed rest influence the outcome of acute low back pain?" Brit J Gen Pract 45:481–484.

Winblad, S. (1975). "Arthritis associated with Yersinia enterocolitica infections." Scand J Infect Dis 7(3):191–195.

Wolfe, F., D. G. Simons, et al. (1992). "The fibromyalgia and myofascial pain syndromes: A preliminary study of tender points and trigger points in persons with fibromyalgia, myofascial pain syndrome and no disease." J Rheumatol 19(6):944–951.

Worcester, J. N., Jr. and D. P. Green (1968). "Osteoarthritis of the acromioclavicular joint." Clin Orthop 58:69–73.

Yu, D. and J. G. Kuipers (2003). "Role of bacteria and HLA-B27 in the pathogenesis of reactive arthritis." Rheum Dis Clin North Am 29(1):21–36, v–vi.

Yunus, M. B. (1992). "Towards a model of pathophysiology of fibromyalgia: Aberrant central pain mechanisms with peripheral modulation." J Rheumatol 19(6):846–850.

9

Physical Factors Affecting Health in Agriculture

Anders Thelin and Kelley J. Donham

NOISE AND HEARING LOSS

Introduction

For a long time safety programs have focused on protection of the fingers, feet, arms, and lungs, but only lately have the ears been realized as a critical organ. Hearing has been called "our forgotten sense." Programs to protect hearing have generally not been implemented before the 1960s or 1970s.

Occupational hearing loss can be caused not only by continuous exposure to noise but also by head injuries, explosions, thermal injuries such as slag burns, or exposure to ototoxic substances. Long lasting exposure to noise in excess of 85 dB is by far the most common reason for hearing loss.

Work-related hearing loss among farmers emerged along with the implementation of new technology. The hearing damage caused by high-intensity and/or steady noise is sometimes called a "second order consequence of technology." Farmers and lumberjacks were, for a period, working with tractors, harvesters, chainsaws, and other engines without any protection (Figure 9.1). They were engaged in maintenance jobs in private workshops and exposed to high-intensity impulse sounds without any prevention equipment or knowledge of the risks. Different kinds of fans, dryers, and mills also generate dangerous noise. Work with livestock may be related to noise exposure. Noteworthy is manual feeding of pigs.

Along with other human beings in the western industrialized society, farmers are exposed to noise from a number of nonoccupational sources, such as driving cars, motorbikes, and lawnmowers. They may be affected by the sound from radios or other personal listening devices that are turned high so they can be heard over the noise of the tractor engine—noise that may last through long workdays. Farmers hunt and shoot more than people in general. Impulse sounds from gunshots are a significant risk factor for developing hearing loss.

Very few studies (McBride and others 2003) of individual exposure (dosimetry) in farming or forestry have been reported. It is also difficult to give a true noise map of a farm. The farm work environment in terms of noise is diversified and constantly changing. The best method, therefore, is to describe typical jobs and the related noise burden.

In a study from 1977 it was shown that only 10% of farmers had normal hearing (Thelin 1981). Fourteen per cent had a severe hearing disability. Other studies from this period also indicate that most farmers had hearing losses (Blatherwick 1969; Tomlinson 1970). Later studies (Marvel and others 1991; McBride and others 2003) indicate that the situation may be improving because of improved machinery design. However, in most countries there still are no hearing conservation programs for farmers.

Noise, Definitions, and Measurements

Sound is waves or alternating periods of compression and rarefaction within an elastic medium, such as air. Noise is unwanted sound. Sound level may be measured as sound pressure level in decibel (dB) units. Zero dB is the faintest sound the average normal hearing young adult can detect.

Figure 9.1. Working with old tractors without any hearing protection.

The dB scale is a logarithmic scale, which means that adding or subtracting sound levels must be done using absolute levels. If for example, you add two equal noise sources of 80 dB, the following formula should be used:

Lptot = 10 lg µ10 Lpn/10
Lptot = 10 lg [10 80/10 + 10 80/10]
Lptot ≈ 83 dB

Thus, adding two equal sources of noise gives a raise of the sound pressure of 3 dB. If one of the sources is much stronger than the other (more than 10 dB) the above formula demonstrates that the weaker sound is insignificant compared to the stronger one and the weaker one may be neglected. In prevention practice the stronger source should always be removed first.

The term *sound level* or *noise level* is used when measuring with a sound level meter. The noise meter filters the sound with respect to the wave length the ear detects most easily (2000–5000 Hz). An A-filter is used blocking most noise outside of this range. The sound level is given in dB(A) with respect to the intensity of the ear. Mostly an A-filter is used and the sound level is given in dB(A).

However, the human ear can detect sound in the frequency interval 20 Hz to 20,000 Hz. Lower frequencies are infrasound and higher are ultrasound. Ultra- and infrasound may affect human beings in other ways than affecting the ear. Speech is sound between 125 Hz and 8,000 Hz

(practically, 500–3,000 Hz) mostly with the sound level of 20 to 60 dB(A). Noise at the level of 120 dB is unpleasant, and 140 dB is painful. A chainsaw generates approximately 110 dB and a tractor between 80 and 100 dB outside and much lower inside a modern cab. There are large variations according to the age and quality of maintenance of the tractor.

To estimate the risk of hearing damage it is necessary to measure the sound level, the exposure time, and the noise frequency. Noise with more high-frequency sounds may be more dangerous than low-frequency noise. The importance of checking the impulse sounds has been stressed lately (Bruel 1976). The first generation sound meters had no capacity to track impulse sounds. Modern sound meters have capacity to catch also very short impulse sounds. A *dosimeter* is a sound meter, which surveys the continual sound level and integrates the total exposure over time.

How Noise Affects the Auditory System

Chronic noise exposure gradually impairs hearing. A noise-induced hearing loss (NIHL) should be considered when the hearing threshold is reduced 25 dB or more. NIHL generally progress insidiously. An individual may not realize he or she has an NIHL until perception of speech is affected.

Exposure to loud noise for a short period of time may produce a temporary hearing loss (a temporary threshold shift, or TTS). A TTS can last for several hours or a day and might be experienced as a ringing in the ears (tinnitus) and difficulty hearing. Chronic noise exposure (months to years) may generate a permanent threshold shift (PTS) from which there is no recovery. As a TTS may mimic a PTS, an audiometric test should not be performed within a 24-hour period after significant noise exposure.

NIHL pathoanatomically results from trauma to the sensory cells of the cochlea. The sensory epithelium of the cochlea consists of one inner row of stereociliated hair cells and three outer rows of stereociliated hair cells. The hair cells may become distorted or even disrupted due to vascular, chemical, and metabolic changes. Other structures of the Organ of Corti can also be affected. Finally retrograde degeneration of cochlear nerve fibers occurs, progressing centrally.

Individual susceptibility to noise is variable

due to genetic differences. Some individuals are able to tolerate high noise levels for long periods of time. Others can rapidly lose hearing in the same environment.

Prolonged exposure to sounds louder than 85 dB is generally regarded as potentially injurious. The level 85 dB is however arbitrary and was first established mainly for economic considerations (further noise reduction might be too expensive). The 85 dB level probably does not protect susceptible people. Thus the goal should be to reduce noise to below 85 dB.

Exposure to hazardous noise tends to have its maximum effect in the high-frequency regions of the cochlea. Hearing loss from noise is usually most severe around 4000–6000 Hz. As the injury expands, more frequencies are affected. Most qualities of continuous noise affect the same frequencies.

The inner ear is partially protected by the activity of the middle ear muscles (stapedius and tensor tympani). Impulse sounds like gunshots penetrate to the cochlea before the normally protective acoustic reflexes have had time to react. Impulse noise exceeding 140 dB may cause immediate and irreversible hearing loss.

Symptoms and Disablement

Commonly the first sign of NIHL is difficulty in comprehending (not just hearing) speech at a dinner party or a meeting where many people are talking at the same time or there is competing background noise. Patients with NIHL hear vowel sounds better than consonants. As IHL affects high-frequency loss, they have more difficulty understanding people with higher-pitched voices (women, children).

The first loss of hearing capacity frequently develops around 4000–6000 Hz and is often not realized by the individual. Over time, the loss expands laterally to a wider range of frequencies. The majority of human speech is largely within the range of 500 to 3000 Hz. A significant loss of capacity in this area leads to a communication disablement. Later in life this is often combined with a presbycusis (normal aging changes) leading to even greater difficulties in perception. The progress unfortunately often proceeds even if the exposure is reduced or dismissed.

A number of other diseases and conditions may also affect the hearing and must be ruled out in evaluating hearing impairment. Presbycusis is a slow progressive deterioration of hearing, which is normally associated with age. Presbycusis affects the higher frequencies mostly, with the most significant loss in the highest area. Hereditary hearing impairment is a common condition and may have very different character. Otosclerosis is regarded as a form of hereditary hearing loss. Sudden deafness or sudden sensorineural hearing loss is a momentary mostly one-sided hearing loss of unknown etiology. The character of the loss may be very different, as well as the prognosis. Ménière's disease, tumors, and some metabolic disorders like diabetes mellitus may also affect the hearing capacity—as well as, of course, infections.

Persons who have been active in hunting and shooting and farming often have more impairment on the left side as the source of the noise is lateralized. Otherwise, the hearing loss in NIHL mostly is bilateral and symmetric.

No medical or surgical treatment is available to heal the effects of NIHL. Focus therefore must be on prevention and those with an established impairment must take steps to reduce their exposure (see below). Hearing amplification may be helpful but must be carefully fitted and optimized to be used. Sometimes hearing aids can be of selective use—as, for example, at a lecture or in a group—but of no need in a standard work situation (driving tractors or working with a chainsaw). The disturbed capacity to realize warning sounds in traffic or in a work situation may be of some risk. Warning sounds in tractors or feeding systems or other technical systems in farming may be replaced by warning lights.

In many countries occupational hearing loss is compensated. The differentiation between NIHL and presbycusis is, however, a problem and the regulations are different. Efforts have been made to standardize the normal loss of hearing over time and calculate the burden of occupational noise exposure (Standardization 1990).

Tinnitus

Tinnitus is a common complication of NIHL. Tinnitus is a perceived ringing or buzzing sound reported by a patient, but it cannot be measured (Axelsson and Ringdahl 1989; Andersson and others 1999). Only a minority of those with tinnitus have major daily problems. Most of these patients also have psychological or psychiatric problems (Holgers and others 2000; Zoger and others 2001).

The damage of the stereociliated hair cells may not only generate a loss of hearing signals but also generate the disturbing buzzing. However, a large number of other (noncochlea-related) causes of tinnitus have been reported (trauma, multiple sclerosis, some drugs) and the mechanisms are not fully understood (Andersson and others 1999). Tinnitus in absence of hearing loss is probably not related to noise exposure.

Patients with tinnitus may need help. Those with depression and/or other forms of psychological problems may need treatment for those basic health problems. Psychological counseling and changes of lifestyle in combination with cognitive therapy may help the patient control and minimize the difficulties.

Hearing Loss Due to Trauma or Toxins

A blunt head injury may cause a traumatic hearing loss by a fracture of the temporal bone with a cochlear injury. Explosions, burns, or a piece of welder's slag may generate forces or penetrate the ear canal and damage the anatomic structures. It is easy to separate a conductive hearing loss from a sensorineural one by using a tuning fork (512 Hz). Traumatic tympanic membrane perforations usually heal spontaneously. If not, grafting the tympanic membrane is possible. Reconstruction of the ossicular chain is also possible.

Some chemical substances are ototoxic and may injure the cochlea. The majority of these substances are components of drugs (aminoglycoside antibiotics, loop diuretics, antineoplastic agents). A combination of these kinds of drugs with noise exposure may be potentially hazardous. Patients under medication thus should be even more cautious with respect to noise exposure.

Hearing loss may also result from exposure to ototoxic substances in the workplace. Some heavy metals have ototoxic potentials as well as some other substances. Organophosphate insecticides in the farm work environment may have an ototoxic effect (Ernest and others 1995; Teixeira and others 2003). A combination of pesticides and noise thus may be hazardous. More research is needed in this area.

Audiometry

Pure tone audiometry is the standard method to screen hearing and to analyze hearing losses in an occupational health service (OHS). The sensitivity to pure tones are measured at 250, 500, 1000, 2000, 3000, 4000, 6000, and 8000 Hz for air conduction. In case of differential diagnostic questions or other special questions, a bone conduction audiometry might also be performed. Normal range of hearing is 0–20 dB at each frequency. The results mostly are presented graphically (Figures 9.2–9.6).

A pure tone audiometry may also be performed as a Bekesy audiometry. This is a self-administrated audiometry in which the patient responds to signals by pressing and releasing a button. This

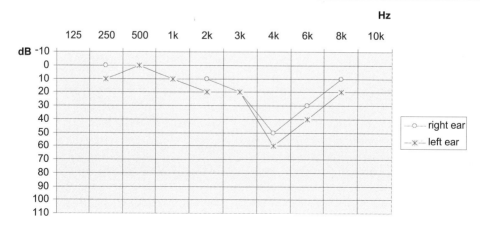

Figure 9.2. Mild noise-induced hearing loss; pure tone audiogram.

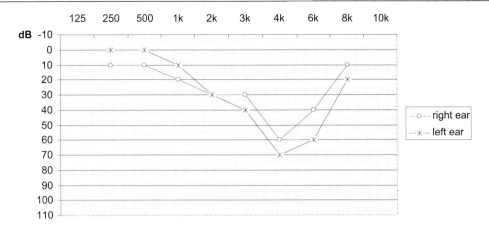

Figure 9.3. Moderate noise-induced hearing loss; pure tone audiogram.

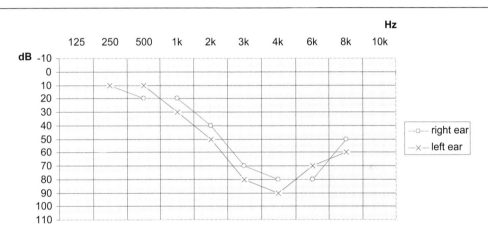

Figure 9.4. Severe noise-induced hearing loss; pure tone audiogram.

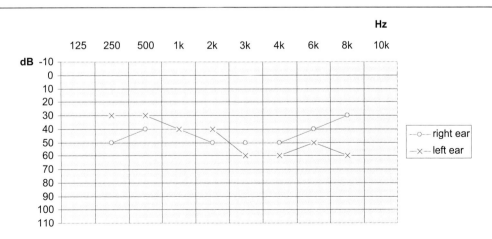

Figure 9.5. Nonorganic hearing loss; pure tone audiogram.

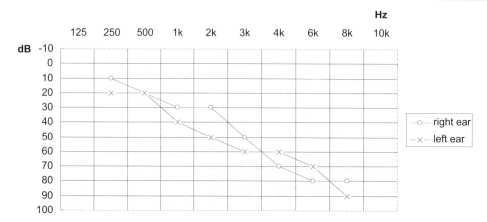

Figure 9.6.   Presbycusis; pure tone audiogram.

method is frequently used in an OHS but is generally regarded not to be as reliable as standard audiometry administered by an audiologist. A Bekesy audiometry may be combined with a computer-supported analysis of the audiogram. Special computer programs to classify hearing losses have been developed (Klockhoff and others 1974).

More in-depth audiometric tests, as, for example, speech audiometry and speech discrimination tests, are generally not conducted in an OHS. Patients needing more testing should be referred to higher-level care facilities such as departments of audiology.

All audiometric tests should be performed with the patient placed in a standardized sound-insulated test booth. Otherwise the results cannot be regarded as standardized and reliable for use in a hearing conservation program or for research purposes.

Other Effects of Noise

The exposure to noise may also have other nonauditory effects (deJoy 1984; Rehm 1985; Abel 1990; Sloan 1991). Noise has been related to increases in diastolic blood pressure (Sloan 1991). Noise may act as a stressor, affecting job performance and increasing injury risk due to interference with perception of speech and auditory signals (Choi 2005; Smith 1989; Jones 1990).

Prevention

NIHL is preventable. Most western industrialized countries have regulations and recommendations to prevent NIHL. Mostly exposure above 85 dB(A)

over an 8-hour working day is not permitted/recommended. A visionary hearing conservation program (HCP), however, might focus measures to reduce average exposure to significantly lower levels. A number of steps and a continuous activity are needed to keep up to the visions.

1. DEFINITION OF NOISE EMISSIONS
A successful HCP includes valid information of noise exposures. The noise must be monitored and defined in terms of personal time-weighted average dose, which includes frequency, time intensity, and type. Personal noise dosimeters must be used to determine these exposures. An individual must wear this instrument through the workday, and it collects, stores, and provides a time-weighted average exposure in relation to the prescribed noise standard (e.g., 85 dB, etc.). Changes in work practices and equipment will require repeated monitoring. In practice this is very difficult to do in farming. The workplace is so varied that it is difficult to develop a noise map, which is relevant for more than a few days. Therefore, those working with noise reduction in farming must rely on information of typical emissions from different types of equipment, confinement systems, and other defined modules of farm work, and estimate total exposures and identify priority areas for reduction.

2. ENGINEERING CONTROLS
Tractors, harvesters, mills, fans, compressors, and other types of equipment used in farming are

common noise sources. Engineers work with information of how the noise is generated and try to reduce the emission. Manufacturers of farm equipment, however, have not had noise reduction as a first priority in the development of new equipment. Proper maintenance of farm equipment is necessary. Worn machinery with defects generally generates more noise. To manage noise emission, an ambitious program for maintenance of equipment will not only save money but will also help the farmer keep down noise exposure.

3. ENCLOSING THE SOURCE OF NOISE

The noise source may be separated from the user or vice versa. Modern tractor cabs have a good internal acoustic environment, relative to old tractors or tractors with no cabs. Fans, dryers, compressors, and mills may all be enclosed. If enclosure is not possible, other techniques are available, such as installing absorbents into the ceiling and walls to reduce echoes.

4. ADMINISTRATIVE CONTROLS

The work might be planned and done so that noise sources such as fans and dryers may be switched off when the worker is in the building. Feeding of animals might be done in such a way that animal activity is not stimulated too much. Alternative work methods might be introduced to replace the noisier methods.

5. WORKER EDUCATION

Farmers, lumberjacks, and employees in farming and forestry must understand the harmful effects of noise. This is a prerequisite for a successful HCP. Education programs might be necessary to help the individuals deal with existing noise hazards and to protect themselves. An OHS program with recurrent audiometric tests provides unique possibilities for personal training based on information of the individual's hearing capacity and how this applies to the work environment.

6. HEARING PROTECTION DEVICES

Many times noise cannot be prevented because of limits in costs and technology. Often the worker must use hearing protection devices to reduce personal noise exposure and hearing impairment. Devices are available in a variety of types and qualities. Earplugs or "aurals" may be premolded, formable, or custom molded. Earmuffs or "circumaurals" have a better protective effect especially in case of high-frequency noise. In special situations a combination of plugs and muffs may be used. Each type of devices has advantages and disadvantages. The user may need support and encouragement to instill a habit of protection. Selection of appropriate, effective, and acceptable hearing protection is vital to compliance and usage.

VIBRATION AND INJURIES RELATED TO VIBRATION

Introduction

Maurice Raynaud first described the white finger phenomenon in his thesis (Raynaud 1862). These intermittent blanching episodes of the fingers and hands are now referred to as primary Raynaud's Disease. Fifty years later a number of reports demonstrated that "white finger" or vascular spasm of the fingers had a relation to the use of handheld vibratory tools (Loriga 1911; Hamilton 1918). The information of the local effects of vibration has facilitated the expanding field of knowledge of a number of vascular, neurological, and musculoskeletal symptoms related to vibration, e.g., the hand-arm vibration syndrome (HAVS) (Stockholm Workshop '86 1986). HAVS is still a relatively new area of study, and there may be cofactors involved in addition to vibration as etiologic agents (Gemne 1997).

Mechanical energy from an oscillating source may transmit vibration energy to other objects. Most structures have their own natural vibration frequency, including the human body. Resonance (resulting in amplification of that energy) results when vibration of a compatible frequency is transmitted to another object with the same natural frequency. Vibration may affect the whole human body as well as local or regional anatomy such as the hands and arms. Vibration energy sources include mechanical tools and self-propelled machines of all types. Vibration-induced injuries are common to farming and many other occupations.

Farmers driving tractors and harvesters in agriculture are exposed to whole-body vibrations, as are farmers and lumberjacks driving tractors or other machines in forestry (Seidel and Heide 1986; Seidel 1993). Driving trucks and other loading equipment may impose the same kind of exposure.

Vibration from steering wheels in farm tractors and handlebars of snowmobiles used in ranching or reindeer herding are transmitted to hands and arms (Anttonen and Virokannas 1994). Grinders and chain saws are well known vibration sources, among other handheld machines. Frequent problems in HAVS have been reported previously among forest workers (Taylor 1974; Pyykkö and others 1986) and also among farmers (Thelin 1981; Peplonska and Szeszenia-Dabrowska 2002).

Whole-Body Vibration

A number of health problems have been attributed to whole-body vibration, including musculoskeletal, neurologic, and circulatory system disorders. A number of experimental studies have been conducted on the effects of whole-body vibrations on the spinal column. The human low back has a resonance frequency of 5 Hz, which is a predominant frequency in many vehicles (Panjabi and others 1986; Panjabi 1996). The design of the driver's seat as well as the driver's position affect the transmission of vibrations to the human low back (Pope and others 1989; Magnusson and others 1993). Research to date has not consistently demonstrated a significant vibration-induced injury to the low back (Bovenzi and Hulshof 1999; Lings and Leboeuf-Yde 2000).

Other studies indicate that whole-body vibration probably can contribute to disorders of female reproductive organs such as menstrual disturbances, abortions, and stillbirths. Animal experiments suggest whole-body vibration may cause harmful effects on the fetus (Seidel 1993).

A kind of "vibration sickness," characterized by diffuse symptoms, such as gastrointestinal problems, decreased visual acuity, muscular pain, and balance disorders, has also been reported among persons exposed to whole-body vibrations. These reports have not been consistent. Many questions remain unanswered regarding the effects of whole-body vibrations.

The International Organization for Standardization (ISO) has established guidelines for whole-body vibration exposure (ACGIH 1996). Despite inconsistent information of negative effects of whole-body vibration it seems wise to make efforts to reduce exposure.

The previous paragraphs have provided an overview of vibration injury. The following paragraphs provide more pathophysiology of these conditions.

VIBRATION WHITE FINGERS (VWF)

The clinical picture of VWF is typical. It consists of episodes of patchy, sharply delineated blanching of the skin, mostly in those parts of the fingers that have been most strongly exposed (Gemne 1992). The fingers turn white in a cold environment with decreased manipulative dexterity. Redness, some swelling, light pain, and paresthesias occur with the return of blood circulation as a consequence of a reactive hyperemia. Taylor and Pelmear (1975) developed a classification system for cold-induced peripheral vascular and sensorineural symptoms (Taylor 1975). This was later modified (Gemne 1987) (Table 9.1). Very severe symptoms are unusual. A progress to fingertip gangrene is possible but very uncommon (Thulesius 1976).

Spasm of the digital arteries elicited by exposure to cooling of the hand or the whole body results in reduced blood flow. The specific pathophysiology of VWF is still obscure. A number of mechanisms have been discussed, such as centrally mediated increased vasoconstrictor tone, local hypersensitivity to cold, occlusive vascular disease, local nerve injury, increased viscosity of the blood, and low blood pressure (Ekenvall 1987; Gemne 1992). Patients with advanced symptoms

Table 9.1. Classification of Cold-Induced Raynaud's Phenomenon in the Hand-Arm Vibration Syndrome: The Stockholm Workshop Scale (Gemne and others 1987)

Stage	Grade	Description
0		No attacks
1	Mild	Occasional attacks affecting only the tips of one or more fingers
2	Moderate	Occasional attacks affecting distal and middle (rarely also proximal) phalanges of one or more fingers
3	Severe	Frequent attacks affecting all phalanges of most fingers
4	Very severe	As in stage, with tropic skin changes in the fingertips

The staging is made separately for each hand.

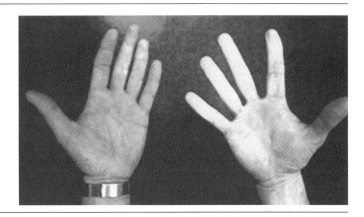

Figure 9.7. Vibration white fingers. (This figure also is in the color section.)

of VWF have an increased basal vascular tone in the finger arteries but no narrowing of the vascular lumen (Ekenvall and others 1987). It is possible that vibration causes an imbalance between alpha-1 and alpha-2 adreneric receptors (Ekenvall and Lindblad 1986). Hereditary factors as well as biochemical conditions and previous traumatic injuries may contribute to the development of white fingers.

DIFFUSELY DISTRIBUTED NEUROPATHY
The cause of the diffusely distributed neuropathy observed in some vibration-exposed persons may be damage to nerve fibers and/or nerve receptors by direct mechanical influence of vibration. Disturbed blood circulation has also been discussed (Bovenzi and others 1997) as an etiologic factor.

PAIN IN HAND AND ARM
Vibration may contribute to an excessive strain on joints or structures around the joints, resulting in pain and restriction of movements. Vibration that may contribute to osteoarthritis in the wrist (Fam and Kolin 1986; Bovenzi and others 1987) and in the shoulder region has also been discussed (Stenlund and others 1992, 1993). Although the latter has not been confirmed by epidemiological data, vacuolization in the small bones of the hands and the forearms have been reported, lending evidence to the negative effects of vibration injury (Kumlin 1973; Karjalainen 1975; Gemne and Saraste 1987; Kivekas and others 1994).

CARPAL TUNNEL SYNDROME
The symptoms of carpal tunnel syndrome (CTS) (see Chapter 8 on musculoskeletal disorders) are caused by a compression of the median nerve as it passes through the carpal tunnel of the wrist. Most people with carpal tunnel syndrome have not been exposed to vibration. However, tool vibration may require a stronger grip on the tool handle, which may exacerbate carpal tunnel syndrome. Static work during wrist flexion may also be a major factor, contributing to the development of CTS in workers with handheld tools (Silverstein and others 1987; Hagberg and others 1992).

OTHER EFFECTS OF VIBRATION
Some studies indicate that vibration may have some negative effects on muscular strength (Necking and others 1992; Färkkilä and others 1986). A hearing deficit has been reported that is not ascribed to noise exposure but related to localized vibration. The mechanisms of these effects are largely unknown (Pyykkö and others 1981; Iki and others 1986).

Other symptoms related to hand-arm vibration include vibration-induced, diffusely distributed neuropathy (to be distinguished from the polyneuropathy of other origin). It is reported as numbness and a light reduction of tactile sense. Wrist pain may be related to strain of joints and tendons more than to vibration exposure or associated to a possible osteoarthritis.

Diagnosis of Vibration-Induced Disorders

Obtaining an informed and accurate history and clinical picture are of primary importance in diagnosing vibration-induced injury. A cold provocation test sometimes demonstrates the symptoms. Measuring the presence of vasospasm by recording finger systolic blood pressure in response to cold is a more objective test (Pyykkö and others 1986). Neurological disturbances may be demonstrated following exposure to vibration and low temperatures (Ekenvall and Lindblad 1986; Lundborg and others 1987; Lundström and others 1992). Carpal tunnel syndrome is verified by electrodiagnostic methods. Plain x-ray may demonstrate changes of joints and skeletal structure.

Vibration-induced white finger may need to be differentiated from other causes. Primary Raynaud's phenomenon is a common condition especially among females (6% of the general female population, perhaps about 13% in Sweden (Leppert and others 1987)). Other causes include trauma of the fingers or the hands, frostbite, occlusive vascular disease (which may be related to smoking), connective tissue disorders, drug intoxication, exposure to vinyl chloride monomer, and several neurological disorders. To judge causality of vibration, attention must be paid to exposure history, including duration, intermittency, and vibration characteristics (Gemne 1997).

Treatment of HAVS

The spasm of the digital arteries and the white fingers usually dissolves in minutes or within 1 hour. This may be facilitated by warming of the hands or the whole body. Swinging the hands or placing them in warm water may reduce the intensity of an attack. Nifidepin, a calcium antagonist, has been used to prevent attacks, but it is not very effective. In severe cases thymoxamine, stanozolol, or prostaglandin E may be useful, but they are not used very often and the effect has been questioned. Surgical sympathectomy has been tried but results have not been positive; in fact, the procedure may aggravate symptoms. Biofeedback has been tried, but results have not been confirmed to date.

The diffusely distributed neuropathy and wrist pain have no special treatment. In case of carpal tunnel syndrome, surgery is a common practice.

Prevention of HAVS

Different steps may be taken to reduce or remove the risk of vibration-induced injuries. Tools and other equipment should be designed in such a way that the risk of injuries is low. The redesign of the traditional chainsaw in the 1970s is a good example of how a risk can be reduced. In the early seventies more than 30% of lumberjacks had VWF. Within a decade the risk of VWF using a chainsaw was almost gone (Pyykkö and others 1978, 1986). The maintenance of tractors, other machines, and tools must properly and continuously be done to diminish the risk of damage by vibration exposure. The period of exposure should be shortened if other methods to reduce or avoid exposure cannot be received. Special gloves to keep the hands warm and absorb some of the vibration energy may be useful in some cases (Griffin 1998). More information or individualized training should be performed.

DISORDERS DUE TO HEAT AND COLD

Stress from physical work of farmers in hot or cold weather is unfortunately often taken for granted, and adverse effects on health and work performance are not always recognized. Although there are few general studies regarding heat and cold stress in occupational activity, they are almost completely lacking in a farming context. Work in hot and cold environments has been driven by the military and space industries, and we are forced to apply information from those sources to agriculture.

Human Thermal Balance

The human body continually works to adapt to its physical surroundings. The following factors interplay to create the thermal balance necessary for health comfort and life: surrounding air temperature, radiant energy, humidity, wind velocity, the insulation and vapor resistance of clothing, and physical activity.

The human thermal balance may be described by an equation (Gagge and Nishi 1977):

$$M - W = R + C + K + E + Hres$$

where M = heat generated by the body, W = mechanical effect, R = radiation, C = convection, K = conduction, E = evaporation, Hres = heat transportation by respiration.

This equation describes a stabilized situation and a necessity over time to keep the body at a constant temperature. For short periods the thermal balance may not be preserved. For example, heavy work may raise body temperature by .5–1.0°C. In case of imbalance the equation must be written thus:

$$S = M - W - R - C - K - E - Hres$$

where S is the heat effect stored or lost. If heat production is larger than heat export, the body temperature will rise. If heat export is larger than heat production, the body temperature will decline. The main ways of heat export is by the skin and through radiation, convection, and evaporation. The interrelation of these depends on the microclimate.

Metabolic processes and physical activity generally generate enough heat to maintain the thermal balance. Only in extreme situations are special efforts activated to generate heat. Generally only for short periods and with heavy work, the body converts muscle energy to mechanical work (e.g., lifting a 50 kg bag of feed). The major component of cellular energy is transformed to heat. To compare the energy produced between different persons, the energy production rate can be expressed in relation to the body surface. Average energy production at rest is approximately 58 W/m^2.

Convection, radiation, and evaporation are the three main ways of heat transfer from (or to) the human body. For convective heat transfer dry-bulb temperature and wind velocity are determining factors. The hotter the air the less heat is exported. If the air temperature exceeds skin temperature, the heat flow is reversed and the surrounding air heats the body.

The objects surrounding the worker dictate radiant heat transfer. Heat is exported from the body when the temperature of the surrounding objects is below +35°C. As high as 60–70% of body heat loss is lost by radiation. Convection as well as radiation depends on the amount of the body surface area that is in contact with the surroundings.

Evaporation is the most efficient way of heat export. Evaporation is affected by humidity, wind velocity, and temperature. Internal physiological factors (circulation, rate of sweat secretion, sweat secretion thresholds) also affect evaporative cooling. The unacclimatized person may come under heat stress much earlier than a well-adapted person. The wet-bulb temperature is a good indicator of humidity, air temperature, and wind speed relative to evaporation conditions. A wet-bulb temperature of +24°C or more impedes an unadapted person and +33°C is regarded as the maximal wet-bulb temperature for safe work for a well-adapted person.

A good capacity to sweat and good climatic conditions to export heat by sweat are prerequisites for work in a hot surrounding. The capability to sweat varies among people and can be developed by acclimatization and physical conditioning. An unacclimatized person may perspire 600 g/h during work. A well-conditioned and acclimatized person can perspire 1000 g/h. Loose-fitting clothing that wicks, low humidity, and high wind velocity all promote the cooling effect.

Sweat consists of water and salts. The losses of water as well as salts must be compensated for or work capacity is quickly reduced. A loss of 1–1.5 liters for a normal-sized person impedes capacity and endurance.

Acclimatization to a hot microclimate requires work in a hot surrounding for a few hours daily for a period of 7–9 days. The acclimatization is lost in a period of weeks if not maintained.

The adaptation to work in a cold surrounding is not as difficult as adapting to heat. A well-trained person acclimatized to work in a cold surrounding tolerates larger variations in local temperature before feeling uncomfortable. There is no evidence for the general concept that the body increases metabolism to compensate for low temperature (Burton 1969).

Cooling of some parts of the body results in reduced sweat production, and localized heating in the same areas generates a raise of sweat production. For example, the forehead is a strategic area for this effect. Physical work generates a natural adaptation to work in cold surroundings. However, heavy physical work will not exclude the risk of cold injuries to the face and extremities when working in very cold and windy weather.

Disorders Due to Heat

A continual exposure to high temperature and humidity extremes is not systematically related to defined diseases of the cardiovascular, respiratory, or excretory systems. However, heat stress may cause several specific heat illnesses. Heatstroke is the most dangerous heat reaction, but heat exhaustion, heat cramps, and heat syncope are more common and less serious and occur prior to heatstroke. Skin disorders and infertility are other reactions related to heat exposure. A number of workers in different occupations may be exposed to heat extremes indoors (steelworkers) or outdoors in hot climate (farmers, ranchers, and construction workers).

Some health conditions, especially conditions with reduced sweat production capacity or reduced evaporation capacity, impede the ability to work (or stay) in a very hot climate. Such conditions are obesity, some cardiac diseases, use of alcohol or medications that inhibit sweating (such as atropine that may be used by insecticide applicators as a prophylactic), conditions with reduced cutaneous blood flow or dehydration, use of drugs that increase the generation of body heat, cancer diseases, and infections. Older persons do not acclimatize as well as younger, and women generate more heat performing the same task as men.

HEATSTROKE

Heatstroke is a life-threatening emergency characterized by high body temperature and a lost or strongly reduced sweating capacity. The classic form appears among persons with reduced capacity to adjust to high temperatures. Exertional heatstroke results from heavy work in a hot environment, especially among persons who are not acclimatized.

A heatstroke is characterized by disturbed mental activity, coma, and/or convulsions (Shibolet and others 1967). Blood pressure goes down and the skin is hot and dry. The core temperature may exceed 41°C. The regulation of the water and electrolyte balance is disturbed (low levels of serum potassium, calcium, and phosphorus). Thrombocytopenia, fibrinolysis, and consumptive coagulopathy (O'Donnell 1975) may result and permanent damage to the liver and kidneys may result. Death may follow.

Rapid reduction of body temperature is the first treatment measure. Spraying or immersing the body in cool water should start immediately while waiting for transportation to a hospital. This action should be ongoing until body temperature has declined to 39°C. The patient should be continuously observed and monitored for hypovolemic and cardiogenic shock. Capacity to treat complications should be at hand.

HEAT EXHAUSTION

A prolonged exposure to heat in combination with inadequate intake of water and salt can cause heat exhaustion characterized by cardiovascular changes, fatigue, and a feeling of being exhausted. Thirst, headache, weakness, and even confusion are common symptoms. The tempera-ture exceeds 38°C, the skin is moist and pulse rate increased. Loss of water and salt causes an imbalance. A progression to heatstroke is a risk and is indicated by raise of temperature and decreased sweating.

After a loss of approximately 1.5% of body weight (~1 liter) the tolerance to heat is reduced, the pulse rate increases, and the risk of a rapid increase of body temperature is obvious. Cardiovascular failure is at stake if the loss of water and salt is not compensated for, and the mental capacity is reduced leading to defective judgment and increased risk of traumatic injuries.

The risk of heat exhaustion is controlled by continuous fluid replacement. Unfortunately, thirst is not an adequate indicator of the dimension of water loss. Thirst is reduced even with a small intake of water. More water and salts than indicated by the thirst must be consumed.

HEAT CRAMPS

A high consumption of water without replacement salts in hot weather may lead to heat cramps. Low levels of sodium alter muscle reactions, resulting in weakness, slow muscle contractions, or severe muscle spasms. Dizziness, malaise, and vomiting may be associated symptoms.

Normally the body reserve of salt is sufficient to compensate for losses when working in a hot environment. However, sodium depletion can occur in extreme situations with long working periods in dry and hot weather. In such situations, sodium compensation with salt is recommended. Salt tablets or a mild salt solution may be given, or perhaps better, one of the balanced salts solutions marketed as sport drinks. Patients with heat cramps should be moved to a cool environment and given a salt solution in addition to water. A few days of rest before return to work is recommended.

HEAT SYNCOPE

Heat syncope is caused by disturbed blood distribution. A larger part than normally of the circulating blood volume is aggregating in the lower parts of the body because of cutaneous vasodilation. The result is a reduction of systolic blood pressure and cerebral hypotension with sudden unconsciousness. Long periods of standing in combination with strenuous work in hot environment predispose to heat syncope.

The patient should be placed in a recumbent position and moved to a cooler place. Recovery is usually rapid without any following symptoms. Persons who are not acclimatized or have a preexisting medical condition are at increased risk.

Skin Disorders Related to Heat

Intertrigo is common among obese persons. The skin in body folds, groin, and the axilla become erythematous and macerated. Obese persons have more problems in hot environments. Good hygiene is necessary (daily bathing with soap). To avoid further problems, weight reduction is highly recommended.

Heat **rash (miliaria)** is caused by inflammation and obstruction of the sweat glands and causes sweat retention. Symptoms vary but the usual appearance is multiple small red inflamed papules with erythema where clothing fits tightly or areas that do not dry easily, such as under arms and at the belt line (See Chapter 4 for details of this condition).

Heat **urticaria** is a condition with elevated (swollen) plaques stimulated by heat. It can be localized or generalized. Antihistamines may help. Corticosteroids are not recommended. The condition is self-limited and will dissipate in 24–48 hours, but may leave the patient with a temporary pruritis at the site of the lesions.

Nephrolithiasis and Hot Environment

Some studies indicate that a chronic dehydration increases the risk of kidney stone disease (Embon and others 1990). Chronic dehydration may be seen among workers in a hot environment and with low water intake. One study indicated an almost fourfold risk for kidney stones in a group of workers in a glass plant in comparison with a control group working (Borghi and others 1993). Uric acid stones are most likely, and adequate fluid intake is recommended to avoid problems.

Infertility

Work in a hot environment for long periods is a well-known and important risk factor for male infertility (Lahdetie 1995). Working in hot environments contributes to an observed loss of semen quality over time (Auger and others 1995; Bonde and others 1996). Hot environment has also been suggested as a risk for cancer of the testis (Haughey and others 1989).

Disorders Related to Cold

Hypothermia

If the body's core temperature is reduced to below 35°C, systemic hypothermia will occur with a cascading group of symptoms. Shivering starts or is more intensive. Work capacity is reduced as a result of deteriorating muscle capacity and because of reduced absorption of oxygen. Physical exhaustion gradually increases. As body temperature declines further, confusion and apathy will begin along with possible hallucinations. Shivering declines and stops due to hypoglycemia. A paradoxical cold reaction may occur, which is a feeling of hotness and may be characterized by undressing, which obviously aggravates the condition.

Below 33°C, muscle function is even more reduced and the cardiac activity is affected due to slowed repolarization. Consciousness is reduced and below 30°C the individual is generally not communicable. The pupils are dilated and breathing is weak. Cardiac activity is hard to detect and the risk of cardiac arrest or ventricular fibrillation is evident. Death may follow.

In all cases of hypothermia, the first action should be to avoid more cooling. If the person is conscious and alert, physical activity may be stimulated to assist personal heat production. If the person is wet, he or she must be transported to a dry environment and wet clothing should be changed. If the patient is unconscious, emergency transportation to a hospital is urgent. It may be dangerous to start warming an unconscious person because of the risk for cardiac fibrillation. Unconscious patients with body temperature as low as 28°C generally recover completely without any permanent disability if complications can be managed.

The circumstances under which a person with hypothermia is found must be analyzed. Preexisting medical conditions, problems with alcohol or drug abuse, depression, and possible suicidal action as well as dementia must be considered. In other cases the circumstances are generally obvious and related to accidents or unusual and unexpected weather conditions.

Localized hypothermia (localized tissue damage) may develop in exposed parts of the body—such as the nose, ears, fingers, and feet—if the temperature is reduced below 15°C due to cooling, humidity, and immobilization. This is related to localized ischemia and development of small superficial thrombosis. Actual freezing of the tissues may also occur, resulting in significant damage.

The degree of a localized cold injury is related to the speed of cooling, depth, and surface area affected. In case of superficial damage, only the epidermis and dermis are affected. The skin is white and without circulation but is not adherent to underlying structures. Warming rapidly restores normal conditions.

Firm skin adherent to underlying structures characterizes deep damage. Rewarming should be performed as soon as possible. This can be done using warm water (40–42°C). Higher water temperatures risk superficial heat damage because the skin has temporarily lost sensitivity.

Prolonged local cold exposure sufficient to damage blood vessels and permanently disrupt blood supply to local areas is called frostbite. There are variable amounts of tissue necrosis with frostbite, which can be mild or extensive with significant loss of skin, digits, or even appendages. In case of a severe frostbite, hospitalization is recommended. Profound tissue damages may have developed and the period of healing may be long and complicated with ulceration or gangrene.

Chilblains or pernio is a superficial skin lesion due to inflammation and related to a temporary exposure to cold. Prolonged exposure may lead to chronic pernio or "blue toes." The acral parts of the toes may be erythematous and edematous with small ulcerating lesions.

Immersion foot is an extensive frostbite involving the whole foot. It is characterized by a prolonged course with chronic complications. After prolonged cold exposure, the foot is cold to the touch, swollen, cyanotic, or waxy white. Some days later hyperemia occurs and the foot turns red, hot, more swollen, and painful. Localized hemorrhage and lymfangitis appear and in some cases thrombophlebitis and gangrene may follow. A month later intense paresthesias sometimes occurs. Chronic localized cold sensitivity and hyperhiderosis may be present for many years with immersion foot, as well as with other localized cold injuries.

CLIMATE AND PHYSICAL AND MENTAL CAPABILITY

Physical performance may be kept unaffected by climate if the temperature balance of the body can be preserved. A maximal performance may be accomplished even in hot climate, but the endurance is reduced due to loss in circulation capacity. This is related to a need of higher blood flow in muscles and skin related to heat dissipation. The energy needed to perform the physical work, however, is the same.

In the case of a low body temperature, maximal performance is reduced. Normally easy tasks may be difficult, leading to fatigue and exhaustion.

Acclimatization and individual physical differences in tolerance to extreme temperatures are important variables in worker productivity. Other working conditions—such as social relations, organizational problems, etc.—also affect how workers will tolerate work in hot or cold climates (Jokl 1982; Enander 1984).

A temperature of 30°C has been shown to reduce mental and physical performance in workplaces (Caplan and Lindsay 1946). The decrease of performance is due mainly to discomfort rather than excess physiologic strain. The Effective Temperature (E.T.) scale is recommended to evaluate climatic effects in workplaces. Wind velocity has great impact on the performance. Studies (Zenz 1975) indicate the following:

- A 5% drop in performance at 29°C (E.T.)
- A 10% drop in performance at 30°C (E.T.)
- A 17% drop in performance at 31°C (E.T.)
- A 30% drop in performance at 32°C (E.T.)

The comfort zone for light work with low air velocity, acceptable clothing, and no radiant heat is given by three parameters: dry-bulb temperature, wet-bulb temperature, and water vapor pressure.

Prevention of Injuries Related to Heat and Cold

Agriculture is a very different and a very diversified type of job with respect to climate and work comfort. The differences are large between work in settings located in tropical zones and ranching in areas with very cold winters, like central Canada, or reindeer production, as in the Scandinavian countries. By tradition farmers have developed strategies to adjust working methods and appropriate clothing to work in local conditions. Introducing new techniques may change the prerequisites. Tractor cabs, for example, may provide protection against dust exposure, but without air

Heat index chart.
Risk zones and related heat disorders.

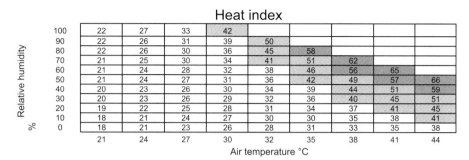

Relative humidity %	21	24	27	30	32	35	38	41	44
100	22	27	33	42					
90	22	26	31	39	50				
80	22	26	30	36	45	58			
70	21	25	30	34	41	51	62		
60	21	24	28	32	38	46	56	65	
50	21	24	27	31	36	42	49	57	66
40	20	23	26	30	34	39	44	51	59
30	20	23	26	29	32	36	40	45	51
20	19	22	25	28	31	34	37	41	45
10	18	21	24	27	30	30	35	38	41
0	18	21	23	26	28	31	33	35	38

Heat index

Air temperature °C

Figure 9.8. Heat index chart—risk zones and related heat disorders.

Table 9.2. Cooling Effect of Wind

The effect of wind given as reference temperature without wind.

	Air temperature degree Celsius					
Wind m/s	0	−5	−10	−15	−20	−25
2	−2	−7	−12	−17	−23	−28
7	−11	−17	−25	−32	−38	−45
11	−16	−23	−31	−38	−46	−53
16	−18	−26	−34	−42	−49	−57
20	−19	−28	−36	−43	−52	−59

conditioning the local in-cab climate may be very hot. If adaptations are not made to accommodate working in cold or hot environments, work quality and quantity are generally reduced. Injury rates are raised. However, few studies are available to quantify these effects (Jokl 1982).

The U.S. National Weather Service has developed guidelines (Figure 9.8) to predict exposure risks according to high temperature and humidity (Bross and others 1994). The cooling effect of low temperature depends on wind velocity; the wind chill index combines these two factors (Table 9.2).

In industries like farming and forestry, it is difficult to control heat exposure and cold exposure by engineering controls. However, in extreme climate situations, worker acclimatization is the first priority. Adequate water should be supplied and workers should be fit, healthy and not be hindered by concurrent medical problems, or taking medications that might complicate extreme temperature exposures.

REFERENCES

Abel, S. M. (1990). "The extra-auditory effects of noise and annoyance: An overview of research." J Otolaryng suppl 1:1–13.

ACGIH (1996). "Threshold Limit Values for Chemical Substances and Physical Agents in the Work Environment." American Conference of Governmental Industrial Hygienists.

Andersson, G., L. Lyttkens, et al. (1999). "Distinguishing levels of tinnitus distress." Clin Otolaryng 24(5): 404–410.

Anttonen, H. and H. Virokannas (1994). "Hand-arm vibration in snowmobile drivers." Arctic Med Res 53 Suppl 3:19–23.

Auger, J., J. M. Kunstmann, et al. (1995). "Decline in semen quality among fertile men in Paris during the past 20 years." N Engl J Med 332(5):281–285.

Axelsson, A. and A. Ringdahl (1989). "Tinnitus—A study of its prevalence and characteristics." Br J Audiol 23(1):53–62.

Blatherwick, F. J. (1969). "Tractor noise—A farm hazard." Alberta Medical Bulletin 34:21–28.

Bonde, J. P., A. Giwercman, et al. (1996). "Identifying environmental risk to male reproductive function by occupational sperm studies: Logistics and design options." Occup Environ Med 53(8):511–519.

Borghi, L., T. Meschi, et al. (1993). "Hot occupation and nephrolithiasis." J Urol 150(6):1757–1760.

Bovenzi, M., P. Apostoli, G. Alessandro, O. Vanoni (1997). "Changes over a workshift in aesthesiometric and vibrotactile perception thresholds of workers exposed to intermittent hand transmitted vibration from impact wrenches." Occup Environ Med 54:577–587.

Bovenzi, M., A. Fiorito, et al. (1987). "Bone and joint disorders in the upper extremities of chipping and grinding operators." Int Arch Occup Environ Health 59(2):189–198.

Bovenzi, M. and C. T. Hulshof (1999). "An updated review of epidemiologic studies on the relationship between exposure to whole-body vibration and low back pain (1986–1997)." Int Arch Occup Environ Health 72(6):351–365.

Bross, M. H., B. T. Nash, F. B. Carlton, Jr. (1994). "Practical Therapeutics—Heat Emergencies." Am Fam Physician 50:389.

Bruel, P. V. (1976). "Do we measure damaging noise correctly?" Bruel & Kjaer Tech Rev nr 1:3.

Burton, A. C. E. O. G. (1969). Man in a Cold Environment. New York, Hafner.

Caplan, A. and J. Lindsay (1946). "An experimental investigation of the effects of high temperature on the efficiency of workers in deep mines." Bull Inst Min Metall (London) 480:481.

Choi, S. and C. Peck-asa, et al. (2005). "Hearing loss as a risk factor for agricultural injuries." AJ Ind Med 48:293–301.

deJoy, D. M. (1984). "The nonauditory effects of noise: Review and perspectives for research." J Auditory Res 24:123–150.

Ekenvall, L. (1987). Vibration syndrome. Clinical and Pathogenic Aspects. Stockholm, Karolinska institutet (thesis).

Ekenvall, L. and L. E. Lindblad (1986). "Is vibration white finger a primary sympathetic nerve injury?" Br J Ind Med 43:702–706.

Ekenvall, L., L. E. Lindblad, et al. (1987). "High vascular tone but no obliterative lesions in vibration white fingers." Am J Ind Med 12(1):47–54.

Ekenvall, L., B. Y. Nilsson, et al. (1986). "Temperature and vibration thresholds in vibration syndrome." Br J Ind Med 43(12):825–829.

Embon, O. M., G. A. Rose, et al. (1990). "Chronic dehydration stone disease." Br J Urol 66(4):357–362.

Enander, A. (1984). "Performance and sensory aspects of work in cold environments: A review." Ergonomics 27(4):365–378.

Ernest, K., M. Thomas, et al. (1995). "Delayed effects of exposure to organophosphorus compounds." Indian J Med Res 101:81–84.

Fam, A. G. and A. Kolin (1986). "Unusual metacarpophalangeal osteoarthritis in a jackhammer operator." Arthritis Rheum 29(10):1284–1288.

Färkkilä, M., S. Aatola, J. Starck, O. Korhonen, I. Pyykkö (1986). "Hand-grip force in lumberjacks: Two year follow-up." Arch Occup Environ Health 58:203–208.

Gagge, A. P. and Y. Nishi (1977). Heat exchange between human skin surface and thermal environment. Handbook of Physiology. D. K. H. Lee. Bethesda, American Physiological Society.

Gemne, G. (1992). Pathophysiology and pathogenesis of disorders in workers using hand-held vibrating tools. Hand-Arm Vibration. A Comprehensive Guide for Occupational Health Professionals. P. T. Pelmear, W.; Wassermann, D. New York, Van Nostrand Reinhold.

—— (1997). "Diagnostics of hand-arm system disorders in workers who use vibrating tools." Occup Environ Med 54(2):90–95.

Gemne, G. and H. Saraste (1987). "Bone and joint pathology in workers using hand-held vibrating tools." Scand J Work Environ Health 13(4 special issue):290–300.

Griffin, M. J. (1998). "Evaluation of the effectiveness of gloves in reducing the hazards of hand-transmitted vibration." Occup Environ Med 55:340–348.

Hagberg, M., H. Morgenstern, et al. (1992). "Impact of occupations and job tasks on the prevalence of carpal tunnel syndrome." Scand J Work Environ Health 18(6):337–345.

Hamilton, A. A. (1918). "A study of spastic anemia in hands of stonecutters. Bull U.S. Bureau of Labor Statistics No 236." Ind Accidents and Hygiene Series 19(53–66).

Haughey, B. P., S. Graham, et al. (1989). "The epidemiology of testicular cancer in upstate New York." Am J Epidemiol 130(1):25–36.

Holgers, K. M., S. I. Erlandsson, et al. (2000). "Predictive factors for the severity of tinnitus." Audiology 39(5):284–291.

Iki, M., N. Kurumatani, et al. (1986). "Association between vibration-induced white finger and hearing loss in forestry workers." Scand J Work Environ Health 12(4 Spec No):365–370.

Jokl, M. V. (1982). "The effect of the environment on human performance." Appl Ergonomics 13(4): 269–280.

Jones, D. (1990). "Recent advances in the study of human performance in noise." Environ Int 16: 447–458.

Karjalainen, P., E. M. Alhava, J. Valtola (1975). "Thenar muscle blood flow and bone mineral in the forearms of lumberjacks." Br J Ind Med 32:11–15.

Kivekas, J., H. Riihimaki, et al. (1994). "Seven-year follow-up of white-finger symptoms and radiographic wrist findings in lumberjacks and referents." Scand J Work Environ Health 20(2):101–106.

Klockhoff, I., B. Drettner, et al. (1974). "Computerized classification of the results of screening audiometry in groups of persons exposed to noise." Audiology 13(4):326–334.

Kumlin, T., M. Wiikeri, P. Sumari (1973). "Radiological changes in carpal and metacarpal bones and phalanges caused by chain saw vibration." Br J Ind Med 30:71–73.

Lahdetie, J. (1995). "Occupation- and exposure-related studies on human sperm." J Occup Environ Med 37(8):922–930.

Leppert, J., H. Aberg, et al. (1987). "Raynaud's phenomenon in a female population: Prevalence and association with other conditions." Angiology 38(12):871–877.

Lings, S. and C. Leboeuf-Yde (2000). "Whole-body vibration and low back pain: A systematic, critical review of the epidemiological literature 1992–1999." Int Arch Occup Environ Health 73(5):290–297.

Loriga, G. (1911). "Il lavoro con i martelli pneumatici." Boll Inspett Lavoro 2:35–60.

Lundborg, G., C. Sollerman, et al. (1987). "A new principle for assessing vibrotactile sense in vibration-induced neuropathy." Scand J Work Environ Health 13(4):375–379.

Lundström, R., T. Stromberg, et al. (1992). "Vibrotactile perception threshold measurements for diagnosis of sensory neuropathy. Description of a reference population." Int Arch Occup Environ Health 64(3):201–207.

Magnusson, M., M. H. Pope, M. Rostedt, T. Hansson (1993). "The effect of backrest inclination on the transmission of vertical vibrations through the lumbar spine." Clin Biomech 8:5–12.

Marvel, M. E., D. S. Pratt, et al. (1991). "Occupational hearing loss in New York dairy farmers." Am J Ind Med 20(4):517–531.

McBride, D. I., H. M. Firth, et al. (2003). "Noise exposure and hearing loss in agriculture: A survey of farmers and farm workers in the Southland region of New Zealand." J Occup Environ Med 45(12): 1281–1288.

Necking, L. E., L. B. Dahlin, J. Fridén, G. Lundborg, R. Lundström, L. E. Thörnell (1992). "Vibration-induced muscle injury. An experimental model and preliminary findings." J Hand Surg [Br] 17: 270–274.

O'Donnell, T. F., Jr. (1975). "Acute heat stroke. Epidemiologic, biochemical, renal, and coagulation studies." JAMA 234(8):824–828.

Panjabi, M. (1996). "What happens in the motion segment?" Bull Hosp Jt Dis 53:149–153.

Panjabi, M., G. B. Andersson, L. Jorneus, E. Hult, L. Mattsson (1986). "In vivo measurements of spinal column vibrations." J Bone Joint Surg Am 68(5): 695–702.

Peplonska, B. and N. Szeszenia-Dabrowska (2002). "Occupational diseases in Poland, 2001." Int J Occup Med Environ Health 15(4):337–345.

Pope, M. H., H. Broman, T, Hansson (1989). "The dynamic response of a subject seated on various cushions." Ergonomics 32(10):1155–1166.

Pyykkö, I., M. Färkkilä, et al. (1986). "Cold provocation tests in the evaluation of vibration-induced white finger." Scand J Work Environ Health 12(4 Spec No):254–258.

Pyykkö, I., O. Korhonen, et al. (1986). "Vibration syndrome among Finnish forest workers, a follow-up from 1972 to 1983." Scand J Work Environ Health 12(4 Spec No):307–312.

Pyykkö, I., E. Sairanen, et al. (1978). "A decrease in the prevalence and severity of vibration-induced white fingers among lumberjacks in Finland." Scand J Work Environ Health 4(3):246–254.

Pyykkö, I., J. Starck, M. Färkkilä, M. Hoikkala, O. Korhonen, M. Nurminen (1981). "Hand-arm vibration in the aetiology of hearing loss in lumberjacks." Br J Ind Med 38:281–289.

Raynaud, M. (1862). De l'asphyxie locale et de la gangrène symétrique des éxtremités. Paris, L. Leclerc, Libraire-Éditeur.

Rehm, S., E. Gros, G. Jansen (1985). "Effects of noise on health and well-being." Stress Medicine 1:183–191.

Seidel, H. (1993). "Selected health risks caused by long-term, whole-body vibration." Am J Ind Med 23(4):589–604.

Seidel, H. and R. Heide (1986). "Long-term effects of whole-body vibration: A critical survey of the literature." Int Arch Occup Environ Health 58: 1–26.

Shibolet, S., R. Coll, et al. (1967). "Heatstroke: Its clinical picture and mechanism in 36 cases." Q J Med 36(144):525–548.

Silverstein, B. A., L. J. Fine, et al. (1987). "Occupational factors and carpal tunnel syndrome." Am J Ind Med 11(3):343–358.

Sloan, R. P. (1991). Cardiovascular effects of noise. Noise and Health. T. H. Fay. New York, New York Academy of Medicine.

Smith, A. P. (1989). "A review of the effects of noise on human performance." Scand J Psychology 30: 185–206.

Standardization, I. O. (1990). ISO-1000: Acoustics: Determination of Occupational Noise Exposure and Estimation of Noise Induced Hearing Impairment, International Organization for Standardization.

Stenlund, B., I. Goldie, et al. (1993). "Shoulder tendinitis and its relation to heavy manual work and exposure to vibration." Scand J Work Environ Health 19(1):43–49.

—— (1992). "Diminished space in the acromioclavicular joint in forced arm adduction as a radiographic sign of degeneration and osteoarthrosis." Skeletal Radiol 21(8):529–533.

Stockholm Workshop '86 (1986). "Symptomatology and diagnostic methods in the hand-arm vibration syndrome." Scand J Work Environ Health 13 (4).

Taylor, W. P., P. L. Pelmear (1975). Vibration White Finger in Industry. London, Academic Press.

Taylor, W. P., P. L. Pelmear, J. Pearson (1974). "Raynaud's Phenomenon in forestry chain saw operators." The Vibration Syndrome. W. Taylor. London and New York, Academic Press.

Teixeira, C. F., L. G. Augusto, et al. (2003). "[Hearing health of workers exposed to noise and insecticides]." Rev Saude Publica 37(4):417–423.

Thelin, A. (1981). "Work and health among farmers. A study of 191 farmers in Kronoberg County. Sweden." Scand J Soc Med Suppl 22:1–126.

Thulesius, O. (1976). "Vibration-induced white fingers." Läkartidningen 73:2269–2270.

Tomlinson, R. W. (1970). A Tractor Noise Limit According to Predicted Driver Hearing Loss. Bedford, National Inst of Agr Eng, West Park, Silsoe.

Zenz, C. (1975). Occupational Medicine. Chicago, Year Book Medical Publishers, Inc.

Zoger, S., J. Svedlund, et al. (2001). "Psychiatric disorders in tinnitus patients without severe hearing impairment: 24 month follow-up of patients at an audiological clinic." Audiology 40(3):133–140.

10
Psychosocial Conditions in Agriculture

Kelley J. Donham and Anders Thelin

INTRODUCTION AND OVERVIEW

The Mental Health of Farmers

Recent reports of increased depressive symptoms and suicides in British farmers have highlighted causal links to acute agricultural crises. Thousands of cattle and sheep were killed to control bovine spongioform encephalopathy and hoof and mouth disease disasters (Eisner and others 1999; Gregoire 2002).

These dramatic crises, however, do not effectively describe the chronic stress often associated with production agriculture—stress that, if not managed, may lead to periodic depression. This chapter will explore the sources of those stresses, their physical and mental health effects, and methods of control.

Comparative studies of the general mental health in urban and rural settings in industrialized countries reveal similar patterns, with a possible exception for depressive disorders and suicides. However, in rural areas people may have considerably greater inaccessibility to medical services (Bentham and Haynes 1986; Shucksmith and others 1996).

There are greater differences when comparing the mental health of urban to farming populations. High risks of suicides among farmers have been reported, especially from the U.S., (Stallones 1990; Zwerling, Burmeister et al. 1995), from the British islands (Booth and others 2000; Thomas and others 2003) and from Australia (Page and Fragar 2002). Studies from other regions have not shown this increase (Thelin 1991; Stallones and Cook 1992; Pickett and others 1993; Liu and Waterbor 1994; Boxer, Burnett et al. 1995).

Substance abuse generally is a sporadic problem among some farmers. However, the overall rate of alcohol-related problems is low, including alcohol abuse and alcohol-related diseases (Walker 1988; Hsieh and others 1989; Thelin 1991; Paxton and Sutherland 2000).

From Peasant to Entrepreneur

The social life of farmers has traditionally been structured by cultural traditions, marriage and extended family relationships, and interdependence in work and daily living (neighboring) (Figure 10.1). In many areas of the world, farming villages, schools, and churches are still important social structures, providing the framework for the agricultural populations. In industrialized western countries the conditions in rural areas have slowly changed over a period of more than 100 years but have accelerated since the industrial revolution and more recently the globalization of our economy. Farming as a way of living has been increasingly isolated to the individual, relatively independent from the former social structure (Thelin 1995; Swisher and others 1998) (Figure 10.2). Neighbors and the rest of the family are mostly integrated into their personal activities. Although farming is still the dominant form of production (regarding the total numbers involved) in most countries, the farming is more and more industrialized, and the farmer today is a specialist, no longer doing everything from building the family home to slaughtering and food preparation. In some areas corporative farms and farming companies today are new actors in farm business. This is related to the fact that modern farming is a very capital-

Figure 10.1. Traditional farming engaged many persons. (Source: Nodiska museets bildbyrå, Stockholm, Sweden)

Figure 10.2. Modern farming often is one man's business. (Source: Natur fotografernas bildyrå, Stockholm, Sweden)

intensive business. It is hard to start farming without a large inheritance or assistance from family.

Basic changes in the way of living and working have a strong impact on the psychosocial work environment. These changes create new sources of stress, and the coping mechanisms imbedded in the former social structures are diminished. The new sources of stress are added to the traditional stressors of harvesting problems: bad weather, machinery breakdowns, and sick animals. These changes and resultant stressors affect the work environment as well as the family environment.

Farming in a Changing World

Beginning in the 1960s, it was noticed that modern living with rapid turnover of goods and service as well as an intensive competition might be a threat to people's health. Thomas Holmes and Richard Rahe developed The Social Readjustment Scale to estimate an individual's psychological stress relative to socioeconomic change (Holmes and Rahe 1967). They found that rapid social changes in people's life were related to cardiovascular diseases.

These findings in human beings have also been identified in studies of apes and other animals as disturbing social structures (e.g., moving apes from one established group to another) related to arteriosclerosis and cardiac infarction (Henry and Stephens 1977; Manuck and others 1995). From an evolutionary perspective these observations indicate that human beings might be prone to new health risks in situations with rapidly developing new social structures.

The effects of different environmental factors like noise, pollution, pesticides, and crowding are relatively well known. The impact of different psychosocial conditions is not so well elucidated. However, the science of how psychological factors generate physical health problems has grown during the last decades. Today it is accepted that chronic stress can exacerbate many general health problems, such as cardiovascular diseases, diabetes, obesity, and depression (Quick and others 1997; Levi and Levi 2000).

STRESS AND STRESS-RELATED HEALTH DISORDERS

Sources of Stress

DISEASES

Health defects, disease symptoms, or fear of diseases are potential stressors. The conceptualization of health defects may generate anxiety and fear of losing control. Diseases in the family may be a stressor of the same magnitude as a personally perceived illness. Destructive feedback mechanisms may be established as stress may generate physical symptoms, which causes more anxiety, fear, and stress. Caring for an ill or disabled family member is a heavy burden for many persons.

FAMILY

Supportive spouses and general good family relationships can buffer stress. On the other hand, conflicts in the family negatively affect all members in the family (Kiecolt-Glaser and others 1996; Golding 1999; Testa and Leonard 2001). Dissolving family relationships (in most cases) is extremely stressful for all involved including the children (Amato 2000; Ellis 2000).

In farming communities the context of family life has changed over time. The extent of change varies in different countries. The social conditions are naturally very different in a village community in comparison to a modern one-man farm operation with his wife working at another job several kilometers away. In certain rural areas there is a deficit of women as they leave the rural setting to seek other employment. Opportunities for establishing a family are disturbed in these areas. The percentage of farmers living alone is larger than among other professionals, at least in some areas (Holmberg and others 2004).

The image of a farmer as a strong independent operator of big machines and manager of large herds has limitations. He/she may be a lonesome person without another person to relate to all day long. If he/she has a spouse, when that spouse comes home they have experiences and contacts to talk about with others unattached to the farm, further isolating the farming partner. Often, members of the previous farming generation, such as the parents of the farmer, are living close by and are still involved with the farming operation. Intergenerational issues in family farming can be an extreme stressor. The generations may differ in how the farm should be managed. The operation of the farm, the farm property, or the estate in the old way may no longer be realistic. Difficulties adjusting to changes in social position in intergenerational farms may cause controversy among all family members involved (Thelin 1995).

Men and women traditionally have had different roles. Especially women on farms may have several roles and expectations. These multiple roles may be a source of conflicts and stress especially when the family includes children (Bird 1999; Stohs 2000). Changes in gender roles have resulted in more equity and in many countries women have major or primary roles as principal or coprincipal operators of the farm. But still the women have major responsibility for children and the household (Lundberg 1998). Women on farms with children often have a job outside the farm. These women still feel a responsibility, obligation, or pressure to assist in farming and may be strongly stressed by these conflicting roles (Walen and Lachman 2000).

ECONOMY

Most farmers are entrepreneurs. To start and run a modern farm in western industrialized countries a large capital investment is a prerequisite. Economic resources of farmers are very unequal. Some have a large family capital in terms of a farm they have inherited. Others have a significant property with money from other sources. However, a great number of farmers permanently have marginal cash resources (Simkin and others 1998). They have to work longer and harder than the average wage earner and they are very exposed to changes in market prices and the effects of weather on the outcome of their investment. Most farmers live with a constant worry about their economic survival, creating stress. Occurrences like a salmonella outbreak or a major change of the farm politics may result in a breakdown economically and mentally (Simkin and others 1998).

RELATION TO GOVERNMENTAL AUTHORITIES AND NEIGHBORS

During recent years, farmers around the world have been criticized for polluting the environment with toxic chemicals and inhumane treatment of animals in types of housing used and husbandry practices.

Many persons living in cities have lost contact with the land that their parents and grandparents had, and they do not understand rural conditions. Sometimes farmers are exposed in the media as ruthless profit makers. In many countries a repressive legislation has been introduced and the farmers have to open their farms for different types of inspections and controls. These actions may be a result of a notice from a distressed neighbor or just a person passing by. In some countries the farmer has to pay for the inspection. These conditions are present more or less among industrialized countries, but they have generally led to a cultural gap between the farming community and the rest of society that is noted by a mutual lack of understanding, mistrust, and more of an adversarial than cooperative relationship.

The changing legal and political framework within which farmers are required to operate is a source of stress to many farmers. More than half of the British farmers are concerned about the amount of record keeping and paperwork involved, and many have difficulties understanding and completing the forms. Farmers are anxious about incurring penalties for mistakes in completing forms or failing to comply with obscure regulations (Simkin and others 1998; Booth and Lloyd 2000). Some farmers have a feeling of inferiority, of being a burden to the rest of society. Farming in general has lost status relative to other professions. However, the farming population is generally more highly educated than the general population in many areas of industrialized countries. Farmers are proud and independent people, but often feel unappreciated. Different types of subsidies increase this feeling of inferiority.

PSYCHOSOCIAL WORKPLACE FACTORS

Previously mentioned loneliness and social isolation are of sporadic concern in modern life on farms, particularly in sparsely populated areas and areas where it is difficult to attract spouses and social institutions are waning. However, several studies indicate that many farmers have well-functioning social networks because the art and practice of "neighboring" (social interaction, mutual caring, and cooperative work) is still alive in many agricultural communities (Simkin and others 1998; Thelin and others 2000). Social isolation is not a widespread stressor in farming. In fact, residents of our urban areas may be more isolated and lonely than our farming population.

Bad weather or machinery breakdown in time of harvesting or sick animals entail temporary stress. On the other hand, nice weather and healthy animals may alleviate stress. These normal stressful variations in the lives of most farmers probably have little negative impact on the longtime outcome of their mental health. Solving problems may contribute to the feeling of control and thus act as a positive health factor.

Farming still requires long hard work. Seventy percent of British farmers work over 10 hours a day and 20% over 15 hours (Simkin and others 1998). Dairy farmers had the longest working days. Many Swedish farmers do not have vacations and some do not have a single day away

from the job during a full year (Holmberg and others 2003). Few possibilities for relaxation and deprivation of sleep are related to stress. However, little is known about the health effects of the long working days and shortness of vacations among farmers.

Many family farmers do not have any employees. They do not have to realize any role conflicts (except possibly for family relations); they have no problems with career development, organizational changes, or many other stressors prevalent in industrial or business enterprises.

PHYSICAL WORKPLACE FACTORS

Environmental pollution, noise, heat, cold, vibration, and work injuries are physical factors in the workplace or living environment of farmers that may induce stress. Gaseous odors or particulate emissions from livestock operations may induce physical stress (Berding and Matthies 2002). This stress has been measured in neighbors of large confined animal feeding operations (Thu 1997). (Many of these neighbors may be farmers.) This situation creates neighbor conflict and can create severe stressors in some farm communities.

As reviewed in Chapter 9, many agricultural operations are very noisy. Noise has been a documented stressor in other settings, such as in those living near airports (Evans and Johnson 2000; Evans and others 2001). Therefore, it is logical that noise could be a stressor in agriculture.

Although many male farmers are not concerned about injury risks, generally, many women are highly concerned about potential injury to their family members, which certainly adds to their risk load. In fact some women are sufficiently concerned that they would advise their children to seek employment outside of agriculture (Thu 1997).

Stress Physiology

The Hungarian-Canadian researcher Hans Seyle (1936) is the father of the modern concept of stress. His first paper "A syndrome produced by diverse nocuous agents" was published in 1936. Seyle noted that a number of different environmental insults could generate the same physical reactions among different individuals. He started to use the term **stress** to refer to the response that the individual makes to environmental insults. He used the word **stressor** as a term for the stimulus causing stress. During his career he developed a model for how the body defends itself in stressful situations and he emphasized that stress is a general physical reaction caused by a number of environmental factors or stressors.

Stress as a general physical response or a syndrome became known as **the general adaptation syndrome (GAS).** It is divided into three stages. The first is the **alarm reaction** characterized by general mobilization through activation of the sympathetic nervous system. The body's systems are activated to maximize strength and prepare them for a fight-or-flight response.

The alarm reaction successively passes on to a second phase of GAS called **the resistance stage**. The objective physical symptoms disappear and the organism adapts to the stressor. The length of this stage depends on the severity of the stressor and the body's adaptive capacity. Although the person gives the outward appearance of normality the body's physical reactions are not normal. This condition is also called **allostasis**. Seyle believed that these reactions in the long run could generate a number of diseases, which he called *diseases of adaptation*.

The capacity to resist stress is individual and has limitations. Sooner or later the ability to resist will expire and the persistent stress will generate endstage effects. This final stage is called the **exhaustion stage**. Seyle thought that exhaustion frequently resulted in depression and sometimes even death.

It was obvious to Seyle that stress reactions in some situations were valuable. The capacity to mobilize extra resources when an individual is in difficulty is an important adaptive mechanism. This good kind of stress was termed **eustress** while the opposite was called **distress**. Distress results in diseases.

Seyle's concept of stress was criticized from different point of views. Cannon described the physiological reactions as an action to stabilize the body to maintain the homeostasis and thought Seyle's concepts as too stereotypical (Cannon 1939). Skinner found that the stress reaction is unique to every individual (Skinner 1985). Others (Mason 1971) noticed that Seyle largely ignored psychological factors, especially the element of emotion. An alternative model focusing on the psychological factors has been developed by Lazarus and Folkman (1984) and has had a great

impact among psychologists. These concepts will be reviewed in detail later in this chapter.

THE NERVOUS SYSTEM AND STRESS

One of the functions of the nervous system is to integrate the different systems of the body and maintain homeostasis. The system has a high level of flexibility and capacity of rapid adjustment to new situations. The nervous system provides internal communication and external communication to and from the environment. The external information may be of physical or psychological character. The mix of information is processed in different centers of the brain and combined or associated with previous experiences, archetypical perceptions, and emotional loadings. Threatening situations and/or situations with a disturbing emotional tension are identified and initiates mobilization and integration of signals. This mobilizing is mediated by the **autonomic nervous system (ANS)**. The ANS has two divisions, the **sympathetic nervous system (SNS)** and the **parasympathetic nervous system (PNS)**. The signals of the SNS and PNS are mediated to autonomic ganglia from where they are relayed to the effector organs. The SNS ganglia are connected in a chain on either side of the vertebral column. The PNS ganglia are located near or in the effector organs. The SNS ganglia have close interconnections, and thus activation of SNS often has a broad effect while the PNS ganglia have few connections and mostly an effect restricted to a single organ or organ system.

Physiologic results of SNS mobilization and activation include increased rate and strength of cardiac contractions and respiration, decrease of gastrointestinal activity, constriction of blood vessels in the skin, stimulation of the sweat glands, and dilation of the pupils. The adrenal medulla is also stimulated to produce more adrenalin (epinephrine) and noradrenalin (norepinephrine). Noradrenalin is also produced by the brain stem in the locus coeruleus. These catecholamines have complex effects acting as neurotransmittors and as hormones. Adrenalin and noradrenalin act similarly on some organs and different on others. Two basic actions may be mentioned: 1) influence of the tone and contraction of muscles, and 2) influence on the metabolism of carbohydrate and fat. The actions are mediated by receptors in the cell membranes of the target cells. At least two kinds of adrenergic receptors are recognized, α-receptors and β-receptors. Contractile action on smooth muscle is mediated by α-receptors while metabolic effects and excitation of the myocardium is mediated by β-receptors.

Effects on the skeletal muscles are also mobilized in case of stress (Lundberg and others 1994). Areas in the neocortex, the amygdala, and the brain stem are engaged in preparing the muscles for physical activity by elevating muscle tone. Central activity in the amygdala area also may upgrade the level of muscular tone.

THE ENDOCRINE SYSTEM AND STRESS

There is a chain of "stress hormones" that are activated during stress. First corticotrophin-releasing hormone (CRH) from the hypothalamus simulates the pituitary to release **adrenocorticotropic hormone (ACTH)**, which in turn stimulates the adrenal gland (cortex) to release **cortisol** (the most important of the stress-related hormones). The axis from hypothalamus via the pituitary to the adrenal cortex is called the **HPA-axis** and is of fundamental importance in the activation of stress reactions (Figure 10.3).

Cortisol is so closely associated with stress that the concentration of cortisol circulating in the blood can be used as an index of stress. Cortisol can easily be analyzed in blood as well as in saliva. There is a normal variation in the concentrations of cortisol in the blood over the day, with highest levels in early morning and lower levels in the evening. There are large differences in the level of cortisol between different individuals, as well as in the variation over the day.

Cortisol has a number of physiological effects. Gluconeogenesis is stimulated and the availability of glucose is raised. Lipolysis is also stimulated, resulting in a release of fatty acids to the circulation. In conjunction with noradrenalin release, vasoconstriction is elicited, resulting in raised blood pressure. Water excretion by the kidneys is increased. Cortisol has a number of effects on the immune system, inhibiting antibody production and local inflammation. This antiphlogistic (inflammation-inhibiting) action is frequently used in different therapeutic applications.

Additional hormones are released during stress as the anterior pituitary is stimulated (Figure 10.4). They are all regulated from the hypothalamus by centrally produced releasing factors. The

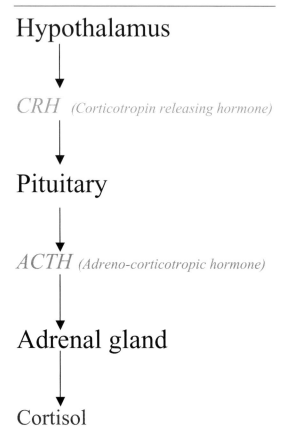

Figure 10.3. The HPA-axis.

sex hormones (luteinizing hormone (LH), follicle-stimulating hormone (FSH), growth hormone (GH), tyreotropin (TSH), and prolactin (PL)) are all involved in the stress complex. Oxytocin is a hormone known more for its effects of uterine contraction and milk let-down during birthing and "mothering." Oxytocin also has effects with regard to the individual's ability to cope with stress. Oxytocin stimulates relaxation, social learning, and trust (Uvnäs-Moberg 1997a). Thus oxytocin may be regarded as an antistress hormone. The production of oxytocin is stimulated by pleasant skin contact, such as massage and slight and nice touch (Uvnäs-Moberg 1997b).

Acute and Chronic Stress

Stress and stress reactions have a common physiologic basis, but are manifested differently among individuals and their personal situations. It is not possible for an individual to take a full control over

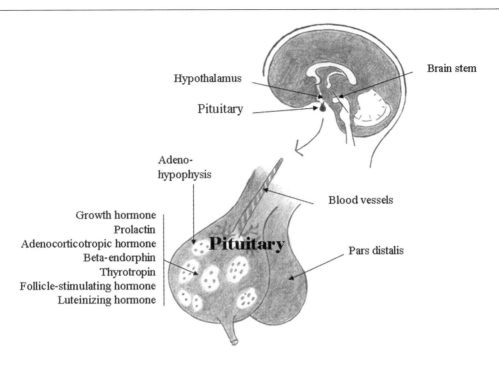

Figure 10.4. Anterior pituitary hormones.

these reactions, but it is possible to learn some coping skills. The immediate reaction to stress is general and more of a "reflex" reaction. The ability to "switch off" the mobilization or cascade of stress events is varied among individuals and depends on many factors, including individual psychological physiological variation, cultural background, and how life and work are organized.

The alarm reaction is mostly short lasting and the activity of the ANS may quickly be adjusted to normal (physiological effects dissipate within 5 hours). This kind of stress should be regarded as a functional resource in arduous situations.

Chronic or long-lasting stress (as well as frequently repeated acute stress reactions) can be dysfunctional and dangerous. The effect of this kind of stress (the resistance stage, according to Seyle) is different and related to a number of diseases. The behavior of people living with long-lasting stress changes: aggressiveness and irritation is elevated with, or finally replaced by, anxiety, depression, and exhaustion. A number of expressions have been used to describe this condition (burnout, chronic fatigue syndrome, neurasthenia, etc.).

Associated with these clinical behaviors are

modifications in the functioning of the ANS and the PNS (Henry and Stephens 1977; Folkow 1997). Elevated tension in the somatic muscles occurs, but the individual is often not aware of the chronic muscle mobilization. However, this muscle tension may result in pain such as neck, shoulder, or back pain (Veiersted and others 1993). The ability of the adrenals to produce elevated levels of cortisol may wane, and cortisol production may "burn out," resulting in a change in the normal daily variation of cortisol concentrations (McEwen 1998; Björntorp and others 2000).

A long-lasting high level of cortisol also affects the brain, resulting in a declining number of functioning cortisol receptors. A size reduction of the hippocampus area has also been demonstrated, probably indicating declining capacity of learning and inadequate coordination of normal stress reactions (Sapolsky 1990).

Sex hormone production is also down-regulated in the case of long-lasting stress. The result is a reduced interest in sex and lower fertility. The production of growth hormone (GH) is likewise affected by stress. The down-regulation may be significant, and stress-related dwarfism has been reported (Sapolsky 1998). The level of growth hormone may be associated with a number of other symptoms related to stress.

Chronic Stress-Related Physiological Reactions

HIGH BLOOD PRESSURE

Animal studies have shown that repeated stressful situations and chronic continuing stress modifies the walls of the arteries, which promotes a higher blood pressure (Swales 1994; Folkow 1995). This reaction in the arterial walls is associated with high activity of the SNS and high levels of cortisol production.

In modern western societies the mean blood pressure of the general population escalates with advancing age. However, two recent controlled studies of populations living in low-stress social conditions had no change in blood pressure over time. The comparison populations (low stress) in these studies included nuns living in a monastery in Italy, and another study of members of a Panamanian Indian tribe (Hollenberg and others 1997; Timio and others 1997). Swedish men living in rural areas have lower blood pressure than men living in the cities (Dahlöf 1992), and Swedish farmers have lower blood pressure than nonfarming rural men, suggesting that both rural living and farming have protective effects. This finding supports the lower mortality of farmers from cardiovascular disease. However, exercise, lower smoking rates, diet, and healthy worker effect may also be related to these findings (Thelin 1981; Stiernström and others 2001).

DYSLIPIDEMIA

Persons under stress with high activity of the HPA-axis system often have low levels of HDL (high density lipoprotein or the "good" cholesterol) and elevated levels of LDL (low density lipoprotein or the "bad" cholesterol) as well as elevated levels of free fatty acids (FFA). The mechanisms behind these changes of the blood lipids are not fully understood.

ABDOMINAL OBESITY

The stress-related hormones, especially cortisol, promote accumulation of abdominal fat, whereas sex hormones and GH (which are inhibited by chronic stress) have the opposite effect (Björntorp 1996).

INSULIN RESISTANCE

It is well known that cortisol decreases the cells' sensitivity to insulin-raising levels of blood glucose. Increased blood glucose stimulates the islets of Langerhans (the beta cells) in the pancreas to produce more insulin. Chronic effects result in the continuous stimulation of the beta cells in the pancreas to the point they cannot keep up the high demand for insulin production over time. The beta cells finally fail, resulting in diabetes type II.

MENTAL EFFECTS

Communication in the brain is mediated by a number of substances, including noradrenalin, serotonin, and dopamine, among others. Cortisol influences the activity and amount of several hormones, which in turn affects the amount and activity of these neurotransmitters, resulting in imbalanced brain chemistry. An imbalance of these brain chemicals is associated with conditions such as depression and anxiety disorders. High levels of cortisol are associated with changes in female sex hormones and decreased thyroid-stimulating hormone; both are related to depression (Sapolsky and others 1986; Gold and others 1988).

IMMUNOLOGICAL REACTIONS

Cortisol has several depressive effects on the immune system. Long-lasting stress with high activity in the HPA-axis suppresses the activity of a variety of different cells in the immune system, and thus the risk of infections and other diseases is raised (Chrousos 1995). (It has been long recognized that exogenous cortisol can be used therapeutically in a number of autoimmune diseases, but also puts the patient at risk for infections.)

Stress and Disease

CARDIOVASCULAR DISEASE

Previously mentioned, chronic stress can result in higher blood pressure, which contributes to cardiovascular disease in two important ways: 1) As a precipitating factor in heart attack or stroke, and 2) as a cause in the development of the degenerative changes of the vessels, resulting in heart disease or cerebral infarction. The risk of serious disturbances of cardiac blood flow is significant if stress is combined with arteriosclerosis of the cardiac arteries (Rosenfeld and others 1978). Acute stress episodes can serve as a trigger of heart at-

Table 10.1. Depression Has Many Faces

Depressed mood
Markedly diminished interest or pleasure in
  everything
Significant weight loss or weight gain
Insomnia or hypersomnia
Psychomotor agitation or retardation
Fatigue or loss of energy
Feeling of worthlessness or guilt
Diminished ability to think or concentrate
Recurrent thoughts of death

tacks (Gullette and others 1997) and chest pains, especially if chronic changes in cardiac vessels has already occurred (Krantz and others 2000; Orth-Gomer and others 2000).

The combination of physiologic changes associated with chronic stress (high blood pressure, dyslipidemia, abdominal obesity, and insulin resistance) is described as **syndrome X or metabolic syndrome**. The metabolic syndrome is strongly related to the development of **arteriosclerosis** and cardiovascular diseases.

MENTAL DISORDERS
The prevalence of episodic depression in the general population is around 5%. Those with deep depression have a significant risk of suicide (Table 10.1). Depression is related to other chronic diseases, including cardiovascular disease (Anda and others 1993; Pratt and others 1996).

Several models have been presented to help explain how stress and depression are interrelated (Owens and Nemeroff 1991; Sapolsky 1998). **Depression** is associated with changes of the production or efficiency of at least three neurotransmittor substances: serotonin, noradrenalin, and dopamine. Pharmacological treatment of depression targets these substances. It is also possible that psychotherapy may affect neurotransmittors. A combined psychological and chemical approach to treating depression (both affect the hypothalamus), affects the transmitter substances. Combined modalities for treating depression are commonly recommended.

Anxiety disorders like phobias, panic attacks, and generalized anxiety are conditions also related to stress. The imbalance of neurotransmitters is again the likely cause.

Insomnia is regarded as a condition of raised physiological activity, a result of repeated or prolonged stress (Bonnet and Arand 1997). Sleeping disturbances are related to the stress hormones. High levels of CRH, ACTH, and cortisol are associated with sleeping disturbances (Kryger 1994).

Drug addiction has been shown in animal studies to be established more rapidly and with lower levels of the drug if the animal is stressed (Piazza and Le Moal 1998). Administration of cortisol supports known addictive mechanisms (Robinson and Berridge 1993). Persons who are more sensitive to stress and individuals who are exposed to stress during long periods have a somewhat higher risk of developing addiction.

Memory may be affected when stress is prolonged. When the concentration of cortisol is permanently high, the hippocampus can be negatively affected, resulting in difficulties in memorizing and enhancing the process of disremembering (McGaugh 1989). The number of nerve cells, dendrites, and synapses decrease, and the size of the hippocampus decreases (Bremner and others 1995; Magarinos and McEwen 1995; Sheline and others 1996). The fact that persons with Cushing's syndrome produce high levels of cortisol and develop a kind of dementia also supports the impression of cortisol as a key factor in this process (Starkman and others 1992).

Persons who have realized very traumatic experiences (e.g., a serious farm injury) sometimes develop a chronic stress syndrome: **post traumatic stress disorder (PTSD)**. This is characterized by flashbacks, anxiety, and facial expression of fear, adhedonia, startle, and SNS-activation. Many types of experiences are related to the development of this disorder (Resnick and others 1992; Roth and others 1997; Schlenger and others 2002). PTSD increases the risk of other medical disorders and produces a long-lasting suppression of the immune system (Kawamura and others 2001). MRI studies of brains of combat veterans with PTSD have demonstrated a significant reduction of the volume of hippocampus (Bremner and others 1995).

Chronic pain disorders may be associated with stress as the spinal cord decreases its ability to control pain. Repeated pain signals may be transformed into persisting hyperexcitability (Melzack and Wall 1965; Boureau and others

2000; Flor 2001). The descending inhibition of the spinal cord may be inhibited by a decreased production of serotonin (Yunus 1992). Reduced female sex hormones, among other hormones, may also increase the experiences of pain (Crofford and others 1994; Anderberg 1999).

Traumatic experiences during childhood modify the sensitivity to pain during life (Anderberg and others 2000). Experiments on animals have shown that stress exposure early in life—as, for example, separation from the mother—results in more stress reactions as adults with a higher activity in the HPA-axis system (Plotsky and Meaney 1993). Stress reactions are more easily activated. Probably a significant number of persons with chronic pain syndromes have easily activated stress reactions because of previous traumatic experiences.

Abdominal disorders, such as hyperbowel activity and abdominal pain, may be caused by stress. Furthermore, **gastric ulcers** and dyspepsia were regarded as diseases with a significant association with stress (Brady and others 1958; Weiss 1968). The detection of the bacterium *Helicobacter pylorus* has somewhat changed that picture but stress probably still has an important role; further exploration is needed to determine why some persons with the bacteria will develop ulcers while others seem to be able to keep the microorganism under control without any symptoms (Pare and others 1993; Levenstein and others 1995; Melmed and Gelpin 1996). (Although stress may be a risk factor for peptic ulcers, the bacterium *Helicobacter pylorus* has been recognized as a more direct cause.)

Irritable Bowel Syndrome (IBS) is a very common health problem, episodically affecting as high as 30% of the population. Clinical symptoms and signs include recidivating pain, abdominal distension, diarrhea, and possibly constipation. A significant proportion of IBS is related to stress (Richter and others 1986; Delvaux 1999). Patients with IBS have a modified activity in the HPA-axis, with raised levels of cortisol in the mornings and lower levels in the evenings. They also report more anxiety, have more depressive episodes, and lack social support (Patacchioli and others 2001). Psychological training programs have a positive effect, with reduced frequency of symptoms and a better general well being (Rome 1999).

OTHER DISEASES RESULTING FROM STRESS

Cohen and coworkers (1991, 1993) have observed that the level of psychological stress is associated with greater infectivity with cold viruses. Their studies, along with others, have shown that stress with a raised activity in the HPA-axis system is related to a more rapid contraction of **infectious diseases** (Bonneau 2001) and diseases of the immune deficiency system like HIV infection (Cole and others 2001).

Stressful events and pain can trigger asthma attacks (Sandberg and others 2000). Actions to control exposure to stressful events has successfully reduced the risk for further asthma attacks (Schmaling 1998).

Autoimmune disorders like **rheumatoid arthritis** and **ulcerative colitis** may also be affected by stress. Stress can affect the progress of the disease by increasing sensitivity and reducing coping efforts and can affect the process of inflammation (Zautra 1998).

Measuring Stress

Measuring the level of stress may be of interest from an individual perspective as well as from an organizational perspective. Researchers have used a variety of methods and most of them fall in three main categories: laboratory methods, psychological measures, and self-reported surveys. Lab methods and psychological approaches have been used mostly in research. There are only a few clinical methods to evaluate individuals.

LABORATORY METHODS

The acute stress reaction may be assessed by measuring heart rate or blood pressure or the combination of them both: the rate pressure product (RPP). Muscle activity may be recorded by electromyography (EMG).

Measurements of catecholamines (noradrenalin and adrenaline) are not reliable as there are large individual variations. The activity in the HPA-axis may be evaluated by assessing the concentration of cortisol in saliva. However, there are also large temporal and individual variations with this hormone.

The level of blood glucose is related to stress. Some of the blood glucose is combined with hemoglobin and assessed as HbA1c. HbA1c is a measure of the average blood glucose concentration over several weeks and is related to long-lasting stress. This lab test is easy to do and the reliability is high.

Psychological Individual Methods

Holmes and Rahe (Holmes and Rahe 1967) developed The Social Readjustment Rating Scale, noting a significant relation between high scores and cardiovascular disease. The antecedent factor may be chronic stress induced by social change.

The "type A" personality (Friedman and Rosenman 1959), especially the hostility factor (Almada and others 1991), is significantly related to cardiovascular disease (Rosenman and others 1975).

There are a number of scales available to test for depression and anxiety. One of the oldest and most used is The Beck Depression Inventory (Beck and Beamesderfer 1974). Another is The Self-rating Depression Scale (Zung 1965). This scale is short (20 items), simple to administer, and has been used in many studies.

Daily life events and hassles, not only traumatic life events, are related to stress and health outcome. Several Perceived Stress Scales are available, which are quite good at predicting headache and IBS (Kanner and others 1981; Cohen and others 1983; DeLongis and others 1988; Fernandez and Sheffield 1996; Searle and Bennett 2001).

Methods for general individual stress testing have also been developed. However, these kinds of scales are fewer because stress is a complex condition. One scale used in Scandinavia is the Stress Profile (Setterlind and Larsson 1995).

To evaluate chronic pain conditions, analogue techniques have commonly been used. Pain is rated by observable behavior of the person. Visual analogue pain rating Scales such as the McGill Pain questionnaire (Melzack 1975) has been available since the 1970s. Different scales have been developed to test for occupational stress in terms of risk for burnout (Maslach 1976).

Screening for Stress on an Organizational Level

Much of the stress people perceive is generated by the interrelationships of individuals (e.g., families and the workplace) (Henry and Stephens 1977; Syme 1989). Family farming often involves the overlap of family relations and workplace relations, creating a complex social environment.

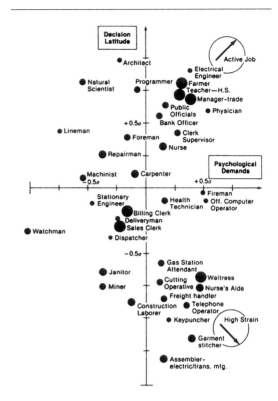

Figure 10.5. The Karasek-Theorell model, the occupational distribution of psychological demands and decision latitude. (Karesek 1989; reprinted by permission of Baywood Publishing Co.)

Robert Karasek and Töres Theorell noticed that most adverse reactions of psychological strain (fatigue, anxiety, depression, and physical illness) occur when the psychological demands of the job are high and the worker's decision latitude (job discretion and job control) is low (Figure 10.5). From this concept, the working life can be described along continuous scales: low strain-high strain, and passive-active. In a four-field diagram, different types of jobs could be identified along these lines as more or less active and more or less high strain (Karasek and Theorell 1990) (Figure 10.5).

Farmers are regarded to have an active job, which may be an explanation for their reduced cardiovascular risk. A number of studies support a view that people in passive jobs have high strain and endure higher risks for cardiovascular disease and depression (Karasek 1981; Karasek and others 1988; Netterström and others 1999).

A number of screening instruments are avail-

	Often	Sometimes	Seldom	Never/ almost never
	1	2	3	4
1. Do you have to work very fast?	☐	☐	☐	☐
2. Do you have to work very intensively?	☐	☐	☐	☐
3. Does your work demand too much effort?	☐	☐	☐	☐
4. Do you have enough time to do everything?	☐	☐	☐	☐
5. Does your work often involve conflicting demands?	☐	☐	☐	☐
6. Do you have the possibility of learning new things through your work?	☐	☐	☐	☐
7. Does your work demand a high level of skill or expertise?	☐	☐	☐	☐
8. Does your job require you to take the initiative?	☐	☐	☐	☐
9. Do you have to do the same thing over and over again?	☐	☐	☐	☐
10. Do you have a choice in deciding HOW you do your work?	☐	☐	☐	☐
11. Do you have a choice in deciding WHAT you do at work?	☐	☐	☐	☐

Figure 10.6. Karasek-Theorell questionnaire (reprinted by permission from Töres Theorell).

able to check the level of job demand, decision latitude, and control over the work situation. Karasek and Theorell's first questionnaire (1990) has been supplemented with other items to include the dimensions of social support and perceived job security (Siegrist 1996) (Figure 10.6).

High functioning personal social support reduces the risk of illness and death. Berkman and Syme (1979) identified four main sources of social contacts of importance to reduced mortality: 1) marriage and life together, 2) contacts with relatives and friends, 3) membership of churches, and 4) membership of other formal and informal groups. Good interpersonal support from colleagues, especially from supervisors, may also alleviate stress. Several scales measuring social support are available (Berle and others 1952; Broadhead and others 1988; Sherbourne and Stewart 1991).

The ability to cope varies between individuals and is of great importance for long-term health status. Antonovsky determined that the ability to cope is related to a "sense of coherence" (SOC), or understanding the complicated structures that surround all of us. (Antonovsky 1979). Antonovsky has developed an SOC-scale, which has been widely used in different situations (Antonovsky 1993).

Treatment and Prevention of Stress Reactions

Historically, stress management programs have tended to focus on solutions at the individual level. Recent research indicates that stress can be effectively addressed by improving the relationships of the individual within the workplace. In the case of farmers, stress may be addressed by improvement of relationships of the individual to his/her family, neighbors, and the local community, and by managing the developing negative perception of farmers from the general public and news media.

Health professionals and especially occupational health physicians and nurses must listen to their patients for signs of stress. Very often a patient presents with a physical health problem. Only after establishing a confidential and trustful relationship between the patient and provider is it possible to determine stressors that might involve personal situations. Sometimes it is difficult to help the patients realize that their reported symptoms are indicators of deeper problems, such as family conflict or economic insolvency.

INDIVIDUAL THERAPEUTIC METHODS

Behavior modification may be needed to help some persons with chronic stress. For example, individuals with a type-A personality or with excessive anxiety may be assisted with **cognitive therapy.** Cognitive therapy is based on the principle that people's beliefs, personal standards, and feelings of self-efficacy strongly affect their behavior (Bandura 1986).

Cognitive therapy includes a variety of strategies to change thoughts. The patients are trained to think about and evaluate their own behaviors. Patients with chronic pain, for example, very often exaggerate their perceived pain by dwelling on it and promoting dysfunctionality. Cognitive therapy for coping with pain (Meichenbaum and Turk 1976; Turk 2001) relies on repeated weakened "doses of pain or stress" in an attempt to build up immunity against pain. Cognitive therapy for low back pain and other symptoms have shown greater effectiveness than traditional behavior modification methods or a placebo (Saunders and others 1996; Compas and others 1998; Kole-Snijders and others 1999; Gardea and others 2001).

Writing or talking about traumatic events (**emotional disclosure)** may help some people (Pennebaker and others 1989, 2001).

Relaxation has been used to treat different stress-related conditions, such as headache, tensions, migraine, and chronic pain conditions (Carlson and Hoyle 1993; Lehrer and others 1994; Sartory and others 1998). **Progressive muscle relaxation** (Jacobson 1938), **meditative relaxation** (Benson and others 1974), **mindfulness meditation** (Kabat-Zinn 1993), guided imagery and yoga are all examples of successful relaxation techniques.

Biofeedback electronic instruments are used to measure biological responses. Biofeedback can be used to decrease muscle tension as a component of stress management (Compas and others 1998). Objective measures of responses are immediately available to the patient and enable them to learn to modify a response. Electromyograph (EMG) biofeedback is the most commonly used method. Thermal biofeedback is based on the principle that skin temperature varies in relation to levels of stress (Blanchard and others 1990).

Personal control is a fundamental element that enables people to cope with stress (Rodin and Langer 1977). However, the fundamental capacity to exercise control may be fixed during childhood and adolescence and thereafter not be easily modified (Antonovsky 1987).

Studies indicate that **physical activity** may be a significant way to cope with stress (Steptoe and others 1989). Physical activity modifies a number of biological reactions to stress in a positive way. Blood pressure, depression, sleeping disturbances, cognitive symptoms, and chronic pain are affected by physical activity. All kinds of exercise is helpful, but running or walking in an open landscape or doing things together in a social context may be more effective than individual exercise with an exercise bicycle or lifting weights.

Good sleep is excellent therapy for stress. Some studies indicate that 8 hours of rest and sleep is not enough (Kryger 1994) and many people need at least 10 hours for relaxation and sleep. To obtain restful sleep, relaxation before going to bed is important. Noise and conflicts should be removed from the bedroom.

FAMILY, NEIGHBORS, AND COMMUNITY

Personal **conflicts must be solved** to counteract stress, and often third-party assistance is necessary. Some individuals with a strong type-A personality may need individual therapy, as they may be conflict-prone.

People who have high levels of social support are more able than other people to cope with stress (House and others 1988). The importance of social networks has been demonstrated in a number of studies (Berkman and Syme 1979) (Figure 10.7).

Health care professionals may help individuals to be more self-disclosing or to support others who are in need of support. Support groups with others in similar circumstances can be very valuable for some people (Gottlieb 1996).

Farmers gain stress reduction from working together. Contacts with neighbors develop the so-

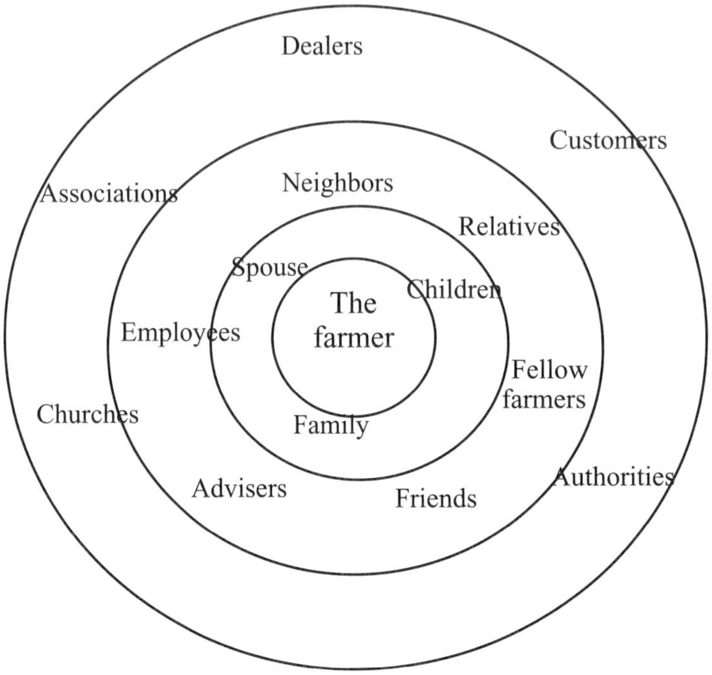

Figure 10.7. Social networks are a resource in coping with stress.

cial networks and may promote new techniques and new methods to reduce job overload. The interaction with others may be formal or informal and not necessarily only with other farmers. Contacts with customers and food consumers may be stimulating and rewarding.

Farmers must be socially astute when working with pesticides and odorous material like manure. Otherwise they will enhance neighbor conflicts that can increase their stress. It is not enough to follow regulations and laws. Especially farming near towns, the farmer must be proactive and develop a public relations practice to avoid conflicts. Practicing proper environmental protection and conservation methods are important, as well as communicating with neighbors to establish relationships and promote understanding of modern farming.

ALCOHOL-RELATED HEALTH PROBLEMS

Several studies from different countries have reported low prevalence of alcohol-related diseases among farmers (Walker 1988; Hsieh and others 1989; Thelin 1991; Paxton and Sutherland 2000). This may be related to selection processes. Farmers are entrepreneurs with a wide personal and direct responsibility and work ethic. Too much

drinking deprives their ability to optimize production and increases the risk of injuries and breakdowns.

However, some farmers are alcoholics and need help to control their drinking. Overconsumption of alcohol produces direct health risks (Rehm and others 1997), such as pancreatitis, liver cirrhosis, mental dysfunction, and injuries, as well as indirect hazards like psychological impairments and social disruptions.

Farmers who excessively drink face the potential of neglecting their land, buildings, and animals. Neglected animals may draw criticism from neighbors or legal action, increasing the farmer's stress level and worsening the farmer's health.

Most authorities agree that both genetic and environmental influences play a role in shaping alcohol abuse. "Alcohol dependency syndrome" is the medical model term for alcohol abuse (Edwards 1977; Edwards and others 1977). Cognitive and physiological theories have generated other models. One of these is the stress reduction hypothesis (Greeley and Oei 1999). The observation behind this model is that alcohol decreases the stress response. Studies indicate that those who had the highest risk of developing problem drinking also had the strongest stress-response–dampening effect (Sher and Levenson 1982).

People with alcohol-related problems are more open to seeking treatment today than in the past. There are a number of treatment models and settings. Furthermore, many alcoholics are able to quit drinking without formal treatment (Sobell and others 1996).

Alcoholics Anonymous (AA), founded by two alcoholics in 1935, is a well-known treatment program used around the world. According to AA, alcoholics never recover. They are always in the process of recovering and they should abstain for the rest of their lives. AA and other 12-step programs like the Minnesota Model attract large numbers of people (Room and Greenfield 1993). Good temporary results are reported (Cunningham and Breslin 2001), but many cannot maintain lifelong abstention and drop out of AA.

Psychotherapy and cognitive therapy have also been proposed to help people with alcohol-related problems. These techniques may be operated in groups, but individual programs may be preferable. It is not known how effective these kinds of programs are. In an analysis of a number of studies 25% were reported to remain abstinent for at least 1 year (Miller and others 2001).

Disulfiram (Antabuse) is a substance that interacts with alcohol to produce a number of unpleasant effects, such as flushing of the face, chest pains, nausea, sweating, headache, and difficulty breathing. It is not possible to drink alcohol while on disulfiram. This treatment requires a highly motivated and compliant patient, and it appears there is only partial long-term success (Krystal and others 2001).

Although most treatment programs aim for total abstinence, it is possible to transform some problem drinkers to moderate drinkers (Davies 1962; Alden 1988; Sobell and others 1996).

Most relapses in treatment programs occur within 90 days after the end of the program (similar to antismoking programs) (Hunt and others 1971). Programs with long-term goals incorporating relapse prevention tend to have the highest rates of success (McLellan and others 2000).

SUICIDES AND SUICIDE PREVENTION

Suicide rates vary a great deal among countries and ethnic groups and with age (Ruzicka 1995). In Europe, Hungary has the highest rate and Greece the lowest. Declining rates as well as increasing rates have been reported. Youth suicides usually gain a great deal of publicity, but the incidence is higher among adults and highest among old people. Individuals over age 85 have the highest suicide rate (Statistics 2001).

Alcohol consumption is associated with suicide, probably because of alcohol's disinhibiting effects and/or because the person is also depressed. A chronic alcoholic has a disintegrating personality, and suicide may be an act to terminate a condition of deprivation, disability, and exhaustion.

However, the main cause of suicide is depression. Most estimates indicate that about 80% of all suicide victims are profoundly depressed. Studies of suicide victims consistently show low concentrations of serotonin in the brain (Asberg and others 1986). Low levels of serotonin metabolites are also found in the cerebrospinal fluid of people who have attempted suicide, as well as high levels of cortisol (Roy 1992). Those attempting suicide that have low levels of serotonin metabolites are 10 times more likely to die of a subsequent suicide attempt compared to those who had normal levels.

Studies indicate that the suicide incidence is raised in groups of farmers who experience extensive economic problems, threats of losing their farms, or threats to their position as an independent entrepreneur (Zwerling and others 1995; Thomas and others 2003). The farmer may face economic problems that may be so overwhelming that they can see no alternative to suicide. Farmers often have access to firearms. Suicides by farmers are commonly gunshots in comparison with other groups (Booth and others 2000). Different intervention approaches may be used to reduce the number of suicides. After risk factors have been identified, some measures may be taken in relation to persons under risk as well as in groups with elevated risks. Individuals who are depressed, who just have been informed of a serious disease (cancer or MS), or who have been confronted with large social difficulties may be contacted and supported to come under treatment and supervision. Firearms may be taken away.

Farmers' support programs may be helpful. To be effective contacts must be personal and persistent. Telephone hot lines have little or no effect on the rate of suicide for people who call (Dew and others 1987). Some farmers with mental and social problems may need help to close down their business and to find other ways of living.

REFERENCES

Alden, L. E. (1988). "Behavioral self-management controlled-drinking strategies in a context of secondary prevention." J Consult Clin Psychol 56(2): 280–286.

Almada, S. J., A. B. Zonderman, et al. (1991). "Neuroticism and cynicism and risk of death in middle-aged men: The Western Electric Study." Psychosom Med 53(2):165–175.

Amato, P. R. (2000). "The consequences of divorce for adults and children." J Marriage and the Family 62:1269–1287.

Anda, R., D. Williamson, et al. (1993). "Depressed affect, hopelessness, and the risk of ischemic heart disease in a cohort of U.S. adults." Epidemiology 4(4):285–294.

Anderberg, U. M. (1999). "[Stress can induce neuroendocrine disorders and pain]." Lakartidningen 96(49):5497–5499.

Anderberg, U. M., I. Marteinsdottir, et al. (2000). "The impact of life events in female patients with fibromyalgia." Eur Psychiatry 15:295–301.

Antonovsky, A. (1979). Health, Stress and Coping: New Perspectives on Mental and Physical Well-Being. San Francisco, CA, Jossey-Bass.

—— (1987). Unraveling the Mystery of Health: How People Manage Stress and Stay Well. San Francisco, CA, Jossey-Bass.

—— (1993). "The structure and properties of the sense of coherence scale." Soc Sci Med 36(6):725–733.

Asberg, M., P. Nordstrom, et al. (1986). "Cerebrospinal fluid studies in suicide. An overview." Ann N Y Acad Sci 487:243–255.

Bandura, A. (1986). Social Foundations of Thought and Action: A Social Cognitive Theory. Englewood Cliffs, NJ, Prentice-Hall.

Beck, A. T. and A. Beamesderfer (1974). "Assessment of depression: The depression inventory." Mod Probl Pharmacopsychiatry 7(0):151–169.

Benson, H., J. F. Beary, et al. (1974). "The relaxation response." Psychiatry 37(1):37–46.

Bentham, G. and R. Haynes (1986). "A raw deal in remoter rural areas?" Family Practitioner Services 13:84–87.

Berding, V. and M. Matthies (2002). "European scenarios for EUSES regional distribution model." Environ Sci Pollut Res Int 9(3):193–198.

Berkman, L. F. and S. L. Syme (1979). "Social networks, host resistance, and mortality: A nine-year follow-up study of Alameda County residents." Am J Epidemiol 109(2):186–204.

Berle, B. B., R. H. Pinsky, et al. (1952). "A clinical guide to prognosis in stress diseases." J Am Med Assoc 149(18):1624–1628.

Bird, C. E. (1999). "Gender, household labor, and psychological distress. The impact of the amount and division of housework." J Health Soc Behav 40:32–45.

Björntorp, P. (1996). "The regulation of adipose tissue distribution in humans." Int J Obes Relat Metab Disord 20(4):291–302.

Björntorp, P., G. Holm, et al. (2000). "Hypertension and the metabolic syndrome: Closely related central origin?" Blood Press 9(2–3):71–82.

Blanchard, E. B., K. A. Appelbaum, et al. (1990). "A controlled evaluation of thermal biofeedback and thermal biofeedback combined with cognitive therapy in the treatment of vascular headache." J Consult Clin Psychol 58(2):216–224.

Bonneau, R., D. A. Padgett, J. F. Shereidan (2001). Psychoneuroimmune interactions in infectious disease: Studies in animals. Psychoneuroimmunology. R. F. Ader, DL; Cohen, N. SanDiego, CA: Academic Press.

Bonnet, M. H. and D. L. Arand (1997). "Physiological activation in patients with Sleep State Misperception." Psychosom Med 59(5):533–540.

Booth, N., M. Briscoe, et al. (2000). "Suicide in the farming community: Methods used and contact with health services." Occup Environ Med 57(9): 642–644.

Booth, N. J. and K. Lloyd (2000). "Stress in farmers." Int J Soc Psychiatry 46(1):67–73.

Boureau, F., T. Delorme, et al. (2000). "[Mechanisms of myofascial pain]." Rev Neurol (Paris) 156 Suppl 4:4S10–14.

Boxer, P. A., C. Burnett, et al. (1995). "Suicide and occupation: A review of the literature." J Occup Environ Med 37(4):442–452.

Brady, J. V., R. W. Porter, et al. (1958). "Avoidance behavior and the development of gastroduodenal ulcers." J Exp Anal Behav 1:69–72.

Bremner, J. D., P. Randall, et al. (1995). "MRI-based measurement of hippocampal volume in patients with combat-related posttraumatic stress disorder." Am J Psychiatry 152(7):973–981.

Broadhead, W. E., S. H. Gehlbach, et al. (1988). "The Duke-UNC Functional Social Support Questionnaire. Measurement of social support in family medicine patients." Med Care 26(7):709–723.

Cannon, W. B. (1939). The Wisdom of the Body. New York, Norton & Co.

Carlson, C. R. and R. H. Hoyle (1993). "Efficacy of abbreviated progressive muscle relaxation training: A quantitative review of behavioral medicine research." J Consult Clin Psychol 61(6):1059–1067.

Chrousos, G. P. (1995). "The hypothalamic-pituitary-adrenal axis and immune-mediated inflammation." N Engl J Med 332(20):1351–1362.

Cohen, S., T. Kamarck, et al. (1983). "A global measure of perceived stress." J Health Soc Behav 24(4):385–396.

Cohen, S., D. A. Tyrrell, et al. (1991). "Psychological stress and susceptibility to the common cold." N Engl J Med 325(9):606–612.

—— (1993). "Negative life events, perceived stress, negative affect, and susceptibility to the common cold." J Pers Soc Psychol 64(1):131–140.

Cole, S. W., B. D. Naliboff, et al. (2001). "Impaired response to HAART in HIV-infected individuals with high autonomic nervous system activity." Proc Natl Acad Sci U S A 98(22):12695–12700.

Compas, B. E., D. A. Haaga, et al. (1998). "Sampling of empirically supported psychological treatments from health psychology: Smoking, chronic pain, cancer, and bulimia nervosa." J Consult Clin Psychol 66(1):89–112.

Crofford, L. J., S. R. Pillemer, et al. (1994). "Hypothalamic-pituitary-adrenal axis perturbations in patients with fibromyalgia." Arthritis Rheum 37(11):1583–1592.

Cunningham, J. A., F. C. Breslin,(2001). "Exploring patterns of remission from alcohol dependence with and without Alcoholics Anonymous in a population sample." Contemporary Drug Problems. 28:559–567.

Dahlöf, C. (1992). "[Middle aged men living in the country have lower blood pressure than men living in the cities]." Läkartidningen 89(21):1874–1877.

Davies, D. L. (1962). "Normal drinking in recovered alcohol addicts." Quarterly Journal of Studies on Alcohol. 24:321–332.

DeLongis, A., S. Folkman, et al. (1988). "The impact of daily stress on health and mood: Psychological and social resources as mediators." J Pers Soc Psychol 54(3):486–495.

Delvaux, M. M. (1999). "Stress and visceral perception." Can J Gastroenterol 13 Suppl A:32A–36A.

Dew, M. A., E. J. Bromet, et al. (1987). "A quantitative literature review of the effectiveness of suicide prevention centers." J Consult Clin Psychol 55(2): 239–244.

Edwards, G. (1977). The alcohol dependence syndrome: Usefulness of an idea. Alcoholism: New Knowledge and New Responses. G. G. Edwards, M. London, Croom Helm.

Edwards, G., M. M. Gross, M. Keller, J. Moser, R. Room (1977). Alcohol-Related Disabilities. Geneva, World Health Organization.

Eisner, C. S., R. D. Neal, et al. (1999). "The effect of the 1996 'beef crisis' on depression and anxiety in farmers and non-farming controls." Br J Gen Pract 49(442):385–386.

Ellis, E. M. (2000). Divorce Wars: Interventions with Families in Conflict. Washington DC, American Psychological Association.

Evans, G. W. and D. Johnson (2000). "Stress and open-office noise." J Appl Psychol 85(5):779–783.

Evans, G. W., P. Lercher, et al. (2001). "Community noise exposure and stress in children." J Acoust Soc Am 109(3):1023–1027.

Fernandez, E. and J. Sheffield (1996). "Relative contributions of life events versus daily hassles to the frequency and intensity of headaches." Headache 36(10):595–602.

Flor, H. (2001). Psychophysiological assessment of the patient with chronic pain. Handbook of Pain Assessment. D. C. Turk and R. Melzack. New York, Guilford Press:70–96.

Folkow, B. (1995). "Integration of hypertension research in the era of molecular biology. G.W. Pickering-Memorial lecture." J Hypertens 18:5–18.

—— (1997). "Physiological aspects of the 'defence' and 'defeat' reactions." Acta Physiol Scand Suppl 640:34–37.

Friedman, M. and R. H. Rosenman (1959). "Association of specific overt behavior pattern with blood and cardiovascular findings; Blood cholesterol level, blood clotting time, incidence of arcus senilis, and clinical coronary artery disease." J Am Med Assoc 169(12):1286–1296.

Gardea, M. A., R. J. Gatchel, et al. (2001). "Long-term efficacy of biobehavioral treatment of temporomandibular disorders." J Behav Med 24(4):341–359.

Gold, P. W., F. K. Goodwin, et al. (1988). "Clinical and biochemical manifestations of depression. Relation to the neurobiology of stress (1)." N Engl J Med 319(6):348–353.

Golding, J. M. (1999). "Intimate partner violence as a risk factor for mental disorders: A meta-analysis." J Fam Violence 14(99–101).

Gottlieb, B. H. (1996). Theories and practices of mobilizing support in stressful circumstances. Handbook of stress, medicine, and health. C. L. Cooper. Boca Raton, FL, CRC Press.

Greeley, J. and T. Oei (1999). Alcohol and Tension Reduction. Psychological Theories of Drinking and Alcoholism. K. B. Leonard, HT. New York, Guilford Press:14–53.

Gregoire, A. (2002). "The mental health of farmers." Occup Med 52(8):471–476.

Gullette, E. C., J. A. Blumenthal, et al. (1997). "Effects of mental stress on myocardial ischemia during daily life." JAMA 277(19):1521–1526.

Henry, J., P. M. Stephens (1977). Stress, Health and the Social Environment. New York, Springer Verlag.

Hollenberg, N. K., G. Martinez, et al. (1997). "Aging, acculturation, salt intake, and hypertension in the Kuna of Panama." Hypertension 29(1 Pt 2): 171–176.

Holmberg, S., A. Thelin, et al. (2003). "The impact of physical work exposure on musculoskeletal symptoms among farmers and rural non-farmers." Ann Agric Environ Med 10(2):179–184.

—— (2004). "Psychosocial factors and low back pain, consultations, and sick leave among farmers and rural referents: A population-based study." J Occup Environ Med 46(9):993–998.

Holmes, T. H. and R. H. Rahe (1967). "The Social Readjustment Rating Scale." J Psychosom Res 11(2):213–218.

House, J. S., K. R. Landis, et al. (1988). "Social relationships and health." Science 241(4865):540–545.

Hsieh, H. H., S. C. Cheng, et al. (1989). "The relation of rural alcoholism to farm economy." Community Ment Health J 25(4):341–347.

Hunt, W. A., L. W. Barnett, et al. (1971). "Relapse rates in addiction programs." J Clin Psychol 27(4): 455–456.

Jacobson, E (1938). "Progressive relaxation: A physiological and clinical investigation of muscle status and their significance in psychology and medical practice." Chicago, University of Chicago Press.

Kabat-Zinn, J. (1993). Mindfulness meditation: Health benefits of an ancient Buddhist practice. Mind/Body Medicine: How to Use Your Mind for Better Health. D. Goleman and J. Gurin. Yonkers, NY, Consumer Reports Books:259–275.

Kanner, A. D., J. C. Coyne, et al. (1981). "Comparison of two modes of stress measurement: Daily hassles and uplifts versus major life events." J Behav Med 4(1):1–39.

Karasek, R., D. Baker, et al. (1981). "Job decision latitude, job demands, and cardiovascular disease: A prospective study of Swedish men." Am J Public Health 71(7):694–705.

Karasek, R. A. (1989). "The political implications of psychosocial work redesign: A model of the psychosocial class structure." Int J Health Serv 19(3):481–508.

Karasek, R. A. and T. Theorell (1990). Healthy Work: Stress, Productivity and the Reconstruction of Working Life. New York, Basic Books.

Karasek, R. A., T. Theorell, et al. (1988). "Job characteristics in relation to the prevalence of myocardial infarction in the US Health Examination Survey (HES) and the Health and Nutrition Examination Survey (HANES)." Am J Public Health 78(8): 910–918.

Kawamura, N., Y. Kim, et al. (2001). "Suppression of cellular immunity in men with a past history of post-traumatic stress disorder." Am J Psychiatry 158(3):484–486.

Kiecolt-Glaser, J. K., T. Newton, et al. (1996). "Marital conflict and endocrine function: Are men really more physiologically affected than women?" J Consult Clin Psychol 64(2):324–332.

Kole-Snijders, A. M., J. W. Vlaeyen, et al. (1999). "Chronic low-back pain: What does cognitive coping skills training add to operant behavioral treatment? Results of a randomized clinical trial." J Consult Clin Psychol 67(6):931–944.

Krantz, D. S., D. S. Sheps, et al. (2000). "Effects of mental stress in patients with coronary artery disease: Evidence and clinical implications." JAMA 283(14):1800–1802.

Kryger, M., T. Roth, W. C. Dement (1994). Principles and Practice of Sleep Medicine. Philadelphia, W.B. Saunders Company.

Krystal, J. H., J. A. Cramer, et al. (2001). "Naltrexone in the treatment of alcohol dependence." N Engl J Med 345(24):1734–1739.

Lazarus, R. S. and S. Folkman (1984). Stress, Appraisal, and Coping. New York, Springer.

Lehrer, P. M., R. Carr, et al. (1994). "Stress management techniques: Are they all equivalent, or do they have specific effects?" Biofeedback Self Regul 19(4):353–401.

Levenstein, S., C. Prantera, et al. (1995). "Patterns of biologic and psychologic risk factors in duodenal ulcer patients." J Clin Gastroenterol 21(2):110–117.

Levi, L. and I. Levi (2000). Guidance on Work-Related Stress. Spice of Life, or Kiss of Death? Luxembourg, Office for Official Publications of European Communities.

Liu, T. and J. W. Waterbor (1994). "Comparison of suicide rates among industrial groups." Am J Ind Med 25(2):197–203.

Lundberg, U. (1998). Work and stress in women. Women, Stress, and Heart Disease. K. Orth-Gomér, M. Chesney, and N. K. Wenger. Mahwah NJ, Erlbaum.

Lundberg, U., R. Kadefors, B. Melin, G. Palmerud, B. Hassmén, P. Engström, A. Elfsberg, I. Dohns (1994). "Psychophysiological stress and EMG activity of the trapezius muscle." Int J Behavior Med 1:354–370.

Magarinos, A. M. and B. S. McEwen (1995). "Stress-induced atrophy of apical dendrites of hippocampal CA3c neurons: Involvement of glucocorticoid secretion and excitatory amino acid receptors." Neuroscience 69(1):89–98.

Manuck, S. B., A. L. Marsland, et al. (1995). "The pathogenicity of behavior and its neuroendocrine mediation: An example from coronary artery disease." Psychosom Med 57(3):275–283.

Maslach, C. (1976). "Burned-out." Human Behavior 5:16–22.

Mason, J. W. (1971). "A re-evaluation of the concept of "non-specificity" in stress theory." J Psychiatr Res 8(3):323–333.

McEwen, B. S. (1998). "Protective and damaging effects of stress mediators." N Engl J Med 338(3): 171–179.

McGaugh, J. L. (1989). "Involvement of hormonal and neuromodulatory systems in the regulation of memory storage." Annu Rev Neurosci 12: 255–287.

McLellan, A. T., D. C. Lewis, et al. (2000). "Drug dependence, a chronic medical illness: Implications for treatment, insurance, and outcomes evaluation." JAMA 284(13):1689–1695.

Meichenbaum, D. and D. C. Turk (1976). The cognitive-behavioral management of anxiety, anger, and pain. The Behavioral Management of Anxiety, Depression, and Pain. P. O. Davidson. New York, Brunner/Mazel.

Melmed, R. N. and Y. Gelpin (1996). "Duodenal ulcer: The helicobacterization of a psychosomatic disease?" Isr J Med Sci 32(3–4):211–216.

Melzack, R. (1975). "The McGill Pain Questionnaire: Major properties and scoring methods." Pain 1(3):277–299.

Melzack, R. and P. D. Wall (1965). "Pain mechanisms: A new theory." Science 150(699):971–979.

Miller, W. R., S. T. Walters, et al. (2001). "How effective is alcoholism treatment in the United States?" J Stud Alcohol 62(2):211–220.

Netterström, B., F. E. Nielsen, et al. (1999). "Relation between job strain and myocardial infarction: A case-control study." Occup Environ Med 56(5): 339–342.

Orth-Gomer, K., S. P. Wamala, et al. (2000). "Marital stress worsens prognosis in women with coronary heart disease: The Stockholm Female Coronary Risk Study." JAMA 284(23):3008–3014.

Owens, M. J. and C. B. Nemeroff (1991). "Physiology and pharmacology of corticotropin-releasing factor." Pharmacol Rev 43(4):425–473.

Page, A. N. and L. J. Fragar (2002). "Suicide in Australian farming, 1988–1997." Aust N Z J Psychiatry 36(1):81–85.

Pare, W. P., M. I. Burken, et al. (1993). "Reduced incidence of stress ulcer in germ-free Sprague Dawley rats." Life Sci 53(13):1099–1104.

Patacchioli, F. R., L. Angelucci, et al. (2001). "Actual stress, psychopathology and salivary cortisol levels in the irritable bowel syndrome (IBS)." J Endocrinol Invest 24(3):173–177.

Paxton, R. and R. Sutherland (2000). Stress in Farming Communities: Making Best Use of Existing Help. Newcastle, North Tyneside & Northumberland NHS Trust.

Pennebaker, J. W., S. D. Barger, et al. (1989). "Disclosure of traumas and health among Holocaust survivors." Psychosom Med 51(5):577–589.

Pennebaker, J. W. and A. Garybeal (2001). "Patterns of natural language use: Disclosure, personality, and social integration." Current Directions in Psychological Science 10:90–93.

Piazza, P. V. and M. Le Moal (1998). "The role of stress in drug self-administration." Trends Pharmacol Sci 19(2):67–74.

Pickett, W., J. R. Davidson, et al. (1993). "Suicides on Ontario farms." Can J Public Health 84(4):226–230.

Plotsky, P. M. and M. J. Meaney (1993). "Early, postnatal experience alters hypothalamic corticotropin-releasing factor (CRF) mRNA, median eminence CRF content and stress-induced release in adult rats." Brain Res Mol Brain Res 18(3):195–200.

Pratt, L. A., D. E. Ford, et al. (1996). "Depression, psychotropic medication, and risk of myocardial infarction. Prospective data from the Baltimore ECA follow-up." Circulation 94(12):3123–3129.

Quick, J. C., J. D. Quick, et al. (1997). Preventive Stress Management in Organizations. Washington DC, American Psychological Association.

Rehm, J. T., S. J. Bondy, et al. (1997). "Alcohol consumption and coronary heart disease morbidity and mortality." Am J Epidemiol 146(6):495–501.

Resnick, H. S., D. G. Kilpatrick, et al. (1992). "Vulnerability-stress factors in development of post-traumatic stress disorder." J Nerv Ment Dis 180(7):424–430.

Richter, J. E., W. F. Obrecht, et al. (1986). "Psychological comparison of patients with nutcracker esophagus and irritable bowel syndrome." Dig Dis Sci 31(2):131–138.

Robinson, T. E. and K. C. Berridge (1993). "The neural basis of drug craving: An incentive-sensitization theory of addiction." Brain Res Brain Res Rev 18(3):247–291.

Rodin, J. and E. J. Langer (1977). "Long-term effects of a control-relevant intervention with the institutionalized aged." J Pers Soc Psychol 35(12): 897–902.

Rome, I. (1999). "A multinational consensus document on functional gastro-intestinal disorders." Gut 45(suppl. 2).

Room, R. and T. Greenfield (1993). "Alcoholics Anonymous, other 12-step movements and psychotherapy in the US population, 1990." Addiction 88(4):555–562.

Rosenfeld, J., M. R. Rosen, et al. (1978). "Pharmacologic and behavioral effects on arrhythmias that immediately follow abrupt coronary occlusion: A canine model of sudden coronary death." Am J Cardiol 41(6):1075–1082.

Rosenman, R. H., R. J. Brand, et al. (1975). "Coronary heart disease in Western Collaborative Group Study. Final follow-up experience of 8 1/2 years." JAMA 233(8):872–877.

Roth, S., E. Newman, et al. (1997). "Complex PTSD in victims exposed to sexual and physical abuse: Results from the DSM-IV Field Trial for Posttraumatic Stress Disorder." J Trauma Stress 10(4):539–555.

Roy, A. (1992). "Hypothalamic-pituitary-adrenal axis function and suicidal behavior in depression." Biological Psychiatry 32:812–816.

Ruzicka, L. (1995). Suicide mortality in developed countries. Adult mortality in developed countries: From description to explanation. A. C. Lopez, G; Valkonen, T. Oxford, Clarendon Press.

Sandberg, S., J. Y. Paton, et al. (2000). "The role of acute and chronic stress in asthma attacks in children." Lancet 356(9234):982–987.

Sapolsky, R. (1990). Effects of stress and glucocorticoids on hippocampal neuronal survival. Stress: Neurobiology and Neuroendocrinology. K. G. Brown MR, Rivier C. New York, Marcel Dekker Inc.

———. (1998). Why Zebras Don't Get Ulcers. New York, WH Freemanm & Company.

Sapolsky, R. M., L. C. Krey, et al. (1986). "The neuroendocrinology of stress and aging: The glucocorticoid cascade hypothesis." Endocr Rev 7(3):284–301.

Sartory, G., B. Muller, et al. (1998). "A comparison of psychological and pharmacological treatment of pediatric migraine." Behav Res Ther 36(12):1155–1170.

Saunders, T., J. E. Driskell, et al. (1996). "The effect of stress inoculation training on anxiety and performance." J Occup Health Psychol 1(2):170–186.

Schlenger, W. E., J. M. Caddell, et al. (2002). "Psychological reactions to terrorist attacks: Findings from the National Study of Americans' Reactions to September 11." JAMA 288(5):581–588.

Schmaling, K. (1998). Asthma. New York, Guilford.

Searle, A. and P. Bennett (2001). "Psychological factors and inflammatory bowel disease: A review of a decade of literature." Psychology and Health Medicine. 6:121–135.

Setterlind, S. and G. Larsson (1995). "The stress profile: A psychosocial approach to measuring stress." Stress Medicine 11:85–92.

Seyle, H. (1936). "A syndrome produced by diverse nocuous agents." Nature 32:138.

Sheline, Y. I., P. W. Wang, et al. (1996). "Hippocampal atrophy in recurrent major depression." Proc Natl Acad Sci U S A 93(9):3908–3913.

Sher, K. J. and R. W. Levenson (1982). "Risk for alcoholism and individual differences in the stress-response-dampening effect of alcohol." J Abnorm Psychol 91(5):350–367.

Sherbourne, C. D. and A. L. Stewart (1991). "The MOS social support survey." Soc Sci Med 32(6):705–714.

Shucksmith, A., D. Roberts, D. Scott, P. Chapman, E. Conway (1996). Disadvantage in Rural Areas. Salisbury, Rural Development Commission.

Siegrist, J. (1996). "Adverse Health Effects of High-effort/low-reward Conditions." J Occup Health Psychol 1:27–41.

Simkin, S., K. Hawton, et al. (1998). "Stress in farmers: A survey of farmers in England and Wales." Occup Environ Med 55(11):729–734.

Skinner, J. E. (1985). Psychological Stress and Sudden Cardiac Death: Brain Mechanisms. Boston, Martinus Bijhoff.

Sobell, L. C., J. A. Cunningham, et al. (1996). "Recovery from alcohol problems with and without treatment: Prevalence in two population surveys." Am J Public Health 86(7):966–972.

Stallones, L. (1990). "Suicide mortality among Kentucky farmers, 1979–1985." Suicide Life Threat Behav 20:156–163.

Stallones, L. and M. Cook (1992). "Suicide rates in Colorado 1980–1989: Metropolitan, non-metropolitan and farm comparisons." J Rural Health 8(2):139–142.

Starkman, M. N., S. S. Gebarski, et al. (1992). "Hippocampal formation volume, memory dysfunction, and cortisol levels in patients with Cushing's syndrome." Biol Psychiatry 32(9):756–765.

Statistics, N. C. F. H. (2001). Health, United States. Hyattsville, Government Printing Office.

Steptoe, A., S. Edwards, et al. (1989). "The effects of exercise training on mood and perceived coping ability in anxious adults from the general population." J Psychosom Res 33(5):537–547.

Stiernström, E.-L., S. Holmberg, et al. (2001). "A prospective study of morbidity and mortality rates among farmers and rural and urban nonfarmers." J Clin Epidemiol 54(2):121–126.

Stohs, J. H. (2000). "Multicultural women's experience of household labor, conflicts, and equity." Sex Roles 42:339–362.

Swales, J. D. (1994). Textbook of hypertension. Oxford, Blackwell Sci Publ.

Swisher, R. R., G. H. Elder, Jr., et al. (1998). "The long arm of the farm: How an occupation structures exposure and vulnerability to stressors across role domains." J Health Soc Behav 39(1):72–89.

Syme, L. (1989). Control and Health. Stress, Personal Control and Health. A. Steptoe and A. Appels. Chichester, England UK, Johan Wiley & Sons:3–18.

Testa, M. and K. E. Leonard (2001). "The impact of husband physical aggression and alcohol use on marital functioning: Does alcohol "excuse" the violence?" Violence Vict 16(5):507–516.

Thelin, A. (1981). "Work and health among farmers. A study of 191 farmers in Kronoberg County. Sweden." Scand J Soc Med Suppl 22:1–126.

—— (1991). "Morbidity in Swedish farmers, 1978–1983, according to national hospital records." Soc Sci Med 32(3):305–309.

—— (1995). "Psychosocial factors in farming." Ann Agric Environ Med 2:21–26.

Thelin, A., E. L. Stiernstrom, et al. (2000). "Psychosocial conditions and access to an occupational health service among farmers." Int J Occup Environ Health 6(3):208–214.

Thomas, H. V., G. Lewis, et al. (2003). "Mental health of British farmers." Occup Environ Med 60(3):181–5; discussion 185–186.

Thu K., P. Lasley, P. Whitten, M. Lewis, K. J. Donham, C. Zwerling, R. Scarth (1997). Stress as a potential risk factor for agricultural injuries: Comparative data from the Iowa Farm Family Health and Hazard Survey (1994) and the Iowa Farm and Rural Life Poll (1989). J Agromedicine 4:181–192.

Timio, M., G. Lippi, et al. (1997). "Blood pressure trend and cardiovascular events in nuns in a secluded order: A 30-year follow-up study." Blood Press 6(2):81–87.

Turk, D. C. (2001). Physiological and psychological bases of pain. Handbook of Health Psychology. A. Baum and T. A. Revenson. Mahwah, NJ, Erlbaum:117–131.

Uvnäs-Moberg, K. (1997a). "Physiological and endocrine effects of social contact." Ann N Y Acad Sci 807:146–163.

—— (1997b). "Oxytocin linked antistress effects—The relaxation and growth response." Acta Physiol Scand Suppl 640:38–42.

Veiersted, K. B., R. H. Westgaard, et al. (1993). "Electromyographic evaluation of muscular work pattern as a predictor of trapezius myalgia." Scand J Work Environ Health 19(4):284–290.

Walen, H. R. and M. J. Lachman (2000). "Social support and strain from partner, family, and friends: Cost and benefits for men and women in adulthood." J Soc Personal Relationships 17:5–30.

Walker, L. W. J. (1988). "Self-reported stress symptoms in farmers." J Clin Psychol 44:10–16.

Weiss, J. M. (1968). "Effects of coping responses on stress." J Comp Physiol Psychol 65(2):251–260.

Yunus, M. B. (1992). "Towards a model of pathophysiology of fibromyalgia: Aberrant central pain mechanisms with peripheral modulation." J Rheumatol 19(6):846–850.

Zautra, A. (1998). Arthritis: Behavioral and Psychosocial Aspects. New York, Guilford Press.

Zung, W. W. (1965). "A Self-Rating Depression Scale." Arch Gen Psychiatry 12:63–70.

Zwerling, C., L. F. Burmeister, et al. (1995). "Injury mortality among Iowa farmers, 1980–1988: Comparison of PMR and SMR approaches." Am J Epidemiol 141(9):878–882.

11
Acute Agricultural Injuries

Murray Madsen, Kelley J. Donham, LaMar Grafft, and Anders Thelin

INTRODUCTION

The focus of this chapter is acute occupational trauma in agriculture resulting from a wound or immediate damage by sudden, one-time application of external force (CDC 2003). Chronic and repetitive motion injuries are covered in Chapter 8, "Musculoskeletal Diseases in Agriculture." Protection and prevention are discussed briefly in the section on "Hazards and Injuries" in this chapter and are covered in more detail in Chapter 14, "Prevention of Illness and Injury in Agriculture."

Acute traumatic injuries in agriculture are associated with a wide variety of agents. The consequences of these injuries are often severe. This chapter provides a general description of fatal and nonfatal injuries in agriculture and discusses unique attributes of select farm and agriculturally related trauma. This chapter provides 1) injury statistics and epidemiological data, 2) rescue and medical treatment considerations, and 3) illustrative injury descriptions. These sections help provide an understanding of how injuries occur as well as assist in anticipating medical implications and prevention opportunities.

The injury statistics and epidemiology section provides a comparative picture of the prevalence and types of agricultural injuries representative of Western industrialized countries. The rescue and medical treatment section presents unique attributes and medical challenges of managing these injuries for the best immediate and long term functioning of the injured farm victim. The injury scenarios section enhances understanding of agents, events, causes, treatment, prevention, and

rehabilitation from acute traumatic injuries in agriculture. These scenarios are augmented with special considerations for those who are first-on-the-scene, first responders and rescue personnel, emergency medical technicians and transport personnel, emergency medical treatment, primary care, secondary/tertiary care providers, and providers of rehabilitation services.

AGRICULTURAL INJURY STATISTICS AND EPIDEMIOLOGY

Farmers, ranchers, farm workers, and their family members participate in a wide range of tasks due to the nature of farm and ranch work. It is common for the young, the elderly, and the primary producer to operate and repair field and farmstead equipment; maintain buildings, structures, and their mechanical and electrical systems; and tend to the multitude of tasks associated with livestock. The circumstances in which they are injured, sometimes fatally, are no less variable.

Quantifying injuries for agriculture, especially nonfatal injuries that are less well reported even in developed countries (Takala 1999), is not straightforward. Many countries do not have data. Often important pieces of desired information have not been collected. And, among sources there are inconsistencies that are a function of the source, collection, and coding methodologies, inclusions/exclusions, and even the basic definitions of agriculture and injury (Murphy 1992). It is generally believed that data sources (where they exist) are not comprehensive enough and underestimate the actual number of both fatal and nonfatal injuries for agriculture (Leigh and others 2001), although perhaps not to the degree that would redirect prevention priorities. The pro and con of these sources and their data will not be detailed. Data presented here—with a North American focus, in part because of unprecedented, comprehensive reports for the U.S. and Canada during the 1990s—is augmented with a sprinkling of other data to provide perspective and illustrate features of an injury picture for Western industrialized nations.

Although patterns of fatal and nonfatal agricultural injury vary by geography and the associated type of agriculture, the circumstances and nature of injuries are often similar. At highest risk for injury are farm and ranch owner-operators, especially elder farmers and ranchers (Myers 1997, 1998, 2001). The family farmer is the most rapidly aging work force in the U.S. (Reed and others 2001). Injury events related to tractors, notably tractor overturns and runovers, are particularly lethal. Collisions between farm equipment and motor vehicles on roadways also contribute to agriculturally related fatalities, though it is often the motor vehicle occupants who are injured. Entanglements in moving parts of machinery too often result in amputation and long-term disability. Nevertheless, working with animals remains the leading cause of nonfatal injuries.

Agriculture consistently ranks as one of the highest injury risk industry sectors (NASD 2004). Death rates for agriculture in North America (NA) as well as the 15 countries of the European union (EU15) are consistently several times greater than the average rate for all industries combined (Figure 11.1). For this reason, agriculture is often described as one of the most hazardous industries in which to work.

Fatal Agricultural Injuries

U.S. Department of Labor, Bureau of Labor Statistics (BLS), data for the agricultural industry (production agriculture, agricultural services, commercial fishing/hunting/trapping, and forestry except logging) reveals an annual average of 806 deaths 1992–2002 (U.S. Department of Labor 2004). Of these deaths, about 85% (Figure 11.2) were in production agriculture associated with the roughly 2 million U.S. farms producing grain and livestock.

Canadian Agricultural Injury Surveillance Program (CAISP) data (Brison and Pickett 2003), unlike BLS Census of Fatal Occupational Injuries (CFOI) data, includes deaths where the victim was a third party, e.g., an occupant of the car in a

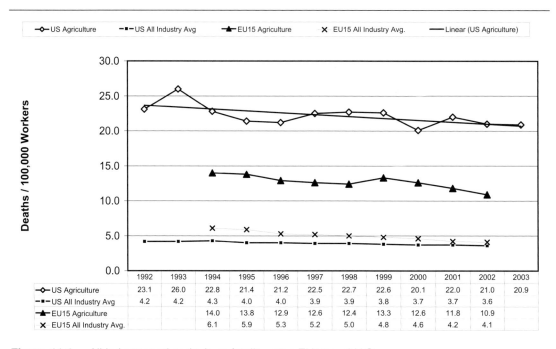

Figure 11.1. All industry and agriculture fatality rates EU15 and U.S.

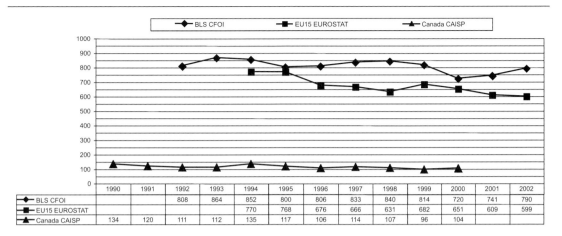

	1990	1991	1992	1993	1994	1995	1996	1997	1998	1999	2000	2001	2002
BLS CFOI			808	864	852	800	806	833	840	814	720	741	790
EU15 EUROSTAT					770	768	676	666	631	682	651	609	599
Canada CAISP	134	120	111	112	135	117	106	114	107	96	104		

Figure 11.2. Agriculturally related work fatalities.

crash with farm machinery. For the U.S., approximately 100 fatal crashes per year occur between "farm equipment other than trucks" and motor vehicles on roadways (U.S. Department of Transportation 2004).

CAISP data show an average 115 agricultural deaths per year during the decade of the 1990s. EUROSTAT, the statistical office with information on the pre-2004 15 European Union (EU15) member-states, shows agricultural mortality averaging 674 deaths per year for the period 1994–2002. (Figure 11.2)

Data reported by the Chicago-based National Safety Council (NSC) show production agriculture was the source of 10.3% of all fatal occupational injuries 1998–2002. The mean fatality rate for the entire U.S. agriculture industry 1992–2002 was 22.3 deaths per 100,000 workers compared to 3.9 deaths per 100,000 workers for all industries combined. The average portion of all work deaths attributed to agriculture in EU15 countries was 12.3%. The average EU15 agricultural death rate approached 12.7 per 100,000 workers while the average for all industries in the EU15 was 5. Canada (1990–2000) experienced 11.6 deaths per 100,000 of farm population based on a single interim year (1996) count of 851,405 farm residents on 246,923 farms (Brison and Pickett 2003). Australia had an annual mean of 152 agricultural fatalities 1989–1992 with a fatality rate comparable to the U.S. of 20 deaths per 100,000 in agriculture and 5.5 work deaths per 100,000 workers overall (Fragar and Franklin 2000; Industries

AIVO 2003). In New Zealand, the total 159 agricultural production workers who died represented 21% of all work-related fatalities (Horsburgh and others 2001).

Tractors most frequently, then other machinery, and least frequently nonmachinery are the common categories of agents or causes for fatalities. The nonmachinery portion of U.S. agricultural fatalities includes a mix of animals, falls, electric current, and drowning among others (Figure 11.3). About half the fatalities on U.S. farms involve machinery. In Canada, 74% of 1086 fatalities 1990–2000 were machinery-related, 29% were associated with "machines other than tractors," and 45% were "tractor-

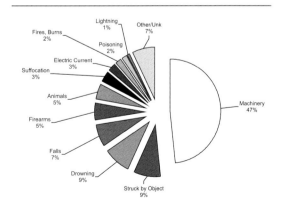

Figure 11.3. Deaths from nontransport unintentional injuries on U.S. farms, 1987–1997 (N = 8338).

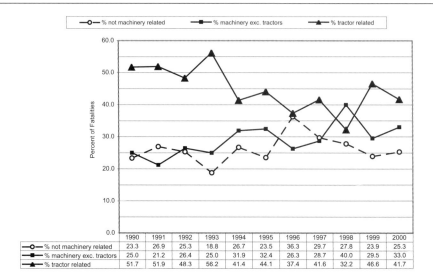

Figure 11.4.   Agricultural deaths by agent, Canada, 1990–2000.

	1990	1991	1992	1993	1994	1995	1996	1997	1998	1999	2000
—O— % not machinery related	23.3	26.9	25.3	18.8	26.7	23.5	36.3	29.7	27.8	23.9	25.3
—■— % machinery exc. tractors	25.0	21.2	26.4	25.0	31.9	32.4	26.3	28.7	40.0	29.5	33.0
—▲— % tractor related	51.7	51.9	48.3	56.2	41.4	44.1	37.4	41.6	32.2	46.6	41.7

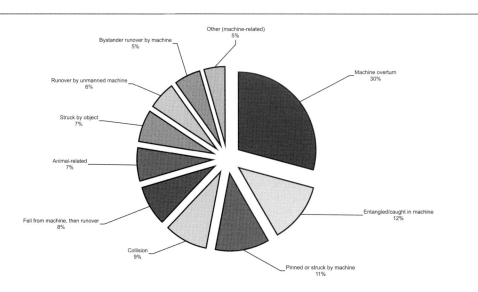

Figure 11.5.   Work-related agricultural fatalities, Canada, 1990–2000.

related" (Figures 11.4 and 11.5) (Brison and Pickett 2003). The trend in Canadian tractor fatalities continued downward across the decade and was nearly offset by the upward trend in the percentage of machine-related fatalities. Together, two categories of machinery, "tractors and other vehicles" and "machinery other than tractors," account for about two-thirds of the agricultural fatalities in the U.S. 1998–

2002 and three-fourths of Canada's agriculturally related fatalities 1990–2000.

Agricultural workers in Great Britain 1991/92–2000/01 were most often fatally injured in falls from heights, struck by a moving vehicle, struck by moving/flying/falling objects, trapped when something collapsed or overturned, contacted moving machinery parts, slipped/tripped/fell on the same level, contacted electric current, or were

Table 11.1. Tractor-Related Fatalities

Percentage	Period	Reference
55% of farm-related incidents in Finland	1988–2000	(Rissanen and Taattola 2003)
50% of agricultural incidents in Columbia	1996	(International Labour Office 2000)
45% of work-related agricultural deaths in Canada	1990–2000	(Brison and Pickett 2003)
37% of production agriculture work fatalities in New Zealand	1985–1994	(Horsburgh and others 2001)
36.5% of agricultural work fatalities (21 states in U.S.)	1996–1997	(National Safety Council 1998, 1999)
14.8% of unintentional injury deaths on farms in Australia	1989–1992	(Fragar and Franklin 2000)

injured by an animal (Field Operations Directorate 2001–2002). In a similar period for Ireland, machinery (53.4%), livestock (13%), collapse of objects or buildings (8.5%), manure slurry–related and other unclassified causes (6.9%), and electrocutions (4.4%) accounted for 247 fatalities (McNamara and Laffey 2003).

Tractor-related events continue to dominate both the agricultural and machinery-related fatality picture (Table 11.1). They represent the single largest portion of the agricultural industry and production agriculture death toll in the U.S. and, consequently, the major portion of deaths related to machinery (Figure 11.6, refer to Figure 11.1).

Equally well studied is the prominence of overturn events in the profile of tractor-related fatalities, running nearly 50% in the U.S. and Canada over the decade of the 1990s (Figure 11.7). Tractor runover fatalities include operators and passengers who fell from tractors and were subsequently run over by trailing implements, operators run over by unmanned tractors, and bystanders run over as machines moved. About one-fourth of the tractor-related deaths in the U.S. and Canada occur when someone is run over by

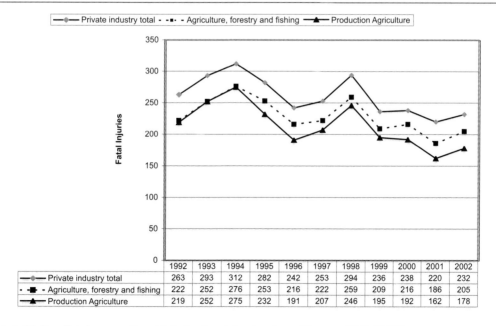

	1992	1993	1994	1995	1996	1997	1998	1999	2000	2001	2002
Private industry total	263	293	312	282	242	253	294	236	238	220	232
Agriculture, forestry and fishing	222	252	276	253	216	222	259	209	216	186	205
Production Agriculture	219	252	275	232	191	207	246	195	192	162	178

Figure 11.6. Fatal occupational injuries involving tractors. (*Source:* BLS CHOI)

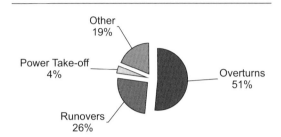

Figure 11.7. Tractor fatalities by event, U.S., 1988–1998. (Source: NSC Injury Facts)

the tractor or tractor/implement combination (Figures 11.7–11.9).

Entanglements occur in the moving parts of machines, including

- Power take-off (PTO) and driveline between the tractor and implement
- Secondary drives that transfer power to various parts of a machine
- Crop-gathering, -processing, -transfer, and -discharge mechanisms

Entanglements also occur in stationary installations with moving machinery parts, such as augers

and conveyors for handling commodities, livestock feed, and manure. It is unclear what portion of PTO-related data truly "belongs" to the tractor.

Considering the nature of injury for the 8,077 deaths in the U.S. agriculture industry 1992–2001, NSC tabulations based on BLS data for farms with 11 or more employees placed 70% in a miscellaneous "all other" category for the nature of fatal injury or illness. Nearly one-fourth of these fatalities were due to "multiple injuries," 4% stemmed from cuts, lacerations, and punctures. Remaining categories in descending order for cause of death were heat burns, amputations, and fractures (NSC 2003).

In summary for the U.S., machinery-related incidents contribute the large majority of deaths in agriculture. The majority of machinery deaths result from tractor-related events. The majority of tractor fatalities are overturns, followed by runovers, roadway crashes, and PTO entanglements. Entanglements lead the nontractor, machinery-related fatalities. The nonmachinery fatalities contain a broad mix of events.

Any global, regional, or national characterization of fatal farm injuries must take into account the substantial differences in agricultural practices generally, and influences such as the impact

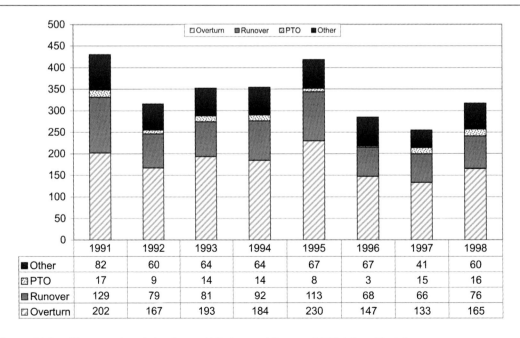

	1991	1992	1993	1994	1995	1996	1997	1998
■ Other	82	60	64	64	67	67	41	60
▨ PTO	17	9	14	14	8	3	15	16
▧ Runover	129	79	81	92	113	68	66	76
▱ Overturn	202	167	193	184	230	147	133	165

Figure 11.8. Tractor-related deaths on U.S. farms. (*Source:* NSC Injury Facts)

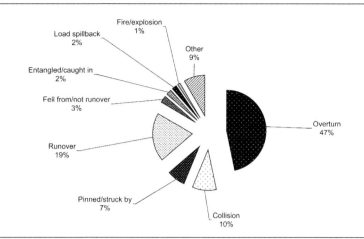

Figure 11.9. Work-related tractor fatality events, Canada, 1990–2000 (N = 489).

of mechanization as well as health and safety regulations. For example, tractor overturn fatalities in the regions of Europe, Scandinavia, and Australia/New Zealand are dramatically lower because of the proportion of their tractor fleet equipped with rollover protective structures (ROPS) or ROPS cabs. Between 1959 and 1990, tractors with ROPS rose from 3–93% in Sweden amid a 275% increase in the number of tractors in use (including new tractors with ROPS cabs) while tractor overturn fatalities dropped from 12 to 0.3 per 100,000 farm tractors (Pope 2000).

Nonfatal Agricultural Injuries

The exposures and events associated with fatal agricultural injuries are somewhat different from the nonfatal injuries. However, obtaining comprehensive statistics on nonfatal events is even more difficult than for fatal injuries.

There is no comprehensive national surveillance system for the U.S. that provides routine nonfatal agricultural injury data (Murphy 2003). NSC applies a customized injury-to-death ratio to estimate disabling injuries. NSC's definition for disabling injury includes injuries with some permanent impairment and those that prevent normal work for at least a full day. NSC estimates typically range from 130,000–150,000 disabling injuries per year for agriculture (3300–4000 per 100,000 workers), about 4% of the estimated total for all industries (NSC 1993, 1994, 1995, 1996, 1997, 1998, 1999, 2000, 2001, 2002, 2003, 2004).

Population-based nonfatal agricultural injury

Table 11.2. NIOSH TISF Injury Study Results

Data Year	Lost-Time Work Injuries*	Injuries/100 Workers
1993	201,081	6.5
1994	121,937	4.7
1995	195,825	6.8

*Half-day or more of restricted work activity (Myers 1997, 1998, 2001).

estimates from regional studies within the U.S. have shown an overall injury rate of approximately 10% per year, ranging widely from 0.5–16.6 nonfatal agricultural occupational injuries per 100 person-years (McCurdy and Carroll 2000). The U.S. National Institute for Occupational Safety and Health (NIOSH) Traumatic Injury Surveillance of Farmers (TISF) sampled 20,000–25,000 U.S. farms in each of three consecutive years to capture their injury experience in 1993, 1994, and 1995 (Myers 1997, 1998, 2001). Injury rates from other studies are within range of the NIOSH TISF findings (Table 11.2).

EUROSTAT reported over 300,000 agricultural work injury incidents, annually resulting in 4 or more days off work in EU15 countries 1994–2001 (European Union 2004). Data for these more severe agricultural injuries (4-or-more-day absence) in EU15 member-states indicate their contribution to the "all work" injury toll (Table 11.3). A 1992 baseline study for Sweden reported 11.5% of farms experienced a work-related injury event.

Table 11.3. Agricultural Work Injuries (4-day, or More, Absence), EU15, Select Member-States, and Norway

		2001	2002
EU15 (pre-2004)	All work	4,702,295	4,441,531
	Ag, hunting, forestry	318,135	284,919
Germany	All work	1,309,331	1,186,803
	Ag, hunting, forestry	125,166	116,462
United Kingdom	All work	384,069	387,522
	Ag, hunting, forestry	11,458	7,698
Finland	All work	60,176	60,067
	Ag, hunting, forestry	6,679	6,553
Sweden	All work	56,168	55,153
	Ag, hunting, forestry	1,748	1,803
Ireland	All work	26,362	21,107
	Ag, hunting, forestry	1,824	1,808
Norway	All work	76,735	71,295
	Ag, hunting, forestry	1,080	918

Created from EUROSTAT EUROPA data (EUROSTAT 2004).

The 1996 agricultural injury rate based on worker compensation records for Finland was 7.4% (Rautiainen 2002). An estimated 3077 injury events occurred among the 141,500 Irish farms in 2001 (9.7% of sampled farms) and the average 1997–2001 was 1782 injury events per year (McNamara and others 2003).

Enterprise type is indicative of nonfatal injury risk, as it is with fatalities. Aggregated U.S. information from the 3-year NIOSH TISF suggest approximately three-eighths of injuries were associated with beef, hog, or sheep operations; two-eighths with cash grain and field crop operations; one-eighth with vegetable, fruit, or nut operations; one-eighth with dairy operations; and the remaining one-eighth with other operations (Figure 11.10).

The mix of these enterprise types differs by region and so does the task or activity engaged in at the time of the event (Table 11.4). Livestock handling ranks above the others in U.S. data. A study of 1996 data for Finland found 45.8% of agricultural worker compensation injuries associated with animal production, 17.6% with crop produc-

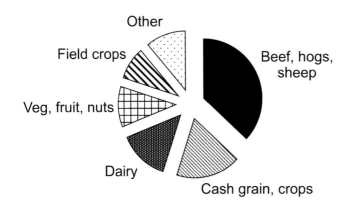

Figure 11.10. Nonfatal injuries by type of farming (U.S. 1993–1995).

Table 11.4. Frequency Rank for Farming Activity When Injured, U.S. and U.S. Regions

Activity When Injured	U.S.	Northeast	South	Midwest	West
Livestock handling	1	2	1	1	2
Farm maintenance	2	1	3		3
Field work	3	3	2		1
Crop handling	4		3	2	
Machinery maintenance				3	

Created from data in NIOSH TISF (Myers 1997, 1998, 2001).

Table 11.5. Distribution (%) of Lost-Time Injuries by Agent of Injury, U.S.

Injury Agent	1993	1994	1995
Livestock	18.1	20.0	20.0
Machinery other than tractors	17.2	19.3	21.3
Working surface	8.5	8.1	8.5
Hand tool	11.4	3.4	7.6
Tractor	5.5	4.7	4.1
Plant/tree	4.3	2.5	3.9
Power tool	3.7	3.0	2.9
Truck/automobile	3.1	4.7	1.5
Other vehicle	2.4	3.3	2.7
Liquids	0.6	1.3	0.3
Pesticide/chemical	0.7	0.7	0.5
Other	24.6	29.1	26.8
Estimated total lost-time injuries	201,081	121,937	195,825

(Myers 1997, 1998, 2001).

Table 11.6. Frequency Rank of Nonfatal Injury Agents, U.S. and U.S. Regions

Injury Agent	U.S.	Northeast	South	Midwest	West
Machinery other than tractors	1	1	1	1	1
Livestock	1	2	1	2	2
Hand or power tools		3	2		
Working surfaces	3				

Created from data in NIOSH TISF (Myers 1997, 1998, 2001).

tion, and 20.5% associated with other farm work (Rautiainen 2002).

The major agents of nonfatal injury on U.S. farms in the NIOSH TISF were livestock, machinery other than tractors, and working surfaces (Table 11.5). The importance of these agents also varied by region (Table 11.6).

NSC estimated farms with over 10 employees,

about 3% of U.S. farms (Murphy 2003), experienced 40,153 nonfatal occupational injuries and illnesses involving days away from work during 2001, an incidence rate of 7300 cases per 100 person-years and more than twice the estimated injury rate for all farms (NSC 2003). This is likely due in no small part to U.S. Occupational Safety and Health Administration (OSHA) regulatory ju-

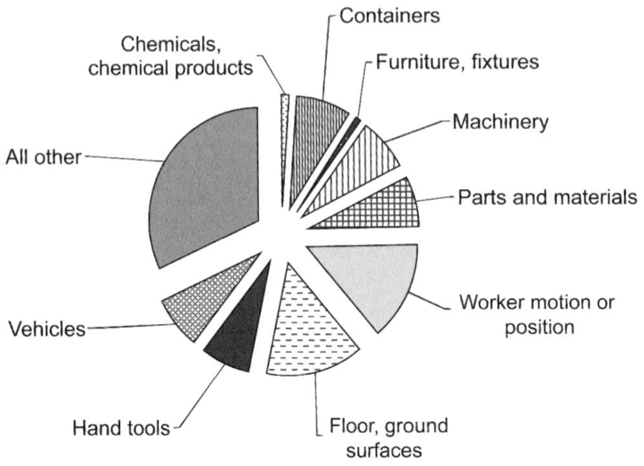

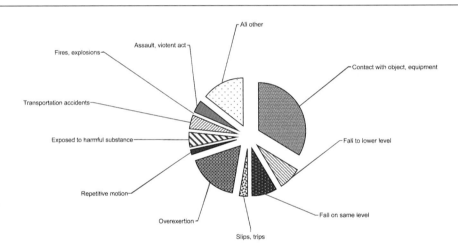

Figure 11.11. Agricultural injury agents, U.S. farms with over 10 employees.

Figure 11.12. Agricultural injury event or exposure, U.S. farms with over 10 employees.

risdiction resulting in improved reporting, the establishment of formalized safety programming, and the inclusion of illness incidents. Annual appropriations prohibit U.S. OSHA from spending to "prescribe, issue, administer, or enforce any standard, rule, regulation or order" on farms with 10 or fewer employees (National Academies Press 1998). The agents of injury, events, and exposures for workers on these large farms differ from those of the NIOSH TISF: Machinery is a much smaller component and ergonomic and other conditions contribute more to the total (Figures 11.11, 11.12).

In Canada 1990–2000, hospitalized agricultural injuries were most commonly caused by animals, machinery, and falls (Figure 11.13). Animals were the leading cause of all types of injuries except open wounds (Brison and Pickett 2003). The primary events in the Canadian machinery-related incidents were

- Entangled/caught in machine, 32.4%
- Pinned or struck by machine, 19.5
- Unspecified person fell from machine (not run over), 8.2
- Struck by object falling from or propelled by machine, 5.5
- Overturn of machine, 5.0
- Operator fell from machine (not run over), 4.8

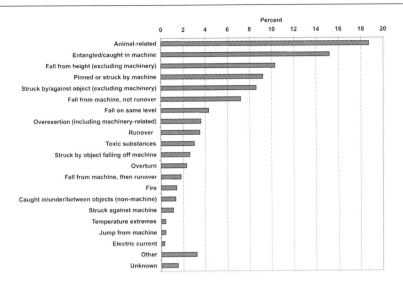

Figure 11.13. Hospitalized agricultural injuries by event, 1990–2000 Canada (14,987 cases).

- Runover of unspecified person, 2.9
- Runover of dismounted operator, 2.7

The types of injury diagnoses in the NIOSH TISF were often multiple, such as amputation with closed injury to the head or chest, complex wound lacerations, and degloving with broad wound contamination. Medical attention was sought in over 8 of 10 such lost-time injuries. These injuries occurred most frequently, in descending order but nearly the same magnitude, to the worker's leg, knee, or hip; back; fingers; or arm and shoulder. Among workers on the small fraction of U.S. farms with 11 or more employees the pattern is similar (Figure 11.14). However, the U.S. national perspective on body part most frequently injured is different than for individual regions (Table 11.7).

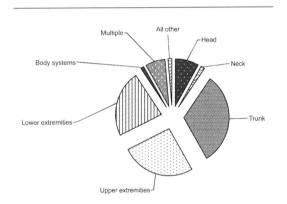

Figure 11.14. Body part injured, U.S. farms with 11 or more employees.

Table 11.7. Frequency Rank for Body Part Injured, U.S. and U.S. Region

Body Part Injured	U.S.	Northeast	South	Midwest	West
Ankle, leg, knee, hip	1	2	1	1	1
Back	2		2	2	2
Fingers, hand	3	1	2		3
Arm, elbow, shoulder	4	1	3	3	3
Toes, foot		2	3		

Created from data in NIOSH TISF (Myers 1997, 1998, 2001).

Table 11.8. Frequency Rank for Nature of Injury, U.S. and U.S. Region

Nature of Injury	U.S.	Northeast	South	Midwest	West
Sprain, strain	1	1	2	1	1
Fracture	2	2	1	2	3
Laceration	3	3	3	3	3
Bruises					2

Created from data in NIOSH TISF (Little 1998; Myers 1997, 1998, 2001).

Sprains and strains were the most frequent diagnosis for the nature of injury in the U.S. NIOSH TISF, followed by fractures and lacerations. Again, the U.S. national perspective is different than for individual regions (Table 11.8). In aggregate, machinery accounted for about three-fourths of amputations, one-half the crushing injuries, and one-fourth of fractures and lacerations. Livestock accounted for about one-fifth of all fractures, lacerations, and crushing injuries (Myers 1998, 2001; Myers and Hendricks 2001).

Collectively, agricultural machinery injury events in Canada had a pattern of diagnoses similar to that for the U.S. (Figure 11.15). Among hospitalized agricultural injury patients from Canada April 1990–March 2000 the most frequent primary diagnoses were fracture, lower limb (15.5%); fracture, upper limb (13.2%); open wound, upper limb (9.1%); and fracture, spine or trunk (8.8%). In Sweden 1992, the body parts injured most frequently included, in descending order, 1) the knee/lower leg/ankle, 2) hand, 3) head excluding eyes, 4) shoulder/upper arm/elbow, 5) foot, and 6) eyes. In Finland 1996, the body parts injured most frequently, also in descending order, were 1) lower limb from ankle to hip, 2) back and spine, 3) upper limb from wrist to shoulder, and 4) fingers (Rautiainen 2002).

Nonfatal injury data are often gathered by survey. More objective and quantitative data may be found in health insurance claims, worker's compensation claims, and medical or hospitalization records. The medical and nonmedical costs associated with injury, for an individual or group, can be an indicator of severity. Another indicator of severity is length of hospital stay.

Of 14,987 machinery-related agricultural injury cases in Canada 1990–2000 grouped by primary diagnosis, the median length of hospital stay was 3 days, the mean 5.9, and the standard deviation 11.6 days (Brison and Pickett 2003). The top five injuries by length of stay in this list were 1) fracture of the spine and trunk (6 days); 2) internal injury of chest, pelvis, and abdomen (5 days); 3) injury to blood vessels (5 days); 4) fracture of a lower limb (4 days); and 5) an open wound to a lower limb (4 days).

In the same Canadian study, hospitalized

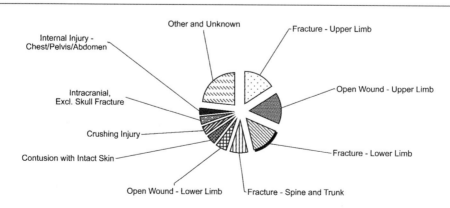

Figure 11.15. Machinery injury diagnoses, Canada hospitalizations (1990–2000). (Brison and Pickett 2003)

machinery-related agricultural injury incidents by primary diagnosis had a median 3-day length of stay, a mean of 6.9 days, and a standard deviation of 12.6 days. Compared to the ranking of all machinery-related agricultural injury cases, the difference is "injury to blood vessels" dropped from third to fifth. Canadian nonmachinery incidents had a median stay of 2 days, a mean of 5.0 days, and standard deviation of 10.7 days. The top five by median length of stay for nonmachinery incidents were 1) injury to blood vessels; 2) fracture of the spine and trunk; 3) fracture of a lower limb; 4) internal injury of chest, pelvis, and abdomen; and 5) burns. The mean stay for this group of primary diagnoses ranged from 4–6 days.

Summarizing nonfatal injuries, agricultural injuries are a substantial contributor to the work injury toll in Western industrialized nations. Livestock and machinery are the two principal agents involved in the main events, which are being struck, caught, run over, entangled, and pinned. Most often these events result in strains and sprains, fractures, and open wounds to the upper and lower extremities or to the trunk of the body. The more severe multiple injuries affect several body parts and systems.

MEDICAL CONSIDERATIONS OF ACUTE AGRICULTURAL INJURIES

The previous section highlights acute traumatic injuries that occur in agricultural communities. There are unique considerations for the medical treatment of these injuries compared to other workplace injuries. Injuries in agriculture stem from a complex mix of people, machines, animals, and the physical environment. This requires an understanding of the situations and risks, as well as the injuries. Therefore, special education regarding rural and agricultural trauma is important in the preparation of rural health professionals. The lack of "on-farm" observations and inexperience with actual farming and ranching further highlights the potential for gaps in understanding. Identifying and treating injuries plus managing recovery and rehabilitation among those engaged in agricultural activities places an important responsibility on rural health care professionals who must understand the environment in which their patients work and the requirements of the tasks they perform (Oehme 2003). The following section of this chapter addresses these issues.

Skills, background, and experience are needed to care for all age groups, both genders, and the variety of injuries (Doty 1998). Best outcomes for patients depend on a team of responders and professional health care providers. Members of this team include those who are

- First on the scene, who may be family members, co-workers, or bystanders.
- First responders and emergency medical service (EMS) personnel, such as law enforcement, firefighters, or rescue squads.
- Emergency medical technicians (EMT) and others providing first aid, stabilization, extrication, and transportation to the emergency room.
- Emergency room care providers.
- Secondary and tertiary care providers.
- Rehabilitation specialists.

Special considerations for managing acute agricultural injuries include the fact that they are confounded by the "triad of T's": 1) excessive **time** until treatment, 2) excessive **trash** or wound contamination, and 3) excessive **trauma** to tissues and organs.

Regarding time, farmers often work alone and may be trapped or caught in a machine away from anybody and not noticed missing or not found for several hours following an injury. Often the first sign to a family member that something might be wrong is when the person does not return for the midday or evening meal. In the U.S. (and similarly in most Western industrialized countries), over 50% of farmer spouses are employed off-farm, so the delay in finding a victim may be longer. Additional time is required for emergency medical services to reach the scene, and even more time to extract and transport the patient to the emergency room. This elapsed time increases the adverse effects of blood loss, shock, tissue damage, and wound contamination. Military and trauma surgeons have identified the "golden hour" of time from injury to stabilization and primary medical care, which is highly related to the survivability of patients suffering severe traumatic injuries.

The agricultural workplace is contaminated with trash, soil, manure, and microbes, as well as a number of chemicals that can contaminate a wound. Since antibiotics are frequently used on livestock and farm orchards, there are antibiotic-resistant organisms in the farm environment that may complicate a wound infection.

Agricultural machinery-related injuries may be extremely traumatic, with extensive tissue damage. Although these types of injuries may have a low probability of severe hemorrhage, they decrease the potential long-term functionality of the affected limb or body part. The severely devitalized tissue, combined with wounds that are often contaminated with animal manure or soil, can create an environment that provides ideal growth for anaerobic organisms like *Clostridium tetani* (cause of tetanus) or *Clostridium perfringens* (a common cause of gangrene).

Improved technology and facilities, including emergency communications, advanced treatment at the site, and improved transportation to high-level care facilities, have dramatically reduced deaths among roadway crash victims and battlefield casualties. However, these improvements have not accrued to benefit the farm person who has suffered a serious farm injury because of the inherent time lag in locating, rescuing, and transporting the victim to a definitive care facility. Furthermore, there may be a lack of appropriately trained personnel in rural areas to manage these injury situations.

Properly informed, prepared staff and appropriate referral along the chain of providers helps assure the best outcome for the patient. A basic understanding of farming, its people, their inherent occupational injuries, and special considerations for treating their injuries is vital for personnel in all links of the chain to assure as near-complete a recovery as possible for farm injury victims.

FIRST ON THE SCENE/FIRST AID—GENERAL

Initial management of major injuries does not come from specialists at a modern hospital but rather from a relative or neighbor at the scene some distance from the nearest medical assistance (National Academies Press 1998). Understanding rescue procedures and first aid methods is extremely important for persons likely to be first on the scene. Familiarity with the victim coupled with lack of formal training can create problems for both the patient and for the safety of the rescuer.

Confronted with a situation involving a family member or close friend, the person first on the scene may rush to rescue without considering their personal risk. The fact that nearly 60% of confined space victims are would-be rescuers points out the magnitude of this hazard. A classic example is a situation where two brothers died trying to rescue their father from a manure pit (Donham and others 1988). Similar situations have been repeated many times resulting in many unnecessary deaths of family members and other first-on-the-scene responders. Farm family members and other persons in the rural environment who might encounter farm injury situations, including feed and seed salesmen, veterinarians, and milk truck drivers among others, should have special training that includes at least the following basic components (Baker and Lee 1995):

1. An understanding of the basic types and circumstances of acute agricultural injuries
2. An understanding of how to assess a situation to avoid harm to the rescuer
3. An understanding of how to prevent further harm to the victim (the first responder should know how to shut off equipment)
4. An understanding of how to summon the rescue and emergency medical services and give them accurate directions to the injury scene
5. An understanding of first aid and cardiopulmonary resuscitation (CPR) (www.redcross.org)

This knowledge base, and the general operational approach at the scene of an injury event as outlined below, will help assure the best outcome for the victim and safety of the responders (Agricultural Safety Program 2005).

DESIGNATE A LEADER
A worker who has had first aid training or the senior person at the scene should assume leadership. This person should direct the rescue until the emergency squad arrives and should update the squad on treatment administered.

ASSIGN A SPECIFIC PERSON TO CALL FOR HELP
The emergency dispatcher will need to know exact location and condition of field (muddy, steep, rough), type of equipment involved in the incident, number of victims, and the extent of their injuries.

ASSESS THE RESCUE SITUATION

Evaluate the situation and develop a rescue plan. Stabilize equipment to minimize the chance of collapse or further injury. Actions to help should not exceed the limitations of the rescuers and available equipment.

ESTABLISH A HAZARD ZONE

Allow the rescuers room to work at freeing the victim. Only the rescuers should be in the area. This area may contain hazards such as fire, toxic or flammable gases, and structural damage.

PROVIDE EMERGENCY FIRST AID

Restore breathing and circulation if necessary. If bleeding, apply pressure to related pressure points. Administer any additional first aid treatment.

STAY CALM

Calm the victim by keeping one rescuer near the victim at all times.

PRESERVE TISSUES IF AMPUTATION OCCURS

Surgical reattachment may be possible. Locate the appendage and wrap it in a moist towel, and then put it in a plastic sack labeled with the patient's name. If saline solution or tap water is not available, contact lens fluid or even a soft drink may be used (Schnitz and Fischer 1996). Keep it on ice but do not let it freeze, and make sure it gets transported to the emergency room with identification of the patient. Avoid clamps on the injured stump to protect vessels, nerves, and soft tissue and to improve the prognosis for reattachment.

The assessment skills and judgment of a person first on the scene are extremely important. Immediate recognition of the type and seriousness of the injury is crucial. If there is only one person first on the scene, that person will have to make a decision to render first aid immediately or go directly for help. This decision depends on the severity of injury, and the responder's ability to administer first aid correctly. The four most life threatening conditions requiring immediate attention are blocked air passages, respiratory arrest, circulatory failure, and severe bleeding (Little 1998; St. John Ambulance 1990). Ideally, every farm family member and rural business or service representative should be trained in basic first aid and CPR (http://www.redcross.org/services/hss/courses/).

General First Aid Supplies

Every farm should have a basic first aid kit readily available. One person should be designated to check it regularly and restock supplies as needed. Readily available means several kits placed in strategic locations, such as on the most frequently used tractors, other harvesting machines like combines, and in the work shop, milk house, and home. First aid materials should be kept in an enclosed container that keeps out dirt and water. There should be a basic kit and a more advanced kit. The basic kit should contain at a minimum the supplies in Tables 11.9 and 11.10 (ISU Extension 1999).

Health professionals in rural areas can perform a vital public service by facilitating first-on-the-scene training courses in their community. This training can complement high school or community college agricultural programs, 4-H programs, and agricultural business programs, and can occur wherever farm families gather. There is an extensive amount of reference and training mate-

Table 11.9. Basic Farm First Aid Kit

1. A basic first aid manual (e.g., *The American Red Cross First Aid Manual*)
2. A card with the phone numbers for the local emergency rescue service, the family physician, and the Poison Control Center
3. 8 ounces of sterile water and an antiseptic soap, such as Betadine, to wash wounds
4. An antiseptic or antibiotic spray for wounds (preferably in a pump bottle)
5. Bandage materials
   a. 2 large (36″) triangular bandages for use as slings, splints, or to control bleeding
   b. 4 safety pins to be used to make a sling
   c. Sterile compress bandages or gauze sponges (4 2″ by 2″, and 4 4″ by 4″)
   d. Pressure bandages (8 6″ by 10″)
   e. Roll cling gauze (2 2″ rolls)
   f. Assorted adhesive bandages
6. Tape
   a. 2 2″ rolls of adhesive tape
   b. 2 2″ rolls of elastic tape
7. Bandage scissors, heavy duty to cut clothing material
8. Eyewash cup, with sterile eyewash solution
9. Flashlight with fresh batteries

Table 11.10. Advanced Farm First Aid Kit (Contains All Basic Items in Table 11.9 Plus These Items)

10. Splints (1/4″ by 3″ by 12–15″)
11. Road flares
12. Plastic bags (one garbage-sized, 2 kitchen-sized, and 2 bread wrapper–sized for proper storage and transportation of amputated limbs or other tissues)
13. Small package of activated charcoal to mix with water as a general antidote

rial available through the U.S. Cooperative Extension Service and other resources such as the Farmedic Program of Cornell University (http://www.farmedic.com/fos/).

FIRST RESPONDERS AND EMERGENCY MEDICAL TREATMENT, RESCUE/EXTRICATION, AND TRANSPORT

Rural areas and communities of industrial countries often have a fire and rescue squad comprised primarily of volunteers. These professionals are well-trained in the basics of firefighting. In many cases, they have been cross-trained in basic first aid and CPR to handle emergency or rescue situations. These services are often called as first responders to an injury scene or extrication situation.

In addition to first responders, many rural communities have some level of EMS. These services are usually located within an ambulance service that may be a local governmental agency or a private enterprise, or affiliated with a hospital. The personnel of these services are more highly trained in medical care compared to first responders. They may be staffed with either volunteers or paid employees. They complete certificate training programs that contain set standards to progress through various steps of EMS training. These professionals have the ability to work under a set of standard protocols and under the guidance of a physician.

Training

Continuing education is important for both first responders and EMTs. However, resources for initial and continuing education for them may be limited by distance to the program venue, their lack of available time if they are volunteers, and insufficient financial resources to pay for the training.

Although these professionals may be well trained in basic firefighting and emergency medicine techniques, few of these individuals are trained in the specifics of farm rescue and emergency medical treatment. There are many unique features of farm injuries for which regular training should be supplemented. There are well-recognized training programs to prepare rural service providers. The most recognized program in the U.S. is Farmedic. This program was initiated in 1981 and now operates out of Cornell University in New York State. The main goal of the Cornell Farmedic Training Program is "to provide rural fire/rescue responders with a systematic approach to farm rescue procedures that address the safety of both patients and responders" (http://www.farmedic.com/fos/). Over 22,000 responders have been trained by this program and certified instructors now exist in many regions of the U.S.

Special training is needed to ensure rescuers are aware of further risks to the victim and to themselves as rescuers, and how to mitigate these risks. Furthermore, trained agricultural rescuers perform confidently in special farm fire and rescue situations such as how to get a person untangled from a combine header or power take-off shaft. Among other unique farm injury scenarios, they learn how to rescue a person from a silo and how to extinguish a silo fire properly. They also learn about special rescue equipment and how to use it in specific farm rescue scenarios.

Rural farm fire/rescue services are equipped with standard tools such as air bags, reciprocating saws, abrasive disk saws, hydraulic spreaders and cutters, and air chisels that are adequate to handle most automobile crashes. However, farm equipment is sturdier and often made with tougher and thicker steel. This makes it more difficult to cut or open this equipment, compared to automobiles, with standard tools. Often it is best, and more expedient, to take the equipment apart rather than destroy it to release the victim. Since rescuers may or may not know how, it can be important to have a person at the scene that is familiar with the mechanics of the equipment, such as a neighboring farmer or an employee of a local farm machinery dealer. Dismantling the machinery may also be preferred by the victim because this equipment is expensive, insurance may not cover this type of destruction, and the farmer will most assuredly be hoping to be able to use it again in the future.

Other Special Considerations of Farm Rescue

The American Academy of Orthopaedic Surgeons book, *Rural Rescue and Emergency Care* (Hensinger and others 1993), describes four phases of rescue: locating, assessing, stabilizing, and transporting.

FINDING THE PATIENT

Finding the scene of a farm injury is often complicated. Even though many rural residences in the U.S. now have emergency (911) addresses with specific GPS coordinates, those coordinates will locate only the center point in the road in front of the driveway to the farmstead. The injury scene may be far from the farmstead or public roadway. And cell phones do not work everywhere. A first responder may need to designate a rendezvous point to meet the rescue team and guide them to the incident location. Direct access to the patient may also be complicated by inclement weather, lack of roads, and difficult surface conditions such as creeks and steep slopes.

ASSESSING AND STABILIZING THE PATIENT

The patient may be trapped or entangled in the machine. Freeing him or her may be complicated by the need to make immediate decisions regarding life threatening issues simultaneous with concerns about causing further injury by moving a person. For example, preventing the risk of a fire or explosion, contact with electricity from a downed wire, or contamination of the patient by various agricultural chemicals must be weighed against moving a victim with a possible spinal cord injury. It is essential for the rescuer to assess these hazards and separate the patient from them while mitigating the possible risk of further harm to the victim. Extended incidents may require rescue personnel to provide emergency medical care for more than 2 hours, assume additional patient care responsibilities, exercise well-developed assessment techniques for late-appearing signs, and cope with increasing stress and deteriorating patient condition (Hensinger and others 1993).

TRANSPORTING THE PATIENT

Fire/rescue vehicles may need to be equipped with four-wheel drive capability to reach the injury scene and then get the patient back to the main road for transport to the hospital. It may also be important to have a four-wheel drive vehicle or tractor at the scene to assist if rescue vehicles become stuck and need to be pulled out.

EMS personnel need excellent assessment skills to communicate with the local health care facilities. To assure best outcome for the patient they should be transported to the closest facility where they can be stabilized. Depending on the degree and nature of the injuries that decision may mean transport to a trauma center. If the distance and emergency nature of the injuries require a helicopter ambulance, that request should be made as early as possible following notification of the incident since the distance to these secondary or tertiary care facilities is usually much greater than the nearest primary care facility.

Sometimes the person first on the scene and/or the victim will choose self-transport. Transporting a person with what appears to be a simple injury, such as a broken arm or leg, can involve complications. Fractures should be properly immobilized and splinted before transport. Moving a person with a broken bone can damage the bone, cause additional bleeding, and otherwise adversely affect blood vessels, nerves, or other surrounding tissues. A first responder without first aid equipment may have to use his or her creativity, using rolled newspapers, magazines, or other materials to fashion a splint and stabilize the fracture.

The transport team or person must be prepared to provide information that may be important to the emergency room personnel as they perform immediate care. The following questions can be used as a guide:

1. Was the patient's skin or clothing contaminated with any of the following materials that would make it necessary to get his or her clothes off and wash the skin?
   - Pesticides (what type—insecticide or herbicide, cholinesterase inhibitor)
   - Anhydrous or aqua ammonia fertilizer
   - Fuels or fluids, e.g., gasoline, diesel, motor or hydraulic oils, or battery acid
   - Animal wastes or soil contamination of an open wound

2. Is there a possible foreign body contamination of wound?
3. Was there amputation of a body part that may be a candidate for surgical reattachment, and is the body part being transported with the patient?
4. What was the mechanism of the injury such that it might cause injuries that are not externally obvious?

EMERGENCY ROOM—GENERAL

The first goal of the emergency room personnel is to create no further harm, stabilize the patient, and make appropriate referral to the next level of care or release to the patient's home. Emergency room personnel should ascertain the information mentioned in points 1–4 above and provide care accordingly. If the patient has been exposed to fuels or chemicals, as a result of a tractor or machine overturn, for example, the patient's clothes must be removed and the body washed thoroughly. If an amputation has occurred, the patient should be assessed for possible reattachment surgery following stabilization. The body part must be obtained from the injury scene, kept on ice, and brought to the emergency room. Also consider that the patient may have secondary injuries, such as pig bites on victims trapped around livestock, or injuries to the contralateral side, self induced while trying to escape.

Possible wound infections resulting from antibiotic-resistant organisms of animal origin indicate avoiding classes of antibiotics that are commonly used on the farm (early beta-lactams (e.g., penicillin), tetracyclines, macrolids, sulfonamides) and starting with newer generations of antibiotics while waiting for culture and sensitivity assessments. Considering antibiotic coverage for anaerobic organisms, reasonable choices to start antibiotic therapy for a farm injury could include 1) Amoxicillin, or a 2nd or 3rd generation cephalosporin; 2) an aminoglycoside (e.g., gentamycin); or 3) a combination of a cephalosporin and an aminoglycoside (e.g., cefuroxime and gentamycin). Consideration should be given to delaying wound closure until it is certain the infection is under control or allowing the wound to heal by secondary intention.

One should make sure the patient is current on his or her tetanus immunization. The current recommendation in the case of either severe or contaminated wounds is revaccination with tetanus toxoid if the last immunization was longer than 5 years ago. If immunization status is unknown or if the patient has had fewer than three total doses of tetanus antitoxin over his/her lifetime, he or she should immediately receive a dose of tetanus toxoid and an additional 250 IU of human origin tetanus antitoxin (Chin 2000).

PRIMARY CARE—GENERAL

Many of the special considerations for agricultural injuries in the emergency room carry forward into primary care treatment. All of the previous considerations need attention to assure they have been dealt with appropriately. If they have not been dealt with in the emergency room, they must be dealt with in primary care.

One can assume that the patient will be very anxious about assuring the farm work gets done, and he/she will want to get back to work sooner than medically advised. Farmers are often stoic individuals and resist extensive care. The health care provider must consider the patient's concern that there often is no one else to do the work at home. There is no one to milk the cows, or feed the pigs, or plant, or harvest, creating a very anxious patient who is sometimes too motivated for his or her own good to get back to farming. Therefore, it would be good to practice some "social medicine" in these instances. Call the farmer's neighbors or relatives and help make arrangements for temporary care of the farm. The Scandinavian countries are advanced in this regard, as they have a special service that provides relief work for farmers in time of sickness or vacation. Unfortunately, these services are not widespread among industrialized countries.

If the patient is to go home directly from primary care, the provider should assume the patient will be out working on the farm sooner than ideal. Therefore, the provider should consider some accommodations to either prepare an earlier, safe return to work or some means that may discourage the patient from working until healing is more assured. Examples might include (if a fracture is involved) making an extra heavy walking cast that can be protected from water and will hold up while performing work around the farm, or making a cast that will impede working to the extent that it will encourage the patient to stay off work for a while. Furthermore, if a laceration or surgery was involved, consider the use of tension sutures as a precautionary measure to help hold the suture line intact even under stress.

SECONDARY/TERTIARY CARE—GENERAL

If the patient has injuries that require secondary or tertiary treatment, such as reconstruction, reattachment, or amputation, special consideration should be given to the patient's continued functioning as a farmer. The physician should discuss with the patient, his/her family, and occupational and rehabilitation therapists the options for reconstruction, considering the patient's desire to continue farm work and the procedures that may result in maximum occupational function for the particular patient. Certain kinds of prosthesis may provide much more function for the farmer. For some patients, an appropriate prosthesis may be better than a partially functioning reattachment.

In general, for amputations, there must be soft tissue padding over bony ends, and amputation should be as distal as possible, especially in children to preserve growing ends. Retention of a small amount of proximal humerus or amputation through the elbow, and similarly a small length of proximal radius and ulna, may be of little value. It is important to preserve the knee. Above-knee amputations should be in the midsection or distal end of the femur. Retaining the carpus can be important in hand injuries. Amputation of a thumb is best repaired microsurgically, and there is functional value if opposition between the digits and thumb can be achieved (Kostuik 1989).

Return to an adequately functioning extremity often requires protecting repaired structures, controlling swelling, and restoring motion through occupational therapy. Unlike other industrial workers who may be reassigned to other tasks or not return to work, the farmer may not have an option and will most likely have a strong desire to return to farming.

REHABILITATION—GENERAL

There often is a gap in continuity of services once the injured farmer leaves the hospital. Provision of rehabilitation services is very rare in rural communities and on the farm to facilitate the person achieving his/her maximum potential. Not only does the recovering injured farmer have physical problems to deal with, but he/she and the family may have mental health issues, too. The health care providers in the patient's community are now the primary professionals to assist the patient in returning to full potential. As mentioned, in most cases the patient will have a strong desire to get back to farming. In many instances, farming may be the only option for employment in the community. Furthermore, there is often an extremely strong cultural bond to the farm that is not broken easily.

Good communication is necessary between the higher-level health care providers and the local provider. This communication should include the nature of the injuries and the expected limitations and prognosis. The details of rehabilitation and assistive technology application are often left to the community. Fortunately, in the U.S., there is quite a bit of assistance for rehabilitation within the agricultural community. The most applicable organization in the U.S. for rehabilitation of farmers is the AgrAbility Network (http://www.agrabilityproject.org).

Created in 1991, AgrAbility links the federal and state Cooperative Extension Service with nonprofit disability service organizations to provide information, education, and technical assistance to serve persons with disabilities who are employed in agriculture. Twenty-eight states have USDA-funded AgrAbility Projects (http://www.agrabilityproject.org/stateprojects.cfm). A health care provider who has a farming patient with a disability may contact the AgrAbility project. That organization has rehabilitation specialists who will visit the farm, assess the patient and the farming operation, and assist the patient in achieving the potential of getting back to the business of farming. There are a number of other private organizations that deal with disabilities in rural America. Although not specific for agriculture, they may be of help in your region (www.AgrAbilityproject.org).

Injuries on the farm create a great deal of stress and mental health issues, which create barriers to full recovery. Chronic injuries present even greater stress for victims. Many may face the loss of work capacity for an extended time period, chronic pain, and learning to deal with lifetime disability. Injured farmers may suffer from posttraumatic stress disorder, which can be disabling in itself. Family disruptions may occur, especially with the loss of a child to a farm injury. Often a spouse assumes the role of total care provider for the injured person while at the same time having to increase his/her role in the farm work—plus, the spouse may be working off the farm to help make ends meet.

Apart from the serious psychological and social impact on patients and relatives, such injuries invariably have profound economic affects. In many instances a death or serious injury to the principal operator of the farm leads to economic failure of that enterprise.

Mental health services in rural areas are not readily available. Too often, there is a stigma in the rural community against admitting to the need for and seeking out mental health services. There are two organizations that are available in North America to help farmers deal with these issues. SHAUN, Sharing Help Awareness United Network, was founded by a woman who suffered the death of a son due to a farm accident and could not find emotional/mental assistance for her or her family. Her organization was developed to provide a service that fills that void. It provides professional counseling and trained peer counseling. SHAUN can be contacted on the Internet at www.shaunnetwork.org. A second organization, AgriWellness, deals with mental health issues generally in the agricultural setting. AgriWellness has professional and peer counselors available to assist farmers and families with mental health issues, regardless of whether there was an injury. AgriWellness can be contacted via the Internet at www.agriwellness.org.

HAZARDS AND INJURIES

This section describes some of the common injury scenarios encountered by farmers, farm workers, and their family members. These scenarios are intended to give the reader a better understanding of the mechanisms of injury, as well as the circumstances of the incidents. More than one body part or system is often involved. They are based on actual cases but some of the details have been fictionalized to emphasize teaching points. The scenarios are not all-inclusive for the types of injuries sustained but do provide an overview of some of the more common incidents as well as some of the confounding factors encountered in emergency situations.

Machinery

As mentioned earlier in the chapter, machinery is one of the main agents, or causes, of fatal and nonfatal injuries on farms and ranches. As a general category, machinery is a catchall for a huge variety of equipment that is used on a daily basis, plus equipment that is used seasonally or only occasionally.

Tractor Injury Incidents

Tractors are used on almost every farm and ranch on a daily basis. They are used on uneven terrain to push and pull implements and accomplish daily farming tasks. They have a high center of gravity, for visibility and clearance, and can quickly overturn onto the driver or run over an operator or bystander.

DESCRIPTIVE INFORMATION

Tractors are often described by their uses and configurations, such as utility or row-crop, narrow or wide front, two- or four-wheel drive, wheeled or tracked. They are powered by engines from 20–500 horsepower (14–350 kilowatts), and weigh 700–23,000 kilograms before additional ballast weight is added. They are typically operated at relatively slow travel speeds in the field but can go 30 mph (48 km/h) or faster on the road. They also perform as a mobile or stationary source providing mechanical, hydraulic, and electrical power.

The operator's station may be open or enclosed in a cab, with or without a framework affording protection for the operator in the event of a tractor overturn. Older operator enclosures, sometimes called "soft cabs" or weather enclosures, do not have rollover protective structures (ROPS) integrated into their design. In some countries, tractors are required to be equipped with ROPS or ROPS cabs. Most Western European countries have required ROPS or ROPS cabs on tractors since the late 1970s. They have been aggressively marketed for after-sale installation since they became commercially available in North America in the mid 1960s. Since 1985, the industry practice in the U.S. has been to provide a ROPS or ROPS cab on all new tractors.

Injury Overview

Tractors and other self-propelled machines including skid-steers, compact utility haulers, all terrain vehicles (ATVs), extendable-reach fork-lifts, trucks, and similar equipment are collectively involved in a majority of deaths to U.S. farmers. Tractor-related events contribute to about half that toll. Foremost among these events are overturns and being crushed by tractors and other mobile equipment. Approximately half the tractor fatalities involve overturns, and about one-quarter involve runovers of the operator, a passenger, or a bystander.

An operator may be thrown against objects during an overturn, thrown to the ground and crushed as the overturn continues, or be pinned to the ground with a good share of the machine's weight coming to rest on their torso or extremities. Being run over by a tractor, other vehicle, or mobile machine can cause internal crushing injuries as the heavy machine's tires or tracks traverse portions of the body. Unexpected or uncontrolled movement of a machine can squeeze a helper or the operator between the machine and an implement or fixed structure. Unsecured objects lifted high by a loader or forklift can spill or roll into the operator's station causing thoracic, cervical, and lumbar injury (Friesen and Ekong 1988).

Illustrative Scenarios

OVERTURN

A 76-year-old male was using a tractor with a rotary cutter attached to the rear 3-point hitch to mow a roadside ditch. Mowing along the shoulder of the road, the wheels on the downhill side of the tractor dropped into a washout and the tractor tipped onto its side in a fraction of a second. The tractor did not have a ROPS, and rolled 180°. The operator was thrown clear of the tractor. The driver of an approaching car stopped and immediately called for emergency assistance. He then began to assist the slightly dazed farmer crawling out of the ditch. The farmer complained of neck, back, hip, and leg pain along with upper-left quadrant abdominal tenderness. To the person first on the scene, he seemed to be breathing rapidly and with some difficulty.

OPERATOR BYPASS START RUNOVER

The 45-year-old farmer was in a hurry to start the morning's chores. His tractor would not start even though he had parked it in the machine shed after using it to feed cattle the evening before. He parked his pickup next to the tractor and con-nected jumper cables from the truck to the tractor battery to boost-start the tractor. Yet, when he turned the ignition key the tractor still would not start. Grabbing the pliers from his pocket, he placed their handle across the terminals of the starter, bypassing the nonfunctioning solenoid switch. The engine cranked and started. In a quick moment the tractor, which he had inadvertently left in gear, moved forward, running over him before he could get out of the way. Luckily, the rear wheel got only his legs.

BYSTANDER RUNOVER

John, who was 12, got home from school and started chores. Every day was the same, doing his chores and watching his 3-year-old brother, Justin, who was dropped off by the babysitter. The cattle needed hay and John thought Justin would be okay playing in the hay storage area. John used the tractor with a front-end loader to pick up a large round bale and headed for the feedlot. He didn't realize Justin was following him, despite being told to stay put. John placed the bale carefully into the manger and shifted the tractor into reverse. He continued to look forward, lowering the loader carefully to avoid hitting the manger. As the tractor began to move rearward and before he turned to look behind, the right side of the tractor lifted slightly. His gaze shifted to the right, ahead of the rear wheel, to see what he had run over just as his young brother's feet and legs came into view from under the tire.

Other scenarios include operators or extra riders falling from the operator station and being run over, helpers being run over hitching an implement to a tractor while the operator slowly backs up, or operators trying to do it alone and being crushed when they lift an object too high in a loader and it rolls down on them. The rollover and runover scenarios can occur with any of the many self-propelled farm machines, tractors, or tractor-implement combinations.

First on Scene, First Responder, and Transport

An operator on a tractor that does not have a ROPS when it overturns has a high probability of dying outright from the injuries. These incidents most frequently occur when the tractor overturns to the side, which is more common than to the rear, and involve the operator being crushed by the tractor (Rural Family Medicine 2003d). Survivors generally have broken bones as well as internal injuries. If the tractor is equipped with a ROPS cab, and the operator does not have the seat belt fastened, the injuries are usually less severe and occur from being thrown around inside the cab.

A first responder may find the engine of a tractor still running with the victim pinned under a portion of the tractor. The victim may also have injuries from caustic or hot fluids. Unless there is immediate danger, such as fire, do not move a tractor overturn victim. The engine of the tractor should be turned off to prevent possible fires or burns to the victim. Equipment with gasoline engines can be shut off by switching the ignition key to the off position, disconnecting the coil or spark plug wires, grounding the magneto, or discharging a CO_2 extinguisher into the air intake. Most gasoline engines and many diesel engines have a key switch to turn them off, but some diesel engines may be different and require pulling a knob or some other action. Other alternatives include manually operating the throttle, shutting off the fuel supply at the fuel tank or in the fuel line, removing the fuel filter to interrupt fuel supply, cutting the fuel supply line (not the lines to the injectors, which are under high pressure), or discharging a CO_2 or halon extinguisher into the air intake (Tyson 1994). If the knowledge to turn off the engine is not possessed by a person at the scene, one person should contact a local farmer or dealer to find out how to turn off the engine.

A shear-type degloving injury may occur when a tractor or other wheeled piece of equipment runs over an extremity causing soft tissue to be pulled loose with underlying tissue and vascular disruption.

Emergency and Post-Crisis Treatment

Explore for fractured skull (usually depressed); thoracic, cervical, and lumbar injury; and potential paraplegia or quadriplegia. It could be a neck fracture if the victim complains of pain or numbness in the neck or down the arms or legs. Chest injury possibilities include mechanical asphyxia, flail chest, pneumothorax, hemothorax, or subcutaneous/mediastinal emphysema (Hensinger and others 1993). Patients suffering from acute and complex spinal cord fractures and dislocations are managed according to established principles.

These include immediate hemodynamic and respiratory support in an attempt to prevent or reduce posttraumatic ischemia and infarction of the spinal cord. Steroids, manitol, and diuretics have been used, although not consistently. Spinal cord injuries are devastating because of their effect on the body and its systems.

Abdominal injuries may include laceration of the liver or spleen, rupture of hollow organs, or other penetrating wounds. Fractures with associated internal bleeding and lacerated rectum should be considered in the pelvic region.

Degloved tissue separated from its perfusing vessels can become gangrenous. Three types of degloving injuries may occur: skin removed from the limb, concealed degloving, and a combination of both. It is especially important to recognize the concealed type since immediate defatting and reapplication of skin as a graft is most efficacious (Letts 1989).

Recovery Care, Management, and Rehabilitation

These types of injuries call for extra-special mental as well as physical rehabilitation. Farmers who suffer an injury that leaves them paraplegic or quadriplegic often still want to farm. Farm rehabilitation specialists do have means to help these people with special lifts to enter tractors, special operational adaptations so they can drive, and much more.

Spinal cord injuries are more than devastating to body functions. They entail serious psychological, economic, and social impacts on the patient and family, and a high cost to society (Kahn 1998).

Contributing Factors and Prevention

Table 11.11 categorizes the contributing factors to these tractor-related injury events according to the man-machine-environment paradigm, and lists preventive strategies.

Skid-Steer Injury Incidents

Descriptive Information
Skid-steers have continued to grow in their capabilities, popularity, and complexity since their advent in the 1950s. They are compact, self-propelled machines with mechanical arms on one or both sides of the machine to lift and lower attachments that carry loads and perform other

Table 11.11.   Analysis of Contributing Factors and Prevention of Tractor-Related Injury Scenarios

	Man	Machine	Environment	Prevention
Tractor overturn	The farmer was elderly (76 years old).	The tractor did not have a ROPS.	There was steep terrain and washout with limited vision due to tall grass.	Install ROPS on tractor or use newer tractor with ROPS. Inspect the area to be mowed on foot, before mowing. Consider age-appropriate tasks for the senior farmer.
Bypass start runover	The farmer was anxious and focused on getting the tractor started and getting chores done; he had a positive previous experience in bypass starting.	Problems were a faulty solenoid switch and a dead battery.	Cold weather compounded the anxiety.	Replace lockouts on the tractor so it cannot be started when in gear. Install a bypass start guard on the starter. Install a new solenoid switch and battery.
Bystander runover	A 3-year-old playing with a 12–year–old were unsupervised because their parents were working elsewhere.	There were no rear view mirrors on the tractor.	Because of the height of the feed bunk, the operator had to look forward while backing out so the front end of the loader would clear.	Install a safe play area on the farm. Arrange for parental supervision, or day care. Discuss and assign age-appropriate tasks.

work tasks. Today they are commonly called *skid-steers* rather than *skid-steer loaders* because of the expanding range of attachments, such as augers, backhoes, blades, jackhammers, and buckets. Common agricultural tasks farmers perform with skid-steers include moving manure from animal facilities, moving large hay bales and silage, utility tasks such as digging post holes, and minor earth- or rock-moving projects. Skid-steers are ballasted with counterweight, enabling them to lift heavy loads. They are able to pivot around in their footprint because the tracks, or

pair of tandem wheels on either side of the frame, rotate in opposite directions in response to the operator's controls.

The steering control of the skid-steer is hydraulically operated as are its liftarms and many attachments. The operator is positioned at controls between or adjacent to the liftarm(s), and behind the attachment connected at the front of the liftarms. Operators enter and exit the operating station through the opening at the front of the machine, over or around the attachment, or from the side of machines designed with a single liftarm

(NIOSH 2003). Old-model skid-steers without rollover protection, falling object protection, and side screens that block access to the scissoring movement of the liftarm(s) alongside the frame, have not totally vanished from use. Occasionally operators remove protective structures for access to buildings with low doorways and minimal floor-to-ceiling clearances.

INJURY OVERVIEW

Like tractors, skid-steers can overturn causing crushing injuries similar to those in a tractor overturn. Skid-steers are inherently less stable than many tractors on steep slopes because of their short and relatively narrow wheel base. Bystanders can be run over during rapid forward and backward shuttle operations. During operations requiring lifting and lowering of the liftarms, an operator's head, neck, or shoulders placed outside of the side frame can be crushed in the scissoring action between the lowering liftarm and the side frame of the skid-steer if the protective side screens have been removed. Since an operator may enter from the front and step over the controls, inadvertent actuation of the controls can occur, causing the bucket, if left in the raised position, to drop suddenly. Heavy material could fall backward into the operator's station from a raised, loaded bucket, injuring the operator. Due to the short wheel base, these machines tend to rock fore and aft forcefully when traveling over rough terrain, or when lowering and raising or dumping the loader/load. If the operator is not wearing a seat belt, and/or forward pitch protection is missing, the operator can be thrown out, risking injury from a runover or being caught in the scissors action of the loader.

Illustrative Scenarios

PINCH OR CRUSH

A 26-year-old pork producer was injured when he was caught between the frame of the skid-steer and the side liftarm of the loader. He was using the loader to load manure from a hog confinement building. The protective cage had been removed to permit operation under the low doorway of the building. The farmer was operating the foot pedals that raise and lower the bucket while leaning to the side to observe the bucket's position for dumping and to avoid hitting the building. He inadvertently lowered the bucket, which crushed him between the liftarm and the side frame of the skid-steer.

HYDRAULIC OIL BURNS

A 38-year-old farmer was using a skid-steer to pick up brush and stumps as clean-up continued following a severe wind storm. He was gathering the debris and loading it into a dump truck. Lifting a log into the truck, a damaged hydraulic line ruptured and the bucket began to fall. Co-workers heard the screams as hot hydraulic fluid sprayed onto the operator's face and hands.

FIRST ON SCENE, FIRST RESPONDER, AND TRANSPORT

Always proceed with the basics: scene safety, shut off equipment, stabilize equipment, and check the patient. An advanced rescue team, paramedic ambulance, and helicopter should be dispatched to the scene prior to your arrival. The injuries sustained in this pinching scenario could include broken bones and internal damage. Even a closed fracture such as of the thigh or pelvis, which often results in damage to the bladder, can involve significant blood loss, and falling blood pressure can mean many tissues no longer receive adequate oxygen (Lycholat 2003). In the event that the person is conscious and talking, be cognizant of the fact that enough internal damage could have been done so that release of the pressure could cause massive internal bleeding and sudden death.

Depending on the extent of the burns, response to the hydraulic fluid shower would be slightly less extensive. Hot oil may be 200°F or more. Decontamination of the patient is important, as is cooling the affected area similar to any burn.

EMERGENCY AND POST-CRISIS TREATMENT

The farmer pinched by the skid-steer liftarms could have significant internal injuries, including broken ribs, punctured and collapsed lung, and ruptured liver or spleen. Assessment of breathing ability should be performed immediately.

RECOVERY CARE, MANAGEMENT, AND REHABILITATION

These types of injury may take an extended recovery period, and all the precautions and assurance of continuity of care as mentioned previously should be put in place.

Table 11.12. Analysis of Contributing Factors and Prevention of Skid-Steer Injury Scenarios

	Man	Machine	Environment	Prevention
Pinch and crush	Behavior of removing the protective cage	Inherent design of the machine— lift apparatus is adjacent to the operator's station. Machine lacks protective cage. Having no interlock system prevents the lift arms from actuating when a person is not in the seat.	Low and narrow building makes it more convenient to operate the machine without the protective cage.	Remodel the building. Replace the protective cage. New machines have an interlock that prevents actuation of the lift arms if no one is in the seat.

CONTRIBUTING FACTORS AND PREVENTION

The primary contributing factor in this case was the missing protective cage that encases the operator, making it impossible for the operator to get caught between the liftarm and the side frame of the skid-steer. Even with an enclosure, it is possible to be crushed between the working equipment attached to the loader arms and the front of the skid-steer when the arms lower. Operating without the protective enclosure and its side screens, and not wearing a seat belt while using the skid-steer or otherwise being secured in the operator station, is an extremely hazardous combination. Proper maintenance is vital for all mechanical equipment. Safety devices need to be kept in place—except when repairs are being made—should function properly, and must be well maintained.

These scenarios are examples of skid-steer–related injury events. Table 11.12 categorizes the contributing factors in the pinch and crush events, according to the man-machine-environment paradigm, and lists preventive strategies.

Other Self-Propelled Machinery Injury Incidents

DESCRIPTIVE INFORMATION

Self-propelled agricultural machines are specialized machines for harvesting grain, tilling or planting fields, applying pesticides, or gathering stones and roots. Combines, windrowers, potato and beet harvesters, and sprayers are common self-propelled machines, and the prevalence of others such as sugar cane harvesters varies with specific commodities cultivated around the world.

Crop gathering, processing, transfer, and discharge mechanisms are powerful, designed to minimize blockage, and aggressively move large volumes of material rapidly. Many of these mechanisms have high inertia and continue to operate for a period of time after power to them has been disengaged. The cutting knives of a forage harvester, for example, can continue to rotate for up to a minute or more after power to them is disengaged. The gathering mechanism of a corn combine head pulls the cornstalk down through the snapping rolls and stripper plates, which remove the ear of corn from the stalk and start the ear on its way to the threshing area. The stalk feeds through from 10–15 feet per second (3.0–4.5 meters per second), much faster than the reaction time of a person (Rural Family Medicine 2003b); there is no time to let go.

INJURY OVERVIEW

In addition to injuries from this equipment overturning, or being runover by the equipment, people can become caught in the mechanisms and be drawn into the machine. Clothing or extremities can be entangled in drive mechanisms that are running. In less severe incidents the fingers of a hand may be crushed or severed as they pass between a belt and pulley, chain and sprocket, or meshing teeth of a gear set.

Inertia causes crop engaging mechanisms to continue in motion, sometimes for minutes after power to them has been disengaged. Contact with these mechanisms at that time can still produce the same effect on body parts as it does on the crop itself. Lacerations, amputations, and degloving injuries may be the result.

Illustrative Scenarios

CUTTING MECHANISM
The 57-year-old hired man had just taken the self-propelled chopper to the field to chop corn for silage. The machine normally gathered the four rows without hesitation, but corn rootworms had damaged the corn, causing the stalks to be twisted by the wind. As he guided the machine slowly through the field, he became increasingly frustrated each time the throat of the chopper, the area that the corn stalks pass through just before they enter the rotating blades to be diced, became plugged. Initially he would shut the machine off and then physically dislodge the stalks. This time, he thought he would save time and dislodge the material with the chopper running. As he grabbed the pile of stalks, they pulled into the machine; he could not let go quickly enough.

SNAPPING ROLL ENTANGLEMENT
The 78-year-old farmer was increasingly troubled with the effects of aging. His hearing and eyesight were not as keen as they had been for the bulk of his life, and the last couple of years he had some trouble with his balance. His combine was older, but still in good shape. He had made a few modifications to make it easier for him to use. This morning, he went to the field to harvest corn. There had been a frost, and as the frost melted it made the stalks tough. He plugged the snapping rolls—again. Two snapping rolls run parallel to each other, one on each side of a corn row. They intermesh and turn toward each other. When they contact a corn stalk, the stalk is suddenly pulled downward. The ear cannot fit between the snapping rolls, and is "snapped" off, the first in a series of processes in corn harvesting.

When the snapping rolls plugged, the farmer got off to remove the corn stalks. He did not shut the combine off, and there was no interlock safety mechanism. As he grabbed the stalks, they suddenly were pulled into the machine, and so was he.

FIRST ON SCENE, FIRST RESPONDER, AND TRANSPORT
Consider contacting a local equipment dealer mechanic. Disengage power; shut off the engine; and stabilize the machine, the scene, and the victim. Assess the injury and call for emergency transportation. Depending on how the person is caught, it may be possible to free him or her by cutting entangled clothing, disconnecting the drive mechanism, or completely dismantling part of the machine. Entanglements in crop-engaging mechanisms may involve the most serious injuries and most complex extrication. The victim's body may block access. Portions of machines may be too strong to be moved with extrication tools such as the "jaws of life." It may be necessary to use common wrenches and sockets to dismantle the machine. Loosening shaft bearings and adjusting devices can relieve pressure, and disassembly of the involved parts can release the patient.

EMERGENCY AND POST-CRISIS TREATMENT
Expect fractures, lacerations, avulsions, crushing injuries, and amputations in these situations.

RECOVERY CARE, MANAGEMENT, AND REHABILITATION
The same general considerations are present in these injuries as mentioned in the introduction to this section.

CONTRIBUTING FACTORS AND PREVENTION
The age of the second farmer was a contributing factor. Cognition, balance, and reaction time all were likely decreased at 78 years of age. The rootworm infestation in the corn, making it difficult to harvest, and the weather, making the stalks difficult to feed into the chopper, were definitely environmental considerations in the two scenarios.

Frustration when working around equipment needs to be controlled in order to reduce the risk of doing something foolish, such as approaching the equipment while it is running. Frequent breaks, proper maintenance, and working with crops at the right stage all help reduce incidences that provoke frustration.

PTO-Powered Implement Injury Incidents

DESCRIPTIVE INFORMATION

Many agricultural machines are made to be pulled, or carried, and powered by a tractor. Mechanical power to these implements is transmitted to them by a rotating driveline called a power take-off (PTO) shaft or implement input driveline (IID) (Society of Teachers of Family Medicine Rural Interest Group 2003). Machines that use the tractor PTO for power do a myriad of jobs on farms and ranches, including harvesting, tilling, grinding feed, and spreading manure.

The PTO stub shaft, typically at the rear of the tractor, is normally surrounded by a guard called a *master shield* or *PTO shield*. The IID connects the PTO stub shaft on the tractor to the implement. There are generally at least two articulating universal joints, allowing for power transmission through the shaft when alignment of the stub shaft of the tractor and the machine are not in a straight line. A driveline shield covers the universal joints and the rotating shaft, preventing loose things from wrapping around the shaft while it is turning. Older machinery may not have been manufactured with this type of guarding system. If the machine has not been maintained properly, guarding may be ineffective, or it may have been removed.

The PTO and driveline are often referred to by the speed at which they rotate. At a standard engine operating speed, a tractor's PTO shaft will rotate at either 540 or 1000 revolutions per minute (RPM). By design, 540 RPM and 1000 RPM tractor PTO stub shafts are made to connect only with the proper coupling for the speed intended for the implement. Some tractors have shiftable PTO speeds, allowing them to operate drivelines at other RPMs without changing tractor speed.

INJURY OVERVIEW

Clothing, long hair, or a person's extremities can be entangled in the PTO driveline, the moving machinery that transmits power to different components of a machine, or the crop-engaging/processing mechanisms themselves. A person becomes entangled in a rotating shaft when any loose or pliable material, like clothing or hair, contacts the shaft and wraps around it. Consecutive wraps "lock" the material onto the shaft, which keeps turning and continuing to wrap up the loose material. At 540 RPM, the IID rotates 9 times per second, at 1000 RPM, almost 17 revolutions per second. A string 10 feet (3 meters) long that catches on a 4-inch (10-centimeter) diameter IID would be completely wound around it in 1 second. A similar shaft at 1000 RPM would wind up over 16 feet (5 meters) of string in a second.

Injuries from rotating shaft entanglements can be some of the most traumatic injuries in agriculture. Head injuries include closed and open skull fractures. Spinal fractures and dislocations are a real possibility. Flail chest, sucking chest wounds, pneumothorax, hemothorax, tension pneumothorax, pulmonary contusion, and myocardial contusion should be considered. Other injuries to anticipate include skin tears, amputations, degloving injuries, fractures, tourniquet-like wrapping of clothing, scalping, and avulsion of skin and external male genitalia (Hensinger and others 1993).

It can be inferred from the previous description of the PTO mechanism that it is extremely important to keep the complete guarding systems in place from the tractor, through the driveline, to the implement, and on the implement itself. Guards and shields must be kept in place, function properly, be inspected often, and be well maintained to afford the protection they were designed and built to provide.

Illustrative Scenarios

PTO ENTANGLEMENT

A 13-year-old boy was baling hay with his father along to supervise his work. His father, after making certain everything was operating and his son was in control of the job, left to do other work. While the youngster was baling, he noticed a large mound of hay in one of the windrows. As he approached it, he worried that driving over the mound would plug the baler, so he stopped just before the hay went in. The boy got off the tractor and started to kick the hay into the baler a little bit at a time, like he had seen his father do on numerous occasions. The tines on the baler that are there to pull hay into the baler, suddenly caught the hay pulling it into the baler and causing the boy to loose his balance. As he fell backward, his shirt sleeve was caught by the PTO shaft.

Belt Drive Entanglement

A 42-year-old hired man was tending the grinder at the silo. For this operation, whole ear corn had been picked and was run through the stationary chopper to grind and blow it into the silo. The hired man noticed during the grinding process that the drive belt on the rear of the machine was vibrating. Thinking the belt was too loose, he climbed onto the back of the implement and attempted to stand on the tightener pulley arm to force it down another notch. His foot was twisted between the tightener and belt, and he was thrown to the ground.

First on Scene, First Responder, and Transport

Stabilizing the patient is very important because of the potential for massive and multiple injuries in this scenario. Extrication from a PTO or other drive shaft usually involves cutting the shaft with a saw. It is far easier to cut through a universal joint than through a multiple layered shaft. A powered reciprocating saw with a metal cutting blade is a good tool to use in this case.

Emergency and Post-Crisis Treatment

Expect long bone compound fractures, arm and/or leg amputations, scalpings, skull and rib fractures. Rapid spinning of the body around a shaft and through narrow spaces between the shaft and surrounding structures can cause blunt force trauma and can tear ligaments loose. Traumatic brain injury may be closed head or penetrating head injuries. Diagnosis of closed head injuries risks delay due to outward appearances. Some symptoms may not appear until well after the injury (Traumatic Brain Injury Resource 2003).

Recovery Care, Management, and Rehabilitation

PTO injuries are some of the most devastating in agriculture. The recovery care and management usually takes specialists in vascular surgery and orthopedics. Decisions will have to be made about the most practical and functional (not necessarily aesthetically best) repair that can be accomplished, considering the patient's likely strong desire to return to farming.

Contributing Factors and Prevention

Even though it may seem expedient to do so, making adjustments or attempting to feed material into a running machine is a recipe for disaster. Although equipment can be observed while it is running to determine necessary adjustments, these adjustments should be performed only while the equipment is shut off. The young man baling hay in the field, and apparently his father, should stop the equipment, spread the mound of hay, and then slowly pull forward over the area so the machine has time to process the extra material.

Other (Non–PTO-Powered) Implement Injury Incidents

Descriptive Information

Tilling and shaping the soil, planting seeds and seedlings, applying nutrients and pesticides, and transporting harvested material to a storage facility are usual tasks in production agriculture. The configurations of tillers, planters, chemical application equipment, and wagons are as varied as those of PTO-powered implements. They range from 1 meter to 35 meters (3–120 feet) or more wide in their working or field configuration and are frequently as long from the tractor hitch point rearward as they are wide.

Although some very wide machines are transported on roadways, they are more typically transformed to a narrower, often taller configuration for transport. Nonetheless, these machines in transport may take up a substantial portion of a rural road as they are towed behind a slow-moving tractor among motor vehicles that are much faster with drivers who are less aware of the hazards and potential actions of the farm equipment operator.

Injury Overview

Injuries from this equipment are quite varied and can include trauma similar to tractor rollovers and runovers, crushing and pinching by hydraulically operated or falling equipment parts, injection from fluids under high pressure (3000 psi—20MPa—hydraulic systems), burns from anhydrous ammonia injection equipment, and engulfment and suffocation in large grain transport trailers.

Illustrative Scenarios

Hydraulic Injection

A 39-year-old farmer had been tilling his fields all day. When he stopped to refuel, he noticed hydraulic oil dripping from the connections at the rear of the tractor. He knew there was a leak and wanted to see whether it was caused by the connections or a hose. As he picked up one of the hoses to check it, he felt something like a hot wire puncture his hand. The almost invisible oil stream penetrated his finger, forcing oil deep into the soft tissues surrounding the tendons and ligaments.

COLLISION

A 69-year-old farmer was driving her tractor with a forage wagon in tow on the two-lane state highway. As she approached her driveway, she turned on her left signal, not realizing that it was obscured by the large wagon. Just as she was making the turn, an automobile struck the side of the wagon. The force of the impact caused the farmer to lose control and the tractor and wagon veered off the road and into the ditch. Fortunately her tractor was equipped with a ROPS. The automobile was totaled.

FIRST ON SCENE, FIRST RESPONDER, AND TRANSPORT

In a hydraulic injection scenario, the victim and family members must be aware that this is an emergency (Vaughn 2003). If appropriate care is not administered in a relatively short period, loss of fingers or a hand may result. The victim should be taken to the nearest emergency room, and the emergency room physician made aware of the possibility of hydraulic injection.

Approaching a roadway collision is not much different when it involves farm equipment than it is with two motor vehicles. The primary difference is that farm implement–related incidents may involve manure, a spray rig loaded with pesticides, or fertilizers such as anhydrous or aqua ammonia. Anhydrous ammonia vapors in air form ammonium hydroxide, a caustic alkaline solution. Contact with skin may result in tissue freezing, dehydration, and severe chemical burns to skin, eyes, mouth, throat, and lungs (Hensinger and others 1993). These scenarios have implications for the health of the victims, bystanders, rescuers, and the environment. Along with caring for victims who may have their clothes saturated with toxic or caustic materials, a hazardous materials response team may be called to control the environmental consequences.

EMERGENCY AND POST-CRISIS TREATMENT

Factors important to high-pressure injection injuries are the type of fluid, amount injected, pressure of injection, degree of spread, and time between injection and treatment (Mizani and Weber 2003). Classically, the mode of injury is one of inadvertent injection of the finger or hand. No immediate clues to the actual severity of the injection are evident. Acute inflammation coupled with tissue distension may rapidly reduce circulation. Once injected, the material disperses along the paths of least resistance. Bone, tendons, and ligaments deflect the injectate into surrounding soft tissue, where it travels along fascial plains, tendon sheaths, and neurovascular bundles (Hensinger and others 1993; Johnston 1998).

For the first hour, injection injuries may appear innocuous and inconsequential. However, the injury soon becomes extremely painful as inflammation induces swelling. Because of the anatomical structure of the hand there is limited tolerance for swelling before damaging pressure occurs to nerves and vessels. Entrapment syndrome may soon develop, which leads to loss of blood and nerve supply to portions of the hand. Immediate and aggressive surgical decompression, debridement, and drainage are necessary to preserve vital structures and prevent permanent disability, which may include amputation. Decompression should be wide, and the wounds should be left open or loosely closed for drainage. Salvage is significantly higher if surgical intervention occurs within 10 hours of the injection.

Antibiotics and corticosteroids are staples of medical treatment following initial surgery for hydraulic injection injury. Follow-up surgeries may be needed in many instances to remove granulation tissues caused by remaining oil. Furthermore, surgery may be needed for scar revisions or tendon repair. Late amputations are unfortunately common, resulting in an overall amputation rate of around 50% for these types of injuries (Johnston 1998).

Thorough irrigation of body parts exposed to anhydrous ammonia should be continued until the patient reaches the nearest burn unit. Rapid transport via air ambulance may be needed to save a patient's sight (Hensinger and others 1993).

RECOVERY CARE, MANAGEMENT, AND REHABILITATION

The extended care for hydraulic injection injury or anhydrous ammonia burn largely depends on the extent of the injury. Rehabilitation plans will depend on the extent of the injury as well. Extended care for the victims of the roadway crash should follow the general principles discussed in the introduction to this section.

CONTRIBUTING FACTORS AND PREVENTION

While traveling on public roadways, it is vital to assure that the appropriate slow moving vehicle emblem is visible along with flashing amber lights during the daytime, although they may not be required by law, and at night. Mirrors mounted on a tractor that transports large loads will help operators see traffic that approaches from the rear and help avoid a collision from the rear when turning left. The general public is unaware of the behaviors of the driver of a farm vehicle and often does not anticipate when a tractor operator may be turning. Furthermore, they do not realize the closing speed, short time to react, and long braking distance required when traveling at 100 km per hour (60 mph) approaching farm equipment ahead moving 5–50 km per hour (3–30 mph). These hazards are seldom taught in driver education classes in rural or urban areas and are not part of the knowledge necessary to pass a driver's examination.

Routine maintenance of all hydraulic hoses and components will reduce the chance of a hydraulic injury. There may be pressure in hydraulic lines even though the flow control is not actuated. Checking for leaks in a pressurized hydraulic line should never be done with a bare hand: Use a piece of cardboard. Routine equipment maintenance, special operating precautions, and personal protective equipment can help prevent anhydrous ammonia burns.

Auger Injury Incidents

DESCRIPTIVE INFORMATION

Augers function to move grain or feed on farms in a wide range of applications. They often may be a component of a mobile farm machine. Combines, grinder mixers, and grain wagons can all contain augers. Augers are also employed as a fixed part of some grain storage facilities and are common as portable machines that can be moved from place to place. They are powered by one of several means, such as an electric motor, gasoline engine, a tractor's hydraulic system, or a tractor PTO (Rural Family Medicine 2003a).

The design of an auger involves a spiraling band of steel (flighting) winding around a center shaft. The powered rotation of the shaft causes the flighting to push the grain in one direction. In most augers, the shaft with attached flighting is fitted inside a metal tube (enclosed auger) with very little clearance between the flighting and inside of the tube so material can efficiently be moved up inclines without leaking back. Grain moves into the intake area of the auger, an area where the flighting extends beyond the tube to let the grain or feed be pulled into the flighting, and on up the tube. The intake area should be equipped with guarding that must be kept in place to prevent fingers, hands, and toes from being pulled into the auger. A person's clothing or extremities can be quickly pulled into the auger just like the grain if caught on the flighting or otherwise screwed between the flighting and tube.

Grain or feed storage bins often have an open (uncovered) auger that moves grain or feed to the intake area for the unloading auger. The auger will move the grain or feed as long as it is immersed in the material.

INJURY OVERVIEW

A person inadvertently getting an appendage into an auger usually experiences severe lacerations and extensive soft tissue damage. Amputation of fingers, hands, arms, toes, and feet are common. Adults may experience digit, hand, or foot amputations, while in children, loss of arms or legs are more common as their size allows movement of the appendage further into the auger tube. Children who manage to escape amputation are frequently left with severe degloving injury, where the skin has been completely detached, requiring multiple skin grafts with associated bone and muscle damage.

Illustrative Scenarios

AUGER ENTANGLEMENT

A 20-year-old employee of a grain farm was unloading grain from wagons into an auger at the edge of the field. The auger moved the grain into a semi-trailer to be sent to town. On the final load of the day, she wanted to make sure that all of the grain was out of the hopper of the auger, so she reached down with her gloved hand and started to brush the grain into the auger flighting. In an instant, the flighting caught her glove and pulled her hand into the auger tube.

First on Scene, First Responder, and Transport

The limb caught in an auger will have massive lacerations. Although there is a crushing aspect to these wounds that may inhibit excessive blood loss, there still is a risk of extensive bleeding. The pressure of the tube on the appendage inside helps to provide a tourniquet effect reducing blood loss until the pressure is released. Severe lacerations and/or amputation can cause substantial bleeding and resultant hypovolemic shock. Indirect pressure on the axillary or brachial artery or femoral or popliteral artery can control bleeding until the extremity is exposed, enabling application of direct pressure (Hensinger and others 1993).

Some first responder/rescue teams have cut away the upper and lower end of the auger, leaving the affected appendage in the section of auger and transporting the patient to the emergency room in that manner. At the emergency room the first responders can free the arm with medical staff on hand to handle the extensive bleeding and shock that may follow.

Emergency and Post-Crisis Treatment

Expect massive lacerations, avulsions, and amputations. Keep in mind that the act of twisting and cutting a limb can also injure joints and attachments of the affected limb.

Recovery Care, Management, and Rehabilitation

Auger injuries often require extensive reconstruction, but they are usually not as severe as PTO injuries. Long-term care and management should follow the general guidelines offered earlier in this chapter for agricultural injuries.

Contributing Factors and Prevention

A machine that is running should not be approached too closely. Without intake screen guarding with properly sized openings over the auger intake, risk for this type of injury is quite high. The act of brushing the grain into the auger

likely cost this woman her hand. A broom could have been kept by the auger to sweep the excess grain so that she would not have had to use her gloved hand. Environmental components of this injury included the concern for wet grain caking in the auger after an evening rain. The auger hopper could have been covered to eliminate this possibility. At worst, a bucket of grain would have been wasted compared to incurring a lifetime disability.

Livestock

Descriptive Information

Working with large farm animals can be hazardous at any time. Moving, sorting, and treating animals that weigh anywhere from 2–10 times more than a person may result in the person being stepped on, run into, kicked, bitten, butted, or gored. These animal-induced injuries usually happen because the animals are frightened, forced to do something they would rather not do, protecting a violation of their territory, or are acting to protect their young. Although animal-related injuries are about as frequent as machine-related injuries, they are often less serious and are a relatively infrequent cause of fatal injury. Animal husbandry during breeding and birthing periods can be particularly hazardous. Bulls, boars, studs, and rams may be aggressive in their natural tendencies and territorial. Getting between a newborn calf and a cow or piglet and a sow stimulates protective instincts, and new mothers often protect their young more aggressively. When compounded with the need to check animals frequently to prevent problems and assist with birthing, the element of human exhaustion is added to the equation.

Injury Overview

Injuries to farmers by livestock are fairly common. Because of the need to handle and treat animals, it is normal to be close to them. Size counts, and the difference between a bull weighing over 1,000 kg (2200 lb) and a 100 kg (220 lb) person can quickly result in an injury. Sudden reflex behavior, fight-or-flight and escape responses, and erratic movements by animals—or people near them—can also lead to injury incidents (Fretz 1989). Reflex behavior by an animal is its first line of defense. When you throw in the protective nature of mothering newborn young, or an adult male with females, you have an equation with an injury in the answer. Lacerations occur while performing minor surgeries on animals, such as dehorning and castration. Hands, arms, and legs can become entangled in ropes and halters. Injuries vary from minor bruises and broken arms/fingers to loss of fingers. Crushing injuries with bruised or broken bones from kicks, butts, or inadvertent crushing against a solid wall are quite common. Large animals such as cattle and horses account for the biggest share of animal-related injuries in the U.S., although injuries certainly do occur from a wide range of animal species, considering agriculture globally.

Illustrative Scenarios

BUTTED BY COW

A 57-year-old farmer was butted by a cow as he tried to check a group of pregnant Holsteins. The man was checking the cows when he noticed one with a newborn calf. As he approached, the cow looked at him nervously. When he bent down to help the calf to its feet, the cow hit him in the back with her head. He fell, and the cow continued to butt him along the ground until he was able to crawl and roll under a barbed wire fence.

FIRST ON SCENE, FIRST RESPONDER, AND TRANSPORT

Stabilization of this scene includes removing any animals that may be in the area or occupying them so they are not a hazard for the responders. Fractures of long bones and ribs are quite common, since the cow has butted and perhaps stomped on the patient. Spinal and cervical immobilization is important.

EMERGENCY AND POST-CRISIS TREATMENT

Expect long bone and rib fractures coupled with internal injuries. Skull fracture or severe concussion can result from a kick to the head. If the victim was rendered unconscious or otherwise unable to escape in the presence of animals, such as hogs, he may also suffer loss of soft tissue that was eaten away (Hensinger and others 1993). Hogs are carnivorous.

RECOVERY CARE, MANAGEMENT, AND REHABILITATION

Relative to machinery-related injuries, animal-related injuries are usually less serious and are most often closed injuries. However, there is a risk of infection from antibiotic-resistant organisms or anaerobic organisms causing tetanus or gangrene. Following the general principles for dealing with agricultural injuries as stated previously in this chapter will help limit complications and shorten recovery.

CONTRIBUTING FACTORS AND PREVENTION

Working around new mothers is very hazardous. Well-designed animal handling facilities for moving, sorting, and loading animals are extremely important. Any animal pen where humans may also be working should have a secure quick escape. For handling newborn calves on pasture, one farmer's solution for safety was a portable pen that could be set around a calf by a tractor with a front-end loader. The farmer would then climb into the pen on the side away from the cow, tend the calf, and then exit and release the newborn.

Farmstead

The farmstead is the workplace, storage place, workshop, and home. It's a place for recreation and often a playground for farm children. The farmstead is a meeting place for many hazardous exposures and the everyday activity of the whole family. It is a nexus of people and activities, machines, and materials, which can be a source of injury for adults working on the farm, a dangerous playground for youngsters, and an environment of tripping or falling hazards for elders. Among the hazards are those associated with maintenance work, hand tools, and machines in the workshop; falling loads such as tires and stacked bales; entanglements in moving parts of conveying equipment; falls on the same level or from heights; entrapment in a confined space; and electric current.

Grain and Silage Storage Injury Incidents

Chapter 3, "Agricultural Respiratory Diseases," describes the health hazards associated with grain and silage storage structures. This section will review only the issue of grain engulfment.

DESCRIPTIVE INFORMATION

Grain bins, trailers, and large wagons are places of risk from grain engulfment and suffocation. Augers and other material handling devices are designed to move large volumes of material quickly and efficiently. Flowing grain hazards result when grain is being removed from the bottom of a bin. A void space shaped like an inverted cone of flowing grain moves downward very rapidly into the intake at the bottom of the bin. A person standing in grain above an auger can sink and be buried in a matter of seconds, unable to escape from the flowing grain, resulting in suffocation if not soon found and rescued.

There are four common scenarios to grain engulfment. The first and most common engulfment involves a worker who enters a bin with the bottom unloading equipment running and is pulled under the surface, much like with quicksand. The second is an avalanche of caked grain that suddenly releases sideways and covers a person when he/she is in the silo, bin, or storage structure. A third involves falling through a crust or bridge of caked grain that spans over a cavity in the grain below. A fairly recent fourth engulfment hazard is operating a high-capacity grain vacuum and sinking into the void in the grain created as the nozzle removes grain from underfoot and the surrounding material flows inward.

INJURY OVERVIEW

Injuries are fairly rare in grain engulfment. Either victims are suffocated with a blocked airway in a matter of minutes or they are alive but trapped, unable to free themselves. There have been rare instances where an individual has been buried for up to an hour and has survived.

Illustrative Scenarios

GRAIN ENGULFMENT

A 16-year-old boy had the task of moving grain from the 20,000-bushel grain bin to a smaller holding bin that could more easily be used for grinding daily hog feed. He started the auger but no grain came out. He tried opening and closing the slide door covering the grain well where the unloading auger intakes grain. He also tried starting and stopping the auger, but nothing worked. He assumed that some spoiled grain had blocked the slide door deep within the center of the bin. He shut off the auger and took a 7-meter-long rod into the top of the bin. Poking the rod down through the grain, he hoped, would dislodge the suspected blockage. After several attempts without success, he decided to use the same method, but left the auger running so he could tell when the blockage broke free. After poking the rod down a few more times, the grain suddenly started flowing, covering him to well above his knees. He struggled to leave the rapidly moving grain, but could not pull his legs out of the grain. In seconds he was buried.

FIRST ON SCENE, FIRST RESPONDER, AND TRANSPORT

There are some special considerations for first response and rescue of flowing grain engulfment victims. If they are buried as deep as or deeper than their waist, the pressure is too high to pull them out. Pulling the grain away from the person is difficult because it will just flow back into the void from where it was removed. Portable structures to surround the victim have been employed with success, but they can be too large to get through openings for access to the victim. If the person is completely engulfed, the victim must be removed as quickly as possible. Cutting triangular-shaped flaps on opposing sides of the bin and allowing the grain to flow away from the victim is the most common recommendation to rescue a person in this situation.

Victims may become unconscious if they have not been able to breathe for four minutes or more. A very pale or bluish complexion, especially around the lips, is evidence that respiration has stopped. Listen for a heartbeat, feel for a pulse at the neck or wrist, and look for the chest to move. CPR must be started immediately if there is no pulse. Grain or some other obstruction may be blocking the airway if the victim's chest does not rise (Homeopathic Online Education ACA 2002).

EMERGENCY AND POST-CRISIS TREATMENT

Expect an airway that is blocked with grain. Keep in mind that the temperature of the grain can be quite cold, even during late spring or early summer. Hypothermia may be a consideration in an unresponsive patient.

RECOVERY CARE, MANAGEMENT, AND REHABILITATION

If victims survive the initial event, they seldom have long-lasting problems. The exception would be if the person had insufficient oxygen for a period of time resulting in cerebral hypoxia. Recovery and long-term management depend on the degree of cerebral damage that might have occurred.

A helper was moving a 20-foot-long grain auger into position to be connected to the tractor so it could be relocated to fill another grain bin. As he moved it to the side for better alignment, the upper end of the elevator contacted the electrical power distribution wire supplying electricity to the bin's aeration fan. Current immediately flowed through the elevator and helper to the ground.

CONTRIBUTING FACTORS AND PREVENTION
There was a small amount of material that was spoiled and plugged the discharge area. The big mistake was entering the grain bin with the unloading auger running, one of many factors contributing to the potential for engulfment (Kingman and others 2004). The switch to activate the auger should be locked and tagged whenever a person is inside so that it cannot be turned on. When entering a bin, rescuers should wear a safety harness that limits the depth to which they could be buried.

Other Farmstead Injury Incidents

DESCRIPTIVE INFORMATION
From materials stored unsafely, to structural problems with buildings, to low electrical wires and exposed liquid petroleum (LP) gas lines, farmsteads provide a vast array of potential injury situations.

INJURY OVERVIEW
Injuries can include electrical current from contact between an overhead electrical power line and a grain auger, long metal ladder, or section of irrigation pipe. Crushing and pinching injuries result from items falling on an individual. Broken bones occur in slips, trips, and falls from a different level or on the same level. Explosions or fires happen with spilled fuels, flammable materials improperly stored, and inadequate repair procedures.

Illustrative Scenarios

CAUGHT UNDER FALLING DUAL WHEEL
A 5-year-old boy was extremely active and loved to play outside. In his mind, the farmstead was a perfect place to imagine being an explorer, with all of the equipment he could crawl in and around. Dad had recently removed the rear dual wheels from the tractor and leaned them against a tree until he could take the time to put them away. The boy spotted them and immediately started to crawl up the side of the tire. These duals were filled with a calcium chloride solution to make them heavier and give added traction to the tractor. As he got almost to the top, the tire shifted against the tree trunk and fell onto him.

FIRST ON SCENE, FIRST RESPONDER, AND TRANSPORT
Response to an incident can involve removing a heavy item that has fallen on the patient. Keep in mind that severe internal injuries could have occurred and removal of the item may result in severe internal bleeding. The necessary equipment should be obtained to lift the object rather than roll it off the victim. Be prepared to lift several hundred pounds and to support it so it does not fall back onto the patient. An inflatable lift bag may be the best choice for this procedure. A long board and cervical stabilization are mandatory.

The amount of electric current, length of time in contact, area of contact, and the path it takes through the body will determine the type of injury. A current of 50–100 milliamperes causes sustained contractions (Edlich and Farinholt 2002). There may be more than one entry and exit. Cardiopulmonary arrest and ventricular fibrillation or asystole may occur (Chasmar 1998). Extrication is safe only after power has been shut off or the victim is thrown clear, which itself can lead to traumatic injuries such as fracture or brain hemorrhage.

EMERGENCY AND POST-CRISIS TREATMENT
Expect long bone and rib fractures coupled with internal injuries from a crushing or falling load incident.

Electrical burns are major and immediate transfer to a burn center is recommended. An EKG is important in the initial evaluation of electric shock victims. X-rays can rule out fractures and dislocations. Hypovolemia and myoglobinuric renal failure can occur. Veins often become thrombosed causing massive edema (Chasmar 1998).

RECOVERY CARE, MANAGEMENT, AND
REHABILITATION

There are no special considerations in recovery through rehabilitation for crushing injuries beyond the basic principles stated earlier.

Acute and delayed spinal cord injury can follow electric current contact for days to years after the injury. Severe potassium deficiency may follow a high-voltage injury (Edlich and Farinholt 2002).

CONTRIBUTING FACTORS AND PREVENTION

Items that are supported by leaning them against a wall need to be chained or otherwise securely fastened to prevent them from tipping over.

Overhead electric power distribution lines in the vicinity of storage facilities are a contributing factor. Placing these lines underground, especially near storage facilities, and use of ground fault interruptor circuits are important preventive measures.

Environment

Farmers spend a good portion of their time working outdoors, regardless of the weather. Because of that, they are exposed to heat and cold and all the physical hazards associated with temperature extremes.

DESCRIPTIVE INFORMATION

For 8–16 hours a day, farmers' and ranchers' work must go on in spite of bad weather. Tending to livestock in cold, wet weather can be especially hazardous. The associated mud and ice can create conditions that make a slip or fall more likely.

INJURY OVERVIEW

Exposure to heat can create problems from sunburn to heatstroke. Cold exposures result in frostbite and hypothermia. Slips and falls due to mud, ice, or other causes can result in broken bones.

Illustrative Scenarios

SLIPS AND FALLS

A 70-year-old farmer was filling a large steel grain bin early in the morning during fall harvest. It had been cold enough to frost that night, but the sun was out and the frost had started to melt, leaving the ground a bit muddy. He climbed the ladder to the top of the bin and crawled up the steep slope of the bin roof to the port in the center where the auger emptied the grain into the bin. After looking into the bin, he turned around to go back down, slipped and slid down the roof, falling to the ground 7 meters (23 feet) below. Following about a 2-hour delay, he was found by his wife.

FIRST ON SCENE, FIRST RESPONDER, AND
TRANSPORT

Since the patient is not trapped, immediate access, care, and treatment is possible.

EMERGENCY AND POST-CRISIS TREATMENT

Expect long bone, rib, pelvic, and spinal fractures along with head and internal injuries.

RECOVERY CARE, MANAGEMENT, AND
REHABILITATION

There is nothing particularly unusual regarding this type of agricultural injury from a fall. Following the general guidelines in managing agricultural injuries presented earlier in this chapter is appropriate.

CONTRIBUTING FACTORS AND PREVENTION

Contributing factors in this scenario were the muddy boots and slick bin roof from melting frost. Grain bins are inherently dangerous to check in this fashion because they are tall and their roofs are steep. Newer grain bins may have ladders or roof cleats to reduce the risk.

SUMMARY

Data on deaths and nonfatal injuries help describe the array of agents, events, and bodily harm experienced by persons engaged in agricultural activities, especially for Western industrialized nations. Factors inherent in production agriculture described in Chapter 1 and in this chapter make agricultural injury prevention complex. Several studies identifying other specific risk factors are listed in Table 11.13.

Risk factors such as those in the table provide additional clues to assist in anticipating and preventing acute injuries. Some risk factors, however, (such as being young or old) cannot be controlled. Nonetheless, being aware of these risks prompts recommendations for accommodations in the workplace and work environment, such as assigning tasks appropriate for the age, physical capabilities, and mental faculties of the workers.

Table 11.13. Risk Factors in Production Agriculture

Symptoms of depression	Park and others 2001
Hours per week worked with animals	Ibid; Carruth and others2001
History of a previous injury	McCurdy and Carroll 2000
Hearing impairment	Ibid; Sprince and others 2003
Vision impairment	McCurdy and Carroll 2000
Prescription drug use	McCurdy and Carroll 2000; Xiang and others 1999
Being elderly	Hard and others 1999
Being young (<20 years)	McCurdy and Carroll 2000; MMWR 1998
Being a younger livestock producer (<39 years)	Sprince and others 2003
Being of African descent	McCurdy and Carroll 2000
Being young and of Hispanic descent	McCurdy and Carroll 2000
Back pain	Carruth and others 2001
High debt load	Xiang and others 1999

Tractors and machinery together clearly dominate the fatality picture, standing out among a broad variety of other, relatively smaller, individual agents. Animals and machinery each share about equally as dominant agents of nonfatal injury. These nonfatal injury events and the all-too-frequently contaminated, mangled, crushed, multipart trauma that results, challenge health care providers serving rural, remote, and frontier communities.

Persons first on the scene through rehabilitation specialists, persons throughout the chain of health-care professionals, must be informed, connected, and communicating to achieve the best possible outcome for the patient. Injury scenarios provide context to expand on treatment basics and deepen understanding of acute agricultural injuries; their potential consequences, complications, and uniqueness; and the possibilities for prevention. Awareness of the varied, ever-present hazards and opportunities for injury in agriculture are fundamental to the safety of farmers, ranchers, farm workers, and their families. Similar awareness coupled with the expertise of health-care professionals should enhance both patient care and participation in injury prevention. Health professionals can be more aware of their agricultural patients, can provide more information, and can participate in prevention programs.

REFERENCES

Agricultural Safety Program. 2005. Agricultural Tailgate Safety Training. Training Module: First on the Scene. Ohio State University Extension.

Baker D, Lee R. 1995. Responding to Farm Accidents. Columbia: University of Missouri Extension.

Brison R, Pickett W. 2003. Agricultural Injuries in Canada for 1990–2003. Kingston, Ontario, Canada: The Canadian Agricultural Injury Surveillance Program.

Carruth AK, Skarke L, Moffett B, Prestholdt C. 2001. Women in agriculture: Risk and injury experiences on family farms. Journal of the American Medical Womens Association 56(1):15–18.

CDC. 1998. Youth Agricultural Work-Related Injuries Treated in Emergency Departments—United States, October 1995—September 1997. Morbidity and Mortality Weekly Report. CDC, U.S. Department of Health and Human Services 47(35):733–737.

———. 2003. Acute Trauma (definition).

Chasmar L. 1998. Electrical Burns in Farmers. Principles of Health and Safety in Agriculture. p 377–378.

Chin J. 2000. Control of communicable diseases manual. p 493–495.

Donham K, Yeggy J, Dague R. 1988. Production rates of toxic gases from liquid swine manure: Health implications for workers and animals in swine buildings. Biol Wastes 24:161–173.

Doty B. 1998. Proposed Document Revision and/or Development. Core Curriculum Guidelines: Rural Practice Training for Family Physicians. Rural Health Committee AAFP.

Edlich R, Farinholt H. 2002. Electrical Burns. In: Orgill D, Talavera F, Stadelmann W, Slenkovich N, Vistnes L, editors. eMedicine.

European Union. 2004. Key Facts & Figures About the European Union. European Union.

EUROSTAT, Statistical Office of the European Communities. 2004. http://epp.eurostat.cec.eu.int/portal/

page?_pageid=1090,30070682,1090_33076576&_d ad=portal&_schema=PORTAL

Field Operations Directorate. 2001–2002. Fatal Injuries in Farming, Forestry and Horticulture. In: Executive HS, editor. National Agricultural Centre.

Fragar L, Franklin R. 2000. The Health and Safety of Australia's Farming Community, a Report of the National Farm Injury Data Centre for the Farm Safety Joint Research Venture. Sydney, Australia: National Farm Injury Data Centre.

Fretz P. 1989. Injuries from farm animals. In: Dosman JA, Crockcroft DW, editors. Principles of Health and Safety in Agriculture. Boca Raton, FL: CRC Press, p 365–366.

Friesen R, Ekong E. 1988. Spinal Injuries Due to Front-End Bale Loaders. In: NASD, editor. Canadian Medical Association Journal: NASD, National Ag Safety Database. p 43–46.

Hard DL, Meyers JR, Snyder KA, Casini VJ, Morton LL, Cianfrocco R, Fields J. 1999. Identifying work-related fatalities in the agricultural production sector using two national occupational fatality surveillance systems, 1990–1995. Journal of Agricultural Safety and Health 5(2):155–169.

Hensinger R, Rineberg B, Morrey B, Eilert R, Hogshead H, McCollough N, McGinty J, Sarmiento A, Strickland J, Craig E and others. 1993. Rural Rescue and Emergency Care. Worsing R, editor. Rosemont, IL: American Academy of Orthopaedic Surgeons.

Homeopathic Online Education ACA. 2002. Asphyxia: Cessation of Respiration. The Homeopathic First Responder.

Horsburgh S, Feyer A-M, Langley J. 2001. Fatal Work Related Injuries in Agricultural Production and Services to Agricultural Sectors of New Zealand, 1985–94. Occupational Environmental Medicine 58:489–495.

Industries AIVO. 2003. Farm Injury Newsletter. Issue 16 ed: Australian Centre for Agricultural Health and Safety.

International Labour Office. 2000. 88th Session, 30 May–15 June 2000. Occupational Accidents in Agriculture: Accidents with Tractors and Agricultural Machinery. Report VI (1) Safety and Health in Agriculture: Sixth item on the agenda. Geneva. http://www.ilo.org/public/english/standards/relm/ilc/ilc88/rep-vi-1.htm.

ISU Extension. Rev 1999. Farm emergency and first aid kits. Safe Farm: Promoting Agricultural Health and Safety, ISU Extension. PM-1563k.

Johnston G. 1998. High-Pressure Injection Injuries in Farmers. Principles of Health and Safety in Agriculture. p 375–376.

Kahn M. 1998. Spinal Cord Injuries in Agricultural Workers. Principles of Health and Safety in Agriculture. p 373–374.

Kingman D, Spaulding A, Field W. 2004. Predicting the Potential of Engulfment Using an On-farm Grain Storage Hazard Assessment Tool. Journal of Agricultural Safety and Health 10(4):237–245.

Kostuik J. 1989. Amputations in Farming. Principles of Health and Safety in Agriculture. p 355–356.

Leigh J, McCurdy S, Schenker M. 2001. Costs of Occupational Injuries in Agriculture. Public Health Reports 116(May–June):235–248.

Letts R. 1989. Farm machinery accidents in children. In: Dosman JA, Crockcroft DW, editors. Principles of Health and Safety in Agriculture. Boca Raton, FL: CRC Press, p 357–361.

Little D. 1998. The ABC's of First Aid. Homoeopathic Online Education, A Cyberspace Academy.

Lycholat T. 2003. Traumatic Injuries. BBC Health.

McCurdy S, Carroll D. 2000. Agricultural Injury. American Journal of Industrial Medicine 38:463–480.

McNamara J, Laffey F. 2003. Managing Safety by Contractors; 27 February 2003; Piltown, Co. Kilkenny. Teagasc.

McNamara J, Ruane D, Connolly L, Reidy K, Good A. 2003. A study of the impact of disability in farm households on the farm business in Ireland, April 8–12, 2003; Raleigh, North Carolina.

Mizani M, Weber B. 2003. High-pressure injection injury of the hand; The potential for disastrous results. Postgraduate Medicine online.

Murphy D. 1992. Farm and Agricultural Injury Statistics. Safety and Health for Production Agriculture. 1st ed. St. Joseph, MI: ASAE. p 43–70.

——. 2003. Looking Beneath the Surface of Agricultural Safety and Health. St. Joseph, MI: ASAE. 103 p.

Myers J. 1997. Injuries Among Farm Workers in the United States, 1993. In: Services USDoHaH, editor: NIOSH, US DHHS, CDC. p 142.

——. 1998. Injuries Among Farm Workers in the United States, 1994. Cincinnati, OH: NIOSH, USDHHS. Report nr NIOSH Publication No. 98–153.

——. 2001. Injuries Among Farm Workers in the United States, 1995. Cincinnati, OH: NIOSH, USDHHS. Report nr NIOSH Publication No. 2001-153.

Myers J, Hendricks K. 2001. Injuries Among Youth on Farms in the United States, 1998 edition. NIOSH. p 5–11.

NASD. 2004. Agricultural Injury. 1996 W Number 1 ed: Centers for Disease Control and Prevention. Marshfield Medical Research and Education Center Fact Sheet.

National Academies Press. 1998. Protecting Youth at Work: Health, Safety, and Development of Working Children and Adolescents in the United States (1998).

NIOSH. 2003. Preventing Injuries and Deaths from Skid Steer Loaders. DHHS (NIOSH).

NSC. 1993. Accident Facts, 1993 edition. Itasca, Il: National Safety Council.

——. 1994. Accident Facts, 1994 edition. Itasca, Il: National Safety Council.

——. 1995. Accident Facts, 1995 edition. Itasca, Il: National Safety Council.

——. 1996. Accident Facts, 1996 edition. Itasca, Il: National Safety Council.

——. 1997. Accident Facts, 1997 edition. Itasca, Il: National Safety Council.

——. 1998. Accident Facts, 1998 edition. Itasca, Il: National Safety Council.

——. 1999. Injury Facts, 1999 edition. Itasca, Il: National Safety Council.

——. 2000. Injury Facts, 2000 edition. Itasca, Il: National Safety Council.

——. 2001. Injury Facts, 2001 edition. Itasca, Il: National Safety Council.

——. 2002. Injury Facts, 2002 edition. Itasca, Il: National Safety Council.

——. 2003. Injury Facts, 2003 edition. Itasca, Il: National Safety Council.

——. 2004. Injury Facts, 2004 edition. Itasca, Il: National Safety Council.

Oehme FW. 2003. How do agricultural workers get illness from their workplace? http://www.vet.ksu.edu/links/agromed/oehmeafternoon.htm.

Park H, Sprince NL, Jensen C, Whitten P, Zwerling C. 2001. Health risk factors and occupation among Iowa workers. American Journal of Preventive Medicine 21(3):203–208.

Pope S. 2000. Tractor ROPS. Rollovers, Rebates & Retrofits. Queensland Government, Department of Industrial Relations.

Rautiainen R. 2002. Injuries and Occupational Diseases in Agriculture in Finland: Cost, Length of Disability, and Preventive Effect of a No-claims Bonus. Iowa City: University of Iowa.

Reed D, Rayens M, Browning S, McCulloch J, Garkovich L. 2001. Sustained Work Indicators of Older Farmers. Lexington, KY: University of Kentucky Chandler Medical Center.

Rissanen P, Taattola K. 2003. Fatal injuries in Finnish agriculture, 1988–2000. Journal of Agricultural Safety and Health 9(4):319–326.

Rural Family Medicine. 2003a. Augers. Rural Family Medicine.

——. 2003b. Combine. Rural Family Medicine.

——. 2003c. PTO Entanglement. http://www.ruralfamilymedicine.org/clinical%20topics/pto.htm.

——. 2003d. Tractor Overturns. Society of Teachers of Family Medicine Rural Interest Group.

Schnitz G, Fischer T. 1996. Farm Injuries. The Indian Hand Center.

Society of Teachers of Family Medicine Rural Interest Group. 2003. PTO Entanglement. Rural Family Medicine.

Sprince NL, Zwerling C, Lynch CF, Whitten PS, Thu K, Logsden-Sackett N, Burmeister LF, Sandler DP, Alavanja MC. 2003. Risk factors for agricultural injury: A case-control analysis of Iowa farmers in the Agricultural Health Study. Journal of Agricultural Safety and Health 9(1):5–18.

St. John Ambulance. 1990. Standard First Aid Safety Oriented Modular Course Workbook. Ottawa, Canada: St. John Priory of Canada Properties.

Takala J. 1999. Global estimates of fatal occupational accidents. Epidemiology 10(5):640–646.

Traumatic Brain Injury Resource. 2003. What Is TBI? http://www.traumatic-brain-injury-resource.com/traumatic_brain_injury/whatis.html.

Tyson B. 1994. Rescuing Farm Accident Victims. Athens, GA: The University of Georgia College of Agricultural & Environmental Sciences/Cooperative Extension Service.

U.S. Department of Labor BLS. 2004. Census of Fatal Occupational Injuries (1992–2002).

U.S. Department of Transportation. 2004. Fatal Accident Reporting System (FARS). National Center for Statistics & Analysis.

Vaughn G. 2003. Hand Injuries, High-Pressure. In: Danzl D, Talavera F, Legome E, Halamka J, Plantz S, editors: eMedicine.

Xiang H, Stallones L, Chiu Y. 1999. Nonfatal agricultural injuries among Colorado older male farmers. Journal of Aging Health 11(1):65–78.

12
Veterinary Biological and Therapeutic Occupational Hazards

Kelley J. Donham and Anders Thelin

INTRODUCTION

The modern age of pharmaceutical use in livestock production began in the late 1940s following the discovery of penicillin. Pharmaceuticals that are commonly used today in modern livestock production include antibiotics, biologicals (immunizing products), and a broad array of therapeutics (Bennish 1999). They are used for treatment of disease, as disease preventives, and as growth promotants. A number of these agents are potentially hazardous to the user via direct exposure through accidental needle sticks or indirect exposure through the air or water. Skin contact or aerosol exposure to antibiotics may occur, as they are often mixed into the animal's drinking water or feed as a disease treatment or prevention, or for growth promotion. As the medication is mixed in the water or ground in with the feed and given to animals, the handler may have skin contact with the medication, or an aerosol may be produced that can be inhaled. Antibiotics and other pharmaceuticals also may be administered by hypodermic injection. When administering injections to hundreds of animals, many that are much larger and stronger than the operator, there are multiple opportunities for inadvertent needle sticks to the handler or assistant.

There are similarities between the regulations concerning pharmaceuticals for human use and veterinary pharmaceuticals. The U.S. Food and Drug Administration (FDA) and the U.S. Department of Agriculture (USDA) are responsible for assuring the efficacy and safety of all licensed drugs in the U.S. There are comparable agencies in other industrialized countries. However, there is one important difference. Most pharmaceuticals for use in humans require a prescription from a licensed physician. Veterinary pharmaceuticals are more widely available to livestock producers. Although the degree of availability varies among industrialized countries, many pharmaceuticals for animals do not require a written prescription by a licensed veterinarian. This policy results in the use of many products under little or no professional control, increasing the risk for potentially harmful drugs to be inappropriately used. However, there are some drugs (those not approved for use in food animals or those used in government disease control programs) that must be either administered or prescribed by veterinarians. These products are generally not available over the counter. For example, the antibiotic chloramphenicol (and most antibiotics that have an important use in human health) is not approved for food animal use, and therefore, not available over the counter. Furthermore, several veterinary immunization products that are part of a governmentally regulated animal disease or public health control program, such as brucellosis and rabies vaccines, are not available over the counter. The European community has more strict regulations on veterinary pharmaceuticals than does the U.S. (Koschorreck and others 2002).

Suffice it to say, most health professionals would be very surprised when visiting a local farm, farm store, or animal drug supply firm's catalogue or website to see the broad array of pharmaceuticals available to the lay public, some of which can be harmful if a person becomes accidentally exposed. In addition to the risk of unintended exposures to veterinary pharmaceuticals, there is evidence of intentional usage of these products by livestock handlers for self treatment (particularly antibiotics and pain medications) (Erramouspe and Carlson 2002). There is also evidence that veterinary and human therapeutic agents excreted in fecal material applied to land may have a detrimental effect on plant growth (Jjemba 2002; Jones 1996) and enhance development of antibiotic-resistant organisms.

This chapter deals with several classes of pharmaceuticals that may cause harm to humans who may be unintentionally exposed. The classes of pharmaceuticals involved include veterinary biologicals (immunization products), antibiotics and hormones used as growth promotants, and pharmaceutical hormones used in veterinary obstetrical procedures. The population at risk of exposure includes livestock producers, farm workers, veterinarians, and veterinary assistants.

VETERINARY BIOLOGICALS

Definitions

Biologicals are products developed from a biological process for the purpose of enhancing immunity to infectious diseases. There are several categories of biologicals, and it is important to understand their differences regarding potential health hazards for humans should they be accidentally exposed.

Antisera are products prepared by hyperimmunizing an animal, such as a horse or cow, with a killed or attenuated infectious agent. The blood serum, now highly concentrated in antibodies, is harvested, refined, and packaged as a product to inject into other animals to provide temporary, passive, but immediate immunity. There are also newer processes, such as monoclonal procedures used in antisera production, that result in a more refined and specific product. Unintentional injections with these products are of low health risk, but may cause local inflammation (foreign protein reaction). However, there is no other anticipated problem except for trauma from the needle stick or infection brought subcutaneous from organisms on the skin or needle.

Bacterins are products made from killed bacteria. These products are designed to afford active immunity and usually include substances to enhance their immunogenicity (adjuvants).

Vaccines are products made from either a live bacterial or a viral agent that either has been altered to reduce its virulence (attenuated) or has been killed. It is used to afford active immunity.

Adjuvants are products that are incorporated with bacterins or vaccines to enhance the immunogenicity of the product. Adjuvants include products such as certain oils, components of mycobacterium organisms, or aluminum salts. Adjuvants work by delaying the adsorption from the site of inoculation and increasing the local inflammatory process and thus stimulating the immune system over a longer period of time.

Toxoids are inactivated toxins (e.g., tetanus toxoid) that create active immunity.

Antitoxins are a type of antisera produced to toxins in another animal species or through an in vitro biological process. Antitoxins (e.g., tetanus antitoxin) produce passive immunity. Remembering the definition of the types of biologicals will help the health professional obtain important details of the person's occupational history, which may lead to the cause(s) or risk of illness that may be involved with an injury from an accidental needle stick. The important facts to remember are as follows:

1. Any needle stick may cause an infection because the needle is likely contaminated.
2. Inoculation with antisera will not cause an infection from the product itself, but can cause an inflammation or allergic reaction.
3. Inoculation with a bacterin will not cause an infection from the product itself, but it can cause inflammation (especially since many bacterins contain inflammatory adjuvants).
4. Inoculation with a vaccine may cause an infection in the accidental host because vaccines usually contain live attenuated organisms. Several such veterinary products in this category are listed in Table 12.1.
5. Inoculation with a toxoid or antitoxin will not cause an infection from the product itself, but may cause severe inflammation.

How Are People Injured with Biologicals?

Needle sticks are common in the everyday world of livestock husbandry (Geller 1990; McGreigan 1994; Patterson and others 1988; Plumb 2002;

Table 12.1.   Veterinary Biologicals and Health Risks from Accidental Needle Sticks

Immunization Product	Live Attenuated	Killed Product	Health Risk from Injection of Product
E. Coli bacterin		X	Inflammation, but not infection
Contagious ecthyma	X		Infection, similar to the wild virus
Johne's disease		X	Inflammation, but not infection
Erysipelas	X	X	
	Some products are live agents.	Most are killed products.	The killed products cause inflammation. The live products may cause a local and systemic infection as well as inflammation
Rabies		X	No known infection or inflammation as a result of current products

Samanta and others 1990; Wiggins and others 1989; Wilkins and Bowman 1997). There is no research or surveillance to provide rates, but most livestock producers, workers in livestock production, veterinarians, and veterinary assistants will admit to being accidentally stuck on several occasions. Accidental needle sticks may cause injury in one or more of the following ways.

INFECTION FROM A CONTAMINATED NEEDLE
Needles used in livestock production are often reused many times. Rarely are the needles sterilized between inoculations when injecting many animals in rapid succession. Therefore, it can be assumed that these needles are contaminated with fecal organisms, skin surface organisms, or even infectious organisms. Furthermore, these needles are usually large bore (e.g., 16–14 gauge), dull, and barbed. They can create a more severe traumatic injury when entering and withdrawing from the injection site.

INFECTION FROM THE PRODUCT INJECTED INTO THE ACCIDENTAL RECIPIENT
Some vaccines contain organisms that may be attenuated for the target species, but are not necessarily attenuated for the accidental host. Several such products are actually infectious for humans. A list of these products appears in Table 12.1.

INFLAMMATION (OR OTHER ADVERSE EFFECTS) FROM THE PRODUCT INJECTED INTO THE ACCIDENTAL HOST
Many veterinary products may not be infectious, but the adjuvants within the product may be highly inflammatory. For example, peanut oil or bacterial cell products of mycobacterium species may be used as adjuvants and both of these substances may be highly inflammatory when injected into a person (Jones 1996). Other veterinary products, such as hormones, sedatives, or antibiotics, may be accidentally injected into workers, causing severe reactions. These latter substances are discussed in later sections of this chapter ("Hormones Used in Veterinary Obstetrics" and "Other Veterinary Pharmaceuticals").

A HYPERIMMUNE REACTION TO A PRODUCT FOR WHICH THE ACCIDENTAL HOST HAS BEEN PREVIOUSLY EXPOSED
In the event that the patient has previously developed an immune response to an infectious agent and then is subsequently injected with a biological for that agent, this person may develop a local or systemic anemnestic response to that product.

    The following published case study illustrates how more than one type of reaction can be involved in an accidental inoculation (Spink 1957). A 60-year-old veterinarian presented with a non-pitting swelling of the left hand. The swelling was severe, showing possible risk of entrapment syndrome. He had been vaccinating cattle earlier in the day and had accidentally stuck himself with the needle used to inoculate the cattle. The swelling and pain began within minutes after the needle stick. Several hours later, the patient developed generalized influenza-like symptoms consisting of fever, muscle aches and pains, headache, malaise, and weakness. He had been vaccinating a group of dairy heifers for brucellosis.

The physician was uncertain as to how to treat this patient. He thought he had a localized infection caused by the accidental inoculation, but could not explain the systemic symptoms and unusually severe swelling in the hand.

The product used to vaccinate these cattle was called the *Brucella abortus* Strain 19 vaccine. This is a live product (a vaccine). This strain does not cause clinical signs of infection in cattle, but it does in humans; in fact, it causes a disease very similar to a naturally acquired case of brucellosis. The explanation of the systemic symptoms is that the patient had vaccine-induced brucellosis infection. The excessive swelling of the hand can be explained by a localized acute hyperimmune (anamnestic) response to the vaccine. Although brucellosis used to be extremely common among livestock-exposed populations and those drinking unpasteurized milk, the governmentally sponsored eradication program has significantly reduced the disease in cattle and swine, and thus in humans. However, brucellosis was so common in the past that most livestock producers and veterinarians (born before the mid-1950s) have been exposed to brucellosis. The patient in this case presentation had previously been exposed to the *Brucella* antigen and developed an anemnestic response at the localized site of the needle stick.

Proper treatment of this patient had to include consideration of the *Brucella* infection (1 gm each of penicillin or doxicycline and streptomycin for 3 weeks). Furthermore, the patient's hyperimmune reaction must be treated. Intravenous manitol and an initial injection of prednisone (with a tapered oral dose) were given to reduce the swelling and the risk of entrapment syndrome in the hand. The patient recovered uneventfully over a course of 3 weeks.

A new brucellosis vaccine is now available called RB51. This product, as with strain 19, is used to vaccinate heifers for brucellosis in regions not yet free of brucellosis. RB51, like strain 19, will infect humans. RB51 is susceptible to a range of antibiotics, and treatment with doxicillin and/or streptomycin is appropriate. However, RB51 is resistant to rifampin and penicillin.

This case is illustrative of the multiple etiologies of a needle stick injury in livestock production or veterinary practice. Although brucellosis was the primary vaccine known to cause an anemnestic response, other vaccines may produce a similar, dual response, or just an infection. Examples of the latter include contagious ecthyma of sheep and goats, Newcastle disease of poultry, and erysipelas vaccine for swine. Contagious ecthyma vaccine can cause a localized skin infection very similar to an actual field-acquired infection. Newcastle vaccine is administered via the aerosol route or in the water to large poultry flocks. If the person is not properly protected, he or she may develop a severe conjunctivitis and systemic influenza-like symptoms (Geller 1990). Contagious ecthyma and Newcastle vaccine infections are self-limiting illnesses, and complete recovery usually occurs over a course of 2–3 weeks. As these diseases are caused by viruses, there is no specific treatment; however, treatment of symptoms may be necessary. Regarding biological products for erysipelas of swine (caused by *Erysipelas rhusiopathiae*), some are bacterins and a few are vaccines (live products). It is the vaccines, of course, that are of greater concern from accidental inoculation. They may cause a localized and systemic disease similar to field-acquired erysipeloid (*E. rhusiopathiae* infection). Treatment with specific antibiotics such as streptomycin may be necessary to protect the accidentally inoculated patient.

Most of the other products are killed products (bacterins), and injuries from accidental inoculations are due to the inflammatory nature of the product, plus any environmental organisms that may have been introduced from a dirty needle. Johne's disease bacterin is a particularly inflammatory product when accidentally inoculated (LaVenture and others 1988). This product can be administered only by a licensed veterinarian as there is a specific governmental control program for this disease. Johne's is a mycobacterium infection of cattle that produces a disease very similar to Crohn's disease of humans. *E. coli* bacterins are also inflammatory substances. Treatment of these agents should focus on the inflammatory response and what infection the dirty needle may have introduced. Rarely, if ever, would the inflammatory condition caused by these latter agents risk an entrapment syndrome.

ANTIBIOTICS

Introduction

In recent years, there has been a heightened concern about the increase of antibiotic-resistant infections in the human and animal populations. The use of antibiotics in human medicine, veterinary medicine, and in agricultural production has been increasingly scrutinized (Bennish 1999; Dunlop and others 1998; Gersema and Helling 1986; Johnston 2001; Khachatourians 1998; Laval 2000; Marwick 1999; Randall and Woodward 2002; Sax 2002; Schmidt 2002; Schwarz and Chaslus-Dancla 2001; Teale 2002; Teuber 2001; Torrence 2001; VandenBogaard 2001; VandenBogaard and Stobberingh 1999; Wierup 2000). The greatest scrutiny has been directed toward antibiotic use in livestock production. Over half of all antibiotics produced in the U.S. and in Europe are used in livestock production (Dunlop and others 1998; Schmidt 2002). Furthermore, 90% of the antibiotics in U.S. livestock production are used as low-level feed additives for growth promotion or disease prophylaxis rather than disease treatment. Since the late 1960s, there has been a concern about environmental and public health hazards related to such massive use of antibiotics in livestock production. Concern was primarily focused on the worldwide development of antibiotic-resistant organisms (Cornaglia 1998; Dromigny and others 2002; Hart and others 1998; Skold 2001; Sundluf and Cooper 1996).

Although antibiotics have been used in livestock production for disease treatment since the 1940s, it was not until the early 1960s that antibiotics began being added to animal feeds as growth promotion or prophylaxis (Laval 2000). This came about following research by animal scientists who discovered that low-level feeding of antibiotics to cattle, swine, and chickens not only increased their growth rate by 3–5%, but they also gained weight while consuming less (increased feed efficiency) (Khachatourians 1998). Since the 1960s, market animals, particularly pigs (86% of young pigs and 30% of finishing pigs), chickens, and veal calves have commonly been fed antibiotics throughout a significant portion of their life cycle (Council for Agricultural Science and Technology 1981). In regard to poultry, egg production and egg hatchability was also found to be improved in chickens fed antibiotics (Council for Agricultural Science and Technology 1981). There is less evidence for use of low-level antibiotics for disease prevention, except for certain diseases (e.g., organic arsenicals and sulfa drugs for prevention of bloody scours in swine and

coccidiosis in poultry, respectively) (Corwin 1997).

The Council for Agricultural Science and Technology (1981), a science and policy analysis organization, estimated that using antibiotics in feed resulted in a savings of $802 million per year in all U.S. livestock and poultry production costs. Additionally, they estimate antibiotic use in livestock spares 1.3 million acres of corn in the U.S. that would otherwise be used as feed. There are beliefs on the part of some scientists, some livestock producers, and some of the general public that antibiotics are necessary to produce livestock in the confined, intensive production systems that have emerged in Western agriculture since the 1970s. Certainly there are challenges to these claims. Others claim that focusing on excellent livestock management and sanitation will achieve the same gains as the use of antibiotics without the accompanying risks (Schmidt 2002; VandenBogaard and Stobberingh 1999; Wierup 2000).

It is clear that under controlled experimental conditions, antibiotics are growth promotants. How this works is not clear, but there are at least three theories: 1) a direct effect on the gastrointestinal mucosa that enhances its absorptive efficiency, 2) a change in the gastrointestinal flora that enhances digestive processes, and 3) a reduction of pathogens that might be subclinically affecting the general health status of the animal (Schmidt 2002).

An additional use of antibiotics in agriculture includes horticultural applications. For example, aminoglycosides are sprayed on some fruit crops to reduce the surface bacterial growth (Teale 2002). However, the total use in this application is much less than in livestock production.

What Are the Potential Health Risks of Antibiotic Use in Livestock?

The potential health consequences of antibiotic use in agriculture are directly related to the increased risk for adverse allergic reactions to antibiotics and the development of antibiotic-resistant infections. These risks can be broken down into occupational and community or public health risks.

ALLERGIC REACTIONS—OCCUPATIONAL HEALTH RISKS
Some of the common antibiotics used as growth promotants in animal feeds include penicillin,

Table 12.2. Antibiotics Used in Livestock Production and Human Health

Antibiotic Class	Animal Species	Used for Treatment	Used for Prevention	Used for Growth Promotion	Human Use?	Bacterial Resistance
Aminoglycosides (gentamycin, neomycin, streptomycin)	Cattle, poultry, sheep, swine	Yes	No	No	Yes	Yes
Beta-Lactams (penicillins, • Amoxicillin • Ampicillin cephalosporins, third generation)	Cattle, poultry, sheep, swine	Yes	No	Yes No	Yes	Yes
Ionophores	Cattle, poultry, sheep	No	Yes	Yes	No	Yes
Macrolides (erythromycin, tilmicosin, tylosin)	Cattle, poultry, swine	Yes	No	Yes	Yes	Yes
Polypeptides (bacitracin)	Poultry, swine	Yes	No	Yes	Yes	Yes
Fluoroquinolones (enrofloxacin)	Cattle, poultry	Yes	Yes	No	Yes	Yes
Sulfonamides	Cattle, poultry, swine	Yes	Yes	Yes	Yes	Yes
Tetracyclines	Cattle, poultry, sheep, swine	Yes	Yes	Yes	Yes	Yes

tetracycline, sulfamethazine, and virginiamycin (a more detailed list is seen in Table 12.2) (Dunlop and others 1998; Khachatourians 1998). Of these, penicillin has been most commonly recognized as a sensitizer, with both dermal (Dal Monte and others 1994) and asthmatic symptoms. Dermal exposure can occur when workers grind, mix, and handle feed that contains antibiotics. Furthermore, aerosols of feed are produced during the mixing and grinding process and whenever the concentrates are handled or moved. Inhalation exposure can occur at any phase in the various processes of preparation, transportation, or feeding. Occupational exposures may occur at the grain elevators and feed manufacturers or on the farm. Cases of allergic dermatitis and asthma or other systemic responses to antibiotic exposure are rare, yet the risk is real for a small percentage of the population.

ALLERGIC REACTIONS—COMMUNITY HEALTH RISKS

There are specified withdrawal times when antibiotics must be withheld from animals to assure that the meat, milk, and eggs are free of residues when the food reaches the market. If a residue did exist in a meat or milk product, and a person sensitized to a certain antibiotic consumes the product, that person may suffer an allergic reaction. However, this would be quite rare. Animal products are randomly checked for residues at the processing plant. Should any residues be found, the carcass, milk, eggs, and other products are removed from the human consumption market. If residues are found in the meat product, the farm of origin is prohibited from selling animals to markets until they test residue free. In the case of milk products, every tanker load of milk is tested for the presence of antibiotics before being unloaded at the processing plant. Should a load test positive, the farm of origin is located and that dairy producer is charged for the entire load of milk that is subsequently disposed, and the farm must test residue free before it can again sell milk. The economic repercussions for today's food animal producers are high for antibiotic residue violations. This process may not remove 100% of the risk, but it does make the risk of public exposure to antibiotics in animal food products remote.

ANTIBIOTIC-RESISTANT ORGANISMS—OVERVIEW

The concern about the use of antibiotics as growth promotants causing an increase in antibiotic resistance has been a growing concern since a highly publicized incident that occurred in the late 1960s. A young farm girl in England became ill with a severe *Salmonella* diarrheal disease with systemic symptoms (Gersema and Helling 1986). Her physicians were hampered in their treatment because they could not find an antibiotic that was effective against the organism, and the little girl died. Public health officials then traced the origins of the infecting organism back to veal calves on the farm. Antibiotics had been extensively used in these calves, and they were harboring resistant *Salmonella*—likely the same organism that caused the little girl's death. This case created a national concern that resulted in the appointment of a special commission to study the situation. The now-famous Swan Commission reported that antibiotics should not be used as low-level growth promotants, and that antibiotics should be administered to livestock only by veterinary prescription. This practice generally spread throughout Western Europe, and now the European Union is in the process of adopting similar policies for all its member nations. However, this has not been the case in the U.S., and the issue is increasingly debated between the medical, public health, pharmaceutical manufacturers, and agricultural communities. Resistant infections are an increasing concern in our hospitals, in the general public, and in our animal populations. However, the degree of connection and risk relative to antibiotic use in livestock and resistant pathogens in humans is not well quantified (Khachatourians 1998; Laval 2000).

Several studies have shown that animals fed antibiotics (on farms where none had been previously used) rapidly develop resistant gut flora. Furthermore, producers and animal handlers in contact with these animals rapidly develop a similar gut flora and resistant pattern.

The use of certain antibiotics has shown to increase the degree of shedding of resistant organisms in the feces (Williams and others 1978). This is probably because the antibiotics reduce the population of susceptible organisms in the gut that were competing with the resistant organisms, allowing the resistant organisms to multiply with less competition (Levy and others 1976).

Development and transfer of antibiotic resistance among microorganisms is well described (Figures 12.1–12.3) (Khachatourians 1998; Schmidt 2002). Resistance develops in one of several ways. A spontaneous mutation may occur that affords the organism antibiotic resistance. The mutated organism multiplies by binary fission (Figure 12.1) passing on resistance to succeeding generations. The organism receiving the resistant genetic material may or may not be a pathogen. However, if it is not a pathogen, there are several ways that this resistant genetic material can be transferred to pathogens (see Figures 12.2 and 12.3). Therefore, nonpathogens with resistant genetic material can serve as reservoirs of resistant genetic material for pathogens.

Bacteriophages (viruses that attack bacteria) may pick up a piece of genetic material from a resistant organism it invades and then transfer that resistance factor to other organisms that were not previously resistant (Figure 12.2). The recipient of that genetic material may be a pathogen. Finally, organisms may transfer resistance by conjugation (Figure 12.3). This occurs when two organisms join in sexual reproduction to exchange genetic material. In such cases, a pathogen may pick up genetic material from a resistant organism (that may be a nonpathogen), producing a new strain of resistant pathogens. These are well-documented mechanisms that explain the increasing reservoir of antibiotic resistant genetic material and resistant organisms in the environment, and thus, the greater potential for resistant infections in the human and animal populations.

ANTIBIOTIC-RESISTANT ORGANISMS—
OCCUPATIONAL HEALTH RISKS

As previously mentioned, research has shown that pigs placed on antibiotics in the feed rapidly de-

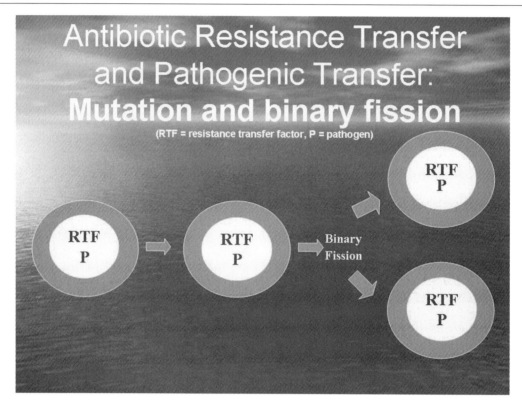

Figure 12.1. Antibiotic resistance transfer and pathogenic transfer: mutation and binary fission.

velop a gut flora contains resistant microorganisms. Additionally, swine producers that have contact with pigs or chickens fed antibiotics rapidly developed a gut flora with resistance patterns similar to the pigs they work with (Levy and others 1976; Padungton and Kaneene 2003). The risk for resistant infections in the animals and workers is intuitive. There are several case reports that have traced resistant occupationally acquired infections in farm workers back to the animals on the farm. Although there is little research data to quantify this risk, health care professionals should always consider the possibility of resistant infections when caring for livestock producers in their differential diagnoses and treatment options.

ANTIBIOTIC-RESISTANT ORGANISMS—PUBLIC AND COMMUNITY HEALTH

There is no question as to the increasing presence of resistant organisms and resistant infections observed by our medical and public health community. The population of resistant organisms (and

genetic material that codes for resistance) is directly related to the total quantity of antibiotics used in society and the way they are used. Overuse and inappropriate use of antibiotics (unnecessary use, wrong antibiotic, wrong route of administration, insufficient dose, and insufficient time of use) in human medicine are important selection factors for resistant organisms (Lathers 2001). Overuse and inappropriate use of antibiotics in agriculture also enhance selection of resistant organisms. The reservoirs for resistant organisms and related genetic material are the gut flora of humans and animals, health care facilities, the water we drink, the air we breathe, and some of the animal food products we eat (Campagnolo and others 2002; McDermott and others 2002; Teuber 2001).

What Are Ways That Resistant Organisms Are Spread Among the Human Population?

Organisms can come into the home through contaminated meat, milk, eggs, vegetables, and fruits (Hamer and others 2002; Teale 2002). Although

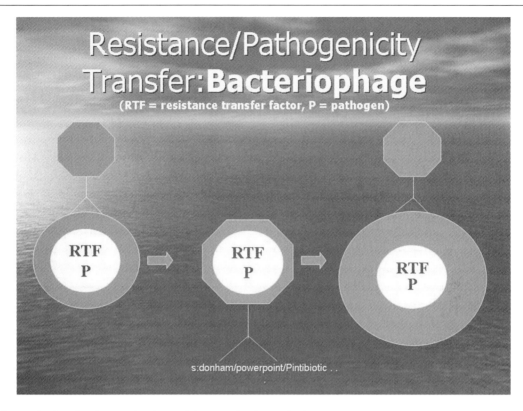

Figure 12.2. Resistance/pathogenicity transfer: bacteriophage.

strong efforts are made by many livestock producers and animal food processors to reduce contamination, meat, eggs, and pasteurized milk may be contaminated with fecal organisms that are present at the time of purchase at the grocery store. Improper cooking, or contamination of the food preparation area, utensils, and the preparer's hands, creates exposure potential to household members. Contaminated water is another potential source of exposure to resistant organisms from livestock origin (Esiobu and others 2002; Kolpin and others 2002). Runoff of manure into surface and groundwater creates a risk of exposure of resistant organisms to the community (Sengelov and others 2003). Recent studies have shown that resistant organisms are present in surface and groundwater around concentrated livestock production facilities (Campagnolo and others 2002) (suggesting a drinking water hazard) and in the exhaust air from swine confinement buildings (suggesting a risk of aerosol exposure) (Zahn 2001).

Antibiotic-Resistant Organisms—Health Effects Summary

As stated earlier, it is very difficult to quantify the human health risks of antibiotics use in livestock. There are scattered case reports of livestock producers acquiring a resistant infection from a farm animal or the environment, presumably by direct contact. However, these cases are found only in rare instances when there is a special interest and/or resources required to conduct the expensive case-finding protocols. It is likely that there are many instances of antibiotic-resistant infections acquired from livestock that result in uneventful recoveries, and therefore are never reported. Another risk, in addition to direct contact, is delayed contact. One example of delayed contact was a patient who reported to her physician for a wound infection. The patient was put on a course of antibiotics and within a few days, developed a severe and highly resistant gastrointestinal infection. The resistant organism was traced to its origin at the farm of the infected patient. The sce-

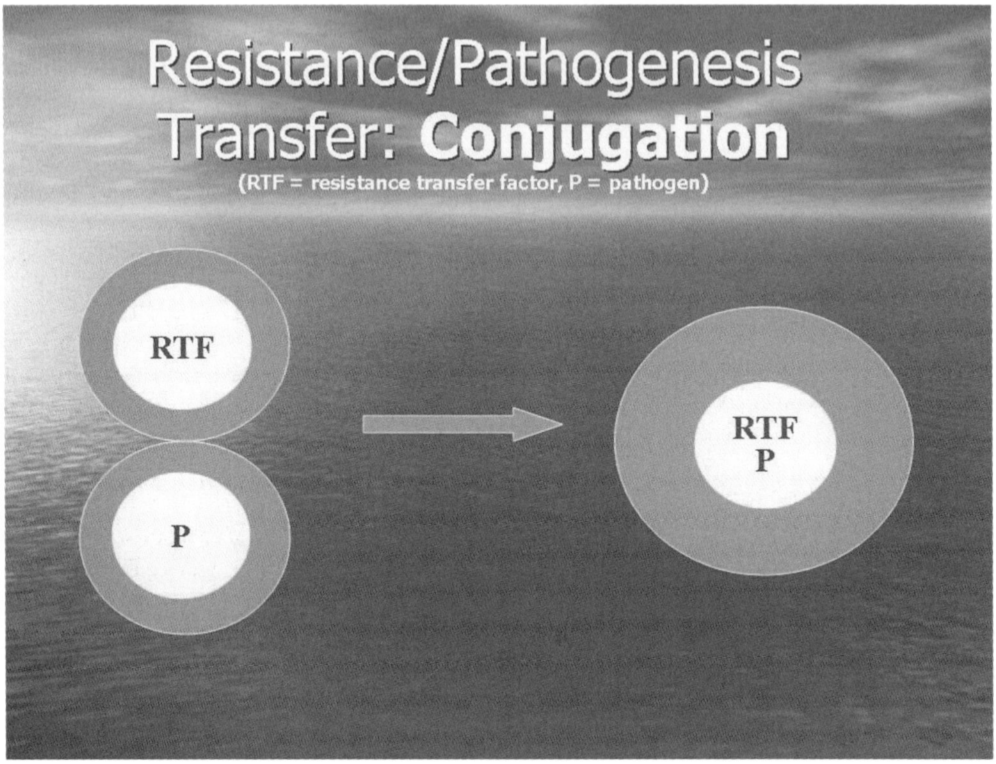

Figure 12.3. Resistance/pathogenesis transfer: conjugation.

nario in this case was that the patient's gut flora already contained the resistant pathogen, having picked it up from cattle on the farm. However, it had been held in check by competition with the remaining gut flora. When the patient was treated with antibiotics for an unrelated infection, the competing gut flora was reduced, allowing the resistant pathogen to grow to a pathogenic level and resulting in a clinically ill patient. It is difficult to know how commonly this happens, but this may be one of the most important occupational or community health risks associated with the use of antibiotics in livestock and humans.

The Presence of Antibiotic-Resistant Organisms in the Farm Environment— Implications for Treatment and Prevention in Farm Patients

Evidence suggests that farm workers, especially around livestock, poultry, or fruit crop production, may be at greater risk for antibiotic resistant infec-

tions relative to the general population. Resistant infections may present in one of several forms, including gastrointestinal infection, wound infections, or urinary bladder infections. Culture and sensitivity prior to initiating antibiotic therapy may be especially important when treating farm patients relative to the general population. A consideration in selection of antibiotics to treat a patient may include selection of an antibiotic of a class not used on the patient's farm, or one not generally used in agriculture (see Table 12.2). Furthermore, one may consider that treatment of a farm patient with a urinary bladder infection (especially a patient who has had a urethral catheter) may have a resistant organism from the person's own intestinal flora. Finally, the health care provider should be alert to the fact that treating a farm patient with a broad-spectrum antibiotic for any infection may allow a resistant pathogen to overgrow in the gut flora and become clinically significant.

HORMONES USED FOR GROWTH PRO-MOTION

Introduction

It was in the early 1950s when Dr. A. H. Trenkle from Iowa State University discovered that animals fed diethylstilbestrol increased the rate of gain, feed efficiency and grade (improved taste and tenderness) of the meat. This discovery has led to further research such that a variety of hormones are now routinely used in livestock production in some countries, but not in the EU. These hormones include estrogens, progesterones, and growth hormone (Sundluf and Cooper 1996). Generally there is no known occupational health hazard associated with their use, but there is a great deal of public angst with the use of these compounds. This section will review the specific hormones used, why they are used, and the potential occupational or community health implications.

Estrogenic Hormones

Estrogenic hormones are commonly given to feedlot steers (castrated male cattle) as they enter the feedlot for market fattening. There are two reasons for their use: 1) hormones increase the rate of gain and feed efficiency, similar and additive to the effect of antibiotics; 2) hormones enhance the flavor and tenderness of the meat product by increasing fat deposits between the muscle bundles (marbling). These effects are obvious economic incentives because it takes less time and less feed to grow a steer to market weight, and the taste and texture of the product are being improved. Meat is graded by the USDA according to the degree of marbling (fat between the muscle bundles) as good, choice or prime (the last being the most desirable). The meat from steers on estrogens may be elevated from their genetic capacity to a higher grade level—for example, from choice to prime. This makes the meat more valuable and increases the prospects of profitability of a beef farm.

The estrogen is administered via hypodermic injection of a time-released pellet on the outside of the ear of the steer as they enter the feedlot. There is little risk of accidental injection, dermal, or aerosol exposure to the worker. The pellets are designed to release the hormone slowly, but in time it will be metabolized when the steer reaches market weight. Therefore, there is a very low risk that any of the hormones reach the food chain.

Diethylstilbestrol is a hormone that has also been used in the past to help maintain pregnancy in women who go into premature labor. However, an epidemiologic study conducted in the early 1980s found that women whose mothers had been administered diethylstilbestrol during pregnancy had an increase of uterine cancer. Therefore, the use of diethylstilbestrol was banned for use in food producing animals as a precaution. However, several new synthetic estrogens have been developed that have been cleared by the FDA for use in food producing animals. Therefore, steers are still commonly injected with estrogenic hormones as they enter the feedlot. There have been no known occupational or consumer hazards with these products. However, there is a certain segment of the population that is against the use of these products. The European community is generally free from using such products and imposes barriers on import of meat from animals that have been administered hormones (or any other growth promotants).

Progesterones are used in feedlot heifers (young, nonpregnant female cows) for similar reasons that estrogens are used in steers. They increase the rate of gain and feed efficiency. They may also be used as an estrous synchronization tool in mature cows. If a cattle producer wants to synchronize the birthing of his calves, he can put the cows on progesterones and they all can be bred within a short period of time once withdrawn from the hormone. This will help the producer manage the planned and timely birthing of calves, which results in an even calf crop in terms of size and maturity. Similar to estrogenic hormones, there are no known occupational hazards or consumer hazards.

Beginning in the 1990s the use of growth hormones in livestock production began to appear. The most commonly used growth hormone in agriculture today is bovine growth hormone (somatotropin) which is used in dairy cattle to increase milk production. The hormone must be given by injection to each cow every 14 days during a portion of the lactation cycle. Therefore, there is an occupational risk for an accidental inoculation of the administrator. However, a small quantity of inoculation on a rare basis is a low health risk. There is a potential for illicit use of the product, as it is an anabolic steroid, and could be abused as an athletic performance enhancer or body building aid. Regarding consumer health, no evidence of the product is found in the milk nor is there any other known abnormality of the milk as a result of growth hormone use. Similar to other hormone usage, this product is not used in the European Union. Furthermore, the EU will not import milk products from animals administered bovine growth hormone.

HORMONES USED IN VETERINARY OBSTETRICS

Introduction

Similar to human medicine, there are a variety of drugs that are used to assist obstetrical problems or other reproductive problems in veterinary medicine. These products are administered by injection, and therefore unintentional inoculation is a concern. Pregnant women who are accidentally inoculated are particularly at risk (Wilkins and Bowman 1997). The two principal hormones to be discussed include oxytocin and prostaglandins.

Oxytocin

Oxytocin is a hormone naturally excreted from the posterior pituitary gland. Two of its primary actions are to cause uterine contraction, and milk let-down from the mammary glands. The product may be used in most mammalian species, but is most commonly used in swine and cattle. The product is used if the animal is having a difficult time delivering, when enhanced uterine contractions may result in a delivery. Furthermore, it is used if an animal (soon after delivery) does not have milk available for the young. An injection of oxytocin will result in almost immediate let-down of milk.

Prostaglandins

Prostaglandins are commonly used in cattle and swine to terminate pregnancy early, to induce parturition in the later stages of pregnancy, and to induce an animal to "cycle," that is, to come into estrus. A dose of this product will cause the corpus luteum to lyse and induce estrus in an animal.

The occupational problem with either oxytocin or prostaglandin is that they can cause abortion in pregnant women who may be accidentally inoculated. If a woman in the later stages of pregnancy is accidentally inoculated with oxytocin, it may cause abortion. Inoculation of a woman with prostaglandins anytime during pregnancy may induce abortion. There are many women working in agriculture today. In the gestation area of swine operations, employees are responsible for administering various hormones. Similarly on dairy farms, women may handle either of these hormones. Women working in these areas should be counseled as to the potential risk for accidental inoculation.

OTHER VETERINARY PHARMACEUTICALS

One concept that must be understood by medical practitioners is that the relative concentration of veterinary pharmaceutical products is often much greater (5–10 times) than analogous products manufactured for human beings. Therefore, even a small accidental injection could administer a harmful dose of certain materials. Anesthetics, analgesics, and injectable antibiotics are commonly used in veterinary medicine, and at least two products are known to have caused serious health risks in humans from accidental inoculation. One such anesthetic/analgesic product used in veterinary medicine is xylazine. A veterinary nurse accidentally inoculated himself with 2 cc of this product, resulting in a dose of 3 mg/kg (the veterinary dose is 0.5–2.0 mg/kg) (Randall and Woodward 2002). The nurse became unresponsive, with slow reacting pupils and a slow heart rate. He was administered fluids, intubated, and remained unconscious for 8 hours. Fortunately he recovered uneventfully over the course of several days.

Tilmicosin (Micotil) is a macrolide antibiotic for use in cattle for respiratory infections. It is particularly effective for gram-positive organisms and some gram-negative organisms as well. Tilmicosin is toxic to the cardiovascular system and potentially fatal for several species, including humans (although the specific mechanism is not known). Intramuscular doses of 20 mg/kg are fatal in most susceptible animals (Plumb 2002). A 1994 publication reported on 36 human exposures to this drug (two were in children) (McGreigan 1994). Twenty-six of the cases were accidental injections, secondary to inadvertent animal movement, and 10 cases were splashes to the face and eyes. The primary effects included pain and swelling at the site of injection, eye pain, burning sensation of the mouth, and bad taste. Twelve of the cases had electrocardiograms, and one of these was abnormal over a limited time period. The patients were treated with standard wound care, and all recovered. However, there have been at least two fatal cases reported, one in a Nebraska rancher and one in a North Dakota rancher. The Nebraska rancher was carrying a syringe filled with mycotil in his pocket, the contents of which he was planning to inject into a cow. He was kicked by another cow near by, driving the needle and contents of the syringe into his inner thigh. He became dizzy with a rapid heart rate. He was transported to a local hospital where he died about 1 hour after the incident (Von Essen 2003).

Extrapolating from animal data, a dose of 700–1400 mg could be fatal to a human being. This corresponds to a 2–4 ml dose of the veterinary product for cattle. There is no known antidote. However, the manufacturer has an emergency number to call in case of exposure (1-800-722-0987). Precautions on use of the product are written in the label directions. Tilmicosin is available only by prescription from a licensed veterinarian. The veterinarian must be certain that precautionary instructions are provided and understood by the user. It is best if the user signs off on precautionary use of tilmicosin before the product is dispensed. Obviously, injection with this material is an emergency (possibly fatal) health hazard.

Finally, it should be mentioned that some veterinary pharmaceuticals can reach illegal markets. As mentioned previously in this chapter, bovine growth hormone is an anabolic steroid and has found illicit channels for use as a performance enhancing drug. Furthermore, ketamine hydrochloride is an analgesic/anesthetic often used in veterinary medicine. This drug has found its way to street recreational use, under the term "angel dust." Ketamine is a controlled substance requiring a veterinary prescription and proper record keeping which is controlled by the FDA or appropriate agency in other countries.

CONTROL AND PREVENTION

Health professionals and veterinarians can promote safety in the use of veterinary pharmaceuticals by gaining knowledge about the harmful aspects of the pharmaceutical products their clients/patients use. Veterinarians must be responsible for assuring their clients are knowledgeable users of products they prescribe or dispense. Furthermore, pharmaceutical companies must assure that they effectively communicate risks of the products they manufacture and sell. A high-quality and readily available "material safety data sheet" should be provided by the pharmaceutical company. Medical and veterinary practitioners should be aware of the harmful aspects of these products. They should also be aware of what products are being used by their patients so that they might anticipate potential hazards and be able to make a rapid diagnosis if they are presented with a potential exposure situation. Both patients and health care providers should know that if there is an exposure situation (needle stick, etc.), the package label should be made available, if possible, so they know the specific product they are dealing with and its potential hazard.

Prevention of needle sticks can be enhanced by having and maintaining good-quality, functional animal handling facilities. Without proper restraint, animals can move unpredictably, resulting in an unintended injection of the operator.

For large agricultural operations, it is the responsibility of the owner/operator to train workers. This includes information on the health risks of all drugs used on the farm, proper injection technique, and the care and cleaning of hypodermic needles.

Pregnant women should be advised of the hazards surrounding the use of oxytocin and prostaglandins.

Finally, health care practitioners, veterinarians, and producers should all be aware of the growing concern of antibiotic-resistant infectious organisms. All involved should support the proper use of antibiotics and avoid unnecessary use. Prudent use of antibiotics in livestock production will help reduce the negative environmental and direct human health risks of antibiotic use (Horrigan and others 2002). Only through these methods can we help assure there will be effective antibiotics for use in combating both human and animal diseases in the future.

REFERENCES

Bennish M. 1999. Animals, humans, and antibiotics: Implications of the veterinary use of antibiotics on human health. Advances in Pediatric Infections 14:269–290.

Campagnolo E, Johnson K, Karpati A, Rubin C, Kolpin D, Meyer M, Esteban J, Currier R, Smith K, Thu K and others. 2002. Antimicrobial residues in animal waste and water resources proximal to large-scale swine and poultry feeding operations. Science Total Environment 299(1–3):89–95.

Cornaglia G. 1998. The spread of macrolide-resistant streptococci in Italy. Alliance for Prudent Use of Antibiotics News Letter 16(1):1,4.

Corwin R. 1997. Economics of gastrointestinal parasitism of cattle. Veterinary Parasitology 72(3–4): 451–457; discussion 457–460.

Council for Agricultural Science and Technology. 1981. Antibiotics in Animal Feed. Ames, IA: Iowa State University. Report nr 88.

Dal Monte A, Laffi G, Mancini G. 1994. Occupational contact dermatitis due to spectinomycin. Contact Dermatitis 31(3):204–205.

Dromigny J, Nabeth P, Perrier Gros Claude J. 2002. Distribution and susceptibility of bacterial urinary tract infections in Dakar, Senegal. International Journal of Antimicrobial Agents 20(5):339–347.

Dunlop R, McEwen S, Meek A, Friendship R, Clarke R, Black W. 1998. Antimicrobial drug use and related management practices among Ontario swine producers. Canadian Veterinary Journal 39(2): 87–96.

Erramouspe J, Carlson R. 2002. Veterinarian perception of the intentional misuse of veterinary medicine in humans. Journal of Rural Health 18(2):311–318.

Esiobu N, Armenta L, Ike J. 2002. Antibiotic resistance in soil and water environment. International Journal of Environmental Health Resources 12(2):133–144.

Geller R. 1990. Human effects of veterinary biological products. Veterinary and Human Toxicology 32(5):479–480.

Gersema L, Helling D. 1986. The use of subtherapeutic antibiotics in animal feed and its implications on human health. Drug Intelligence & Clinical Pharmacy 20(3):214218.

Hamer D, Friedman D, Gill C. 2002. From the farm to the kitchen table: The negative impact of antimicrobial use in animals on humans. Nutrition Reviews 60(8):261–264.

Hart S, Korey E, Stamatis G. 1998. Nearly one-third of *Streptococcus pnemoniae* resistant to newer antibiotics: study in doctor's office patients points to resistance rates in children.

Horrigan L, Lawrence R, Walker A. 2002. How sustainable agriculture can address the environmental and human health risks of industrial agriculture. Environmental Health Perspectives 110(5):A256.

Jjemba P. 2002. The potential impacts of veterinary and human therapeutic agents in manure and biosolids on plants grown on arable land: A review. Agriculture Ecosystems and Environments 93(1–3): 267–278.

Johnston A. 2001. Animals and antibiotics. International Journal of Antimicrobial Agents 18(3): 291–294.

Jones G. 1996. Accidental self inoculation with oil based veterinary vaccines. New Zealand Medical Journal 109(1030):363–365.

Khachatourians G. 1998. Agricultural use of antibiotics and the evolution and transfer of antibiotic-resistance bacteria. Canadian Medical Association Journal 159(9):1129–1136.

Kolpin D, Furlong E, Meyer M, Thurman E, Zaugg S, Barber L, Buxton H. 2002. Pharmaceuticals, hormones, and other organic wastewater contaminants in U.S. streams, 1999–2000: A national reconnaissance. Environmental Science Technology 36(6): 1202–1211.

Koschorreck J, Kock C, Ronnefahrt I. 2002. Environmental risk assessment of veterinary medicinal products in the EU—A regulatory perspective. Toxicology Letters 131(1–2):117–124.

Lathers C. 2001. Role of veterinary medicine in public health: Antibiotic use in food animals and humans and the effect on evolution of antibacterial resistance. Journal of Clinical Pharmacology 41(6): 595–599.

Laval A. 2000. Veterinary use of antibiotics and resistance in man: What reaction? Pathologie et Biologie 48(10):940–944.

LaVenture M, Hurley S, Davis J, Patterson C. 1988. Accidental self-inoculation with mycobacterium paratuberculoris bacteria (Johne's bacteria) by veterinarians in Wisconsin. Journal of the American Veterinary Medical Association 192(9):1197–1199.

Levy S, Fitzgerald G, Macone A. 1976. Changes in intestinal flora of farm personnel after introduction of a tetracycline-supplemented feed on a farm. North England Journal of Medicine 295:583–588.

Marwick C. 1999. Animal antibiotics use raises drug resistance fear. Journal of the American Medicine Association 282(2):120–122.

McDermott P, Zhao S, Wagner D, Simjee S, Walker R, White D. 2002. The food safety perspective of antibiotic resistance. Animal Biotechnology 13(1): 71–84.

McGreigan M. 1994. Human exposures to tilmicosin (MICOTIL). Veterinary and Human Toxicology 36(4):306–308.

Padungton P, Kaneene J. 2003. Campylobacter species in humans, chickens, pigs, and their antimicrobial resistance. Journal of Veterinary Medical Science 65(2):161–170.

Patterson C, LaVenture M, Hurley S, Davis J. 1988. Accidental self-inoculation with Mycobacterium paratuberculosis bacterin (Johne's bacterin) by veterinarians in Wisconsin. Journal of the American Veterinary Medical Association 192(9):1197–1199.

Plumb D. 2002. Veterinary Drug Handbook. Ames, IA: Iowa State University Press. p 804–805.

Randall L, Woodward M. 2002. The multiple antibiotic resistance (mar) locus and its significance. Research in Veterinary Science 72(2):87–93.

Samanta A, Roffe C, Woods K. 1990. Accidental self-administration of xylazine in a veterinary nurse. Postgraduate Medical Journal 66(773):244–245.

Sax H. 2002. Successful strategies against increasing antibiotic resistance. Therapeuishe Umschau 59(1): 51–55.

Schmidt C. 2002. Antibiotic resistance in livestock: More at stake than steak. Environmental Health Perspectives 110(7):A396–402.

Schwarz S, Chaslus-Dancla E. 2001. Use of antimicrobials in veterinary medicine and mechanisms of resistance. Veterinary Research 32(3–4):201–225.

Sengelov G, Agerso Y, Halling-Sorensen B, Baloda S, Andersen J, Jensen L. 2003. Bacterial antibiotic resistance levels in Danish farmland as a result of treatment with pig manure slurry. Environmental International 28(7):581–595.

Skold O. 2001. Resistance to trimethoprim and sulfonamids. Veterinary Research 32:261–273.

Spink W. 1957. The significance of bacterial hypersensitivity in human brucellosis: Studies on infection due to strain 19 Brucella abortus. Annals of Internal Medicine 47:861–874.

Sundluf S, Cooper J. 1996. Human health risks associated with drug residues in animal-derived foods. Veterinary Drug Residues 636(5–17).

Teale C. 2002. Antimicrobial resistance and the food chain. Journal of Applied Microbiology 92(suppl): 85S–89S.

Teuber M. 2001. Veterinary use and antibiotic resistance. Current Opinion in Microbiology 4(5): 493–499.

Torrence M. 2001. Activities to address antimicrobial resistance in the United States. Preventive Veterinary Medicine 51(1–2):37–49.

VandenBogaard A. 2001. Human health aspects of antibiotics use in food animals: A review. Tijdschrift Voor Diergenees Kunde 126(18):590–595.

VandenBogaard A, Stobberingh E. 1999. Antibiotic usage in animals: Impact on bacterial resistance and public health. Drugs 58(4):589–607.

Von Essen S. 2003. Linking options. Clinical Toxicology 41(3):229–233.

Wierup M. 2000. The control of microbial diseases in animals: Alternatives to the use of antibiotics. International Journal of Antimicrobial Agents 14(4):315–319.

Wiggins P, Schenker M, Green R, Samuels S. 1989. Prevalence of Hazardous exposures in veterinary practice. American Journal of Industrial Medicine 16(1):55–66.

Wilkins JI, Bowman M. 1997. Needle stick injuries among female veterinarians: Frequency, syringe contents and side-effects. Occupational Medicine 47(8):451–457.

Williams R, Rollins L, Pocurull D, Selwyn M, Mercer H. 1978. Effect of feeding chlortetracycline on the reservoir of Salmonella typhimurium in experimentally infected swine. Antimicrobial Agents and Chemotherapy 14(5):710–719.

Zahn J. 2001. Evidence for transfer of Tylosin and Tylosin-resistant bacteria in air from swine production facilities using sub-therapeutic concentrations of Tylan(r) in feed. Journal of Animal Science 79:189.

13
Zoonotic Diseases: An Overview

Kelley J. Donham, Danelle Bickett-Weddle,
Gregory Gray, and Anders Thelin

INTRODUCTION

The World Health Organization (WHO) defines *zoonoses* as "those infections which are naturally transmitted between nonhuman vertebrate animals and man."(Marano and Pappiaoanou 2004) However, a more comprehensive definition for zoonotic diseases are those infections common to animals and man (Acha and Szyfres 2003). This latter definition will be used in this chapter, as it covers diseases such as histoplasmosis and tetanus; organisms that infect, but are not transmitted between both humans and animals. This chapter will first review the general concepts of zoonoses and then describe in summary form those infections that are most important in the rural and agricultural setting of industrialized countries (Donham 1985).

This introductory section explores general concepts and epidemiologic principles of zoonotic infections. In particular, it covers these five topics: significance of zoonoses, trends of zoonotic infection, classification, ecological aspects of these diseases, and general characteristics of zoonoses.

Significance of Zoonotic Disease

The number of these diseases is relatively large. Over 200 zoonoses are distinguished worldwide. Approximately 80 zoonoses occur in industrialized countries, and at least 40 of these can be health hazards for agricultural workers (Donham 1985). Thirty-two of these infections will be briefly covered in this chapter. They will be organized relative to the primary species of livestock or environment that poses the greatest risk.

Zoonotic diseases are a major economic problem for the animal industry. The eradication control program for brucellosis costs the United States about $54 million dollars annually. Canada spends over $2.5 billion annually for surveillance and control of bovine spongiform encephalopathy (Unterschultz and others 2004). Similarly, large amounts are also expended to prevent zoonotic infections from reaching the public through potentially contaminated red meat, poultry, and dairy products.

Over 60% of the cataloged human pathogens are zoonoses, and 75% of emerging diseases are zoonoses (Taylor and others 2001; Erhardt 1974). Zoonoses may cause acute health problems, as well as chronic health problems. The latter are often associated with considerable psychological stress. Examples of diseases that are known to be associated with chronic disease include Lyme disease and brucellosis. The number of zoonotic infections in humans is difficult to obtain for several reasons. First, mild infections are common for many of these pathogens and thus often not seen by a physician (especially farmers). Typically, only severe illnesses or deaths are commonly reported. Second, many clinical infections seen by a physician remain unreported, often because they are not specifically diagnosed. Clinical manifestations of many zoonoses are variable and nonspecific, mimicking those of acute viral infections. Lack of physician awareness and lack of diagnostic support for these diseases further increase underreporting. Third, lack of a comprehensive zoonotic illness reporting system adds to underrepresentation. Although countries and states vary in their requirements for reporting infectious diseases, most states include only certain diseases specified by their public health authority. For example, the U.S. Public Health Service routinely collects and compiles data on just 24 zoonoses (CDC 1972, 2005).

For all of the above reasons, only a small percentage of human zoonotic infections is properly diagnosed, treated, and reported. Thus it is thought that the number of zoonotic infections tallied by state health departments is a gross underestimate of the actual number of infections. Even less data exist on the magnitude of diseases transmitted from humans to animals, since the focus for study of zoonoses is directly on human health.

Although the actual annual rates of zoonotic diseases are unknown, trends of individual zoonotic disease frequency can be traced. Some have changed from widespread epidemics to geographically localized and sporadic cases. Examples include bubonic plague, which has caused several pandemics since the Middle Ages and is now isolated to certain local ecosystems in North America, Asia, Europe, and Africa.

Infectious diseases have generally decreased in importance relative to chronic diseases in industrialized countries during the last half of the 20th century. This is thought to be due to better nutrition, improved public health measures, better environmental sanitation, and effective antibiotic therapy. However, since the beginning of the mid-1980s to today, there has been a reemergence of numerous infectious diseases and newly recognized zoonoses in certain areas of the world. For example, in North America and Europe, we have seen the emergence of Lyme disease, Hantavirus infections, West Nile encephalitis, spongiform encephalopathy, and avian influenza. Even with the emergence of these new diseases, zoonoses generally do not occur as epidemics, but as sporadic cases among people in limited geographical areas. They are primarily occupational and environmental hazards, and they occur in populations with specific risk factors that bring them into close association with the natural reservoirs of the disease agents.

Trends in Zoonotic Infections

Future trends likely include an increase of emergence of new or newly introduced infections stimulated by the high degree of mobility of people and agricultural products, and because of livestock production practices and spillover into livestock from wildlife reservoirs. However, there will be a continued decrease in the number of human cases of zoonoses for which there has been an active control or eradication program, such as brucellosis and bovine tuberculosis. The frequency of zoonoses contracted from feral animals will likely increase as participation in outdoor recreational and occupational activities continues to rise.

With some exceptions, vast epidemics of a zoonotic disease are not likely. However, natural or human-induced disasters or changes in agricultural practices may create a different scenario. Either intended (e.g., bioterrorism or agroterrorism) or unintended events (such as the introduction of disease X from infected arthropods escaping from an airplane coming from a country where disease X is endemic) could result in zoonotic disease outbreaks. Additionally, new agents could be introduced into an "immunological virgin" population. For example, migratory waterfowl are vast reservoirs for potentially zoonotic influenza viruses. Should the close association of swine and poultry production allow a recombination of an avian virus within swine that is transmissible to, and among, people, an epidemic of severe influenza could result (Murphy 1998).

Besides influenza, other zoonotic diseases may emerge as important agriculturally related infections such as nipah virus infection, hepatitis E from swine (worldwide), variant Creutzfeldt-Jakob disease (from bovine spongiform encephalopathy or mad cow disease) in the U.K., and morbillivirus virus in Australia (Murphy 1998).

CLASSIFICATION OF ZOONOTIC INFECTIONS

Zoonotic diseases are often classified in one of two ways: according to the major reservoir of the infectious agent or according to the mode of transmission of the infectious agent among natural host species. Zoonoses also may be grouped in a less formal way, according to the major human population at risk. An understanding of these classification systems can increase comprehension of the natural history of these diseases and thus aid in their diagnosis and control (Schwabe 1969).

Terms defining the major reservoir of zoonoses

include *zooanthroponoses, anthropozoonoses,* or *amphixenoses.* A **zooanthroponosis** is a zoonotic disease for which humans are the natural hosts of the infectious agent. Other vertebrate animals acquire infection through contact with humans. For example, dairy farmers infected with *Mycobacterium tuberculosis* can transmit this infection to their cattle.

An **anthropozoonosis**, in contrast, is a disease for which a vertebrate animal species other than a human is the natural host. For example, leptospirosis is primarily a disease of domestic cattle, swine, and numerous wildlife species. If a person becomes infected with leptospirosis, it is almost certain that it was obtained from an animal (or the environment) rather than through contact with another human.

Amphixenosis refers to a disease for which humans and other vertebrate species serve equally well as natural hosts. Infections may be transmitted freely between the two types of hosts. It often is difficult to determine whether human infections are acquired from animals or other humans. Examples of amphixenoses include infections due to certain strains of *Staphylococcus, Streptococcus, Escherichia coli,* and *Salmonella* that are not host-specific.

Terms used to classify zoonotic diseases according to their primary mode of transmission are **direct zoonosis**, **cyclozoonosis**, **metazoonosis**, and **saprozoonosis**. Direct zoonosis requires only one vertebrate species host to maintain the infectious agent. For example the rabies virus is maintained in the wild skunk, bat, raccoon, fox, and coyote populations of North America, and transmission is by direct contact (usually bite wound) from an infected animal to a susceptible animal.

Cyclozoonosis refers to a zoonotic disease that requires two or more vertebrate hosts for maintenance of the infectious agent. The tapeworm, *Echinococcus granulosis,* for example, is maintained in a cyclical transmission pattern involving sheep who ingest tapeworm eggs passed in feces of dogs, which then encyst in sheep viscera. The life cycle is completed when dogs ingest infected tissues of the sheep, and adult tapeworms develop in the dogs' intestines.

An example of a **metazoonosis** is an agent that requires both a vertebrate and an invertebrate host for maintenance of the infectious agent, such as Saint Louis encephalitis. A mosquito may main-

tain the virus temporarily, but it must transmit the virus from an infected vertebrate to a susceptible vertebrate for permanent transmission or amplification of the infectious agent.

A **saprozoonosis** refers to a disease caused by an infectious agent that is maintained in a fomite, e.g., soil, water, or another type of inanimate object. Histoplasmosis, for example, is contracted when animals inhale spores from soil with high concentrations of avian or bat feces, where the organism grows.

In addition to the above two classification methods, zoonoses can be less formally grouped according to the major human population at risk. Zoonoses typically are most common where contact with animals or their environment is maximized—that is, whenever human activities encroach disease cycles as they occur in their natural setting. Several specific types of activities increase the risk for acquiring a zoonotic disease. There are at least 40 zoonoses that are occupational hazards for people who work in agriculture and other occupations involving animal contact. Farmers, veterinarians, packing plant workers, and hair and hide industrial workers are all at risk of acquiring diseases such as brucellosis, ornithosis, anthrax, and contagious ecthyma, among others.

Outdoor recreational activities, such as hiking and swimming, have increased risks of acquiring diseases such as Rocky Mountain spotted fever, Colorado tick fever, leptospirosis, and tularemia. The trend in the United States for people to live on urban fringes and in the country creates an additional risk for acquiring certain diseases such as Rocky Mountain spotted fever and the arthropod-borne encephalidities. Pet ownership also increases the risk for certain diseases such as toxoplasmosis, leptospirosis, ringworm (caused by animal dermatophytes), visceral larval migrans, and cutaneous larval migrans.

Categorizing zoonoses by risk group is not a hard and fast scheme, but it can be a significant help in obtaining a pertinent case history and a list of differential diagnoses for health care providers.

Ecologic Considerations of Zoonoses

Also helpful to clinicians is an understanding of why zoonotic outbreaks occur. This can best be developed by examining the infectious agent as part of its total environment–that is, by understanding the disease as part of an ecological system (Schwabe 1969).

Infection and sporadic disease are a natural part of ecosystems. In natural, undisturbed ecosystems (those in which all inhabitants have evolved balanced interrelationships with each other and their environment), infectious agents typically maintain a steady, low rate of infection in the host population. In this way, infectious agents derive what they need for survival without decimating the host population.

Disease outbreaks are a result of a change of the natural balances of ecosystems. Ecosystems change slowly through time as a result of natural processes, such as erosion, changes in climate, or geophysical events. However, human intervention produces most of the rapid and large-scale ecosystem changes. As ecosystems and the interactions among organisms change, a frequent result is the alteration of the number and types of organisms present. For example if ecologic changes create a more favorable environment for an arthropod vector of a zoonotic disease or a vertebrate host for a zoonotic disease, one might see an epidemic of that zoonotic disease in the community. Introduction of a new feral species into an ecosystem can result in an epidemic if this species is host to an infectious agent, or if the new species is a disease vector. Similarly, introduction of new domestic species into a natural ecosystem can lead to disease outbreaks when these species transmit infectious agents from native vertebrates to humans. For example, agricultural practices involved with raising cattle in the Midwestern United States have led to the potential transmission of rabies from skunks to cattle through bites and then to people through exposure to infected cows.

People who enter environments where zoonoses are part of the natural ecosystems increase their risk of acquiring one of these infectious agents. Examples include cases of rabies in hikers who are bitten by rabid skunks, or Colorado tick fever in campers bitten by ticks.

Alteration of the abiotic components of an ecosystem can result in changes in the population structure, and thus in disease outbreaks. For example, as land in Egypt has been irrigated for agricultural purposes with water made available by the Aswan Dam, the habitat for snails has improved. Snails have increased, providing an intermediary host for the blood fluke *Schistosoma*. Thus, the incidence of schistosomiasis in humans has subsequently increased dramatically in this area.

The last points to be discussed in this overview are the general epidemiologic characteristics of zoonoses. These 10 characteristics, many of which have already been discussed, capsulize the importance of zoonoses as a human health risk:

1. Although there often are only one or a few major host species, zoonotic infectious agents typically have a broad potential host range. For example, the bacterium *Francisella tularensis,* which causes tularemia, has been isolated from over 100 mammalian species and numerous other vertebrates. However, many of these species are "accidental hosts" and are not significant in perpetuation of the disease cycle.

2. Zoonoses often cause severe economic burdens because of both loss of diseased animals and the cost of preventing and treating infections in animals or humans.

3. The majority of zoonoses are anthropozoonoses, being maintained primarily by a vertebrate host species other than humans.

4. Animals may be inapparent carriers of zoonotic pathogens. They may pose a health hazard for humans and other livestock without presenting demonstrable clinical signs.

5. Humans are often accidental, dead-end hosts for zoonotic pathogens. They do not transmit their infection to other people or animals.

6. Human zoonotic infections typically result in morbidity, but rarely in mortality. For example, leptospirosis, brucellosis, histoplasmosis, and Q fever can all cause moderate to severe illness. However, when properly treated, they rarely lead to death.

7. Many zoonoses have nonspecific, variable clinical signs and symptoms, mimicking other diseases, especially influenza.

8. Human zoonotic infections often occur sporadically rather than in epidemics. This is because humans are often dead-end hosts, not transmitting infections to other people.

9. There are specific groups of people who have an increased risk of acquiring infection. These risk groups include persons with greater than average contact with animals:

meat processing plant workers, farmers, veterinarians, pet owners, and people living in rural areas or engaging in outdoor activities.

10. Many zoonotic infections in humans are never diagnosed because of numerous factors, such as their specific and clinical signs and symptoms, lack of physician awareness and lack of adequate diagnostic support. Thus, accurate statistics regarding rates of zoonotic diseases are not readily available.

Summary

Because zoonotic diseases are difficult to recognize, clinicians must have knowledge of potential zoonotic pathogens in their regions, and how to detect them, before they can be recognized. Practitioners can utilize three information sources: local practicing veterinarians, veterinarians in public health sections of colleges of veterinary medicine and in state health departments, and the national-level public health agencies, such as the Centers for Disease Control and Prevention (CDC) in the U.S., and the WHO. In addition, physicians can develop an awareness of environments and activities typically responsible for contraction of these diseases. Such awareness can assist in obtaining a patient history that is likely to lead to correct diagnosis.

Tables 13.1 and 13.2 focus on the 32 zoonoses considered to be of greatest agricultural occupational significance in developed agricultural countries today. Many of these diseases could be contracted from several types of livestock. Table 13.1 divides these 32 diseases into 6 groups, according to according to human risks associated with specific types of farm operations, i.e., swine, dairy cattle, beef cattle, sheep, poultry, and the rural environment (Donham 1985). Categorizing these diseases in this way is meant to help the reader/health professional organize the information in a relative way according to exposure. For example, if the reader/health professional is dealing with swine producers, there should be only 5 or 6 zoonotic disease risks that immediately come to mind, out of the some 200 different zoonotic diseases.

Because the specific information on each of these diseases would require extensive material beyond the extent of this book, the information is summarized in tabular form (Table 13.2). The information on each disease is organized in Table 13.2 in a manner that helps the reader anticipate zoonotic risks according to the particular farmer's exposures, possibly leading to a differential diagnosis and enough direction to obtain further information to obtain a specific diagnosis. The information is outlined according to the following information: 1) natural history of the infectious agent, 2) modes of transmission to humans, 3) typical symptoms and signs, 4) questions pertinent to determining a relevant patient history, and 5) laboratory confirmation. Prevention also will be briefly covered.

Table 13.1. Agricultural Occupational Zoonoses According to Primary Reservoir for Workers

X = Other possible (secondary/species/environment sources) of infection
? = Rare or unknown source of infection
— = Not a source of infection

Swine	Dairy Cattle	Beef Cattle	Sheep	Poultry	Rural Environment
X	Brucellosis	X	X	—	X
Erysipeloid (*Erysipelothrix rhusiopathiae*)	X	X	X	X	X
Swine influenza	—	—	—	X Avian Influenza	—
Nipah Virus	—	—	—	—	—
Streptococcus suis	—	—	—	—	—
X	Q Fever	X	X	—	?
—	Bovine Spongiform Encephalopathy (Mad Cow)	X	?	—	—
X	Ringworm	X	—	—	?
—	Pseudocowpox (Milker's Nodules)	X	—	—	—
X	X	Anthrax	X	—	?
X	X	Leptospirosis	X	—	X
X	X	Rabies	—	—	X
X	X	Foot and Mouth Disease	X	—	—
—	—	—	Contagious Ecthyma (Orf)	—	—
—	—	—	Hydatid Disease (*E. granulosus*)	—	X
—	—	—	Tularemia	—	—
—	—	—	—	Newcastle Virus	—
—	—	—	—	Ornithosis (psittacosis)	X

Table 13.1. Agricultural Occupational Zoonoses According to Primary Reservoir for Workers (*continued*)

X = Other possible (secondary/species/environment sources) of infection
? = Rare or unknown source of infection
— = Not a source of infection

	Swine	Dairy Cattle	Beef Cattle	Sheep	Poultry	Rural Environment
Histoplasmosis	—	—	—	—	—	
Blastomycosis	—	—	—	—	—	
Coccidioidomycosis	—	—	—	—	—	
Arboviral Enchephalidities	—	—	—	—	—	
Lyme Disease	—	—	—	—	—	
Hantavirus	—	—	—	—	—	
Hepatitis E		—	—	—	—	—
Nematodiasis (*Ascaris suum*)		—	—	—	—	—
Salmonellosis		X	X	X	X	X
E. coli		X	X	X	X	X
Pork Tapeworm (*Taenia soleum*)		X	X	X	X	X
Trichinosis (*Trichinella spiralis*)		—	—	—	—	X

Table 13.2. Zoonotic Bacterial Infections

Disease (Common Names)	Etiologic Agents	Health Effects (A) Human (B) Animal	Animal Hosts (A) 1. Zoonthroponosis 2. Anthropozoonosis 3. Amphixenosis (B) Specific Animals Infected	Mechanisms of Transmission (A) Reservoir (B) 1. Directzoonosis 2. Metazoonosis 3. Saprozoonosis 4. Cyclozoonosis (C) Specific mechanisms	Epidemiology (A) Populations at Risk (B) Geographic Distribution	Treatment	Prevention or Control
Bacterial							
Anthrax (Malignant pustule, wool sorter's disease)	*Bacillus anthracis*	(A) *Human*— Localized skin lesions usually on hands or arms —Pulmonary form less common but much more severe with fairly high case fatality rate— gastrointestinal form is least common (B) *Animal*—usually overwhelming bacteremia and septicemia with rapid death in cattle, sheep and goats— less acute in horse, pigs and dogs	(A) 1. (B) Cattle-sheep-goats- horse-pig-dog	(A) Soil —Water (stagnant ponds near incubator areas) (B) 1, 3 (C) Direct contact with infected animals or their carcasses or body parts or animal fertilizers or feeds —Inhalation of spores from hair or hide of infected animals —Consumption of improperly cooked meat or infected animals	(A) Sheep and goat producers —Cattle producers —Veterinarians and other animal health workers —Hair and hide processors —Abattoir workers (B) Worldwide in endemic foci —May be transported to distinct locations with hair and hides from infected animals	1. Penicillin and its derivatives 2. Ciprofloxacin 3. Doxycycline 4. Fluroquinolins	1. Vaccination of animals in endemic areas 2. Personal protection when handling potentially infected animals or tissues 3. Deep burial unopened and covering with lime or infected animal carcasses 4. Vaccination of humans at high risk
Brucellosis (undulant fever, Malta fever, Bangs disease)	*Brucella abortus* *B. suis* *B. melatensis* *B. canis*	(A) *Human*—generalized prolonged influenza like illness, spiking fevers, myalgia, malaise —occasional chronic forms include lesions of heart valves, abscesses of bone, liver, or other body parts	(A) 1. (B) Cattle-swine —Sheep-goats —Less common are dogs, camels, deer, buffalo, and others —wildlife reservoirs in U.S. Bison, elk, wild pigs	(A) Cattle, swine, goats, dogs mainly —Other susceptible animals less important (B) 1. (C) Direct contact with infected animals or their tissues especially placenta and abortion products —Ingestion of unpasteurized milk and cheese products from infected animals	(A) Sheep, goats, cattle and swine producers primarily —Abattoir workers (B) Worldwide, especially in dairying areas —Rare today with industrialized countries due to eradication program	1. Children— Trimethoprim —Sulfamethoxizole—21 days —Aminoglycosides—21 days 2. Adults —Doxicline + Rifampin— 30 days	1. Eradication of disease in the primary livestock species (several countries have established such programs) 2. Personal protection when handling infected animals, especially following abortion —sanitation of the animal environment

Table 13.2. Zoonotic Bacterial Infections (*continued*)

Disease (Common Names)	Etiologic Agents	Health Effects (A) Human (B) Animal	Animal Hosts (A) 1. Zooanthroponosis 2. Anthropozoonosis 3. Amphixenosis (B) Specific Animals Infected	Mechanisms of Transmission (A) Reservoir (B) 1. Directzoonosis 2. Metazoonosis 3. Saprozoonosis 4. Cyclozoonosis (C) Specific mechanisms	Epidemiology (A) Populations at Risk (B) Geographic Distribution	Treatment	Prevention or Control
		(B) *Animal*— Abortions —possible chronic infections of R.E., system urogenital, Bone and other tissues		—Possible airborne transmission			—pasteurization of milk products
Colibacillosis	*Escherichia coli*	(A) Gastroenteritis mainly —Possible wound infections or abscess —Cystitis —0157:H7 strain possible toxic-renal syndrome (B) Gastroenteritis mainly —Possible wound infections or abscess	(A) 1, 2 or 3, depending on specific strain (B) All livestock species, most mammals feral or domestic —0157:H7 strain cattle	(A) Animals are the reservoir for some strains, man for others and both man and animals for other strains (B) 1 (C) Direct or indirect contact with infected animal and waste —Accidental ingestion of organism via hand to mouth contact or ingestion of contaminated food	(A) Livestock and poultry farmers —General public food and water (B) Worldwide	1. Neomycin 2. Chloromycetin 3. Gentamycin	1. Excellent personal and environmental sanitation
Erysipeloid (pork finger, fish finger, swine erysipelas)	*Erysipelothrix rhusiopahtiae*	(A) Skin lesions with pain and swelling —Possible local spread of lesions from primary site —Possible endocarditis (B) Acute-septicemia with high fever, skin lesions and moderate death loss —Chronic-polyarthritis myocarditis	(A) 1. (B) Swine mainly —Also chickens, turkeys, and sheep —Also found in slime layer on fish	(A) Animals—Swine, sheep, chickens, turkeys Environmental—soil, slime layer on fish (B) 1, 3 (C) Direct contact—contamination of a cut or break in skin with soil or infectious materials from animal or tissues	(A) People raising or handling swine, sheep, or poultry —Abattoir workers and meat cutters —Fish processing workers (B) Worldwide	1. Beta Lactams (e.g., penicillin, cephalosporins) —21 days	1. Control by swine vaccination, adding only erysipelas free-animals to the herd 2. Humans-proper treatment of cuts and wounds 3. Environmental sanitation 4. Abattoir workers, protective gloves proper treatment of cuts, environmental sanitation (*continued*)

365

Table 13.2. Zoonotic Bacterial Infections (*continued*)

Disease (Common Names)	Etiologic Agents	Health Effects (A) Human (B) Animal	Animal Hosts (A) 1. Zooanthroponosis 2. Anthropozoonosis 3. Amphixenosis (B) Specific Animals Infected	Mechanisms of Transmission (A) Reservoir (B) 1. Directzoonosis 2. Metazoonosis 3. Saprozoonosis 4. Cyclozoonosis (C) Specific mechanisms	Epidemiology (A) Populations at Risk (B) Geographic Distribution	Treatment	Prevention or Control
Leptospirosis (Weil's disease, swine herd's disease, swamp fever, mud fever)	*Leptospira interrogans* —many different serovars involved	(A) Generalized febrile, influenza-like illness of variable severity —mild cases malaise, myalgia, symptoms of meningitis, vomiting —severe cases, hepatorenal involvement jaundiced, case-fatality ratio 20–40% (B) Abortion —Hepatorenal involvement with jaundice, possible kidney failure —healthy carriers are common	(A) 1. (B) Cattle, swine, and the main livestock species infected —Sheep and goats, less common —Dogs–Rats —Wildlife including squirrels, raccoons, mice, shrew, bandicoot, fox, jackals, hedgehog and others	(A) Cattle, swine, rats mainly, but most other susceptible animals also —water, muddy soil (B) 1 and 3 (C) Direct and indirect contact with urine from infected animals —Contact with abortion products of infected animals —Contact with water contaminated with urine from infected animals	(A) Persons working with cattle or swine —Persons working in rice paddies contaminated from urine of infected animals —Abattoir workers —Persons swimming in contaminated water —Hunters and trappers (B) Worldwide, specific serovars vary with locality	1. Aminoglycoside (e.g., sheptomycin) —21 days 2. Tetracycline (e.g., doxycycline)—21 days	1. Control infection—in livestock with good environmental sanitation, immunization and proper veterinary care 2. Prevent infected animals from urinating in water where humans have contact 3. Personal protection of workers when handling infected animals or tissues 4. Rat control
Lyme Disease	*Borellia burgdonferi*	(A) Variable systemic disease with skin lesions at tick bite site —Acute influenza-like symptoms may be followed by cardiac and neurological symptoms which may be followed by chronic arthritis and neurological symptoms	(A) 1. (B) Several small rodent species, deer, and dogs	(A) Small wild rodents (peramyscus leucopus in N. America) —Several species of Ixodes ticks (B) 2. (C) Tick bites from Ixodes ticks	(A) People working outdoors in endemic regions (B) Western and Eastern Europe, Australia, Asia, N. America	1. Doxycycline 200–400 mg/day for 17 days 2. For chronic cases, ceftriaxone for 6 months	1. Avoidance tick vectors 2. Wear well covered clothing, pant legs tucked into socks or boot tops 3. DEET insect repellant 4. Body and hair inspect when returning from endemic areas

Table 13.2. Zoonotic Bacterial Infections (*continued*)

Disease (Common Names)	Etiologic Agents	Health Effects (A) Human (B) Animal	Animal Hosts (A) 1. Zooanthroponosis 2. Anthropozoonosis 3. Amphixenosis (B) Specific Animals Infected	Mechanisms of Transmission (A) Reservoir (B) 1. Directzoonosis 2. Metazoonosis 3. Saprozoonosis 4. Cyclozoonosis (C) Specific mechanisms	Epidemiology (A) Populations at Risk (B) Geographic Distribution	Treatment	Prevention or Control
Salmonellois	*Salmonella ty-phimurium (2,000 sero types)*	(A) Gastroenteritis of variable severity depending on dose and virulence of specific organism —May produce bacteremia and septicemia (B) Gastroenteritis —Bacteremia and septicemia possible in severe cases	(A) 1, 2 and 3 (depending on the strain) (B) Dairy animals are particularly important in agriculture —All livestock and mammalian species may be infected	(A) Fecal–oral —G.I. tract of most livestock species —Man-respiratory tract, skin, or G.I. tract —Water, soil contaminated with animal waste (B) 1 or 3 (C) Direct or indirect contact with infected animal or their environment —Consumption of unpasteurized milk —Animal food products	(A) Livestock workers—particularly dairy —Those who drink unpasteurized milk or consume improperly prepared or stored animal food products (B) Worldwide-ubiquitous	1. Neomycin 2. Chloromycetin 3. Gentamycin	1. Practice sound animal hygiene and management 2. Detection and eradication of mastitis in dairy herds 3. Pasteurization of milk 4. Sanitary preparation and storage of animal food products 5. Prudent use of antibiotics to prevent resistant infection
Tetanus (Lockjaw)	*Clostridium tetani*	(A) Tonic Clonic Convulsions —Spastic contraction of skeletal muscles —Respiratory failure —Death (B) Hyperirritability of central nervous system —Tonic clonic —Spastic paralysis —Death	(A) 1. (B) Sheep and horses mainly —other animals more resistant	(A) Soil —Large intestine of herbivores and to a lesser extent carnivores (B) 3. (C) Wound contamination with soil or feces containing spores, anaerobic conditions in wound required for organism to germinate and produce toxin	(A) Most agricultural workers who are subject to punctures or cuts, especially in areas where herbivore animals were raised —Infants with umbilical infections of unvaccinated mothers (B) Worldwide	1. Debride, clean wound 2. Antitoxin + tetanus toxoid if previously unimmunized 3. Penicillin or derivatives 4. Anti-convulsants for seizures	1. In humans-proper treatments of open wounds to prevent infection 2. Immunizations (every 10 years) with tetanus toxoid, repeat following severe exposures if last toxoid longer than 5 years

(*continued*)

367

Table 13.2. Zoonotic Bacterial Infections (*continued*)

Disease (Common Names)	Etiologic Agents	Health Effects (A) Human (B) Animal	Animal Hosts (A) 1. Zooanthroponosis 2. Anthropozoonosis 3. Amphixenosis (B) Specific Animals Infected	Mechanisms of Transmission (A) Reservoir (B) 1. Directzoonosis 2. Metazoonosis 3. Saprozoonosis 4. Cyclozoonosis (C) Specific mechanisms	Epidemiology (A) Populations at Risk (B) Geographic Distribution	Treatment	Prevention or Control
Tularemia	*Fracisella tularensis*	(A) Four forms: —Ulceroglandular—most common form, localized wound infection with generalized symptoms, cellulites with regional lymphadenitis —Oculoglandular, severe conjunctivitis with regional lymphadenitis —Pulmonary-pneumonia with severe generalized symptoms —Typhoidal-gastroenteritis, fever, toxemia, ulcers in mouth, pharynx and esophagus (B) Varies according to species, some rodents and lagomorphs are most susceptible, prolonged generalized illness fatal septicemia. Sheep, other rodents and birds have inter-	(A) 1. (B) Infections found in 125 species of invertebrates ticks, mosquitoes, deer flies, horse-sheep flies, fleas —Sheep are most important for agricultural exposure —Other important exposures are Lagomorphs and small rodents (rabbits, muskrats are important reservoirs for hunters and trappers) contaminated water in endemic areas	(A) Sheep —Infected mammals —Ticks and other blood-sucking arthropoids —Contaminated water (B) 1, 2 and 3 (C) Handling infected sheep —Bites from blood-sucking arthropoids, mainly horse and deer flies —Handling infected rodents, Lagomorphs (hunters and trappers) —Consumption of water from a stream or pond —Possibly inhalation —Cat bites	(A) Sheep ranchers, sheep shearers and sheep handlers —Outdoor occupation or recreation where exposure to blood sucking arthropods is common —Hunters, trappers (B) North America —Mexico —European Continent —Turkey —Iran —China —Japan	1. Aminoglycoside (e.g., streptomycin) 7.5–15 mg/ 15g—21 days 2. Alternatives Ciprofloxacin, doxicycline	1. Personal protection (gloves and dust mask) when handling potentially infected sheep or wild mammals 2. Avoid drinking untreated surface water from ponds and streams 3. Thoroughly cook meat from small wild mammals consumption 4. Wear gloves when handling or cleaning small wild mammals

Table 13.2. Zoonotic Bacterial Infections (*continued*)

Disease (Common Names)	Etiologic Agents	Health Effects (A) Human (B) Animal	Animal Hosts (A) 1. Zooanthroponosis 2. Anthropozoonosis 3. Amphixenosis (B) Specific Animals Infected	Mechanisms of Transmission (A) Reservoir (B) 1. Directzoonosis 2. Metazoonosis 3. Saprozoonosis 4. Cyclozoonosis (C) Specific mechanisms	Epidemiology (A) Populations at Risk (B) Geographic Distribution	Treatment	Prevention or Control
		mediate susceptibility with non-fatal generalized illness. Carnivores have low susceptibility, usually with subclinical infection are sheep and arthropods					
Viral							
Arboviral Encephalitis: West Nile Virus (WVN)	Flavivirus	(A) Influenzalike illness of variable severity, WNV and EEE are most severe	(A) 1. (B) A variety of avian species	(A) Several species of mosquitoes	(A) Anybody working outdoors in endemic areas	1. Supportive treatment	1. Mosquito control 2. Protection from mosquitoes
Eastern (EEE)	Togavirus	—Encephalitis and neurological disease most common with WNV and EEE	—Horses for WNV, EEE, WEE, and VEEE	—WNV—crows and related birds—EEE, WEE, VEE, SLV— various wild ground nesting birds	(B) WNV—Africa, E.U., N. America	2. Acyclovir or ribivirin if used early in the disease course	—covered clothing
Western (WEE)	Togavirus	(B) Most animal species are asymptomatic, except WNV is fatal to a wide variety of avian species	(C) Small rodents for CEV	—CV—small rodents e.g., ground squirrels	—EEE, WEE, SLV, CV, America	3. Corticosteroids to reduce brain swelling	—DEET repellant
St. Louis virus (SLV)	Togavirus	—Horses may be severely affected by WNV and Eastern Equine Encephalitis			—VEE—Mexico, Central and S. America various related viruses in most parts of the world temperate to tropical climates		—Avoid being outside at dusk
Venezuelan Encephalitis (VEE)	Alphavirus						
California or LaCross virus (CEV)	Bunyaavirus						

(*continued*)

369

Table 13.2. Zoonotic Bacterial Infections (*continued*)

Disease (Common Names)	Etiologic Agents	Health Effects (A) Human (B) Animal	Animal Hosts (A) 1. Zooanthroponosis 2. Anthropozoonosis 3. Amphixenosis (B) Specific Animals Infected	Mechanisms of Transmission (A) Reservoir (B) 1. Directzoonosis 2. Metazoonosis 3. Saprozoonosis 4. Cyclozoonosis (C) Specific mechanisms	Epidemiology (A) Populations at Risk (B) Geographic Distribution	Treatment	Prevention or Control
Contagious ecthyma (orf)	Pox virus	(A) Skin lesions on hands and arms —Start out as small papules, progress to large vesicles which then ulcerate —Last 4–8 weeks (B) Vesicular lesions in the mouth and on the lips	(A) 1. (B) Sheep mainly —Goats also	(A) Sheep —Goats —The animal environ— e.g., animal sheds, feed bunks, etc. (B) 1 and 3 (C) Direct contact with infected animals or their environment, especially handling and examining infected animals	(A) People raising or handling sheep or goats (B) Worldwide where ever sheep are raised	Symptomatic, wound protection, topical antibiotics to prevent bacterial infections	1. Isolation of infected animals 2. Wearing protective gloves when handling or treating infected animals or working in their environment 3. Practice excellent sanitation of the animal environment
Foot and mouth disease (Aphthous fever, aphthosis)	Picorna virus, rhinovirus subgroup	(A) Not highly infectious for humans —May cause mild influenza-like illness —Vesicles in mouth and on lips and hands (B) Highly contagious in animals —Vesicles in mouth on lips, teats and udders and between the toes	(A) 1. (B) Cattle mainly —Sheep and goats	(A) Cattle, their tissues also (B) 1. (C) Direct contact with infected animal or their environment, especially during handling or examination, or milking	(A) Dairy and beef cow handlers (B) Eradicated from most of North America and Western and Central Europe, Australia, New Zealand and Mexico —Present in South American, Eastern Europe, Asia, some African countries	No specific treatment, only symptomatic	1. Eradication program which consisted of quarantine, identification and destruction and sanitary disposal of infected animals and environmental clean up 2. Personal protective gloves when handling infected animals or tissues
Hepatitis E	Calici virus	(A) Similar to hepatitis A —Mild to moderate febrile illness	(A) 1. (B) Pigs and humans	(A) Pigs feces oral route (B) 1, 2	(A) Anyone with exposure to pigs or contaminated drinking water (B) Mainly in developing countries —Also in US and Mexico	(A) No known treatment	1. Sanitation in pig production and in people working with pigs, e.g., hand washing

370

Table 13.2. Zoonotic Bacterial Infections (*continued*)

Disease (Common Names)	Etiologic Agents	Health Effects (A) Human (B) Animal	Animal Hosts (A) 1. Zooanthroponosis 2. Anthropozoonosis 3. Amphixenosis (B) Specific Animals Infected	Mechanisms of Transmission (A) Reservoir (B) 1. Directzoonosis 2. Metazoonosis 3. Saprozoonosis 4. Cyclozoonosis (C) Specific mechanisms	Epidemiology (A) Populations at Risk (B) Geographic Distribution	Treatment	Prevention or Control
		—Children and pregnant women at an increased risk (25% case fatality)					
Hanta Virus Pulmonary syndrome (HVPS). Renal Syndrome (HVRS)	Bunya virus (various strains)	(A) HVPS—variable severity —Influenzalike then pulmonary edema, 38% case fatality —HVRS—Initial influenzalike followed by hypotension, varying degree of renal failure	(A) 1. (B) Various species of wild rodents (rats and mice)	(A) Species of wild rats and mice —in apparent life-long carriers (B) 2. (C) Aerosol transmission via rodent droppings, food contamination by rodent feces, urine, and saliva	(A) Farmers, loggers, outdoor recreation, cleaning out rodent nests (B) E.U. countries, N. and S. America, Eastern Europe, Asia	1. Symptomatic, often critical care treatment —antiviral (ribavirin) helpful if started early	1. Reduce habitat for feral rodents where humans live 2. Personal protection when cleaning where rodents have been (gloves, respirator) 3. Hypochlorite sanitization solution to clean with
Influenza (Grippe)	Myxovirus	(A) Variable effects depending on virulence of the specific strain —swine and avian strains transmissible to man, avian strain more severe in man (B) Mild to severe upper respiratory illness with generalized symptoms. Wild avian (ducks amd geese) asymptomatic carriers	(A) 1, 2, or 3 (the various interrelationships are not completely understood yet) (B) Swine, horses, poultry, domestic and wild ducks and geese —Swine may serve as mixing vessels creating new human communicable strains	(A) Infected animals —The specific roles of swine, horses, and birds as reservoirs of influenza for man are yet to be determined (B) 1 (C) Direct contact, primarily respiratory droplet from infected animals	(A) Swine handlers primarily —Possibly persons working with poultry or horses (B) Worldwide	1. Supportive/ Symptomatic 2. Antiviral agents e.g., amantadine, rimantadine	Provide excellent sanitation in the animal environment, including ventilation —Vaccinate horses —Swine vaccine has been successful in one European Country —separate housing of poultry and swine —Prevent contact of wild avians and domestic avians and swine

(*continued*)

371

Table 13.2. Zoonotic Bacterial Infections *(continued)*

Disease (Common Names)	Etiologic Agents	Health Effects (A) Human (B) Animal	Animal Hosts (A) 1. Zooanthroponosis 2. Anthropozoonosis 3. Amphixenosis (B) Specific Animals Infected	Mechanisms of Transmission (A) Reservoir (B) 1. Directzoonosis 2. Metazoonosis 3. Saprozoonosis 4. Cyclozoonosis (C) Specific mechanisms	Epidemiology (A) Populations at Risk (B) Geographic Distribution	Treatment	Prevention or Control
Milker's Nodules (paravaccinia)	Paravaccinia subgroup of pox virus	(A) Wart-like nodules on the skin of hands and forearms (B) Nodules on the teats and udders of cows	(A) 1. (B) Cattle	(A) Infected cattle (B) 1 (C) Direct contact with teats and udders of cows with active lesions —Hand milking or washing the udder and teats prior to machine milking are primary exposures	(A) Dairy cow milkers and handlers (B) Europe and the United States	1. No specific treatment 2. Treat/prevent secondary bacterial infections with topical antibiotics	—Separation of infected animals —Protective gloves when milking or treating infected cattle
New Cattle Disease (in poultry synonyms are pseudo fowl pest, pneumoenencephalitis)	Paramyxovirus	(A) Conjunctivitis —Occasionally mild influenza-like illness (B) Disease varies depending on the specific virus strain 3 main forms: —mild respiratory illness —respiratory form with nervous involvement in chicks —severe highly fatal pneumonenencephalitis	(A) 1. (B) Chickens primarily, turkeys also —Many other avian species may be infected but are primarily asymptomatic	(A) Infected avian species, domestic or wild (B) 1. (C) Direct or indirect contact with infected birds, their environment, or their tissues —Direct contact with Newcastle vaccines in water or aerosolized	(A) Poultry workers —Those who administer aerosol vaccines to chicken flocks —Poultry processing plant workers (B) Worldwide	1. Symptomatic 2. Avoid sunlight 3. Anti-inflammatory/antibiotic eye drops	Most developed countries have eradicated the severe form and have programs to keep it out of the country —Outbreaks do occur and a test and slaughter program is invoked

Table 13.2. Zoonotic Bacterial Infections (*continued*)

Disease (Common Names)	Etiologic Agents	Health Effects (A) Human (B) Animal	Animal Hosts (A) 1. Zooanthroponosis 2. Anthropozoonosis 3. Amphixenosis (B) Specific Animals Infected	Mechanisms of Transmission (A) Reservoir (B) 1. Directzoonosis 2. Metazoonosis 3. Saprozoonosis 4. Cyclozoonosis (C) Specific mechanisms	Epidemiology (A) Populations at Risk (B) Geographic Distribution	Treatment	Prevention or Control
							—Effective vaccines are available. They can infect workers. Protective clothing and full face respirators should be used when handling sick birds or vaccines —Practice good sanitation and personal hygiene in poultry processing plants
Nipah virus	Virus of the paramyxoviridae family related to hendra—(a zoonoses of horses in Australia)	(A) – Initial influenza-like progresses to encephalitis —40–70% case fatality	(A) 1. (B) Pigs, fruit bat (*pzeropus spn*)	(A) Fruit bats (*pteropus spn*) (B) Pigs to people by virus in urine and respiratory secretions —Fruit bats to pigs by virus in saliva and urine	(A) Swine farmers and family (B) Outbreaks in Malaysia and Bangladesh	1. No specific treatment known, symptomatic and palliative	1. Eradication of infected swine 2. Limiting exposure of pigs to flying bats
Rabies	Rhabdovirus	(A) Progressive encephalitis with personality changes and hyperactivity to external stimuli resulting in spastic contractions of	(A) 1. (B) Many species of domestic and wild mammals —Mainly cattle as agricultural risk sources —Species of canidae, mustecidae, viveridae	(A) Reservoirs vary depending on geographic location; include skunks, bats, raccoons of N. America, foxes in E.U. (B) 1.	(A) Animal handlers working with bovine species —Agricultural workers in outdoor areas where disease is endemic in wild animal population	—Postexposure immunization for exposed —Thorough washing of bite wounds with soap and water	—Vaccination or removal of reservoir host —Preexposure vaccination of people at high risk

(*continued*)

Table 13.2. Zoonotic Bacterial Infections (*continued*)

Disease (Common Names)	Etiologic Agents	Health Effects (A) Human (B) Animal	Animal Hosts (A) 1. Zooanthroponosis 2. Anthropozoonosis 3. Amphixenosis (B) Specific Animals Infected	Mechanisms of Transmission (A) Reservoir (B) 1. Directzoonosis 2. Anthropozoonosis 3. Saprozoonosis 4. Cyclozoonosis (C) Specific mechanisms	Epidemiology (A) Populations at Risk (B) Geographic Distribution	Treatment	Prevention or Control
		skeletal muscles, usually resulting in dysphasia, respiratory failure and death (B) Variable encephalitis depending on species, but usually behavior changes, paralysis of muscles of mastication, death		(C) Direct contact via bite wound or contamination of preexisting wound with saliva —Aerosol transmission is rare.	(B) Occurs in most areas of the world except Australia, most islands in Caribbean and Hawaii		
Rickettsia Q Fever (Query Fever)	Coxiella burnetti	(A) Generalized febrile illness with pneumonitis —possible endocarditis —possible abortion —case fatality rate <10% (B) Often unapparent —May cause abortion especially in sheep	(A) 1. (B) Cattle, sheep, goats —Many small wild animal species	(A) Cattle, sheep, goats —Ticks —Several species of small wild mammals (B) 1, 3 (C) Inhalation of airborne organisms in dust —Direct contact with infected animals, particularly placenta and placental fluids —Consumption of raw milk	(A) Farmers/farm workers in contact with animals or cleaning up the animal environment or assisting at birthing of calves or lambs (B) Worldwide	1. Tetracycline (e.g., doxycycline) 2. Furoquinalines (e.g., difloxacin or ciprofloxacin)	—Personal protection when handling infected animals (especially during parturition) —Respiratory protection when working in a dusty environment contaminated with the organism —Immunization
Psittacosis (ornithosis, fowl chlamydiosis)	Chlamydia psittaci	(A) Variable depending on strain —Generalized febrile illness with headache, constipation and pneumonitus	(A) 1 (B) Turkeys primarily —Ducks, geese, and chickens also —Psittacine birds —Many species of wild birds who flock (starlings, pigeons etc.)	(A) Subclinically infected poultry, psittacine birds, and many species of wild birds (B) 1 (C) Direct contact with infected birds, their tissues or fecal material via direct contact of mucous membranes, or inhalation	(A) Persons raising and handling poultry particularly turkeys —Poultry processing plant workers —People handling psittacine birds (B) Worldwide	1. Tetracycline (e.g., doxycycline) 2. Erythromycin (14 days)	—Personal protection when handling infected birds, their environment, or their carcasses —Eliminate carrier state by

Table 13.2. Zoonotic Bacterial Infections (*continued*)

Disease (Common Names)	Etiologic Agents	Health Effects (A) Human (B) Animal	Animal Hosts (A) 1. Zooanthroponosis 2. Anthropozoonosis 3. Amphixenosis (B) Specific Animals Infected	Mechanisms of Transmission (A) Reservoir (B) 1. Directzoonosis 2. Metazoonosis 3. Saprozoonosis 4. Cyclozoonosis (C) Specific mechanisms	Epidemiology (A) Populations at Risk (B) Geographic Distribution	Treatment	Prevention or Control
		(B) General latent or subclinical —Under stress symptoms may be seen including depression, emaciation, respiratory distress					feeding birds tetracycline —Screen animals before they enter processing plant —Personal protection for poultry processing
Parasitic Echinococcosus (hydatid disease)	Echinococcus granulosus	(A) Chronic-progressive space-occupying lesions of the liver, lung, or other body organ —Symptoms vary depending on type and extent of tissue involvement —Anaphylactic shock may occur if the echinococcus cyst ruptures (B) Sheep, goats space occupying lesion of liver, lung, brain, or other tissue. Usually fatal over a long period. —Dog (primary host) asymptomatic or mild enteritis	(A) 1 (B) Sheep, goats, and a few wild ruminants are secondary hosts —Several species of candidae are primary hosts	(A) Maintenance depends on transmission cycle between sheep and dog (B) 4 (C) Dog eats infected tissue of sheep—adult tapeworm develops in gut of dog—eggs shed in dog feces—man pick up eggs via fecal-oral route	(A) Sheep herders, sheep handlers, especially those that keep dogs (B) Western United States —Latin America —Mediterranean coast countries —Southern Russia —Middle East —Kenya —Australia —New Zealand	1. Surgery to remove cysts 2. Repeated aspiration and injection of cyst with anti helmintics such as mebeudazole or niclosamide	—Control depends on breaking the dog-sheep cycle —Eliminate parasite from dogs with parasiticides —Avoid dogs with possible infections —Avoid feeding tissues of infected sheep to dogs by sanitary disposal of dead sheep and offal

(*continued*)

Table 13.2. Zoonotic Bacterial Infections (*continued*)

Disease (Common Names)	Etiologic Agents	Health Effects (A) Human (B) Animal	Animal Hosts (A) 1. Zooanthroponosis 2. Anthropozoonosis 3. Amphixenosis (B) Specific Animals Infected	Mechanisms of Transmission (A) Reservoir (B) 1. Directzoonosis 2. Metazoonosis 3. Saprozoonosis 4. Cyclozoonosis (C) Specific mechanisms	Epidemiology (A) Populations at Risk (B) Geographic Distribution	Treatment	Prevention or Control
Pork Tape Worm	Taenia soleum	(A) Minor abdominal discomfort for adult tape worms —Humans may have encysted forms of *T. soleum* in tissue causing variable myositis or other illness (B) No major symptoms reported, but reason to reject meat at the abattoir	(A) 1. (B) Pigs, cattle, humans	(A) Pigs, cattle (B) 4. (C) Consumption of improperly cooked pork or beef	(A) Those consuming improperly cooked or frozen beef or pork —People contact with feces from person infected with T. Soleum (B) Worldwide	1. Niclosamide or mebendazole	1. Prevent feces of human from being consumed by pigs or cattle 2. Detection elimination of "measly" pork or beef at abattoir 3. Proper cooking/freezing of pork/beef
Beef Tape Worm	Taenia saginata						
Trichinosis	Trichanelta spiralis	(A) Flulike illness with fever, periorbital edema (B) Asymptomatic	(A) 1. (B) Pigs, bears, sea mammals	(A) Infected meat of pigs, bears, sea mammals	(A) Those consuming improperly cooked or frozen pork, bear meat, or sea mammal meat	1. Usually no treatment required	1. No feeding of pigs with any food (garbage that may have pork) 2. Rodent control around swine 3. Properly cook/ freeze pork
Fungal Blastomycosis	*Blastomyces dermatiditis*	(A) Variable illness similar to other deep mycoses —Generalized illnesses with pulmonary involvement —S. America form more likely to have skin lesions	(A) 1, 3 (B) Humans, dogs, many other species	(A) Soil (B) 3 (C) Inhalation of soil or bird or bat feces	(A) Anybody working in endemic areas where soil is aerosolized (B) B. dermatiditis— South/ Central Canada, Midwest US, Central and S. America, Western Europe —P. brasiliensis—Latin America	1. Itraconizole 200–400 mg/day for 6 months or ketoconazule or amphotacin B	1. Prevent inhalation of dust (use of respirator) when working in areas such as old farm structures or when disturbing soil in endemic areas
S. America Blastomycosis	*Paracoccidioides brasiliensis*						
Coccidioidomycosis	*Coccidioides immitis*						

Table 13.2. Zoonotic Bacterial Infections (*continued*)

Disease (Common Names)	Etiologic Agents	Health Effects (A) Human (B) Animal	Animal Hosts (A) 1. Zoanthroponosis 2. Anthropozoonosis 3. Amphixenosis (B) Specific Animals Infected / Mechanisms of Transmission (A) Reservoir 1. Directzoonosis 2. Metazoonosis 3. Saprozoonosis 4. Cyclozoonosis (C) Specific mechanisms	Epidemiology (A) Populations at Risk (B) Geographic Distribution	Treatment	Prevention or Control
Dermatophytosis (Ringworm, taenia dermatomycosis)	—*Trichophyton verrucosum* —*T. equinum* —*T. mentagrophytes* —*Microscorum canis* —*M. nanum* —*M. gallinaciae*	(A) Skin infection of variable severity —Crusty inflamed lesions that tend to clear centrally—pustules may develop in the active portion of the lesion —Lesions usually occur on face, arms, and head (B) Similar to man except lesions usually much less inflamed and dry —May be patches of hair loss —May be subclinical infections	(A) 1. (B) Most animal species have their own fungal agents which cause skin infections. The infected animal species important in agriculture includes: cattle, goats, sheep, horse, rat, swine, and chicken / (A) *T. verrucosum* (cattle mainly, sheep and goats also possible —*T. equinum* horses —*T. mentagrophytes* (Rat) —*M. canis* (Dog, cat) —*M. nanum* (Swine) —*M. gallinaciae* (Chicken) (B) The environment (barns, feed bunks, corals, etc.) may also serve as a reservoir, because the organism lives for long periods of time off the host. (C) Close direct contact of bare skin to infected animals or their environ.	(A) Farmers and livestock handlers —Persons who milk infected animals are particularly at risk —Children are at greater risk than adults (B) Worldwide	1. First try topical treatment with imidazoles (chlortrimazole or ketaconazole) 2. Systemic treatment may be necessary with fluconazole or ketaconazole	—Practice excellent—animal health programs—e.g., sound nutrition, excellent environmental sanitation prevent overcrowding —Isolate infected animals —Wear protective gloves and clothing when handling infected animals —Practice good personal hygiene
Histoplasmosis	*Histoplasma capsulatum*	(A) Often subclinical —Variable, depending on dose and immune response of the individual	(A) None of above. Animal relationship to humans comes form the fact that the organism grows particularly well in soil contaminated with fecal material of birds or bats. These / (A) Soil—particularly that contaminated by aged feces of birds or bats. (B) 3 (C) Inhalation by producing aerosols of the organism during disturbance of soil that	(A) Farmers as well as other persons who live and work in endemic areas of infection —immunocompromized people are high risk of disseminated disease	1. Amphotericin B 2. Ketoconazole, dapsone, rifampin, are alternatives	—Wetting down soil and wearing a good particle filtering respirator when working in dusty environ- *(continued)*

(B) Dogs most likely affected with generalized symptoms and pulmonary involvement

—*C. immitis*—South West U.S.

Table 13.2. Zoonotic Bacterial Infections (*continued*)

Disease (Common Names)	Etiologic Agents	Health Effects (A) Human (B) Animal	Animal Hosts (A) 1. Zooanthroponosis 2. Anthropozoonosis 3. Amphixenosis (B) Specific Animals Infected	Mechanisms of Transmission (A) Reservoir (B) 1. Directzoonosis 2. Metazoonosis 3. Saprozoonosis 4. Cyclozoonosis (C) Specific mechanisms	Epidemiology (A) Populations at Risk (B) Geographic Distribution	Treatment	Prevention or Control
		—Usually a febrile illness with influenza-like symptoms, cough, pneumonitis, usually recover 2–3 weeks —Chronic forms may be extremely severe and very difficult to treat, with chronic pneumonitis, liver infections, bone infections or other tissues (B) Often subclinical —Similar to human illness	species may also act to distribute the organism in nature (B) Most animals have subclinical infections —Dogs are the primary animal species that develop illness	contains the organism— e.g., cleaning or razing old chicken coops, working in areas where old bird roosts have been	(B) Worldwide, in specific localities where soil and climatic conditions are favorable for growth of the organism		ment conductive for growth of the organism—e.g., old bird roosts, old poultry house, etc.
Protein Particle Variant Creutzfeldt Jacob disease (VCJD) Transmissile spongiform encephalopathy (TSE) Mad Cow Disease	An abnormal infectious protein particle	(A) An unusual progressive/fatal encephalopathy leading to dementia, ataxia, and death (Variant form of Creutzfeldt-Jacob disease) (B) Progressive encephalopathy leading to altered behavior, ataxia, and death	(A) 1. (B) Cattle (only species known infectious for humans) —Elk, mink, cats, sheep, deer also have TSEs	(A) Infected cattle (B) 1.	(A) People consuming brain and nerve tissue from infected cattle (B) Anima mainly in UK, EU, Japan —Two cases in Canada	1. There is no known treatment	1. Detect infected animals and eliminate them from the food chain 2. Do not eat brain/spinal cord tissues of cattle 3. Do not feed dead herbivore animals to cattle

REFERENCES

Acha P, Szyfres B. 2003. Chlamydioses, Rickettioses, and Viroses. Washington, D.C.: Pan American Health Organization. p 158–159, 236–240.

CDC PHS. 1972. Manual of Procedures for National Morbidity Reporting and Surveillance of Communicable Diseases.

CDC UPHS. 2005. Nationally Notifiable Infectious Diseases.

Donham K. 1985. Zoonotic diseases of occupational significance in agriculture: A review. International Journal of Zoonoses 12:163–191.

Erhardt C. 1974. Ranking of causes of death. In: Berlin J, editor. Morbidity and Mortality in the United States. Cambridge: Harvard University Press. p 25.

Heinen P. 2003. Swine Influenza: A Zoonosis. Veterinary Sciences Tomorrow. Swine virus epidemiology.

Hubbert W, McCulloch W, Schnurrenberger P. 1975. Diseases Transmitted from Animals to Man. Springfield, IL: Charles C. Thomas Co. p 147–148.

Kilbourne E. 1973. The molecular epidemiology of influenza. Journal of Infectious Disease 127:478–487.

Kruse H, Kirkemo A, Handeland K. 2004. Wildlife as source of zoonotic infections. Emerging Infectious Diseases 10(12):2067–2072.

Marano N, Pappiaoanou M. 2004. Historical, new and reemerging links between human and animal health. Emerging Infectious Diseases: Center for Disease Control, U.S. Public Health Service.

Murphy F. 1998. Emerging zoonoses. Emerging Infectious Diseases 4(3).

Schwabe C. 1969. Veterinary Medicine and Human Health. Baltimore: Williams and Willins. p 229–240 of 713.

Taylor L, Latham S, Woolhouse M. 2001. Risk factors for human disease emergence. Philos Trans R Soc Lond B Biol Sci 356:983–989.

Unterschultz J, Hobb J, Lerohl M. 2004. Does one mad cow equal one dead industry? BSE in Canada. University of Alberta.

14
Prevention of Illness and Injury in Agricultural Populations

Kelley J. Donham and Anders Thelin

INTRODUCTION

Creating a more healthful and safe working environment for agricultural populations is a huge challenge relative to other occupations (Fragar 1996; Schenker 1996). Challenges include cultural and belief issues of the people (Jones and Field 2002), socioeconomic issues, and issues of the work environment that create special risks. These challenges vary across different sectors of agriculture. The most numerous of these groups (small independent family farms) are perhaps the most challenging. Health and safety promotion on larger farms with employed farm workers is challenging, but in different ways, as are the challenges for health and safety of migrant and seasonal workers. Prevention issues according to specific exposure categories have been discussed in previous chapters. Personal protective equipment was discussed in Chapters 3, "Agricultural Respiratory Diseases," 6, "Health Effects of Agricultural Pesticides," and 9, "Physical Agent." This chapter deals with the broad aspects of prevention from a philosophical and programmatic perspective.

The Challenge of Prevention in Small Independent Family Farms

If anybody is going to tell me how to be safe on my farm, they had better first come out here and walk a mile in my shoes.

> Myron Zumbach, Corn and
> cattle producer, Coggin, IA

Myron was interviewed by a farm safety specialist from the University of Iowa regarding his opinion on how farms can be made safer. His statement and attitude tell a lot about the challenge of creating safer and more healthful farming operations. Myron is typical of most agricultural producers in industrialized countries. They are proud and do not like interference from outside entities, especially if the outsiders have no understanding or credentials in agriculture. Farm people in most areas of the world share a common culture of pride, independence, work ethic, mistrust of institutions, defensiveness, and a need to be understood and respected. Their work takes priority over most anything. Their work is more than a job. It is their duty and mission in life. To do this work, they accept risks as part of the job. Health and safety is not a high priority of the farmer or rancher, relative to matters of production.

This cultural ideology has been present in the farming community for hundreds of years and perhaps first expressed by Thomas Jefferson in his address to the constitutional convention, justifying a federal system that would allow the farming population not to lose representation relative to an urban aristocracy (Kelsey 1994).

> *. . . Cultivators of the Earth are the most valuable of citizens. They are the most vigorous, the most independent, the most virtuous, and they are tied to their country and wedded to its liberty and interests by the most lasting bonds.*
>
> Thomas Jefferson, 1785

These attributes of the farmer have been referred to as the "agrarian myth." It still permeates farming communities of Western civilization in many areas around the world. Farmers feel they are in a noble profession that "feeds the world." They feel they have "an obligation to society and thus a right to farm with little interference."

Although issue ownership and community-based involvement are important prevention programs, active farm people generally (with some notable exceptions) have not been in charge of their occupational health and safety problems (Loos and others 2001; Moore and McComas 1996). There have been numerous community coalitions that have been formed by outside leaders, via funding from governmental or philanthropic sources (Palermo and Ehlers 2002). However, the likelihood of long-term sustainability or expansion of these efforts is unlikely without outside leadership and resources or incentives. The reasons include time and economic limitations of modern farm life, the culture of agriculture, and the immediate (in the moment) orientation to life's challenges. Furthermore, resources that initiate and sustain effective prevention programs are limited in rural communities. Those of us in the health and safety professions are left to understand and accept this system and culture, working within the farming culture to try to make a difference.

THE FARM WORKER

The Migrant and Seasonal Farm Worker (MSFW)

Most industrialized countries have regulations that help protect MSFW. In the U.S., migrant health clinics for acute care are supported by the government, and they aim to serve this population regardless of immigration status. Furthermore, there are unions and advocacy groups that help draw attention to safety issues and help promote protective policies. However, there are many work hazards in this population that need to be addressed (Nawrot and Wright 1998; Schenker 1996). Many of these issues were discussed in Chapter 2, "Special Risk Populations in Agricultural Communities."

The Non-Minority Farm Worker

The non-minority farm worker may be less protected than the MSFW. There are typically fewer of these workers in a given enterprise, and thus not qualified for legal protection and/or regulations (such as in the U.S. where more that 10 employees are required before federal labor inspection and enforcement apply). They are often part of a small family farm operation. They experience the same hazards, but are not afforded the protection given to workers on large family farms with many employees.

The Nature of Farming; Comparing Farming to Other Types of Industry

Compared to other industries, small family farms have an entirely different relationship between labor and management. Table 14.1 compares farming to other enterprises relative to the nature of the work. Note that as some farming enterprises become larger, they tend to acquire characteristics of both industrial enterprises and family farms.

ECONOMICS

The general economic features of agriculture were discussed in the introductory chapter. Farming has always been an economic challenge for the individual producer. However, the growth of the agricultural supply-driven economy fueled by the global economy that was exacerbated in the early 1970s has created high production, keeping commodity prices nearly level relative to continual rising production costs and creating an ever-narrowing profit margin (Donham 1995b).

The latter has resulted in making choices of available resources, such as using old and (often unsafe) equipment, relying on unpaid family labor, and not allocating resources to safety and health applications (Arcury 1997). A fact that needs to be elucidated and appreciated by the farm community and their insurers is that agricultural injuries have a profound economic impact on the individual family and the local economy. Leigh and co-workers (2001) found that for the U.S., nonfatal farm injuries cost about $9,000 per injury, which totals $4.6 billion per year in 1992 dollars. Another economic analysis resulted in findings that installation of rollover protective structures on tractors in the U.S. would save the public $1.5 billion (Pana-Cryan and Myers 2002). The challenge to health and safety professionals is to use this economic reality to help create behavior change in the farming population and behavior

Table 14.1. Differences in Agricultural Work and Industrial Work

	Industry (Individual Plant)	Farm (Family Farm)[1]
Place of work	Separate from home	Workplace and home the same
Work force	Separate labor and management	Combined labor and management
	Concentrated and easily observed	Remote and variable worksites
	Limited to adults	Children, elderly, and spouses are part of the work force
Work routine	Limited to 40-hour work week	No limit to work hours, and often significant rush times
	Usually trained in work practices	No specific training in work practices
	High degree of routine	Highly variable work tasks
	Work tempo controlled	Periods of rush work during harvesting and planting
	Equipment maintained by special crew	Must repair own equipment
Health requirements	Medically selected (preplacement)	No selection for work
	Medical monitoring	Little medical monitoring
	Hygienic facilities provided	Rare to have hygienic facilities
	Health insurance and workers compensation insurance comes with the job	Variable depending on the country (must provide own insurance in the U.S.)
	Regular vacations and holidays	No regular vacations, holidays, or sick leave (although Scandinavian countries have a relief worker system)
Emergency medical services	Available	Usually a long distance to health services
	Co-workers and others nearby to assist in an emergency	Usually work alone
	Rehabilitation services available	Rehabilitation services are rare
Worker protection regulations and enforcement	Occupational health and safety regulations present and enforced	Few regulations and rarely enforced

[1]Many large family farms have employees. Therefore some of the work organization and social and health benefits for these operations would share some characteristics with the industrial operations.

changes in insurance and other businesses supporting agricultural business to provide incentives for lowered occupational health and safety losses.

The inherent risks, marginal economics, and cultural peculiarities of the farm occupation are challenges to health professionals to be creative and dedicated in finding ways to promote health and safety in agriculture.

TYPES OF PREVENTION STRATEGIES

Organizations and Professionals Who Conduct Agricultural Health

There are several professional fields that are applicable to health and safety in agriculture (see Chapter 1, "Introduction and Overview," Figure 1.2). These include a broad array of environmental, public health, safety, clinical medicine, and veterinary professionals (McNab 1998; Wheat and others 2003; Lee 1994; Lundvall and Olson 2001; Martin 1997; Pistella and others 2001; Stueland and others 1996). Equally (or even more important) are active farmers, ranchers, or others with no particular professional training but with a commitment to the health of their profession. These individuals work in various sectors to provide prevention expertise. They include (in no particular order) the following: 1) government agricultural extension services (Purschwitz 1997), 2) public health, 3) research universities, 4) government-associated farm health and safety associations, 5) insurance companies, 6) governmental regulatory agencies, 7) private not-for-profit organizations, 8) practicing veterinarians, 9) farm organizations, 10) primary health care providers, and 11) farm company occupational health and safety programs. Each of these organizations or professional groups has different methods of prevention. The degree of activity of each of these entities varies by region (Fragar and Houlahan 2002). Table 14.2 provides a list and general description of some of these organizations.

Nongovernmental Organizations Practicing Agricultural Safety and Health

There are numerous nongovernmental organizations focusing on agricultural health and safety around the world. The people in these organizations work diligently to improve the health and safety of people who produce food and fiber for the world's people. Much of the work at these organizations is voluntary, and it arises from the genuine concern for the health and safety of this important population. Table 14.3 lists examples of various nongovernmental organizations, people employed, and their activities relative to prevention in agricultural safety and health.

METHODS OF PREVENTION IN AGRICULTURAL SAFETY AND HEALTH

Murphy (1991) has reviewed the history and development of agricultural health and safety. He points out that contributions to this field have come from industrial safety, industrial engineering, education, psychology, and public health. Illness and injury prevention includes elimination of the hazard, application of safeguards, and change in human behavior. Aherin and co-workers (1992) point to three basic methods that may be applied to achieving the actions mentioned above: 1) enforcement of regulations and standards, 2) engineering out the hazard, and 3) educating/training the individual to change unsafe behaviors. Most health and safety specialists recognize that engineering is the most effective strategy, followed by health and safety regulations and enforcement. Education and training to change behavior is a distant third in regard to effectiveness. Although some agricultural health and safety specialists favor greater use of health

and safety regulations because of their proven effectiveness (Purschwitz 2003), regulations and enforcement have largely been absent from production agriculture (this is true primarily for small independent family farms as opposed to large farms) (Meyers and Bloomberg 1993). The reasons for this have to do with the nature of agricultural production and the nature of the farmer (mentioned above). Farm organizations and farm groups have resisted most attempts at regulations. This is true especially in the United States (relative to Australasia, the EU, and Canada), which has generally given higher importance to individual rights than to protection of groups. For example, the U.S. passed the Occupational Safety and Health Act of 1970, with the goal of providing a safe and healthful work environment for all workers. However, in 1976, following pressure from farm groups, an amendment to the regulation was instituted which made it illegal to spend federal dollars to inspect or enforce regulations on farm businesses unless there were 11 or more employees (Kelsey 1994). Although farmers in many states can request an inspection through the Occupational Safety and Health Administration (OSHA) consultation program, the amendment to the OSHA act effectively eliminated nearly 90% of the farms in the U.S. from mandatory coverage under OSHA (Purschwitz 2003). Other countries have had more success in achieving regulations for agricultural safety and health, but compared to other industries, agriculture remains relatively unregulated. Even if there were rigorous inspection requirements for small farms, the nature of production agriculture makes enforcement very difficult. There are millions of small family farms with few employees or only family employees, often in remote locations (American Academy of Pediatrics 2001).

One advantage of large farms is that they have a greater potential for developing effective health and safety programs compared to small farms. These farms have more resources, possibly affording newer and safer equipment, and many have a labor management structure that could better assure compliance with safety programs.

REGULATIONS AND ENFORCEMENT

Although not extensive, some regulations for agricultural health are in existence in most indus-

Table 14.2. Examples of Governmental and Non-Profit Entities Practicing Prevention and Agricultural Health and Safety

Entity	Methods	Activities	Location	Program Examples
1. Agricultural Extension Services	Education and demonstration	Education presentations to farm groups. Mass communication (newsletter, articles in farm publications, websites). Didactic training	All industrialized countries have them, usually as a part of the agricultural technical advisory service. The U.S. has had dedicated activity since 1943.	Irish Agriculture and Food Development Authority provides safety programs, distributes informational brochures and pamphlets to farmers, and sponsors farm safety day camps. (http://www.teagasc.ie/links.htm)
2. Public health	Surveillance epidemiology regulations/enforcement Industrial hygiene education/behavioral change • Social psychology • Social marketing	Epidemiologic studies based on surveillance results	Public health departments are located in most counties as well as in the federal level.	The Iowa (U.S.) Department of Public Health advanced a law that requires health care providers to report all agriculturally related injuries and illnesses. They publish reports on the results and use the information to set prevention goals. (http://www.idph.state.ia.us/ems/sprains.asp)
3. Government-associated health and safety associations	Education conferences Farm safety inspections	Information dissemination Voluntary farm safety checklist	Operated by the Ontario Workers Compensation Association Guelph, Ontario Canada.	These groups disseminate information about the new slow moving vehicle sign requirements on farm machinery. (http://www.farmsafety.ca)
4. Insurance companies	Collecting actuarial information and promoting risk reduction	Monitoring of injuries and costs of insurance for farmers Awards: rate reduction for claims reduction	Most European companies are quasi-private/government organizations. In the U.S., they are private companies.	The Finnish farmers social insurance company. (MELA).
5. Governmental regulatory organizations	Monitoring, inspections for established regulations	Training of employers regarding the regulations	Located within the state or national level governmental units, such as the Department of Labor or Environment.	The worker protection act or the U.S. Environmental Protection Agency (http://www.epa.gov/oppfead1/safety/workers/workers.htm)
6. Private not-for-profit organizations	Education health services	Advocacy training	Many organizations are located in North America and Australia; there are fewer in the EU.	The AgriSafe Network is headquartered in Spencer, Iowa. (http://www.agrisafe.org/network.htm)

(continued)

385

Entity	Methods	Activities	Location	Program Examples
7. Farm organizations	Education	Safety promotion	The Farm Bureau in the U.S. and farmers unions in the EU are active in health and safety.	The Iowa Farm Bureau of Iowa, U.S., is involved in a joint program Certified Safe Farm with the University of Iowa. (http://www.public-health.uiowa.edu/icash/csf/led)
8. Primary health care providers	Diagnosis and treatment Some preventive activities	Supplying of personal protective equipment Health screenings for farmers	Located worldwide.	Finnish Agricultural Health Service The AgriSafe Network (http://agrisafe.com)
9. Veterinarians	Research Practicing veterinary preventive medicine Epidemiology Research Animal disease surveillance and control	Evaluation, consultation on air quality inside livestock confinement buildings Practice and consultation in control of zoonotic infections, antibiotic usage	Located around the world.	Private practicing veterinarians Veterinarians in employment of Departments of Agriculture, food and drug agencies, public health agencies, universities, and producer organizations American Association of Swine Practitioners
10. Farm company or commodity group health and safety programs	Monitoring health and safety regulations	Education of membership	Most livestock commodity groups and large production agricultural operations have programs for occupational and environmental health.	The National Pork Board (U.S) (http://www.porkboard.org/home/default.asp)

Table 14.3. Nongovernmental Agricultural Health and Safety Prevention Organizations[1]

	Funding	Function	Description	Contact Information
International				
International Association of Agricultural Medicine and Rural Health	Affiliated as an international nongovernmental organization to the World Health Organization, specialized institutions of the United Nations, and the Internal Centre for Pesticide Safety	Professional education Publish a journal	An association for agricultural and rural health professionals. Collaborate with other health and medical, as well as scientific, associations and their representatives in areas such as agriculture, food industry, ecology, environment protection, pedagogy, economy, etc. Maintain elective collaboration with governments, governmental organizations, professional associations and other groups or individuals interested in the field of agriculture or rural health.	http://www.iaamrh.org/ Phone: 00 36 1 450 17 68 (Hungary)
International Commission on Occupational Health (ICOH): Occupational Health in Agriculture Scientific Committee	Nongovernmental, professional society	Professional education, conferences Advocates for guidelines and regulations	The largest and oldest international occupational health professional organization. The Agricultural health scientific working committee holds international conferences and other activities to educate its members.	http://www.icoh.org.sg/committees/oh_in_agr.html Phone: (319) 335-4190 (Committee Chair)
International Social Security Association: Section for Agriculture	Funding as a membership organization of social insurance organizations	Advocacy Member education Research	Undertake the tasks, within the framework of the existing ISSA international sections for accident prevention, of implementing accident prevention for all those employed in agriculture and forestry (including special cultures).	http://agriculture.prevention.issa.int/

(continued)

Table 14.3. Nongovernmental Agricultural Health and Safety Prevention Organizations[1] (*continued*)

	Funding	Function	Description	Contact Information
Pesticide Action Network	Nongovernmental organization	Advocacy Education Research	Network of over 600 participating nongovernmental organizations, institutions, and individuals in over 90 countries working to replace the use of hazardous pesticides with ecologically sound alternatives.	http://www.pan-international.org/
North America AgriSafe Network	Membership organization	Farm population education Professional training Preventive clinical occupational health services	Represent rural-based hospitals, health clinics, and county health departments that provide preventive occupational health services for the farming community.	http://www.agrisafe.org/network.htm Phone: (888) 424-4692
Agriwellness, Inc.	Nonprofit, publicly funded organization serving Iowa, Kansas, Minnesota, Nebraska, North Dakota, South Dakota, and Wisconsin	Advocacy Farm population education Professional training Behavioral health services	Promote accessible mental health services for the farm population. Establish training for mental health professionals in agricultural issues.	http://www.agriwellness.org/ Phone: (712) 235-6100
Canadian Agricultural Safety Association	Main sponsors are Agriculture and Agri-Food Canada and FCC (Farm Credit Canada)	Education Advocacy Research	Improve the health and safety conditions of those that live and/or work on Canadian farms.	http://www.casa-acsa.ca/ Phone: (306) 665-2272
Farm Safety 4 Just Kids	Nonprofit organization based in Earlham, IA, serving the United States and Canada	Advocacy Farm population education	Promote a safe farm environment to prevent health hazards, injuries, and fatalities to children and youth.	http://www.fs4jk.org/ Phone: (800) 423-5437

Table 14.3. Nongovernmental Agricultural Health and Safety Prevention Organizations[1] (*continued*)

	Funding	Function	Description	Contact Information
Farm Worker Health and Safety Institute	Nonprofit membership organization	Advocacy Farm population education Professional training	Formed in 1992 to protect the health and safety of migrant and seasonal workers (MSW) against environmental hazards such as pesticides. Conduct training programs with a focus on teaching MSW to organize around environmental issues and ways to impact public policy. Function as the training arm of MSW organizations in the U.S., Mexico, Central America, and the Caribbean.	http://www.cata-farmworkers.org/fhsi.htm Phone: (856) 881-2507
Institute for Rural Environmental Health	Organization located in the College of Public Health, University of Iowa, U.S.	Research Education Service	A unit of the Department of Occupational and Environmental Health of the University of Iowa College of Public Health.	http://www.public-health.uiowa.edu/oeh/ Phone: (319) 335-4415
Institute of Agricultural Rural and Environmental Health	Nonprofit based at Saskatoon's Royal University Hospital at the University of Saskatchewan, Canada	Farm population education Professional training Research	Conduct and stimulate research, education, and health promotion programs aimed at enhancing the health and well-being of agricultural, rural, and remote populations.	http://iareh.usask.ca/ Phone: (306) 966-8286
Iowa's Center for Agricultural Safety and Health	State funding and grants and contracts	"Helping to keep farm families alive and well in agriculture"	Combines the College of Public Health, the Iowa Dept of Public Health, the Department of Agriculture, and Agricultural Extension.	http://www.public-health.uiowa.edu/ICASH/ Phone: (319) 335-4438

(*continued*)

Table 14.3. Nongovernmental Agricultural Health and Safety Prevention Organizations[1] (*continued*)

	Funding	Function	Description	Contact Information
National AgrAbility Project	Cooperative State Research, Education and Extension Service, an agency of the USDA, which administers the AgrAbility Project; the Project funds both a National AgrAbility Project and several State AgrAbility Projects	Professional training Direct services to disabled farmers	Provide training, technical assistance, and information on available resources to the State AgrAbility Project staffs. Provide direct technical consultation to consumers, health and rehabilitation professionals, and other service providers on how to accommodate disabilities in production agriculture.	http://www.agrabilityproject.org/ Phone: (866) 259-6280
National Association for Rural Mental Health (NARMH)	Membership organization composed of approximately 500 organizations and individuals from the United States and a few foreign countries	Advocacy Professional training	Develop and enhance rural mental health and substance abuse services and to support mental health providers in rural areas. Now NARMH has added the goal of "developing and proactively supporting initiatives that will strengthen the voices of rural consumers and their families."	http://www.narmh.org/ Phone: (303) 202-1820
National Center for Farmworker Health, Inc.	Private, nonprofit corporation located in Buda, Texas	Farm population education Professional training	Improve the health status of MSW families by providing information services and products to a network of more than 500 migrant health center service sites in the United States as well as other organizations and individuals serving the MSW population.	http://www.ncfh.org/ Phone: (800) 531-5120

Table 14.3. Nongovernmental Agricultural Health and Safety Prevention Organizations[1] (*continued*)

	Funding	Function	Description	Contact Information
National Children's Center for Rural and Agricultural Health and Safety	Funded by the National Institute for Occupational Safety and Health and the Federal Maternal and Child Health Bureau	Advocacy Professional training	Enhance the health and safety of all children exposed to hazards associated with agricultural work and rural environments.	http://research.marshfieldclinic.org/children/ Phone: (888) 924-SAFE(7233)
National Education Center for Agricultural Safety (NECAS)	Partnership between Northeast Iowa Community College and the National Safety Council	Farm population education Professional training Research	As a training center reduce the level of injuries, preventable illnesses, and fatalities among farmers, ranchers, their families, and employees.	http://www.nsc.org/necas/ Phone: (888) 844-6322
National Farm Medicine Center	Nonprofit program of the Marshfield Clinic Research Foundation, funded by internal sources and competitive government grants and contracts	Health professional training development in agricultural health, networking, information exchange	Conduct research addressing human health and safety associated with rural and agricultural work, life, and environments.	http://research.marshfieldclinic.org/nfmc/ Phone: (800) 662-6900
National Institute for Farm Safety		Farm population education Professional training	An organization dedicated to the professional development of agricultural safety and health professionals, providing national and international leadership in preventing agricultural injuries and illnesses to the agricultural community.	http://www.ag.ohio-state.edu/~agsafety/NIFS/nifs.htm Phone: (608) 265-0568

(*continued*)

Table 14.3. Nongovernmental Agricultural Health and Safety Prevention Organizations[1] (*continued*)

	Funding	Function	Description	Contact Information
National Rural Health Association	Nonprofit association headquartered in Kansas City, Missouri, with a Government Affairs Office in Washington, D.C.	Advocacy Farm population education Research	Provide a forum for the exchange and dissemination of ideas, information, research, and methods to improve rural health.	http://www.nrharural.org/ National Office: (816) 756-3140 Government Affairs: (703) 519-7910
NIOSH Agricultural Centers Regional programs in Iowa, Ohio, New York, North Carolina, Kentucky, Texas, California Washington, Colorado, Wisconsin (National Children's Center for Rural Agricultural Health and Safety)	Federally funded through the U.S. Public Health Service/CDC/NIOSH	Research Farm population education Professional training	Nine Centers were established by cooperative agreement to conduct research, education, and prevention projects to address the nation's pressing agricultural health and safety problems. The Centers are distributed throughout the nation to be responsive to the agricultural health and safety issues unique to the different regions.	http://www.cdc.gov/niosh/agctrhom.html
North American Agromedicine Consortium	Professional organization	Farm population education Research	Devoted to the programmatic teamwork of land-grant and medical universities and their partners to promote health and prevent disease for farmers and farm workers and their families, others in rural communities, and consumers of food and fiber.	http://www.agromedicine.org/

Table 14.3. Nongovernmental Agricultural Health and Safety Prevention Organizations[1] (*continued*)

	Funding	Function	Description	Contact Information
Sharing Help Awareness United Network	Nonprofit	Advocacy Farm population education	Provide peer support and professional services and advocate counseling for farmers, farm workers, and their families who have experienced death or disability.	http://www.shaunnetwork.org/ Phone: (877) 867-4286
European Union				
Health & Safety Executive: Agricultural and Food Sector (Great Britain)	Government	Farm population education Professional training Research	Protect people's health and safety by ensuring that risks in the changing workplace are properly controlled. The Agricultural and Food Sector is based in the Midlands and is made up of three sections: Health, Education, and Employment; Safety; and Forestry, Strategy, and Evaluation.	http://www.hse.gov.uk/agriculture/index.htm
Norwegian Farmers' Association For Occupational Health And Safety	Farmers's union Government Individual farmer	Farm population education Preventive health services	Establish a national occupational health service through cooperation with local occupational health services. Be available for farmers and all others working in agriculture. Help create a safe and pleasant workplace, minimizing the risk from health hazards and ensuring a safe environment for children.	http://www.landbrukshelsen.no/ Phone: (32) 29 90 30
Finnish Farmers Safety and Preventive Health Association	Government (part of health services) Farmers' union Connected to farmers' insurance	Preventive health services Farm population education	Connected to regional health centers. Provides clinical and on-farm technical safety services, education to farmers and to health care providers.	http://www.lpa.fi/tt_main.asp?path=780;1661

393

Table 14.3. Nongovernmental Agricultural Health and Safety Prevention Organizations[1] (*continued*)

	Funding	Function	Description	Contact Information
Australasia				
Australian Centre for Agricultural Health and Safety	A center within the University of Sydney	Research	Collect and report farm injury data.	http://www.acahs.med.usyd.edu.au/ Phone: (61) 2 6752 8210
Farmsafe Australia, Inc.	Nonprofit	Farm population education Professional training	The mission or role of Farmsafe Australia has been defined as "to lead and coordinate national efforts to enhance the well-being and productivity of Australian Agriculture through improved health and safety awareness and practices."	http://www.farmsafe.org.au/ Phone: (02) 6752 8218 (NSW, Australia)
National Rural Health Alliance, Inc.	Nongovernment organization	Advocacy	Work to improve the health of people throughout rural and remote Australia. Strengthen collaboration between and among rural and remote communities, service providers, policymakers, researchers, non-government agencies, Member Bodies, and other stakeholders.	http://www.ruralhealth.org.au/ Phone: (02) 6285 4660

[1]This is not intended to be an exhaustive list but, rather, examples of organizations within the industrialized agricultural countries. The authors realize there are many other organizations that could be included in this table.

394

trialized countries. Table 14.4 lists some of these regulations. In the U.S., individual states, such as Washington, may choose not to take federal OSHA funds and can choose to write their own regulations and thus inspect and enforce regulations on small operations (Meyers and Bloomberg 1993). Regulations pertaining to agriculture covered under OSHA include safe storage and handling of anhydrous ammonia, signs and tags that remind of hazards, and tractor rollover protection and machinery safety. There are no regulations pertaining to air quality. There are regulations regarding migrant workers, for example, and sanitation facilities in labor camp housing and in the field (OSHA 2005). Furthermore, OSHA in the U.S. (and in most industrialized countries that have similar governmental regulations) enforces regulations that include general safety standards, such as legal responsibility to inform workers of all hazardous substances used in the workplace. There are confined space entry and lock-out/tag-out regulations. Furthermore, under the general duty clause, hazards not otherwise specified may be covered, if there is reasonable scientific evidence that a health or safety hazard exists. The fact that permissible exposure limits for air contaminants are not applicable extends a risk to many agricultural workers. For example, there is a large body of literature that indicates respiratory hazard risks occur in confined livestock production facilities when organic dust exposures exceed 2.5 mg/m^3 and ammonia exceeds 7 ppm (Donham 1995a). Although many facilities have exposures above these levels, there are no regulations that guide owners/operators for their own health, or their family members' or employees' health. Although there are regulations in regard to migrant labor camps, which may be monitored and enforced, other production agricultural facilities are rarely monitored and enforced in the U.S., even if there are more than 10 employees in an operation. The reasons are many, including lack of enforcement personnel, lack of trained personnel, unpopularity with the law on the part of the employers, and lack of concern on the general public. As farms become larger, more agricultural operations will be subject to OSHA regulations. However, at present, it is estimated that only about 10% of the farming operations in the U.S. are now covered. There are currently nearly 1,900,000 small family farm operations with an estimated 3,000,000 persons not covered by the OSHA regulation (USDA 2002, 2004).

There are regulations that provide some protection for children working on farm operations other than the family farm. In 1938, the U.S. Department of Labor promulgated the Fair Labor Standards Act. This act was amended in 1966 and again in 1970 to cover employment of youth in agriculture. The so-called Hazardous Occupations Order for Agriculture (HOOA) aims to protect youth under 16 (Child Labor Bulletin No. 102) (U.S. Dept of Labor 1970). The particular hazardous activities that youth under the age of 16 are not allowed to perform (except for stated exemptions seen below) are included in Table 14.5.

The exemptions to these orders are as follows: Youth 14 and 15 years of age are exempt if they

1. Are employed by their parents (or stand in) on farms owned or operated by their parents
2. Are student learners in a legitimate vocational agriculture program (applicable to items 1–6 in Table 14.5)
3. Have had training in tractor and or machine operation as offered by the 4-H or Vocational Agriculture Training Program (there are separate tractor and machine operation training programs, and the exemptions are applicable, respectively, to items 1 and 2 in Table 14.5)

Additional exemptions exist to allow (with written parental permission) for age 12- and 13-year-old youth to be employed on farms where a parent or parental stand-in is employed.

Furthermore, age 10- and 11-year-old youth may be employed for no more than 8 weeks total duration during a season from June 1–October 15, and for hand harvest crops. Youth must be local permanent residents. Employers must obtain a permit from the Secretary of Labor to hire youth of this age.

Although there are fines if these orders are violated, the HOOA act is not well monitored or enforced, unless there is a complaint. Therefore this act functions more like a voluntary educational program than a regulation. Although there has been relatively little outcome evaluations of this training, the effectiveness of these educational programs is questionable. Evidence from two Ph.D. dissertations have shown no decrease in injuries for youth who have taken the training as

Table 14.4. Regulations in Agricultural Health and Safety

	Regulation Name	Effective Date	Target Group	Description	Document Location
International					
International Labour Organization	C184 Safety and Health in Agriculture Convention, 2001	June 21, 2001	Agricultural employers and workers	Aim to prevent injuries and illnesses arising out of, linked with, or occurring in the course of work by eliminating, minimizing, or controlling hazards in the agricultural working environment.	http://www.ilo.org/ilolex/cgi-lex/convde.pl?C18
International Labour Organization	R192 Safety and Health in Agriculture Recommendation, 2001	June 21, 2001	Agricultural employers and workers	Recommendations supplementing C184 Safety and Health in Agriculture Convention, 2001, concerning occupational safety and health surveillance, and preventive and protective measures.	http://www.ilo.org/ilolex/english/recdisp1.htm
North America					
United States Congress	29 CFR 00 Migrant and Seasonal Agricultural Worker Protection Act	March 31, 1989	Migrant and seasonal agricultural workers	Remove activities detrimental to migrant and seasonal agricultural workers; require farm labor contractors to register; assure protection for migrant and seasonal agricultural workers, agricultural associations, and agricultural employers.	Code of Federal Regulations, Title 29 (Labor), Chapter 5 (Wage and Hour Division, Department of Labor), Part 500
United States Department of Labor, Occupational Safety and Health Administration	29 CFR 1910.142 Federal Migrant Housing Regulations	June 27, 1974	Migrant and seasonal agricultural workers	Migrant agricultural labor housing constructed or substantially renovated since April 3, 1980 must comply with these standards of the Occupational Safety and Health Administration.	

Table 14.4. Regulations in Agricultural Health and Safety (*continued*)

	Regulation Name	Effective Date	Target Group	Description	Document Location
United States Environmental Protection Agency	40 CFR 170 Worker Protection Standard (WPS) for Agricultural Pesticides	October 20, 1992	Agricultural employers and workers	Require an agricultural employer or a pesticide handler–employer to assure that each worker and handler subject to the standard receives the required protections.	http://www.epa.gov/pesticides/safety/workers/PART170.htm
United States Department of Labor	29 CFR 1928 Occupational Health and Safety Standards for Agriculture	1970	Agricultural employers and workers	Occupational safety and health standards applicable to agricultural operations.	http://www.access.gpo.gov/nara/cfr/waisidx_03/29cfr1928_03.html
Mexico: Secretariat of Labor and Social Security	Official Mexican Standard NOM-003-STPS-1999	June 28, 2000; Modification 2003; effective 2004	Agricultural employers	Agricultural activities—use of pesticides and materials of vegetable nutrition or fertilizers—safety and health.	http://www.mexicanlaws.com/NOM-003-STPS-1999.htm
Mexico: Secretariat of Labor and Social Security	Official Mexican Standard NOM-007-STPS-2000	September 9, 2001	Agricultural employers and workers	Establish the safety conditions that the installations, machinery, equipment, and tools utilized in agricultural activities must consist of in order to prevent risks to the workers.	http://www.mexicanlaws.com/NOM-007-STPS-2000.htm
European Union	Occupational Health and Safety (Tractor Safety) Regulations 1986	April 16, 1986		Issued under the "Occupational Health and Safety Act 1985" (CIS 88-1751). Applies to tractors; use of safety devices (machine guards, protective frames, power take-off guards).	http://europe.osha.osha.eu.int/data/products/oshainfo_916
Australia	National Standards for Plant	1994	Farmers and Farm workers	Each state has their own regulations. However, the 1994 National Standards for Plant was an attempt to standardize the states.	http://www.ohs.anu.edu.au/publications/pdf/working_paper_2.pdf

Table 14.5. Hazardous Occupations Orders for Youth in Agriculture—U.S.

Excerpted from the Fair Labor Standards Act, U.S. Department of Labor

The Secretary of Labor has found and declared that the following occupations in agriculture are hazardous for minors less than 16 years of age. No minor under 16 may be employed at any time in these occupations except as exempted.

(1) Operating a tractor of over 20 PTO horsepower, or connecting or disconnecting an implement or any of its parts to or from such a tractor

(2) Operating or assisting to operate (including starting, stopping, adjusting, feeding or any other activity involving physical contact associated with the operation) any of the following machines:

Corn picker, cotton picker, grain combine, hay mower, forage harvester, hay baler, potato digger, or mobile pea viner

Feed grinder, crop dryer, forage blower, auger conveyor, or the unloading mechanism of a powered self-unloading wagon or trailer

Power post-hole digger, power post driver, or nonwalking–type rotary tiller

(3) Operating or assisting to operate (including starting, stopping, adjusting, feeding, or any other activity involving physical contact associated with the operation) any of the following machines:

Trencher or earthmoving equipment

Fork lift

Potato combine

Power-driven circular, band, or chain saw

(4) Working on a farm in a yard, pen, or stall occupied by:

Bull, boar, or stud horse maintained for breeding purposes

Sow with suckling pigs, or cow with newborn calf (with umbilical cord present)

(5) Felling, bucking, skidding, loading, or unloading timber with butt diameter of more than 6 inches

(6) Working from a ladder or scaffold (painting, repairing, or building structures, pruning trees, picking fruit, etc.) at a height of over 20 feet.

Table 14.5. Hazardous Occupations Orders for Youth in Agriculture—U.S. (*continued*)

(7) Driving a bus, truck, or automobile when transporting passengers, or riding on a tractor as a passenger or helper	
(8) Working inside:	A fruit, forage, or grain storage designed to retain an oxygen-deficient or toxic atmosphere
	An upright silo within 2 weeks after silage has been added or when a top unloading device is in operating position
	A manure pit
	A horizontal silo while operating a tractor for packing purposes
(9) Handling or applying agricultural chemicals classified under the Federal Insecticide, Fungicide, and Rodenticide Act (as amended by Federal Environmental Pesticide Control Act of 1972, 7 U.S.C. 136 et seq.) as Toxicity Category I, identified by the word "Danger" and/or "Poison" with skull and crossbones; or Toxicity Category II, identified by the word "Warning" on the label	Specified tasks include cleaning or decontaminating equipment, disposal or return of empty containers, or serving as a flagman for aircraft application.
(10) Handling or using a blasting agent, including but not limited to, dynamite, black powder, sensitized ammonium nitrate, blasting caps, and primer cord	
(11) Transporting, transferring, or applying anhydrous ammonia	

http://www.abe.iastate.edu/Safety/clb102.htm

specified under HOOA (Silletto 1976; Williams 1983). Furthermore, Risenberg and Bear (1980) have found that this training may be associated with an increase in injuries. This observation is one that may be more of an association with exposure (trained youth have more work exposure) than as a direct causal effect. A similar increase in automobile crashes with youth taking drivers training programs has also been noted (Robertson 1983).

ENGINEERING

Most health and safety specialists agree that along with regulations and enforcement, engineering applications to health and safety can be very successful in preventing occupational injuries and illnesses (Meyers and Bloomberg 1993; Myers 2000; Thelin 1990). The engineering approach assumes that human beings will err in judgment and behavior and put themselves at risk. The engineering goal is to remove the hazard to protect the operator in spite of unsafe behaviors (Powers and others 2001; Rains 2000). The hierarchy of safety engineering includes the following in order of priority (Murphy 1991):

1. Eliminate the hazard.
2. Apply safeguard technologies.
3. Employ warning signs.
4. Train and instruct the operator.
5. Employ personal protection equipment.

Items 3–5 above also appear in the educational and or public health approach to prevention and will be discussed in those sections. Elimination of the hazard is the ultimate goal of safety engineering, and there are several essential principles to reaching this goal, including

1. Hazard analysis
2. High product reliability
3. Failsafe design
4. Monitoring and structural safety factors
5. Passive protection

Hazard analysis is a way to trace the sequence of events leading to an injury event. Fault-free analysis is a specific type of hazard analysis involving all the events and structures that link to an injury event. The engineering correction is in the design of a feature in the process or machine

that eliminates the "weak link(s)" in the operation. An example is the situation where one must be sitting in the seat and push down on the clutch before the machine will start. This prevents a runover from the operator trying to start the tractor while not seated in the control position. A failsafe design principle is used in electrical circuits (e.g., ground fault circuit interrupters, GFCI) to prevent electrocutions from electrical hand tools used in agriculture. GFCI may be installed in the service, power outlet, or the appliance itself. The interrupter measures the amount of current leaving the service and returning back to the service. If there is even a slight difference (leakage), GFCI will shut off service to that line immediately, preventing a possible electrical shock to the operator. Passive protection can be fairly effective in eliminating risk of injury in agriculture. Clearly the most important passive protection system in agriculture is the rollover protective structure (ROPS). Installation of this device, for all practical purposes, eliminates the probability of a fatal outcome if the tractor should overturn (Reynolds and Groves 2000). Additional protection can be afforded if seat belts are worn with the ROPS. However, more than 50% of tractors in the U.S. do not have ROPS (Donham 1997). A discussion of legislation and programs to enhance ROPS installation is seen in Chapter 11, "Acute Agricultural Injuries," of this text. Another important example of passive protection is the shield used to prevent bypass start of a tractor. Should the main electrical start switch (usually activated by the key) fail to activate the solenoid switch on the starter of a tractor, one can bypass that bad connection by (standing next to the starter motor, usually in front of the rear wheels of the tractor) and jumping across the connections from the battery where they attach to the starter. However, all too often the tractor is unknowingly in gear, and as the tractor starts, it jumps forward, running over the operator. A shield covering the terminal connections prevents this type of injury (unless the operator removes it).

Although engineering is one of the two most effective means of injury prevention, there are significant limitations to its effectiveness, including the following:

1. Most engineering is applied by the manufacturer on new equipment. However, there is a

lot of old equipment in use without modern safety features. Retrofit equipment may not be available, it is expensive, and retrofitting safety equipment on old machinery may increase the liability of the manufacturer.

2. There are few incentives or programs to promote or facilitate the operator investing in new engineering controls for old machines.

3. Developing and installing new engineering safety equipment on machines is subject to economics and marketing. If the new engineering design and deployment on new machines puts one company at an economic, liability, or marketing disadvantage to competitors, such engineering will not likely happen.

4. Safety features can be defeated. For example, the bypass starting shields can be removed. Warning devices can be turned off.

5. Often, failures in engineering or engineering interventions may not become apparent until injuries occur. This means that many machines are in service that are outdated or need modification.

At any rate, even though there are limitations to engineering controls, they are one of the two most effective strategies available to reduce agricultural injuries.

ERGONOMICS/HUMAN FACTORS ENGINEERING
Human factors engineering is a cross between engineering design, psychology, and human anatomy and physiology. The principles include designing machines so that they can be operated effectively and safely by a person, instead of trying to make the person conform to the machine. Examples of human factors engineering in agriculture include design of the cabin of a tractor, where the gauges, lights, warning devices, controls, and seat are designed so that the operator can readily see and operate the machine in a safe and productive manner. Ergonomics and human factors engineering are similar. However ergonomics often relates more to the physical work that the operator has to do, modifying the processes of work to conform to the limitations of human physiology and anatomy. Limitations of implementing ergonomic fixes include dealing with the beliefs and culture of the people you are working with to achieve adoption of the principle. For example, one well-intended worker for the

Foreign Service working in a developing country obtained funding to buy farmers new long-handled hoes to prevent stooping and excess back strain. Upon returning to the field several months later, he had found the farmers had cut off the handles of all the new hoes, because they believed they could do a better job of weeding the crop with the short hoe (even at the expense of extra strain on their backs).

THE PUBLIC HEALTH APPROACH
The public health approach to agricultural health and safety is being increasingly used in various industrialized countries (Fragar and Houlahan 2002; Chapman and others 1996; Huneke and others 1998). Similar to engineering, the public health approach to injury and illness prevention assumes that to err is human, but there are ways to overcome human error. Public health includes a broad concept of health, including physical and mental health promotion and disease, injury, and disability prevention. Public health utilizes several scientific fields to promote health, including surveillance, epidemiology, health behavior, social marketing, and evaluation. Using program evaluation results to improve interventions is a key principle to the public health approach to injury and illness prevention (Meyers and Bloomberg 1993). Surveillance is critical to understanding the problem and measuring successes (Browning and others 2003; Hubert and others 2001; Hubler and Hupcey 2002; Kumar and others 2000; Stueland and others 1996; Yoder and Murphy 2000). Surveillance data regarding agricultural injuries and illnesses are not widely available (Stiernstrom and others 1998). Available data sources are discussed in Chapters 1 and 11 of this text.

William Haddon (1963) is credited as the first person to apply epidemiologic principles to injury investigation. L. W. "Pete" Knapp (1965) is perhaps the first person to apply epidemiologic principles to agricultural injuries. Injury epidemiology is an ecologic approach to discovery of the event. An understanding of the causal factors and variables in the causes of the injuries will lead to the understanding and interplay of factors contributing to an injury event. In its simplest form, there are contributions from human behavior (Man), the Environment, and the Machine (Mausner and Kramer 1985). "Observational epidemiology" or simple descriptive epidemiology will provide clues to the causes of an event. Analytical epidemiology involves the study of many events, the establishment of rates, and determining statistically significant risk factors based on multiple variables in the events. Once risk factors are understood, interventions can then be specifically designed to control them. The epidemiologic approach rejects the idea that injuries are caused by accidents (uncontrollable "acts of God"). The public health approach rejects the use of the term "accident," because this thinking denies the reality that there are many controllable factors responsible for injury events. Understanding these factors allows for the ability to control injuries. Expanding on the man, machine, environment analysis, Haddon (1963) proposed a matrix to study injury events. He suggested that these factors can be analyzed at various times of the event, i.e., pre-event, event, and post-event. Analysis in this detail can reveal ways to control an injury event.

Social marketing is another tool used in public health to promote health. Social marketing uses and adapts concepts from commercial marketing to promote behavior change (Social Marketing Institute 2005). Commercial marketing attempts to create a change in behavior so that people will buy a certain product or service. Social marketing attempts to sell an idea that will create a specific health behavior change. This concept is just beginning to be used in agricultural health and safety in a more formalized manner. Because of the lack of regulations and enforcement, delivery mechanisms for retrofitting newer engineering controls and evidence of effective awareness-level education programs, agricultural health professionals believe that social marketing may have an important place for prevention of agricultural illnesses and injuries. The workshop entitled "Tractor Risk Abatement and Control: The Policy Conference" (Donham 1997) called for social marketing as an important tool to promote installation of ROPS on tractors and to avoid extra riders on tractors and other machines. A more recent document on tractor safety entitled "National Agricultural Tractor Safety Initiative" (Swenson 2004) also calls for social marketing as a tool to enhance installation of ROPS on tractors. Social marketing has been used effectively in other public health applications, such as sun protection (Australia) and antismoking campaigns (most in-

dustrialized countries). There have been few social marketing programs in agricultural health and safety to date. One social marketing program to decrease pesticide exposures in horticulture workers resulted in little evidence of reduced illness or exposure (Flocks and others 2001). However, there are several additional social marketing programs in the planning process, and evaluation of these programs will help guide such efforts in the future. One disadvantage to social marketing is that it is very expensive. In order to affect a large number of people or a large geographical area, expenditures can easily reach millions of dollars. Additionally, it takes time to see results of social marketing; perhaps 5–10 years to detect a difference. The large amount of money and long-term commitment to a project is usually not available for agricultural health and safety programming.

Surveillance, another important public health approach to prevention in agricultural health and safety, has been initiated by the U.S. National Institute for Occupational Safety and Health (NIOSH). Examples of programs that emphasized surveillance include "Farm Family Health Hazard Surveillance," and "Nurses Using Rural Sentinel Events." Other examples of larger more integrated programs that are public-health based include Iowa's Center for Agricultural Safety and Health (I-CASH 2005) and the Agricultural Health and Safety Section of the California Department of Health Services (CDHS 2005). These programs include the application of surveillance, epidemiology, industrial hygiene, and ergonomics. More details of these programs are seen in Table 14.2.

INDUSTRIAL HYGIENE APPROACH
In 1904, the U.S. Public Health Service established training in Industrial Hygiene (U.S. Dept of Health 1973). Since that time, this area of interest has developed into an internationally recognized profession with the objectives of 1) **recognition** of relationships of hazardous exposures to health in the workplace, 2) **measurement and evaluation** of the hazardous agent(s), and 3) formulating and implementing a plan to **control** or eliminate the hazard. Industrial hygiene is a type of public health approach that is based on the integration of engineering, chemistry, physics, and biology. Methods used to control the problem are multifac-

torial, including identification and removal of the source of hazardous agent(s) by changing work practices or process, ventilation, and/or proper selection and use of personal protective equipment. Examples of use of industrial hygiene principles in agriculture include measurement of dust and gas exposures in livestock confinement buildings and designing and evaluation of oil sprinkling systems to suppress dust (Nonnenmann and others 2004). Another example of the industrial hygiene approach is the substitution of hazardous chemicals with less toxic substitutes, such as the replacement of organophosphates with microbial insecticides, and genetically modified resistant crops. Industrial hygienists have great potential to effect change for use in agricultural health and safety. Most industrial hygienists in agriculture focus in the area of research. Industrial hygienists as private practitioners in agriculture are limited, as small farmers are often unable to pay these professionals, and on larger farms, there is little awareness of the services of these individuals. There is only one known program in the world that provides academic training of hygienists for agricultural application (University of Iowa Agricultural Health and Safety Training Program).

OCCUPATIONAL HEALTH SERVICES FOR FARMERS
Occupational health services generally began to evolve in the late 1940s. These services applied preemployment physical exams, acute medical treatment of injuries and illnesses, surveillance, medical screening, and education of workers on occupational exposure risks and prevention. By the 1950s, many of the larger industries had included occupational health services as a part of their worker health programs. In the late 1970s in Scandinavia, it was recognized that many small business workers, including farmers, did not have adequate access to occupational health services. The staff of the acute care medical service did not have the training or interest to detect and treat illnesses caused by occupational exposures. Therefore, small industry-specific occupational health services began to evolve. It was in 1978 that the farmers' occupational health service (Lantbrukshälsan) began in Sweden (Hoglund 1990). This voluntary program developed with support from the farmers union, the government, and the individual farmer (each paid about 30% of the cost). About 40 clinics were established around the country, serving 40% of all farmers. These clinics were locally managed by trained nurses and staffed with industrial hygiene technicians and physical therapists. Unfortunately, the Swedish program ended in 2002 as a result of financial difficulty. However, the Swedish Farmers Union still employs a person who deals with occupational health issues. It also has a funding mechanism that allows it to conduct agricultural occupational health research.

A program similar to the original Swedish occupational health service for farmers was developed in Finland in the late 1970s, about the same time the Swedish program began. This program was integrated into a total of 350 health care clinics in most rural municipalities (Notkola and others 1990). The staff received special training in agricultural health and safety. The Finnish program had 37,000 members in 2003 (36% of farmers).

In 1987, a program now called the AgriSafe Network was founded in Iowa, by Iowa's Center for Agricultural Safety and Health at the University of Iowa, U.S. (Gay and others 1990). This program was modeled after the Swedish and Finnish systems, being managed by nurses, with assistance in industrial hygiene and medicine from the University of Iowa. In 2002, AgriSafe became a not-for-profit corporation (AgriSafe Network 2005). Its 20 clinics scattered across the state serve farming communities within their region through the provision of clinical screenings, occupational health consultation, fitting and selection of personal protective equipment, and community agricultural health educational programming.

The Norwegian Farmers' Association for Occupational Health and Safety (NFAOHS) was established in 1994. It is funded and managed by the Norwegian Farmers' Union and the Norwegian Farmer and Smallholder's Union, in partnership with the Department of Agriculture. Staffing is similar to that described for the programs in Sweden, Finland, and Iowa. The individual clinics are associated with state-run clinics in the region (NFAOHS 2005). The integrated services offered by these occupational health services have been universally well received by the farmers. The farmers appreciate being seen by health professionals who understand agriculture and empathize with the difficulties and exposures inherent in their occupation.

Most of these services have not been in place long enough to measure long-term results. However, most have shown success in increasing the use of personal protective equipment and in reducing health care costs (Donham 1995b). The Lantburkshälsan organization in Sweden showed a reduction in noise-induced hearing loss in the farm population from 25% to 5%, through the widespread use of hearing protection (Thelin 1990). A reduction in cardiovascular diagnoses and higher scores in a number of wellness measures was seen in farmers using an occupational health service compared to farmers not using such a service (Thelin and others 1999). Furthermore, the AgriSafe clinic network in Iowa has employed the Certified Safe Farm Program, which encourages farms to achieve a certain standard of safety. This program has shown a combined decrease (compared to controls) in self-reported out-of-pocket medical expenses and payouts for insurance companies of 24%. The program has also shown a significant reduction in an acute respiratory and systemic condition called *organic dust toxic syndrome*.

There have been other research programs that have utilized the on-farm safety inspection without the clinical screening component of the occupational health services. Rasmussen (2003) reported on a research project in Denmark that had two elements of an occupational health service (on farm safety audit and in-depth education). The results of this study were a 23% reduction on all injury rates and 40% reduction in more severe (medically treated) injuries. The Farm Safety Walkabout program, developed by Iowa's Center for Agricultural Safety and Health, is designed as a video-guided activity for children and parents to walk about the farm and identify and remove hazards. Legault and Murphy (2000) developed a self-inspection guide for farmers to find and remove hazards around the farm. Two Finnish researchers (Heikkonen and Louhevaara 2003) reported on a project comparing farm inspections by farmer peer groups and trained professionals. Evaluation revealed that the farms visited by the peer group resulted in greater improvements on the farm.

EDUCATION

Education has been the most common type of prevention program employed in agriculture.

However, as previously mentioned, there is little evidence that "information only" programs make significant long-term differences in injury and illness outcomes (DeRoo and Rautiainen 2000; Murphy and others 1996). Wilkins and coworkers (2003) have shown that injury training for principal operators had no relation to installation of rollover protective structures. Westaby and Lee (2003) have shown that safety knowledge is negatively associated with agricultural injuries. There are numerous reports on the evaluation of educational programs, but most cite that evaluation is difficult and often not optimally completed (Kidd and others 1996, 2003), due to lack of outcome data or control groups. Most often, only short-term evaluation is available, such as retention of information and increased use of personal protective equipment (Mandel and others 2000). Numerous prevention programs have been directed toward farm youth (Kidd and others 1997; Liller 2001; Liller and others 2002). Many adhere to the concept that young children are more malleable in their makeup, and behavior change may be easier at an early age, with results carrying on into later life (Reed and others 2001). There is also the assumption that the children will bring safety information home, which helps influence the parents to adopt more safe behaviors (Steffen 1998). However, long-term outcome evaluation of education for youths and adults is extremely difficult to track (Reed and others 2003). Although these programs in combination may have a preventive effect, it is difficult to prove (Cole and others 1997; Elkind and others 2002).

One of the reasons for the difficulties in achieving effectiveness of these programs is the cultural conundrum found in the agricultural population regarding its occupational health and safety. Most farm people are concerned by the injury problems they encounter (Thu and others 1990). However, when an injury happens, the individual readily accepts all blame for the occurrence of the event. These authors' common experience in interviewing injury victims results in statements similar to the following: ". . . I was just careless, I just did a stupid thing. I knew that grain auger did not have a shield over it, and I just got my hand too close when trying to push the remaining grain into the machine." When asked how it could be prevented, they say "we need to get more education out about this problem and be more careful." The point is that farmers know what the major safety hazards are. However, they do not think beyond being careful around hazardous machines. They do not make the transition that "being careful" at first means having and using appropriately guarded machines which protect us from our human condition of "being careless and doing stupid things." They do not think that being safe should include actions to remove the hazard. The latter is the failing of the awareness-level educational models. Clearly the old formula of "education will create attitudes that lead to behavior change" does not necessarily lead to reduced risk of injuries and illness.

On the other hand, education can play an important part in a more integrated program. Community-based programs with farmer involvement, materials, and approaches that have been evaluated (Morgan and others 2002), and programs with multiple modalities appear to have more impact (Brandt and others 2001; Flocks and others 2001; Hjort and others 2003). The following section describes how education can be an effective part of an integrated intervention model.

THE IOWA INTEGRATED MODEL OF PREVENTION PROGRAMMING

The authors suggest that education may most likely be effective in an integrated modality model of agricultural illness and injury prevention. Figure 14.1 illustrates our model of integrated prevention modalities that enhances the probability of achieving the desired behavior change. The model suggests that (in the absence of regulations or engineering controls) the best possible effectiveness of an intervention program includes education based on a combination of applied theories from safety education, social psychology, and public health (Olson and Zanna 1991). Furthermore, this model uses theory from the fields of epidemiology, engineering, industrial hygiene, ergonomics, and regulation. Practicality, convenience, and common sense must also be incorporated in the model to predict best prospects for positive outcomes (McCullagh and others 2002). Inclusion of health and safety in the management of the farm operation is crucial to an integrated model of prevention (Suutarinen 2004). Iowa's Center for Agricultural Safety and Health (http://www.public-health.uiowa.edu/ICASH/) has used this concept in several of its interventions. One recent such program is the Certified Safe Farm (CSF) (http://www.public-health.uiowa.edu/icash/csf/) (CSF 2005). CSF combines a triad of services (clinical screenings, on-farm safety audits, individual and group education) among other occupational health services. This program requires a defined level of performance in removing safety hazards in order to attain "Certified Safe Farm" status. Incentives for certification include potential insurance discounts and discounts on agricultural production inputs such as feed, seed, and farm loans. CSF includes components of all the points displayed in Figure 14.1. The CSF program is connected to the AgriSafe Network (http://www.agrisafe.org/) (AgriSafe

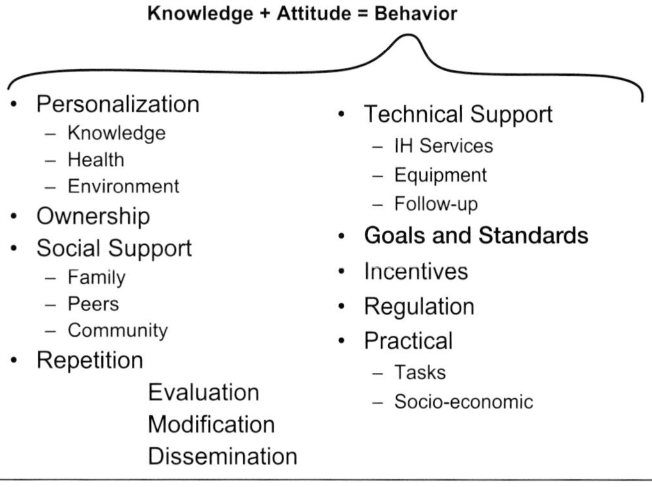

Knowledge + Attitude = Behavior

- Personalization
  - Knowledge
  - Health
  - Environment
- Ownership
- Social Support
  - Family
  - Peers
  - Community
- Repetition

- Technical Support
  - IH Services
  - Equipment
  - Follow-up
- **Goals and Standards**
- Incentives
- Regulation
- Practical
  - Tasks
  - Socio-economic

Evaluation
Modification
Dissemination

Figure 14.1. Promoting safe work behavior.

Network 2005), which provides the clinical services for the program and opportunities of expansion for the CSF through its system. Standards of safety are set, and CSF farm auditors rate and assist the farms they inspect to become certified by providing assistance to obtain the required safety equipment and the specific PPE they need. The farm operation needs to have a health and safety plan with goals that are based on science and input from the farm operators-managers. The health care professional who conducts the clinic screenings spends time with the farm operator to review the screening results and develops a personal/family health and occupational exposure control plan specific for their farm operation. The health professional will contact the client to discuss progress on the plan over the year. The trained CSF program auditors have an agricultural background, which enables them to develop on-farm intervention plans that are practical and within the economic, social, and cultural reference of the farm family. They can assist the farmer to make safe, low-cost changes. They may also connect farmers with resources such as the Farm Bureau who in some states (e.g., Iowa and Virginia) will provide $150 to a farmer to help finance a new ROPS for a tractor (Stone and others 2001).

There are other program examples of the integrated approach in farm health and safety that may be used, such as the occupational health services in Scandinavia and the farm audit and educational program described by Rasmussen and co-workers (2003) in Denmark. The latter program resulted in a decrease of 30% of all farm injuries and 42% of medically treated injuries.

Prevention in Medical or Veterinary Practice
Many farmers think their health care system is failing them when it comes to occupational health concerns. Farmers think their providers do not know, understand, or care about their situations (Walsh 2000). Medical care and veterinary practitioners can have a tremendous positive effect on the health and safety of their farm clients. Without changing practice patterns significantly, there are many things that rural practitioners can do in medical and veterinary practice to help assure they are effective (Rohrer and Culica 1999; Rohrer and others 1998). Veterinarians have significant credibility with the farm community as sources of health information (Thu and others

1990). Following are a few things that can be done:

1. Attain a general working knowledge about agriculture in your community by asking questions of your farm patients, visiting their farms, talking with your local extension agent, and reading farm magazines.
2. Attain specific knowledge about your individual patients/clients, farms/ranches. Ask them questions about their operations in order to determine their specific risks. This will not only help the health professional anticipate possible occupational health issues, but it will build patient/client relations by demonstrating an interest in the individuals' farming operation and health.
3. Based on your intake of information about their operations, you may take an occupational history of the individual and family members.
4. Be a source of information and referral for your patients or clients. Keep timely informational pamphlets and brochures available in the office for them. Have a computer available with key websites in your favorites list that can provide up-to-date access for you and your patients (see the section "Nongovernmental Organizations Practicing Agricultural Safety and Health," earlier in this chapter).
5. Keep up to date with advances in agricultural production and economy by regularly reading farm magazines and newspapers.
6. Subscribe to one or more of the scientific journals that deal primarily with agricultural health issues. Examples include the *Journal of Agromedicine,* the *Journal of Agricultural Safety and Health,* and the *International Journal of Agricultural Medicine and Environmental Health.*
7. Promote and facilitate farm health and safety events and programs in your communities, such as Farm Safety Day Camps (Magazine 2005).
8. Communicate and consult within your interdisciplinary health community (veterinarians, physicians, PAs, nurses, etc.) on patients/clients. For example, if a physician has a patient with an unusual infectious disease, a call to his/her veterinarian may help pin-

point a zoonotic illness that may be present on this person's farm.

9. Get to know and consult with other nonmedical professionals in your community such as your county or area extension agent, your state agricultural safety specialist, and vocational agricultural teachers in your community that can help keep you informed and assist in putting on farm community prevention programs.

10. Facilitate development of farm health and safety organizations in your community, such as Farm Safety 4 Just Kids (FSJK 2005) and AgriSafe Network (AgriSafe Network 2005) (Table 14.4).

11. Become active in a local, state, national, and/or international organization regarding agricultural health and safety (see Table 14.4).

12. If there are migrant or seasonal workers in your practice area, become familiar with cultural issues and language, which may be an asset to your success as a health care provider; learn their language and/or have ready access to translators in your region.

13. Understand the lifestyle of your farm patient population and implement wellness promotion as part of your services to the farming population (McCrone 1999).

SUMMARY

Due to the nature of farming, farm work, and the farm culture, farmers, ranchers, and other practicing production agricultural workers are generally not leading occupational health and safety efforts in most countries. Certainly there are exceptions, which include but are not limited to Farm Safe Australia (Farm Safe Australia 2005), the United Farm Workers Organization in the U.S. (United Farm Workers 2005), and Farm Bureaus in some states of the U.S. The bulk of prevention has been carried out by governmental and voluntary nongovernmental organizations. Although regulations have proven to be one of the most effective ways to reduce health and safety hazards in agriculture, there are few regulations that directly affect independent small farming operations, and those that are present are not regularly enforced (at least in the U.S.). Most regulations in agriculture pertain to large farms, particularly where migrant and seasonal workers are employed. The reasons for lack of regulations are many, but

there has not been a sufficient calling from the farm population, the general public, or political lobbyists.

Engineering interventions are very effective preventive methods. However, they primarily apply to newer equipment and buildings (unless retrofitting is done) because old equipment and structures without effective engineering protection may remain in use for years. It may not be economical for the farmer to upgrade old equipment. New equipment and structures are usually safer, but replacement of old equipment and buildings may take decades.

Educational programs have by far been the main tactic used in agricultural health and safety prevention. These programs have primarily been awareness-level efforts. Although long-term outcome evaluation of educational programs has been limited and hampered by lack of baseline data, there has been little evidence that they have been effective in reducing illnesses and injuries over extended periods. The authors have described a multifaceted approach that combines education with multiple levels of intervention that may have the best chance for effecting prevention. This includes provision of occupational health screenings, on-farm safety audits, and performance incentives.

REFERENCES

AgriSafe Clinic of Spencer Municipal Hospital. http://www.spencerhospital.org/agrisafe.htm. Spencer, IA.

AgriSafe Network. 2005. AgriSafe Network.

Aherin R, Murphy D, Westaby J. 1992. Reducing Farm Injuries: Issues and Methods. St Joseph, MI: ASAE.

American Academy of Pediatrics CoI, Poison Prevention, Committee on Community Health Services. 2001. Prevention of agricultural injuries among children and adolescents. Pediatrics 108(4):1016–1019.

Arcury T. 1997. Occupational injury prevention knowledge and behavior of African-American farmers. Human Organization 56(2):167–173.

Brandt V, Struttmann T, Cole H, Piercy L. 2001. Delivering health education messages for part-time farmers through local employers. J Agromed 7(3): 23–30.

Browning S, Westneat S, Donnelly C, Reed D. 2003. Agricultural tasks and injuries among Kentucky farm children: Results of the Farm Family Health and Hazard Surveillance Project. South Med J 96(12):1203–1212.

CDHS. 2005. Agricultural Health and Safety Section (AHSS).

Chapman L, Schuler R, Wilkinson T, Skjolaas C. 1996. Agricultural safety efforts by county health departments in Wisconsin. Public Health Rep 111(5): 437–443.

Cole H, Kidd P, Isaacs S, Parshall M, Scharf T. 1997. Difficult decisions: A simulation that illustrates cost effectiveness of farm safety behaviors. J Agromed 4(1/2):117–124.

CSF. 2005. Certified Safe Farm Home Page. Iowa's Center for Agricultural Safety and Health. http://www.public-health.uiowa.edu/icash. Iowa City, IA.

DeRoo L, Rautiainen R. 2000. A systematic review of farm safety interventions. Am J Prev Med 18(4 Suppl S):51–62.

Donham K. 1995a. Respiratory dysfunction in swine production facility workers: Dose-response relationships of environmental exposures and pulmonary function. Am J Ind Med 27(3):405–418.

——. 1995b. Agricultural medicine and environmental health: The missing component of the sustainable agricultural movement. Boca Raton, FL: CRC Press Inc. p 583–589.

——. 1997. Tractor risk abatement and control: The policy conference. Iowa Center for Agricultural Safety and Health. Iowa City, Iowa: College of Public Health, U of I.

Elkind P, Pitts K, Ybarra S. 2002. Theater as a mechanism for increasing farm health and safety knowledge. Am J Ind Med Suppl(2):28–35.

Farm Safe Australia. 2005. Farm Safe Australia Home Page.

Flocks J, Clarke L, Albrecht S, Bryant C, Monaghan P, Baker H. 2001. Implementing a community-based social marketing project to improve agricultural worker health. Environ Health Perspect 109(Suppl 3): 461–468.

Fragar L. 1996. Policy issue. Agricultural health and safety in Australia. Australian Journal of Rural Health 4(3):200–206.

Fragar L, Houlahan J. 2002. Australian approaches to the prevention of farm injury. N S W Public Health Bull 13(5):103–107.

FSJK. 2005. Farm Safety for Just Kids Home Page.

Gay J, Donham K, Leonard S. 1990. The Iowa Agricultural Health and Safety Service Program. Am J Ind Hygiene 18:385–389.

Haddon W. 1963. A note concerning accident theory and research with special reference to motor vehicle accidents. Annals New York Academy of Sciences 107:636–646.

Heikkonen J, Louhevaara V. 2003. Empowerment in farmers' occupational health services. Ann Agric Environ Med 10(1):45–52.

Hjort C, Hojmose P, Sherson D. 2003. A model for safety and health promotion among Danish farmers. J Agromed 9(1):93–100.

Hoglund S. 1990. Farmers' health and safety programs in Sweden. Am J Ind Med 18(4):371–378.

Hubert D, Ullrich D, Murphy T, Lindner J. 2001. Texas entry-year agriculture teachers' perceptions, practices, and preparation regarding safety and health in agricultural education. J Agric Saf Health 7(3): 143–153.

Hubler C, Hupcey J. 2002. Incidence and nature of farm-related injuries among Pennsylvania Amish children: Implications for education. Journal of Emergency Nursing 28(4):284–188.

Huneke J, VonEsse S, Grisso R. 1998. Innovative approaches to farm safety and health for youth, senior farmers and health care providers. J Agromed 5(2): 99–106.

I-CASH, College of Public Health. 2005. I-CASH. The University of Iowa.

Jones P, Field W. 2002. Farm safety issues in Old Order Anabaptist communities: Unique aspects and innovative intervention strategies. J Agric Saf Health 8(1):67–81.

Kelsey T. 1994. The agrarian myth and policy responses to farm safety. Am J Public Health 84(7): 1171–1177.

Kidd P, Reed D, Weaver L, Westneat S, Rayens M. 2003. The transtheoretical model of change in adolescents: Implications for injury prevention. Journal of Safety Research 34(3):281–288.

Kidd P, Scharf T, Veazie M. 1996. Linking stress and injury in the farming environment: A secondary analysis of qualitative data. Health Education Quarterly 23(2):224–237.

Kidd P, Townley K, Cole H, McKnight R, Piercy L. 1997. The process of chore teaching: Implications for farm youth injury. Fam Community Health 19(4):78–89.

Knapp L. 1965. Agricultural injury prevention. J Occup Med 7(11):545–553.

Kumar, Adarsh, Varghese M, Mohan D. 2000. Equipment-related injuries in agriculture: An international perspective. Injury Control & Safety Promotion 7(3):175–186.

Lee B. 1994. Agricultural health and safety: Opportunities for nursing research. J Agromed 1(1):75–80.

Legault M, Murphy D. 2000. Evaluation of the Agricultural Safety and Health Best Management Practices Manual. J Agric Saf Health 6(2):141–153.

Leigh J, McCurdy S, Schenker M. 2001. Costs of occupational injuries in agriculture. Public Health Rep 116(3):235–248.

Liller K. 2001. Teaching agricultural health and safety to elementary school students. Journal of School Health 71(10):495–496.

Liller K, Noland V, Rijal P, Pesce K, Gonzalez R. 2002. Development and evaluation of the Kids Count Farm Safety Lesson. J Agric Saf Health 8(4):411–421.

Loos C, Oldenburg B, O'Hara L. 2001. Planning of a community-based approach to injury control and safety promotion in a rural community. Australian Journal of Rural Health 9(5):222–228.

Lundvall A, Olson D. 2001. Agricultural health nurses: Job analysis of functions and competencies. AAOHN J. 49(7):336–346.

Magazine PF. 2005. Farm Safety Day Camps.

Mandel J, Carr W, Hillmer T, Leonard P, Halberg J, Sanderson W, Mandel J. 2000. Safe handling of agricultural pesticides in Minnesota: Results of a county-wide educational intervention. J Rural Health 16(2):148–154.

Martin S. 1997. Agricultural safety and health: Principles and possibilities for nursing education. J Nurs Educ 36(2):74–78.

Mausner J, Kramer S. 1985. Epidemiology—An Introductory Text. Philadelphia, PA: WB Saunders.

McCrone J. 1999. A case control study of the health status of male farmers registered at Ash Tree House. Journal of the Royal Society of Health 119(1):32–35.

McCullagh M, Lusk S, Ronis D. 2002. Factors influencing use of hearing protection among farmers—A test of the Pender Health Promotion Model. Nurs Res 51(1):33–39.

McNab W. 1998. Incorporating farm safety into the health education curriculum. J Sch Health 68(5): 213–215.

Meyers J, Bloomberg L. 1993. Farm industry slips through cracks: Public health principles should be utilized to reform, improve agricultural injury prevention efforts. Occup Health Saf 62(11):56–60.

Moore E, McComas J. 1996. Acceptance of an injury-prevention program in rural communities: A preliminary study. Prehospital & Disaster Medicine 11(4): 309–311.

Morgan S, Cole H, Struttmann T, Piercy L. 2002. Stories or statistics? Farmers' attitudes toward messages in an agricultural safety campaign. J Agric Saf Health 8(2):225–239.

Murphy D. 1991. Safety and Health for production agriculture. Code of Fed Regs 29:4345–4346.

Murphy D, Kiernan N, Chapman L. 1996. An occupational health and safety intervention research agenda for production agriculture: Does safety education work? Am J Ind Med 29(4):392–396.

Myers M. 2000. Prevention effectiveness of rollover protective structures—Part I: Strategy evolution. J Agric Saf Health 6(1):29–40.

Nawrot R, Wright W. 1998. Make it safe: An injury prevention program for Hispanic farm workers and families at work and play. International Electronic Journal of Health Education. 1(4):219–221.

NFAOHS. 2005. NFAOHS Home Page.

Nonnenmann M, Donham K, Rautiainen R, et al. 2004. Vegetable oil sprinkling as a dust reduction method in swine confinement. J Agric Saf Health 10(1): 7–15.

Notkola V, Husman K, Tupi K, Virolainen R, Nuutinen J. 1990. Farmers occupational health programme in Finland 1979–1987. Social Science and Medicine 30(9):1035–1040.

Olson J, Zanna M. 1991. Attitudes and Beliefs, Attitude Change and Attitude-Behavior Consistency. In: Baron R, Graziano WG, Stangor C, editor. Social Psychology. Philadelphia PA: Holt, Rinehart & Winston, Inc. p 192–225, 226–269.

OSHA. 2005. OSHA home page. In: Labor UDo, editor: U.S. Dept of Labor.

Palermo T, Ehlers J. 2002. Coalitions: Partnerships to promote agricultural health and safety. J Agric Saf Health 8(2):161–174.

Pana-Cryan R, Myers M. 2002. Cost-effectiveness of roll-over protective structures. Am J Ind Med Suppl(2):68–71.

Pistella C, Kanzleiter L, Henderson R, Herman J. 2001. Pennsylvania Statewide Agricultural Emergency Response Training Program: A collaborative partnership with agromedicine, academic and community resources. J Agromed 7(3):11–21.

Powers J, Harris J, Etherton J, Ronaghi M, Snyder K, Lutz T, Newbraugh B. 2001. Preventing tractor rollover fatalities: Performance of the NIOSH autoROPS. Injury Prevention 7(Suppl 1): i54–58.

Purschwitz M. 1997. University of Wisconsin agricultural safety and health activities. Wis Med J 96(8):25–29.

——. 2003. Creating a safer and healthier agriculture— Are we asking the right questions? J Agric Saf Health 9(2):87–88.

Rains G. 2000. Initial rollover effectiveness evaluation of an alternative seat belt design for agricultural tractors. J Agric Saf Health 6(1):13–27.

Rasmussen K, Carstensen O, Lauritsen J, Glasscock D, Hansen O, Jensen U. 2003. Prevention of farm injuries in Denmark. Scand J Work Environ Health 29(4):288–296.

Reed D, Kidd P, Westneat S, Rayens M. 2001. Agricultural Disability Awareness and Risk Education (AgDARE) for high school students. Injury Prevention 7(Suppl 1):i59–63.

Reed D, Westneat S, Kidd P. 2003. Observation study of students who completed a high school agricultural safety education program. J Agric Saf Health 9(4):275–283.

Reynolds S, Groves W. 2000. Effectiveness of roll-over protective structures in reducing farm tractor fatalities. Am J Prev Med 18(4 Suppl):63–69.

Risenberg L, Bear W. 1980. Instructional impact on accident prevention in engineering a safer food machine. St Joseph, MI:138–144.

Robertson L. 1983. Injuries: Causes, control strategies, and public policy. Lexington MA, Lexington Books, DC Heath & Company.

Rohrer J, Culica DV. 1999. Identifying high-users of medical care in a farming-dependent county. Health Care Manage Rev 24(4):28–34.

Rohrer J, Urdaneta M, Vaughn T, Merchant J. 1998. Physician visits in a farming-dependent county. J Rural Health 14(4):338–345.

Schenker M. 1996. Preventive medicine and health promotion are overdue in the agricultural workplace. J Public Health Policy 17(3):275–305.

Silletto T. 1976. Implications for agricultural safety education programs as identified by Iowa farm accident survey. [Unpublished doctoral dissertation]. Iowa State University, Ames, IA.

Social Marketing Institute. 2005. Social Marketing.

Steffen R. 1998. Farm safety day campus: Developing a definition using the Delphi method. J Agric Saf Health 4(2):109–117.

Stiernstrom E, Holmberg S, Thelin A, Svardsuff K. 1998. Reported health status among farmers and non farmers in nine rural districts. J Occup Environ Med 40(10):917–924.

Stone J, Hanna M, Guo C, Imerman P. 2001. Protective headgear for midwestern agriculture: A limited wear study. J Environ Health 63(7):13–19.

Stueland D, McCarty J, Stamas PJ, Gunderson P. 1996. Evaluation of agricultural rescue course by providers. Prehospital Disaster Med 11(3):234–238.

Suutarinen J. 2004. Management as a risk factor for farm injuries. J Agric Saf Health 10(1):39–50.

Swenson E. 2004. National Agricultural Tractor Safety Initiative. University of Washington.

Thelin A. 1990. Epilogue: Agricultural occupational and environmental health policy strategies of the future. Am J Ind Med 18:523–526.

Thelin A, Stiernstrom E, Holmberg S. 1999. Differences in the use of health care facilities and patterns of general risk factors in farmers with and without occupational health care programs. Int J Occup Environ Health 5(3):170–176.

Thu K, Donham K, Yoder D, Ogilvie L. 1990. The farm family perception of occupational health: A multi-state survey of knowledge, attitudes, behaviors and ideas. Am J Ind Med 18:427–431.

United Farm Workers. 2005. United Farm Workers Home Page.

University of Nebraska Medical Center. http://www.unmc.edu/. Omaha, NE.

U.S. Dept of Health. 1973. The Industrial Environment—Its Evaluation & Control. Education, and Welfare, PHS, CDC, NIOSH:1–5.

U.S. Dept of Labor. 1970. Child Labor Requirements in Agriculture under the Fair Labor Standards Act. Child Labor Bulletin No. 102.

USDA. 2002. 2002 Census of Agriculture.

——. 2004. Farm Numbers and Land in Farms Final Estimates 1998–2002. National Agricultural Statistics Service.Walsh M. 2000. Farm accidents: Their causes and the development of a nurse led accident prevention strategy. Emerg Nurse 8(7):24–31.

Westaby J, Lee B. 2003. Antecedents of injury among youth in agricultural settings: A longitudinal examination of safety consciousness, dangerous risk taking, and safety knowledge. Journal of Safety Research 34(3):227–240.

Wheat J, Turner T, Weatherly L, Wiggins O. 2003. Agromedicine Focus Group: Cooperative extension agents and medical school instructors plan farm field trips for medical students. South Med J 96(1):27–31.

Wilkins JI, Engelhardt H, Bean T, Byers M, Crawford J. 2003. Prevalence of ROPS—Equipped tractors and farm/farmer characteristics. J Agric Saf Health 9(2):107–118.

Williams D. 1983. Iowa's farm people characteristics and agricultural accident occurrence. Unpublished doctoral dissertation, Iowa State University, Ames, IA.

Yoder A, Murphy D. 2000. Evaluation of the Farm and Agricultural Injury Classification Code and follow-up questionnaire. J Agric Saf Health 6(1):71–80.

Index

Page numbers followed by *t* indicate tables.